Antimicrobial Resistance and Implications for the Twenty-First Century

Emerging Infectious Diseases of the 21st Century

Series Editor: I. W. Fong

> *Professor of Medicine, University of Toronto*
> *Infectious Diseases, St. Michael's Hospital*

Recent volumes in this series:

MALARIA: GENETIC AND EVOLUTIONARY ASPECTS
Edited by Krishna R. Dronamraju and Paolo Arese

INFECTIONS AND THE CARDIOVASCULAR SYSTEM: New Perspectives
Edited by I. W. Fong

REEMERGENCE OF ESTABLISHED PATHOGENS IN THE 21ST CENTURY
Edited by I. W. Fong and Karl Drlica

BIOTERRORISM AND INFECTIOUS AGENTS: A New Dilemma for the 21st Century
Edited by I.W. Fong and Ken Alibek

MOLECULAR PARADIGMS OF INFECTIOUS DISEASE: A Bacterial Perspective
Edited by Cheryl A. Nickerson and Michael J. Schurr

NEW AND EVOLVING INFECTIONS OF THE 21STCENTURY
Edited by I.W. Fong and Ken Alibek

ANTIMICROBIAL RESISTANCE AND IMPLICATIONS FOR THE 21ST CENTURY
Edited by I.W. Fong and Karl Drlica

A Continuation Order plan is available for this series. A continuation order will bring delivery of each new volume immediately upon publication. Volumes are billed only upon actual shipment. For further information please contact the publisher.

I. W. Fong · Karl Drlica

Editors

Antimicrobial Resistance and Implications for the Twenty-First Century

 Springer

I. W. Fong
Director, Infectious Disease
St. Michael's Hospital
30 Bond St., Suite 4179
Cardinal Carter Wing
Toronto, Ontario M5B 1W8
Canada
fongi@smh.toronto.on.ca

Karl Drlica
International Center for Public Health
225 Warren Street
Newark, NJ 07103-3535
USA
drlica@phri.org

ISBN: 978-0-387-72417-1   e-ISBN: 978-0-387-72418-8

Library of Congress Control Number: 2007930198

Printed on acid-free paper.

9 8 7 6 5 4 3 2 1

springer.com

Preface

For many years after the discovery of antibiotics, microbial resistance was largely ignored. Now, however, the prevalence of antibiotic-resistant microorganisms, both in the community and in hospitals, has reached a level that impacts treatment efficacy. New, more potent agents have been introduced, but resistant microbes continue to be selectively enriched. Unfortunately, the problem of drug-resistant microorganisms extends beyond bacteria: it is also of major concern with the management of viral diseases, such as that caused by Human Immunodeficiency Virus, and with parasitic diseases such as malaria. Meanwhile, it is becoming increasingly difficult to identify new compound classes and more active derivatives of existing agents, especially since many pharmaceutical companies have abandoned efforts to find and develop new antimicrobials.

Antimicrobial Resistance and Implications for the Twenty-First Century serves as a status report on resistance. This set of comprehensive, up-to-date reviews by international experts covers problems being observed among a variety of bacteria (*Streptococcus pneumoniae*, enteroccoci, staphylococci, Gram-negative bacilli, mycobacteria species), viruses (HIV, herpesviruses), and fungi (*Candida* species, fusarium, etc.). The chapters explore molecular mechanisms of drug resistance, epidemiology of resistant strains, clinical implications, and future directions, including strategies for restricting the acquisition of resistance. The work is intended for experts and students in the fields of infectious disease, microbiology, and public health. While our goal is to stimulate basic research on resistance, the work should also help international bodies, such as the World Health Organization, formulate effective plans to combat the acquisition and dissemination of resistant strains. We hope that the documentation provided in *Antimicrobial Resistance and Implications for the Twenty-First Century* can be used by public health and medical communities to exert the political pressure needed to limit the indiscriminate use of antimicrobials and to provide the incentives needed to find new antimicrobials and treatment strategies.

Contents

Contributors

Beth A. Arthington-Skaggs
Mycotic Diseases Branch
Centers for Disease Control
   and Prevention
Atlanta, GA
USA

G. Boivin
Research Center in Infectious Diseases
   of the Centre Hospitalier
Universitaire de Québec and
Department of Medical Biology
Université Laval, Québec
Canada

Patricia A. Bradford
Wyeth Research, Pearl River, NY
USA

Buenaventura Buendía
Servicio de Microbiología
Hospital Universitario de la Princesa
Madrid
Spain

Diane M. Citron
David Geffen School of Medicine
   at UCLA
Los Angeles, CA
USA

Adam L. Cohen
Division of Healthcare
Quality Promotion

Centers for Disease Control
   and Prevention
Atlanta, GA
USA

Richard Cooke
Health Protection Agency Laboratory
University Hospital Aintree
Aintree
UK

Peter D.O. Davies
Tuberculosis Research Unit
Cardiothoracic Centre
Liverpool
UK

Charles R. Dean
Novartis Institutes for BioMedical
   Research Inc.
Cambridge, MA
USA

Felipe Rodriguez de Castro
Servicio de Neumología
Hospital Dr. Negrín, Universidad de las
   Palmas de Gran Canaria
Las Palmas de Gran Canaria
Spain

W.L. Drew
Departments of Laboratory Medicine
   and Medicine
University of California
San Francisco, CA
USA

Karl Drlica
Public Health Research Institute
Newark, NJ
USA

Ellie J.C. Goldstein
R M Alden Research Laboratory
Santa Monica, CA
USA

Rachel Gorwitz
Division of Healthcare
Quality Promotion
Centers for Disease Control
   and Prevention
Atlanta, GA
USA

David W. Hecht
Loyola University Medical
Center Maywood
IL and Hines VA Hospital
Hines, IL
USA

Daniel B. Jernigan
Division of Healthcare
Quality Promotion
Centers for Disease Control
   and Prevention
Atlanta, GA
USA

Xinying Li
Public Health Research Institute
Newark, NJ
USA

Tao Lu
Public Health Research Institute
Newark, NJ
USA

Muhammad Malik
Public Health Research Institute
Newark, NJ
USA

Javier Aspa Marco
Servicio de Neumología
Hospital Universitario de la Princesa
Universidad Autónoma de Madrid
Madrid
Spain

Jorge L. Martinez-Cajas
McGill University AIDS Centre
Lady Davis Institute for Medical
   Research
Jewish General Hospital
Montreal, Quebec
Canada

Olga Rajas Naranjo
Servicio de Neumología
Hospital Universitario de la Princesa
Universidad Autónoma de Madrid
Madrid
Spain

John Papp
Laboratory Research Branch,
Division of STD Prevention
Centers for Disease Control
   and Prevention
Atlanta, GA
USA

Steven Park
Public Health Research Institute
Newark, NJ
USA

Jean-Michel Pawlotsky
French National Reference Center
   for Viral Hepatitis B, C and delta
Department of Virology
INSERM U841, Henri Mondor Hospital
University of Paris XII
Créteil
France

David S. Perlin
Public Health Research Institute
Newark, NJ
USA

Marco Petrella
McGill University AIDS Centre
Lady Davis Institute
    for Medical Research
Jewish General Hospital
Montreal, Quebec
Canada

John H. Rex
AstraZeneca Pharmaceuticals
Macclesfield
UK

Jesús Sanz Sanz
Servicio de Medicina
Interna-Enfermedades
    Infecciosas
Hospital Universitario de la Princesa
Madrid
Spain

David Trees
Laboratory Research Branch
Division of STD Prevention

Centers for Disease Control
    and Prevention
Atlanta, GA
USA

Mark A. Wainberg
McGill University AIDS Centre
Lady Davis Institute
    for Medical Research
Jewish General Hospital
Montreal, Quebec
Canada

J.-Y. Wang
Public Health Research Institute
Newark, NJ
USA

Hillard Weinstock
Epidemiology and Surveillance Branch,
Division of STD Prevention
Centers for Disease Control
    and Prevention
Atlanta, GA
USA

Chapter 1
Mechanisms of Resistance by Gram-Positive Bacteria (Streptococci and Enterococci)

Javier Aspa Marco, Olga Rajas Naranjo, Felipe Rodriguez de Castro, Buenaventura Buendía, and Jesús Sanz Sanz

1.1 *Streptococcus Pneumoniae* Resistance to Penicillin: Mechanism and Clinical Significance

Streptococcus pneumoniae (*S. pneumoniae*) is the most commonly identified bacterial cause of meningitis (Schuchat et al., 1997), otitis media and acute sinusitis in adults (Barnett and Klein, 1995; Jacobs, 1996), and community-acquired pneumonia (CAP) at all ages throughout the world (Marston et al., 1997). It is also a frequent cause of bacteremia and one of the most frequent pathogens involved in Chronic Obstructive Pulmonary Disease (COPD) exacerbations.

In the past, approximately 80% of patients hospitalized with bacteremic pneumococcal infections died of their illness (Austrian and Gold, 1964). With effective antimicrobial agents, mortality has decreased, but it remains at nearly 20 percent for elderly adults (Austrian and Gold, 1964; Kramer et al., 1987; Whitney et al., 2000). The principal groups at risk of developing pneumococcal infections are immunocompetent adults with chronic illnesses (cardiovascular, lung, or liver diseases), patients with functional or anatomic asplenia, patients with lymphoproliferative illnesses (chronic lymphatic leukemia, multiple myeloma, and non-Hodgkin lymphoma), and those with congenital deficits of immunoglobulin synthesis. Cigarette smoking is the strongest independent risk factor for invasive pneumococcal disease among immunocompetent, nonelderly adults (Nuorti et al., 1998).

1.1.1 Microbiologic Characteristics of S. pneumoniae

The first description of *S. pneumoniae* dates back to 1870 when it was referred to as a "slightly elongated diplococcus". In 1881, Louis Pasteur (France) and George Sternberg (United States) (Pasteur, 1881; Sternberg, 1881) managed to identify this bacterium by isolating it from human saliva.

S. pneumoniae, an aerobic and facultatively anaerobic bacterium, is a typical extracellular gram-positive coccus characteristically arranged in pairs or short chains when grown in a liquid media. Colonies of encapsulated strains are round, mucoid, and unpigmented; colonies of nonencapsulated strains are smaller, glossy,

I. W. Fong and K. Drlica (eds.), *Antimicrobial Resistance and Implications for the Twenty-First Century.* © Springer 2008

and appear flat. All colonies are alpha-hemolytic on blood agar (the organism produces a green halo of hemolysis) if incubated aerobically and may be beta-hemolytic if grown anaerobically.

The basic structural framework of the cell wall is the peptidoglycan layer, which is similar in composition to that found in other gram-positive bacteria. The other major component of the cell wall is teichoic acid. In addition, virulent strains of *S. pneumoniae* are covered with a complex polysaccharide capsule. This capsule is poorly antigenic and by itself is not toxic. It can act as a barrier to toxic hydrophobic molecules and can promote adherence to other bacteria or to host tissue surfaces. Composed of one of 90 serologically distinct polysaccharides, the virulence contribution of the capsule lies in its antiphagocytic properties. Capsular material purified from the most commonly isolated serotypes is the basis of the current antipneumococcal vaccine. Most resistant strains belong to a small number of serotypes (6, 9, 14, 19, and 23) (Fenoll et al., 1991, 2002), which are prevalent in young children and which are being included in the new conjugate vaccines. Although pneumococci exist in unencapsulated forms, only encapsulated strains have been isolated from clinical material (Musher, 2000).

1.1.2 Historical Overview of Pneumococcal Antibiotic Resistance

S. pneumoniae has been known as an important human pathogen for over 120 years. Before penicillin became widely available, mortality and morbidity estimates for pneumococcal disease were extremely high. The discovery of penicillin in 1929 by A. Fleming from the fungus *Penicillin notatum*, in St. Mary's Hospital, London (Fleming, 1929), marked the beginning of the so-called antibiotic era in which death rates resulting from pneumococcal disease decreased dramatically. A few years later, mutants of pneumococcal penicillin G-resistant strains were identified (in vitro resistance), and in 1943 resistance was demonstrated in animal models (Eriksen, 1945; Schmidt, 1943).

Clinical resistance to penicillin was not observed until 22 years later (Boston, 1965), when two strains with a moderate loss of penicillin susceptibility (minimun inhibitory concentration [MIC] 0.1–0.2 mg/l) and with no apparent clinical significance were isolated from patients (Kislak et al., 1965). Two years later, in 1967, Hansman and Bullen (1967), in Australia, were the first to document and analyze the clinical importance of such resistance by isolating pneumococci with MIC 0.6 mg/l from the sputum of a patient with hypogammaglobulinemia. In 1970, the isolation of a diminished penicillin susceptibility strain from Anguganak (New Guinea) (MIC 0.5 mg/l) was reported (Hansman et al., 1971a). In 1974, the first case with clinical pneumococcal meningitis with a MIC of 0.25 mg/l was described in the United States in a patient with sickle cell anemia. About that time, the incidence rate of intermediate resistance to penicillin in New Guinea and Australia had reached 12%. Between 1977 and 1980, a major outburst of penicillin resistance took place in Dourban (South Africa) with the discovery of patients suffering from pneumonia, meningitis, bacteremia and other invasive infections caused by highly penicillin-resistant pneumococci (MIC 4–8 mg/l).

These organisms were also resistant to chloramphenicol and other antimicrobial agents (Appelbaum et al., 1977; Finland, 1971; Hansman et al., 1971a,b). Since then, especially in the last two decades, penicillin and multidrug resistance have been a common characteristic of *S. pneumoniae* in many countries around the world (Appelbaum, 1992; Martínez-Beltrán, 1994), with a very uneven geographic distribution: levels of pneumococcal resistance have been as high as 60% in certain areas of South America and, both surprising and alarming, up to 80% in some Asian countries (Appelbaum, 1992, 2002; Empey et al., 2001; Felmingham and Gruneberg, 2000; Jones, 1999).

The worldwide spread of nonsusceptible strains of pneumococci has been related to the dissemination of certain highly resistant pneumococcal clones (serotypes 6A, 6B, 9V, 14, 19F, and 23F) (Davies et al., 1999; McDougal et al., 1992). All have been associated with infections in humans and are represented in the currently available pneumococcal vaccines. In contrast, serotypes 1, 3, 4, 5, 7, 11, 15, and 18, rarely carry resistance genes (Chenoweth et al., 2000; Musher, 2000).

1.1.3 Clinical and Microbiological Concepts of Antibiotic Resistance

Microbial resistance causes a great deal of confusion when choosing an empirical treatment for pneumonia. Strictly speaking, the term resistance refers to the in vitro susceptibility of a pathogen to various antibiotics; however, in vitro data do not necessarily translate into in vivo efficacy. Favorable pharmacokinetic/pharmacodynamic (PK/PD) parameters and high concentrations of antimicrobials at the site of infection may explain the good clinical outcomes achieved despite MIC values in vitro that appear to be "resistant" or "nonsusceptible".

In vitro susceptibility testing by the microbiology laboratory assumes the isolate was recovered from blood, and is being exposed to serum concentrations of an antibiotic given in the usual dose. Resistance levels are always defined in relation to these serum concentrations based on the National Committee for Clinical Laboratory Standards (NCCLS) breakpoints—now known as the Clinical Laboratory Standards Institute (CLSI)—and they do not make reference to tissue concentrations at the site of infection. Susceptibility breakpoints proposed by NCCLS (NCCLS, 1997) for *S. pneumoniae* were derived from laboratory and clinical data relating to the treatment of the most serious and difficult to treat clinical infections, such as meningitis, and not to otitis media, sinusitis, or pneumonia. According to these values, *S. pneumoniae* was classified as susceptible (MIC < 0.06 mg/l), of intermediate susceptibility (MIC 0.1–1 mg/l) and resistant (MIC > 2 mg/l). These breakpoints are meant to be considered for infections in which borderline penetration makes these levels significant. In meningitis, for instance, the penicillin concentrations reached in cerebrospinal fluid (CSF) might barely reach the 2 mg/l level established for high-degree penicillin resistance (Musher et al., 2001). In contrast, in pneumococcal pneumonia, the drug levels reached in serum and lung tissue exceed the MIC of the microorganism because there is no anatomical barrier hindering access of antimicrobials to the infection site (Craig, 1998a). Knowing the origin of these definitions helps to resolve the apparent paradox that infections of the respiratory tract due to seemingly β-lactam-resistant pneumococci may still respond well to standard doses of these drugs (Kaplan and Mason, 1998; Heffelfinger et al., 2000). Such inconsistencies have prompted the NCCLS to revise its breakpoints for *S. pneumoniae* susceptibility testing, which differentiate between meningeal and non-meningeal sites of pneumococcal infection (Heffelfinger et al., 2000). (See Table 1.1).

Recently, Weiss and Tillotson (2005) have suggested that the definition of clinical resistance, rather than strictly relying on NCCLS criteria, should be linked to the inability of an antibiotic agent to reach a concentration high enough at the site of infection to inhibit local bacterial replication, and to the need for a second antibiotic prescription 72 h after starting therapy because of the deterioration or nonsignificant improvement of the initial medical condition.

1.1.4 Overview of Bacterial Resistance

The use of antimicrobials, appropriate or not, encourages the development of resistance in bacterial strains. In recent decades, bacteria have demonstrated their almost limitless ability to adapt to different circumstances, specifically to the ecological pressure caused by different antimicrobial agents. In the early 1900s, ethylhydrocupreine (optochin) was examined as treatment of pneumococcal pneumonia. Unfortunately, the trial of optochin was terminated because of drug toxicity, and the emergence of resistance to the drug during therapy was also noted (Moore and Chesney, 1917). This appears to have been the first documentation of the emergence of resistance to an antimicrobial agent in humans. It was the clinical deployment of penicillin, however, that initially brought home the importance of the emergence

Table 1.1 NCCLS Breakpoint definitions of susceptibility and resistance for pneumococcal pneumonia per year (NCCLS)

Antibiotic	NCCLS (1999)			NCCLS (2000)			NCCLS (2002)		
	S	I	R	S	I	R	S	I	R
Penicillin	≤ 0.06	0.12–1	≥ 2	≤ 0.06	0.12–1	≥ 2	≤ 0.06	0.12–1	≥ 2
Amoxicillin	≤ 0.5	1	≥ 2	< 2	4	≥ 8	< 2	4	≥ 8
Cefotaxime/ceftriaxone	≤ 0.5	1	≥ 2	≤ 0.5	1	≥ 2	≤ 1	2	≥ 4
Erythromycin	≤ 0.25	0.5	≥ 1	≤ 0.25	0.5	≥ 1	≤ 0.25	0.5	≥ 1
Levofloxacin/Ofloxacin	≤ 2	4	≥ 8	≤ 2	4	≥ 8	≤ 2	4	≥ 8

S, susceptible; I, intermediate; R, resistant.
The values show MIC in mg/l, in accordance with the recommendations made by the NCCLS (National Committee for Clinical and Laboratory Standard) for each year of reference for Pneumococcal Pneumonia, defining the standards for susceptibility, intermediate resistance and resistance (NCCLS, 1999; NCCLS, 2000; NCCLS, 2002).

of resistant organisms. Alexander Fleming first noted that certain bacteria, such as *Haemophilus influenzae*, were naturally resistant to penicillin, but the real significance of resistance emerged as penicillin was used to treat serious staphylococcal infections. Whereas before 1946 about 90% of *Staphylococcus aureus* isolates in hospitals were susceptible to penicillin, 75% of isolates were resistant by 1952 (Findland, 1955). The first decade of the clinical use of the sulfonamides also saw the widespread emergence of resistance to these drugs in several gram-negative bacteria (Coburn, 1949; Woods, 1962). At that time, the close relationship between the use of low dosage antibiotics and the later appearance of resistance had already been proven and described by Fleming, in relation to *S. pneumoniae* resistance to sulphamides, and later by Chain, with reference to the appearance of penicillin resistance (Florey, 1943; Maclean and Fleming, 1939).

Bacteria can evade antimicrobial action by adopting diverse mechanisms. These mechanisms include enzymatic inhibition, altered porin channels, alterations of outer or inner membrane permeability, modification of target proteins, antibiotic efflux, and altered metabolic pathways. Intrinsic resistance is a naturally occurring phenomenon that occurs in the absence of antimicrobial selection pressure. It occurs in all members of a genus or species and the term implies that not all species are intrinsically susceptible to all antimicrobials. Acquired resistance may be permanent or temporary or "adaptative" (dependent on the growth conditions). Permanent resistance arises either from mutations (chromosomal resistance) or from acquisition of DNA from an outside source by one of three distinct mechanism: (a) transduction (transfer of genetic material via phage); (b) transformation (gene transfer via naked DNA); or (c) conjugation (transfer requiring direct contact between the bacteria). Within this category there are two elements that are self-transferable: plasmids and transposons (Tillotson and Watson, 2001).

1.1.5 Mechanisms of Resistance to Penicillin and Cephalosporins

Different processes are involved in the appearance, evolution, and dissemination of β-lactam-resistant pneumococcal strains. Among those that stand out are (1) the horizontal dissemination of the penicillin-resistant mosaic genes among pneumoccal strains; (2) the geographic spread of resistant clones; and (3) the selective pressure that indiscriminate use of antibiotics exerts upon pneumococci (Doern et al., 1998).

Analysis of penicillin-resistant strains of *S. pneumoniae* isolated from the clinical environment reveals a remarkable number of changes in these bacteria which, directly or indirectly, relates to the appearance of antibiotic-resistant pneumococcus. The main mechanism of resistance in clinical isolates of *S. pneumoniae* was first shown to involve the alteration of penicillin target proteins, the so-called PBPs, by the demostration that PBPs from penicillin-resistant bacteria had radically reduced affinities and/or binding capacities for the antibiotic molecule (Klugman, 1990). Therefore, the antibiotic is neither modified nor destroyed by hydrolysis (β-lactamases), but poorly recognized (Chenoweth et al., 2000; Garau, 2002; Musher, 2000).

Alterations to the penicillin-binding properties of these proteins are brought about by transfer of portions of the genes encoding the PBPs from other strepto-coccal species resulting in mosaic genes, indicating that the origin of these PBP genes were heterologous recombination events in which non-pneumococcal bacteria served as DNA donors (Baquero et al., 1998b; Markiewicz and Tomasz, 1989; Musher, 2000). These alterations can occur without affecting cell wall building functions of the enzymes or pneumococcal virulence (Markiewicz and Tomasz, 1989). Incorporation of one altered, low-affinity PBP gene marks the beginning of a resistant clone, which then expands through cell division until a member of the lineage engages in a second recombinational event that results in the modification of another of the high-molecular weight PBP genes in the recipient pneumococcus. The descendants of such a cell, which now has an increased penicillin MIC, may undergo further recombination events, each increasing the resistance level (Baquero et al., 1998a). Because only portions of the genes are transferred and because there is a range of PBPs that can be modified in a stepwise manner, the level of resistance to penicillin can vary considerably.

The combination of high antibiotic selective pressure and a high proportion of "colonizable" hosts is very effective for the selection of resistance. The presence of sublethal concentrations of a drug exerts selective pressure on a population of pathogens without eradicating it. Under these circumstances, mutant strains that possess a degree of drug resistance are favored and tend to dominate the population. From such populations with low-level resistance, more highly resistant organisms are more readily selected. It thus follows that one tactic to prevent the emergence of resistance is to minimize the time that suboptimal drug levels are present by thoughtful attention to dosing (Burgess, 1999).

The following PBPs have been identified in penicillin-susceptible pneumococci: 1a, 1b, 2x, 2b y 3 (Appelbaum, 2002). The final resistance level will depend on the combined action of the different PBPs or, in other words, the resistance phenotype of each strain depends on the genotype of all the PBPs involved (Baquero et al., 1998b). Loss of susceptibility affects all β-lactams, but to differing degrees depending on the affinity of each drug for the altered PBPs. Each β-lactam has a maximum binding affinity for any particular PBP, which is responsible for its specific antibacterial effect. At low concentrations, each β-lactam has a different preferential activity with respect to each of the PBPs. Ampicillin and amoxicillin provide an example of preferential affinity for two different PBPs: amoxicillin is clearly linked to PBP1a/1b, whilst ampicillin has it for PBP3. Resistance to third-generation cephalosporins requires alterations mainly in PBP2x and 1a, but they scarcely produce a lower susceptibility to penicillin. Conversely, mutations in PBP 2b are determinant for the development of high-level penicillin resistance, but not for cephalosporins. Selective mutations in PBP1a do not significantly modify resistance to penicillin, although they can contribute to the development of highly resistant isolates when they appear in combination with other variants. Gene mutations in PBP1a, PBP2x, and PBP2b should be considered highly penicillin-resistant mutants; hence, a much higher β-lactam concentration is needed to bind to and, consequently, inhibit the enzymatic activity of these enzymes than those in strains with isolated variations in PBP 2x and 2b (Baquero et al., 1998b). PBP3 appears to be important but nonessential for cell growth, since a variety of β-lactams are bound by PBP3 far below their respective MICs, and since the penicillin affinity of PBP3 is not altered in intrinsically penicillin-resistant strains. Mutations in PBP3 could very slightly alter the MIC of penicillin and cephalosporins. This protein seems to be the target of clavulanate acid, although the consequences of this interaction do not seem to be immediately followed by any antibiotic effect.

The penicillin-nonsusceptible *S. pneumoniae* have, to a greater or lesser degree, cross-resistance to all β-lactam antibiotics: carbenicillin, ticarcillin, aztreonam, and first-, second-, and third-generation cephalosporins. This resistance affects all β-lactam antibiotics that target PBP1 and PBP3, but it does not affect imipenem or other carbapenems that preferentially target PBP2b. This seems to be due to the specific location of PBP2 within the cytoplasmic membrane or to only a small quantity of PBP2 existing in a bacterial cell. In terms of MICs, the most active β-lactam antibiotics against *S. pneumoniae* are amoxicillin, cefotaxime, and ceftriaxone; the MICs of these antibiotics are two- to fourfold lower than bencylpenicillin. Amoxicillin has the best MIC:serum ratio of all oral β-lactam antibiotics against pneumococci. Piperacillin is the most active of the extended-spectrum penicillins. The addition of clavulanate, sulbactam, or tazobactam does not affect MICs of the parent β-lactam compound. First- and second-generation cephalosporins are active against penicillin-susceptible pneumococci, but have variable activity against penicillin-intermediate strains. Penicillin-resistant pneumococci (MIC ≥ 2 mg/l) are predictably resistant to first- and second-generation cephalosporins, and 15–30% of penicillin-resistant isolates are also resistant to cefotaxime and ceftriaxone (Lynch and Zhanel, 2005). In Spain, among 229 patients with CAP caused by pneumococcus with lowered susceptibility to penicillin (intermediate resistance and resistance), 73.4% showed diminished susceptibility to imipenem, 88.6% to cefuroxime, 14.8% to amoxicillin, while only 7.9% showed diminished susceptibility to cefotaxime (Aspa et al., 2004).

Penicillin-nonsusceptible *S. pneumoniae* are far more likely to be resistant to non-β-lactam antibiotics (e.g., macrolides, clindamycin, tetracyclines, chloramphenicol, and trimethoprim-sulfamethoxazole) (Aspa et al., 2004; Doern et al., 1998; Zhanel et al., 2003), so it could be considered that penicillin-resistance is a marker of resistance to other antimicrobial agents (Linares et al., 1992). Viridans group streptococci, which are part of the commensal bacteria of the upper respiratory tract, can transfer macrolide, penicillin, and tetracycline resistance genes to *S. pneumococci* (Cerda Zolezzi et al., 2004). This reflects transfer of non-β-lactam resistance determinants via transposons (e.g., transposons carrying ermB often coharbor the *TetM* gene, which confers resistance to tetracyclines, and chloramphenicol-resistance determinants) (Gay et al., 2000).

The close association between β-lactam resistance and macrolide resistance is not because the genes encoding resistance are linked, but because resistant determinants are selected in the same environment, and additive selective determinants confer selective advantage to those strains each time they are exposed to antibiotics. Resistance to fluoroquinolones is generally unrelated to penicillin susceptibility

(Vanderkooi et al., 2005; Schrag et al., 2004), although there is some evidence of an association between resistance to penicillin and resistance to fluoroquinolones (Chen et al., 1999; Ho, et al., 2001; Linares et al., 1999) and, in Spain, because of the high prevalence, also to macrolides (Garcia-Rey et al., 2000; Linares et al., 1999; Aspa et al., 2004). Some studies have also reported higher rates of fluoroquinolone resistance among penicillin nonsusceptible *S. pneumoniae* (Doern et al., 2005; De la Campa et al., 2004).

1.1.6 Relationship Between Pneumococcal Serotypes and Penicillin-Resistance

The emerging resistant populations probably belong to a limited number of *S. pneumoniae* serogroups. Among them, serogroups 6, 9, 14, 19, and 23 are much more frequently found to be resistant (Fenoll et al., 2002; Fenoll et al., 1991). In some geographic areas the increasing incidence of penicillin-resistant pneumococci is very closely related to the higher incidence of the above-mentioned serotypes (Appelbaum, 2002), which are grouped in a series of specific clones: clone 1 belongs to serotypes 23 and 19, and clone 3 to serotypes 14 and 27.

Penicillin-resistant clones most frequently retain a stable combination of genetic background, PBP gene sequence, and capsular serotype. However, capsular switch events, in which a well-defined penicillin-resistant clone served as a recipient of new capsular determinants, are being described with increasing frequency. The mechanism of these spontaneous genetic exchanges is genetic transformation. The capacity of multiresistant pneumococci to undergo capsular switch is an issue of obvious concern, particularly in the setting of selective pressure by conjugate vaccines. Spontaneous capsular switching in vivo can spread mutiresistance to new serotypes in a single genetic event and can generate highly virulent multidrug-resistant bacteria (Tomasz, 1999).

1.1.7 Risk Factors for Infection by Penicillin-Resistant S. pneumoniae

Knowing risk factors is essential for the clinician to suspect resistance to *S. pneumoniae* and to select appropriate empiric antimicrobial treatment. Risk factors for the acquisition of antimicrobial-resistant *S. pneumoniae* have been studied by means of multivariate analyses (Niederman, 2001) (see Table 1.2), and various authors have found that patients with penicillin-resistant pneumococci have a significantly higher incidence of (1) use of β-lactam antibiotics during the previous 3–6 months, (2) hospitalization during the previous 3 months, (3) nosocomial pneumonia, (4) suspected aspiration, (5) episodes of pneumonia during the previous year, (6) alcoholism, (7) noninvasive disease, and (8) an initially critical condition. Additional risk factors include age less than 5 years or greater than 65 years,

Table 1.2 Pneumococcal resistance risk factors

Previous β-lactam treatment (Bedos et al., 1996; Clavo-Sanchez et al., 1997; Nava et al., 1994; Pallares et al., 1987).
Nosocomial infection (Bedos et al., 1996; Pallares et al., 1987)
Previous hospitalization (Aspa et al., 2004; Pallares et al., 1987)
CAP in previous year (Pallares et al., 1987)
Serious illness (Pallares et al., 1987)
HIV infection (Aspa et al., 2004; Bedos et al., 1996)
Alcohol abuse (Clavo-Sanchez et al., 1997)
Comorbidities ($\geq$ 2) (Aspa et al., 2004)
Age < 5 and > 65 years old[a] (Aspa et al., 2004; Clavo-Sanchez et al., 1997; Karlowsky et al., 2003)
Others (children, closed institutions) (Bauer et al., 2001)
Chronic pulmonary disease (Aspa et al., 2004)
Suspected aspiration (Aspa et al., 2004)

CAP, Community-acquired pneumonia
[a]Cephalosporin resistance.

white race, closed communities (military camps or schools), attendance in day care centers, and the presence of chronic underlying disease (mainly COPD). (Aspa et al., 2004; Campbell and Silberman, 1998; Clavo-Sanchez et al., 1997; Ewig et al., 1999; Nava et al., 1994). It has also been shown that pediatric residents in long-term care facilities may be an important reservoir of penicillin-resistant pneumococci (Pons et al., 1996; Ekdahl et al., 1997). Finally, a French multicenter study reported (Bedos et al., 1996) that age less than 15 years, isolation of the organisms from the upper respiratory tract, and human immunodeficiency virus (HIV) infection as risk factors for pneumonia caused by antibiotic-resistant *S. pneumoniae*.

1.1.8 The Clinical Consequences of Penicillin Resistance in Patients with Pneumonia

S. pneumoniae has been consistently shown to represent the most frequent causative agent in CAP. Prospective studies reported an incidence of pneumococcal pneumonia of about 30–40% (Ruiz et al., 1999). Moreover, there is some evidence that most episodes without established etiology are in fact due to *S. pneumoniae* (Menéndez et al., 1999). Mortality from pneumococcal pneumonia is less than 1% for outpatients, but it reaches 50% in hospital-treated episodes and as much as 30% in bacteremic disease and the elderly population (Whitney et al., 2000).

It is well known that resistant pneumococcal strains can spread rapidly. Although the resistance pattern may vary among geographic areas and over time, data from many countries show resistance of *S. pneumoniae* to many antimicrobials. When selecting antibiotic therapy for pneumococcal infection, it is important to remember that *S. pneumoniae* strains resistant to penicillin also have decreased sensitivity to other β-lactam agents and non-β-lactam antibiotics, such as macrolides (Alfageme et al., 2005). The current situation in some countries has become a matter of deep concern. In Spain, according to the latest published studies, 30–50% of the strains presented some type of resistance to penicillin and 24–40% to macrolides. High frequencies of resistance to aminopenicillins and expanded-spectrum cephalosporins were observed only among penicillin-resistant strains (Aspa et al., 2004; Baquero et al., 1999). For *S. pneumoniae* the prevalences of highly resistant strains were 5% for amoxicillin and amoxicillin-clavulanic acid, 7% for cefotaxime, 22% for penicillin, and 31% for cefuroxime (Perez-Trallero et al., 2001, 2005). *S. pneumoniae* strains with a high level of resistance to penicillin and other antimicrobial agents appeared in the United States in the early 1990s (Breiman et al., 1994).

In recent years, there has been a greater concern to know the extent to which antimicrobial resistance may come to influence the morbidity and mortality of pneumococcal infections, specifically pneumonia. Treatment failures due to drug resistance have been reported with meningitis (Catalan et al., 1994; Sloas et al., 1992) and otitis media (Jacobs, 1996; Poole, 1995), but the relationship between drug resistance and treatment failures among patients with pneumococcal pneumonia is less clear (Buckingham et al., 1998; Choi and Lee, 1998; Deeks et al., 1999; Dowell et al., 1999; Feikin et al., 2000b; Turett et al., 1999). Several studies of patients with CAP failed to find an independent association between pneumococcal resistance and outcome when strains with penicillin MICs < 1 mg/l were considered (Feikin et al., 2000a; Friedland, 1995; Pallares et al., 1995). For strains with penicillin MICs of 2–4 mg/l some data suggest that there is no increase in therapeutic failure rates (Choi and Lee, 1998; Deeks et al., 1999), whilst others point to an increase in mortality (Deeks et al., 1999) or in the incidence of complications (Buckingham et al., 1998; Dowell et al., 1999; Turett et al., 1999). Fortunately, current levels of resistance to penicillin or cephalosporins mostly do not surpass MICs of 4 mg/l (Doern et al., 1998; Aspa et al., 2004; Perez-Trallero et al., 2001). Berry et al. (2000) have identified multiresistant pneumococci that are able to cause pneumonia in a rat model. Bactericidal activity (>3 logs of killing) was demonstrated by using amoxicillin-clavulanate against a pneumococcal strain with an amoxicillin MIC of 2 mg/l, but this antibiotic was not able to reliably produce a bactericidal effect when the infecting strain had an amoxicillin MIC of 8 mg/l.

For many of the presented studies, a number of confounding variables, other than the specific MIC of the pathogen, may influence outcome measurements. These include age, underlying disease, severity of illness at presentation, multilobar infiltrates, and immunosupression. In addition, several of the reports fail to specify the drug regimen followed by the patients, which again limits interpretation of the results. Pharmacodynamically, time above MIC ($T > MIC$) is the parameter for β-lactams that best correlates with bacteriological erradication (Craig, 1998a). Variability exists among the various parenteral β-lactams in regard to their abilities to attain a $T > MIC$ of 40% for drug-resistant strains. For example, among the cephalosporins, ceftriaxone, cefotaxime, and cefepime would all produce $T > MIC_{90}$ greater than 40% (range, 87–100%) against *S. pneumoniae*. Cefuroxime just hits the 40% target, whereas ceftazidime and cefazolin attain $T > MIC_{90}$ of only 32 and 20%, respectively. Similarly, among the remaining β-lactams, penicillin and ampicillin at standard dosages achieve $T > MIC_{90}$ far exceding 40%, whereas ticarcillin does not (23%). These data suggest that high-dose penicillin, cefotaxime or ceftriaxone is adequate for the treatment of invasive pneumococcal disease with penicillin MICs up to 2 mg/l but probably inadequate for higher MICs (Craig, 1998a,b).

A valid approach to initial antimicrobial treatment of suspected pneumococcal CAP in the era of drug resistance should consider local epidemiological data, individual risk factors of pneumococcal resistance, and the severity of the disease. Pallares et al. (1995) conducted a prospective study including 504 hospitalized patients with severe pneumococcal pneumonia and found a significantly higher (38%) mortality among patients with penicillin-resistant strains compared with patients with penicillin-sensitive strains (24%). When this excess of mortality was controlled, however, for other predictors of mortality, the risk of death was comparable for both groups of patients. These authors concluded that resistance to penicillin and/or cephalosporins was not associated with increased mortality in patients with pneumococcal pneumonia. In a prospective cohort study of 101 hospitalized patients with CAP caused by *S. pneumoniae*, Ewig et al. (1999) reported a mortality rate of 15 versus 6% among patients with any antimicrobial resistance compared with patients without resistance. Moreover, the authors found that discordant (compared with accurate) antimicrobial treatment was not associated significantly with death (12 versus 10%. RR:1.2; IC95 [0.3–5.3]; p:0.67) and concluded that antimicrobial resistance of *S. pneumoniae* was not associated independently with an adverse prognosis. Similarly, in a retrospective cohort study of bacteremic pneumococcal pneumonia with a high prevalence of HIV infection, only highly penicillin-resistant pneumococci were identified as independent risk factors of mortality (Turett et al., 1999). Finally, in another retrospective cohort study of invasive pneumococcal disease (Metlay et al., 2000), 44 of 192 patients (23%) infected with pneumococcal strains showed some degree of penicillin nonsusceptibility. Abnormal vital signs and laboratory values on admission were not different in relation to the presence of drug-resistant strains. There was no increased mortality in patients with drug resistance strains after adjustment for baseline differences in severity of illness (Fine et al., 1997). Compared with patients infected with penicillin-susceptible pneumococcal strains, patients whose isolates were nonsusceptible had a significantly higher risk of suppurative complications.

Feikin et al. (2000b) examined factors affecting mortality from pneumococcal pneumonia in nearly 6,000 hospitalized patients and found an increased mortality associated with age, underlying disease, Asian race, and residence in a particular (local community). Mortality was not associated with resistance to penicillin or cefotaxime when these factors were entered in a multivariate model adjusted for confounders. When deaths during the first four hospital days were excluded, mortality was signficantly associated with penicillin MICs of 4 mg/l or higher (OR 7.1) and cefotaxime MICs of 2 mg/l (OR 5.9). These data corroborate earlier case reports indicating that high-level resistance may be associated with adverse outcome.

More recently, Yu et al. (2003), in a large-scale study of 844 hospitalized patients with bacteremic pneumococcal pneumonia, found that penicillin resistance was not a risk factor for mortality. With the new NCCLS-2002 breakpoints, these authors showed that discordant therapy with penicillins, cefotaxime, and ceftriaxone (but not cefuroxime) did not result in a higher mortality rate. Similarly, time required for defervescence and frequency of suppurative complications were not associated with concordance of β-lactam antibiotic therapy. In contrast, Lujan et al. (2004) state that discordant therapy prescribed at admission was independently associated with higher (27-fold) mortality in bacteremic pneumococcal pneumonia. Both studies agree on how infrequently patients are found to be receiving discordant therapy using the new breakpoints, leaving open the possibility that the small size of the final sample may have affected the results.

In the authors' study (Aspa et al., 2004), already mentioned, in contrast with previous studies (Metlay et al., 2000), disseminated intravascular coagulation, empyema, and bacteremia were significantly more common in patients with drug-susceptible pneumococcal CAP, which may reflect the biological cost that resistance-determining mutations engender on the fitness of bacteria. Using multivariate survival analysis, factors related to mortality in this population were (1) bilateral disease [Hazard ratio (HR):1.98], (2) suspected aspiration (HR:2.79), (3) shock (HR:5.76), (4) HIV infection (HR:2.06), (5) renal failure (HR:1.86), and (6) Prognostic Severity Index (PSI) score categories IV versus I–III (HR:2.61) and categories V versus I–III (HR:3.24). Different groups of patients with significant mortality/morbidity were also analyzed (ICU, PSI class >III, renal failure, chronic lung disease and bacteremic patients). Only in patients with PSI class >III, the initial antimicrobial choice classified as "other combinations," was associated with higher mortality (Aspa et al., 2006). In conclusion, an independent association between initial antimicrobial regimen and 30-day mortality in community-acquired pneumococcal pneumonia patients could not be demonstrated, except for those with high PSI score.

Overall, several methodologic issues affecting the interpretation of currently available data deserve comment. First, current levels of resistance to penicillin or cephalosporins mostly do not surpass MICs of 2 mg/l, and as a consequence only a few patients receive a truly discordant antimicrobial treatment in the presence of microbial resistance. Therefore, high doses of β-lactams should remain the therapy of choice for pneumococcal pneumonia (Heffelfinger et al., 2000). Second, owing to the small numbers of strains with high-level resistance (MIC $\geq$ 4 mg/l) observed, current studies are underpowered to establish the impact of infections with these strains on outcome. As a result, the Drug-Resistant *S. pneumoniae* Therapeutic Group (Heffelfinger et al., 2000) has recommended that penicillin susceptibility categories be shifted upward so that the susceptible categories include all isolates with MICs of 1 mg/l, the intermediate categories include isolates with MICs of 2 mg/l, and the resistant category includes isolates with MICs of 4 mg/l.

Moreover, the overall mortality for community-acquired pneumococcal pneumonia has been stable at 10–20% for the past 40 years. Mortality is also an imprecise outcome measurement of the impact of antibiotic resistance, yet few studies have recorded other, more sensitive outcome measurements. With regard to the impact of antibiotic resistance, many authors have indicated that optimization of antibiotic therapy could be achieved through the understanding and application of pharmacokinetic and pharmacodynamic principles, which could be used to predict the likely success or failure of antibiotics and to indicate their optimal dosing schedules. At this time, it is generally believed that the current prevalence and levels of resistance of pneumococcus to penicillins in most areas of the world do not indicate the need for significant treatment changes in the management of CAP. For penicillin-sensitive pneumococcal infections, penicillin or/and aminopenicillin could still be used. In the case of infections with isolates of intermediate resistance to penicillin, a higher dosage of penicillin or amoxicillin should effectively treat these infections, judging from the pharmacokinetic/pharmacodynamic parameters described above. Alternative antibiotic therapies are only suggested in cases of infections with strains demonstrating high levels of resistance (Feldman, 2004).

1.1.9 Combination Antimicrobial Therapy for Pneumococcal Pneumonia

Several retrospective studies have suggested that the use of macrolide/β-lactam combination as part of the initial antimicrobial treatment of patients with CAP requiring hospital admission may shorten the hospital stay and reduce mortality rate in comparison with those treated with monotherapy (Stahl et al., 1999; Brown et al., 2003; Martinez et al., 2003; Sanchez et al., 2003; Waterer et al., 2001), even when *S. pneumoniae* is finally identified as the causative organism. However, many aspects of the apparently beneficial effects of the combined therapy remain unclear and/or controversial. There are inconsistencies in reported outcomes and confusing biases that may have influenced these results. For instance, groups receiving the β-lactam/macrolide combination, as opposed to monotherapy, are not comparable with regard to the average prognosis. In a retrospective study on 213 hospitalized patients, Burgess and Lewis (2000) concluded that it may not be necessary to add a macrolide to a non-pseudomonal third-generation cephalosporin in the initial empirical therapy of CAP. Furthermore, Johansen et al. (2000) has reported antagonism for the combination of penicillin–erythromycin with *S. pneumoniae* both in vitro and in animal models of invasive disease, suggesting that β-lactam antibiotics and macrolides should not be administered together unless pneumococcal infection is ruled out. Data from a recent, prospective multicenter study of patients with bacteremic pneumococcal illness (Baddour et al., 2004) suggest that combination antibiotic therapy improves survival, but only among critically ill patients and without demonstratable advantages from regimens including macrolides in comparison with non-macrolide combinations. In the above-mentioned study (Aspa et al., 2006), they found that combination therapy was not better than the use of β-lactams alone. Two meta-analyses evaluating the advantages of adding a macrolide to a β-lactam in the treatment of CAP patients have recently been published, and no additional benefits were found for this combination (Shefet et al., 2005; Mills et al., 2005).

Weiss and Tillotson (2005) have recently addressed the controversy of combination versus monotherapy in the treatment of hospitalized CAP. For the vast majority of patients with CAP (i.e., outpatients and inpatients on medical wards), the type of antibiotic regimen prescribed does not have any significant impact. For patients with severe pneumonia, the controversy remains alive. In the authors' opinion, severe CAP can be defined as the presence of *S. pneumoniae* bacteremia, the necessity of ventilatory support, or patients with a PSI class greater than III. In the meantime, for practical purpose, patients hospitalized with a diagnosis of severe CAP may benefit from a dual antibiotic therapy combining a third-generation cephalosporin and a macrolide. For the majority of hospitalized patients with CAP who are not severely ill, fluoroquinolone monotherapy remains an approved, tested, and reliable option.

1.2 Streptococcal Resistance to Macrolides, Tetracycline, Quinolones, and Clindamycin—Mechanisms and Implications

Resistance to the macrolide antibiotics (e.g., erythromycin, clarithromycin, and azithromycin) has escalated in association with penicillin resistance. In addition, macrolide resistance can develop independently of penicillin resistance. In many parts of the world macrolide-resistant *S. pneumoniae* are more common than penicillin- resistant *S. pneumoniae*. Macrolide resistance reflects both clonal dissemination (vertical) as well as horizontal spread of genetics elements. Taking into account the current levels of *S. pneumoniae* macrolide resistance in hospitalized patients with moderate or moderately severe CAP, empiric monotherapy with a macrolide is not optimal (Lynch and Zhanel, 2005).

Fluoroquinolones are widely prescribed for treatment of both community- and hospital-acquired respiratory infections, and recent guidelines include the newer respiratory fluoroquinolones as potential therapeutic options for CAP. However, rates of fluoroquinolone resistance have increased dramatically within the past decade among both nosocomial and community pathogens. Some fluoroquinolone-resistant pneumococci are also resistant to β-lactams and macrolides (Chen et al., 1999; Bhavnani et al., 2005; Low, 2004).

1.2.1 Macrolides and Related Antimicrobial Agents

Macrolides are composed of a minimum of two amino and/or neutral sugars attached to a lactone ring of variable size. Macrolides commercially available or in clinical development can be divided into 14-, 15-, and 16-membered lactone ring macrolides. These classes differ in their pharmacokinetic properties and in their responses to bacterial resistance mechanism (Leclercq and Courvalin, 1991). Lincosamides (lincomycin and the more active semisynthetic derivate clindamycin), are alkyl derivates of proline and are devoid of a lactone ring. Streptogramin antibiotics are composed of two factors, A and B, that act in synergy and are produced by the same microorganism.

Any discussion of mechanisms of resistance to macrolide antibiotics must include the chemically distinct, but functionally overlapping, lincosamide and streptogramin B families. This type of resistance has been referred to as MLS resistance (Perez-Trallero et al., 2001). Members of the MLS antibiotic superfamily include the macrolides (carbomycin, clarithromycin, erythromycin, josamycin, midecamycin, mycinamicin, niddamycin, rosaramicin, roxithromycin, spiramycin, and tylosin), the lincosamides (celesticetin, clindamycin, and lincosamycin), and the streptogramins (staphylomycin S, streptogramin B, and vernamycin B).

In 1956, a few years after the introduction of erythromycin therapy, resistance of staphylococci to this drug emerged and subsequently spread in France, the United Kingdom, and the United States (Chabbert, 1956; Finland et al., 1956; Garrod, 1957). The MLS cross-resistance phenotype due to modification of the drug target is widely distributed and has, since then, been detected in *Streptococcus* spp. (Gilmore et al., 1982).

1.2.1.1 Mechanisms of Resistance to Macrolides

Most strains of erythromycin-resistant pneumococci have one of two mechanisms of resistance: an altered ribosomal binding site (modification of the target of the antibiotics), or an active efflux pump mechanism (Leclercq and Courvalin, 1991; Sutcliffe et al., 1996). Pneumococci resistant to erythromycin by either mechanism are also resistant to azithromycin, clarithromycin, and roxithromycin.

Modification at the Ribosomal Target Site

Macrolides and ketolides inhibit protein synthesis by binding to ribosomal target sites in bacteria. Several distinct mutations and phenotypes have been detected in laboratory and clinical settings. The dominant mechanism, encoded by the *erm*AM (erythromycin ribosome methylation) gene, methylates a highly conserved region of the peptidyl transferase loop in domain V of 23S mRNA. The result of this rRNA methylation is a conformational change in the ribosome that reduces the affinity of MLS antibiotics for the binding site. Streptogramin A-type antibiotics are unaffected, and synergy between the two components of streptogramin against MLS-resistant strains is maintained (Chabbert and Courvalin, 1971). This phenotype is referred to as the MLS$_B$ phenotype (Leclercq and Courvalin, 1991; Montanari et al., 2001); it is prevalent in Europe (in Spain is the predominant phenotype) and South Africa (Johnston et al., 1998; Montanari et al., 2001; Pallares et al., 2003). *Erm*AM-positive isolates typically display high-level resistance to both macrolides (i.e., erythromycin MIC > 64 mg/l) and clindamycin (MIC > 8 mg/l) that cannot be overcome by increasing the dosage. Hence, clinical failures can be expected with *erm*AM strains.

Active Efflux of Antibiotics

The other known resistance mechanism involves active (proton-dependent) efflux (the M resistance phenotype), which is encoded by the *mefE* gene. This mechanism of macrolide resistance confers resistance to 14- and 15-member macrolides, but it does not affect lincosamides, streptogramins, or 16-member macrolides. Typically, pneumococci containing *mefE* have MICs to erythromycin of 1–32 mg/l, and high concentrations of macrolides achieved in tissue or at sites of infection may overcome this mechanism of resistance.

The prevalence of *erm*B and *mef*E mechanisms varies according to geographic region. Efflux accounts for less than 20% of macrolide-resistant pneumococcal isolates in Europe (Hoban et al., 2001; Klugman, 1990; Lynch and Martinez, 2002) but is the most prevalent mechanism of macrolide resistance in North America (Johnston et al., 1998; Montanari et al., 2001; Pallares et al., 2003). In Spain, according to the latest published studies, macrolide resistance rates have remained between 20 and 40% (Aspa et al., 2004; Baquero et al., 1999), and the prevalence of the M phenotype has been increasing in the last few years (Pallares et al., 2003). The proportion of MLS$_B$/M phenotypes was 96.8/3.2 in the years 1990–1997; 91.3/8.7 in the years 1997–1999, and, 87.6/12.4 between 2000 and 2002.

1.2.1.2 Clinical Implications of Macrolide Resistance

Pneumococci are often susceptible to MLS antibiotics. The first strains resistant to erythromycin appeared in 1976, and their incidence progressively increased to 20–30% by 1986 (Acar and Buu-Hoi, 1988). This evolution may have been due to the epidemic spread of strains of serotypes 6, 14, 19, and 23 and to a rapid dissemination of Tn1545-related transposons in pneumococcci (Leclercq and Courvalin, 1991).

Some authors have reported failure while using a macrolide to treat a pneumococcal infection caused by a macrolide-resistant strain (Fogarty et al., 2000; Kelley et al., 2000; Lonks et al., 2002; Moreno et al., 1995; Waterer et al., 2000). In a recent review (Klugman, 2002), it was stated that there is increasing evidence of bacteriologically confirmed macrolide failure in pneumococcal pneumonia therapy at MICs $\geq$ 4 mg/l. This is particularly important in Europe, where the predominant mechanism of resistance is typically high level (Aspa et al., 2004). To add to the confusion, the emergence of resistance to macrolide agents during treatment has been recently reported (Musher et al., 2002). Approximately 20% of pneumococcal strains are naturally resistant to macrolides not related to antibiotic use. The use of macrolides to treat CAP, even if a successful outcome is achieved, should be limited, because their use has been associated with increasing penicillin resistance among *S. pneumoniae*. In addition, macrolides have been associated with acquired penicillin resistance. Macrolide resistance is present in less than 5% of penicillin-sensitive pneumococci and in 48–70% of isolates with high-level penicillin resistance (MIC > 2 mg/l) (Lynch and Martinez, 2002). In hospitalized patients with moderate or moderately severe CAP, empiric monotherapy with a macrolide is not optimal. Ketolides are semisynthetic derivates of 14-membered macrolides in which the clanidose at position 3 is replaced with a keto group. Because of this and other additional modifications, telithromycin is active against macrolide-resistant pneumococci that contain either the methylase or the efflux mechanism of resistance

(Bonnefoy et al., 1997; Rosato et al., 1998). Telithromycin is active against nearly 100% of macrolide-resistant *S. pneumoniae* (Garau, 2002). Telithromycin was approved by the United States Food and Drug Administration (FDA) for the treatment of mild to moderately severe CAP due to *S. pneumoniae* (including multidrug resistant strains). Given the limited data among seriously ill patients, telithromycin is not approved for severe CAP or bacteremic cases. Nonetheless, telithromycin has a role for treating community-acquired pneumococcal infections resistant to either or both penicillin and macrolides. Although rare, there are clinical isolates of *S. pneumoniae* resistant to telithromycin (MIC $\geq$ 4 mg/l) and other macrolides (Boswell et al., 1998). Post marketing reports of associated hepatotoxicity should also be considered (Clay, et al., 2006, Graham, 2006). At present, the committee of the recent IDSA/ATS guidelines (Mandell, et al., 2007) is awaiting further evaluation of the safety of this drug by the FSA before making its final recommendation.

1.2.1.3 Risk Factors for Macrolide-Resistant *S. pneumoniae*

Risk factors for pneumonia caused by erythromycin-resistant pneumococcus have been assessed in a prospective Spanish study (Aspa et al., 2004). Previous hospital admissions (OR 1.89) and resistance to penicillin (OR 15.85) were identified as independent risk factors. Age less than 5 years, white race, and a nosocomial origin of the infections have also been reported as risk factors for antibiotic resistance (Hyde et al., 2001; Moreno et al., 1995) (see Table 1.3).

1.2.2 Quinolones

In association with increased utilization of fluoroquinolone antibiotics in both hospital and community settings, concern has been raised regarding the potential for emergence of resistance to this class of antibiotics. NCCLS breakpoints for levofloxacin are as follows: MIC = $\leq$ 2 mg/l (susceptible); 4 mg/l (intermediate), 8 mg/l (resistant). For moxifloxacin and gatifloxacin, breakpoints are $\leq$1 (susceptible); 2 (intermediate); >4 mg/l resistant. Breakpoints for ciprofloxacin have not been defined, but most experts consider ciprofloxacin MICs of >4 mg/l, as a marker of resistance.

Table 1.3 Pneumococcal macrolides-resistance risk factors

Age less than 5 years old
Nosocomial pneumonia
White race
Previously hospitalizad
Penicillin-resistance

Aspa et al., (2004), Hyde et al., (2001), and Moreno et al., (1995).

1.2.2.1 Mechanisms of Resistance

Once inside the cell, the antimicrobial action of the fluoroquinolones is mediated through the inhibition of two DNA topoisomerase enzymes, DNA gyrase (also termed "topoisomerase type II") and topoisomerase IV. As a class, topoisomerases are essential in controlling the topological state of DNA by catalyzing supercoiling, relaxing, unknotting, and decatenation reactions which are vital for DNA replication, transcription, recombination and repair. Topoisomerases maintain cellular DNA in an appropriate state of supercoiling in both replicating and nonreplicating regions of the bacterial chromosome. DNA gyrase removes the excess positive supercoiling that builds up ahead of the DNA replication fork as a result of enzymes replicating DNA. Without DNA gyrase-mediated relaxation, this excess positive supercoiling would ultimately arrest DNA replication. After DNA replication, topoisomerase IV activity helps to separate the daughter DNA molecules.

Resistance to quinolones occurs (Hooper and Wolfson, 1993) by mutation in the genes that encode DNA gyrase and topoisomerase IV, or by lowering intracellular drug content through an active efflux mechanism (Drlica and Zhao, 1997; Janoir et al., 1996; Muñoz and De La Campa, 1996).

Altered Target Sites

The complexity in the interactions of the quinolones with topoisomerases is the basis for the differences in antibacterial spectrum among the quinolones and is also the basis for differences in the selection of resistant strains of bacteria. Generally, bacteria develop resistance to fluoroquinolones through chromosomal mutation in the target enzymes of fluoroquinolone action, DNA gyrase and topoisomerase IV. Because of the different preferential targets of various fluoroquinolone molecules, pneumococcal resistance could theoretically be conferred by mutations in either the *gyrA* or *gyrB* genes (encoding DNA gyrase) or in the *parC* or *parE* genes (encoding DNA topoisomerase IV (Appelbaum, 1992; Muñoz and De La Campa, 1996; Pan et al., 1996; Pestova et al., 1999; Tankovic et al., 2003).

The various fluoroquinolone antibiotics differ in their potency and potential for emergence of resistance based on intrinsic activity (e.g. MIC), pharmacokinetics and pharmacodynamics (Ambrose et al., 2003). The overall potency (based on MICs alone) of fluoroquinolones, from most to least active, is as follows: gemifloxacin, moxifloxacin, gatifloxacin, levofloxacin, and ciprofloxacin (Zhanel et al., 2003). The fluoroquinolones also differ in target site affinity. Ciprofloxacin and levofloxacin preferentially inhibit pneumococcal topoisomerase IV rather than gyrase enzymes (Fernandez-Moreira et al., 2000). Gatifloxacin, moxifloxacin, and gemifloxacin have been reported to bind to both targets, but may preferentially bind GyrA (Smith et al., 2004a). According to the mechanisms of action, gatifloxacin, moxifloxacin, and gemifloxacin would be expected to preferentially select for *gyrA* mutations, whereas ciprofloxacin and levofloxacin would select for mutations in *parC* (Fukuda and Hiramatsu, 1999).

Importantly, a "first-step" *parC* mutation in the quinolone resistance determinant region (QRDR) increases the risk for subsequent mutations (e.g., *gyrA*) that confer high-level resistance to fluoroquinolones (Lim et al., 2003). The use of less potent fluoroquinolones (e.g., ciprofloxacin) or low doses of newer fluoroquinolones (e.g., levofloxacin 500 mg qd) may amplify the risk of fluoroquinolone resistance development (Lynch and Zhanel, 2005; Low, 2004, 2005; Tillotson et al., 2001).

Resistance to fluoroquinolones in *S. pneumoniae* arises primarily by spontaneous point mutations in the QRDR of either or both *parC* and *gyrA* (Bast et al., 2000). Low-level resistance to ciprofloxacin typically develops from mutations altering *parC*. High-level fluoroquinolone resistance strains typically have dual mutations in QRDRs affecting both *parC* and *gyrA* (Bast et al., 2000). Mutations in *gyrB* or *parE* are less common, and their impact on fluoroquinolone resistance is less than *gyrA* or *parC* (Piddock, 1999; Lynch and Zhanel, 2005). Mutations in *parC* confer resistance to ciprofloxacin but these isolates remain susceptible to gatifloxacin, moxifloxacin, and levofloxacin (De la Campa et al., 2004). However, once a first-step *parC* mutation exists, the probability for developing a *gyrA* mutation with exposure to fluoroquinolones increases (Smith et al., 2004a; Gillespie et al., 2003). Dual mutations (*parC* and *gyrA*) are associated with high-grade resistance to ciprofloxacin (MICs $\geq$ 16 mg/l) and confer resistance to levofloxacin, gatifloxacin, and moxifloxacin and, most of the time, to gemifloxacin (De la Campa et al., 2004; Brueggemann et al., 2002).

First-step mutations will not be detected by conventional susceptibility testing and may reflect the "tip of the iceberg" for fluoroquinolone resistance. When isolates harbor single-step mutations in *parC* or *gyrA*, treatment with conventional doses of fluoroquinolones may fail to achieve bacterial eradication (Smith et al., 2004b). Emergence of resistant subpopulations occurs more readily when isolates containing preexisting mutations (either *parC* or *gyrA*) are exposed to fluoroquinolones concentrations within a specific range (Allen et al., 2003). Further, secondary mutations are acquired more rapidly than first-step mutations and may result in isolates that are highly resistant to all fluoroquinolones (Smith et al., 2004a). Ciprofloxacin and low doses (500 mg qd) of levofloxacin are less potent than gatifloxacin (Hoban et al., 2001), moxifloxacin (Dalhoff et al., 2001) or gemifloxacin (File and Tillotson, 2004). Theoretically, the use of more potent fluoroquinolones (particularly gemifloxacin or moxifloxacin), as well as high doses of levofloxacin (750 mg qd), may be a means to reduce emergence of resistance (Tillotson et al., 2001).

Mutant Prevention Concentration (MPC) is defined as the drug concentration that prevents the growth of mutants from a susceptible population of >10^{10} cells (Dong et al., 1999). Conceptually, dosing above the MPC may minimize selection of resistant mutants during antibiotic treatment (Smith et al., 2004a; Blondeau et al., 2001). MPC studies found that moxifloxacin and gatifloxacin had lower MPCs than levofloxacin; this likely reflects higher intrinsic activity of moxifloxacin and gatifloxacin for mutants (Blondeau et al., 2001; Li et al., 2002; Allen et al., 2003).

An additional concept, the mutant selection window, is defined as the range of concentration between the MIC and MPC (Zhao and Drlica, 2001, 2002). Within this concentration range, selection of resistant subpopulations may occur. On the contrary, antimicrobial concentrations exceeding the upper boundary of the selection window (MPC) should restrict the expansion of all single-step mutants, just as antimicrobial concentrations above the lower boundary of the window (MIC) are expected to limit the outgrowth of susceptible pathogens.

Efflux Pump

It has been found that *S. pneumoniae* also exhibits resistance to fluoroquinolones through active efflux of the drug. This efflux mechanism is probably a multidrug transporter that contributes a degree of intrinsic resistance to *S. pneumoniae*. Hydrophilic fluoroquinolones, such as ciprofloxacin, appear more prone to efflux than more hydrophobic molecules, such as the newer quinolones, in gram-positive organisms. Therefore, the clinical significance of inhibiting the accumulation of the drug in the bacterial cell is probably less important for the new fluoroquinolones (Chen et al., 1999; Hoban et al., 2001), and results in low level resistance.

1.2.2.2 Clinical Implications of Quinolone Resistance

The fluoroquinolones represent valuable alternatives for the therapy of pneumo-coccal infections, since their activity is not affected by resistance to other antimi-crobials classes (Critchley et al., 2002). However, although they have not reached the high resistance levels that other antibiotic classes present (such as β-lactams and macrolides), resistance to quinolones is increasing progressively. In the last decade the prevalence of pneumococcal infections resistant to fluoroquinolones has increased in Canada (Chen et al., 1999) from 0% in 1993 to 1.7% in 1997. This coincides with the increased use of ciprofloxacin in adults. During the 1990s, pneu-mococcal infections resistant to fluoroquinolones only represented some 2–3% of the strains in North America (Chen et al., 1999; Whitney et al., 2000) and Spain (Linares et al., 1999); however, a tendency towards an increase in resistance is being observed now. In a recent study carried out in the United States (Doern et al., 2005) to assess trends of antimicrobial resistance in *S. pneumoniae*, 2.3% of iso-lates had ciprofloxacin MICs of ≥ 4 mg/l. Trend analysis since 1994–1995 indi-cated that rates of resistance to β-lactams, macrolides, tetracyclines, trimethoprim–sulfamethoxazole (TMP–SMZ) and multiple drugs have either plateaued or have begun to decrease. Conversely, fluoroquinolone resistance among *S. pneumoniae* is becoming more prevalent. Ciprofloxacin MIC of ≥ 4 mg/l as an indicator of dimin-ished fluoroquinolone activity, appeared to have occurred in the United States some-time during the 2-year period between the winters of 1999–2000 and 2001–2002. This change was sustained through the most recent survey period, the winter of 2002–2003. The progressive increase in ciprofloxacin resistance seems to indicate that there are pneumococcal strains with single mutations that remain sensitive to the new fluoroquinolones in vitro (Chen et al., 1999; Ho et al., 2001; Schmitz et al., 2001; Whitney et al., 2000). Their real incidence, however, is not known (Davies et al., 2002), since the usual sensitivity tests are incapable of detecting this group of pneumococci with emerging fluoroquinolone resistance. Although the clinical importance of such low-level resistance is not known, these mutants are ready to undergo the second mutation, which involves DNA gyrase (*gyrA*) and results in high-level resistance (ciprofloxacin MIC 16–64 µg/ml) (Baquero, 2001; Tillotson et al., 2001).

Another, perhaps more refined means for comparing pneumococci with respect to the effect of fluoroquinolones over time is mutation analysis. In 1999–2000, Davies et al. (2002) studied 528 clinical isolates of pneumococci and found that 4.5% of the levofloxacin-sensitive strains (MIC ≤ 0.06–2 µg/ml) had a single mutation. However, variations were noticeable according to the MICs of each strain: no mutation was found in those whose MIC was ≤0.06 µg/ml and a single mutation was found in 7.3% of those that had a MIC of 1 µg/ml and in 71% of those with MIC of 2 µg/ml (all in *parC*). Other authors have shown similar results (Low et al., 2002). Both studies coincide in pointing out that lowering the levofloxacin pneumococcal sensitivity breakpoint to 1 µg/ml would be a pre-emptive measure for detecting strains with a single mutation (a direct precursor to fluoroquinolone clinical resistance) (Davies et al., 2002; Low et al., 2002). Doern et al. (2005) estimated that 21.9% of the 2002–2003 collection of *S. pneumoniae* isolates had mutations in *parC* (21%), *gyrA* (0.3%), or both (0.6%). A similar analysis of the isolates evaluated during their survey yielded an estimated QRDR mutation rate of 4.7%. On the basis of current NCCLS criteria, the following percentages of isolates would have been classified as intermediately resistant or resistant to levofloxacin, gatifloxacin, moxifloxacin, and gemifloxacin: 0.8, 0.7, 0.5, and 0.5%, respectively.

This relatively low resistance rate when compared to β-lactams has caused clinicians to consider fluoroquinolones as relatively safe agents in the treatment of pneumococcal infections. Be that as it may, it is worth pointing out that the appearance of resistance during treatment may occur, reflecting spontaneous mutations in QRDR. Such a change could adversely affect the pharmacodynamics of the drug. In a recent review, Fuller and coworkers found 20 published reports of clinical failures with fluoroquinolones used to treat respiratory tract infections due to fluoroquinolone-resistant *S. pneumoniae*. These authors specifically report 12 treatment failures in CAP and two failures in hospital-acquired pneumonia (Fuller and Low, 2005). More recently, Endimiani et al. (2005) have described a similar case in a patient with CAP. In all these cases, patients were treated with ciprofloxacin or with levofloxacin at low doses (500 mg qd). Failure of treatment of pneumococcal pneumonia due to resistance to levofloxacin has recently been described (Davidson et al., 2002; Zhanel et al., 2002). Pneumococcus acquired resistance during treatment in half of the cases, while the other half showed resistance after the antibiotic treatment (Empey et al., 2001; Ross et al., 2002; Urban et al., 2001). In a Spanish study on 2,822 pneumococcal resistant strains, half of the isolates came from patients who had been treated with a quinolone in the preceding three months (Linares et al., 1999).

Emergence of resistance during treatment is usually due to quinolones that are associated with a high mutation rate as determined in vitro. During epidemic outbreaks, when successive isolates of a strain are not routinely tested, changes within the susceptibility pattern are often overlooked. Sometimes several weeks elapse before the percentage of clinical failures becomes high enough to be noticed by physicians and to suggest reevaluation of the standard treatment (Davidson et al., 2002; Ho et al., 2001; Schrag et al., 2000). Risk factors for clinical failures with levofloxacin include low doses, residence in long-term care facilities, prior use of fluoroquinolones within 3 months, recent or current hospitalization, and first step (*parC*) mutation (Fuller and Low, 2005; Low, 2005; Lim et al., 2003).

Resistance to β-lactam or macrolide antibiotics during therapy rarely develops because resistance to these agents requires acquisition of exogenous DNA that encodes for resistance. In Fuller's work (Fuller and Low, 2005), the authors emphasize that clinical failures due to fluoroquinolone-resistant *S. pneumoniae* have not yet been observed when the more potent fluoroquinolones (moxifloxacin, gatifloxacin, gemofloxacin, or high-dose levofloxacin) are used. The explanation of this fact has already been commented on previously (Tillotson et al., 2001).

Globally, and particularly in North America, most fluoroquinolone-resistant strains of *S. pneumoniae* arise from heterogeneous mutations and are not clonal. However, clonal spread of fluoroquinolone-resistant pneumococci has been documented (Lynch and Zhanel, 2005; De la Campa et al., 2004).

Fluoroquinolone resistance in *S. pneumoniae* has been slow to emerge, with levofloxacin resistance rates in most countries being less than 1% (Sahm et al., 2001; Critchley et al., 2002). As the prevalence of resistant-pneumococci increases, so does the likelihood that treatment will fail if susceptibility testing is not performed (Davidson et al., 2002). Davidson et al. state that current data indicate that recent exposure to a fluoroquinolone should be a contraindication to the use of another fluoroquinolone for the empirical treatment of CAP. Other risk factors for infection with a resistant strain may need to be taken into consideration before one of these agents is prescribed (Ho, 2001). Since fluoroquinolones are being recommended in some situations for the empirical treatment of CAP, it may now be appropriate to consider recommending routine testing and reporting of the susceptibility of pneumococci to these agents. However, it is noteworthy that although routine testing reliably detects high-level resistance to levofloxacin, it may not always detect low-level resistance (Davidson et al., 2002).

Since *S. pneumoniae* can develop resistance to fluoroquinolones during therapy and since cross-resistance to other fluoroquinolones is likely to occur (Chen, 1999), physicians should be aware of the consequences of substituting one fluoroquinolone compound for another if a patient does not have a response to the initial therapy (Davidson et al., 2002).

1.2.2.3 Risk Factors for Quinolone-Resistant *S. pneumoniae*

The following risk factors have been established for quinolone resistance: previous exposure to quinolones, old age, nursing home, nosocomial acquisition of the infection, isolates in sputum, penicillin resistance and COPD (Ho et al., 2001). The nasopharynx is considered the main reservoir of penicillin- and macrolide-resistant pneumococci. It has been suggested that, similarly, the airways of elderly COPD

Table 1.4 Pneumococcal quinolones-resistance risk factors

Previous exposure to quinolones
Age > 65 years old
Nursing home
Nosocomial pneumonia
Isolates in sputum
Penicillin-resistance
Chronic Obstructive Pulmonary Disease (COPD)

Ho et al., (2001)

patients may have the same role for fluoroquinolone-resistant pneumococcal strains (Ho et al., 2001) (Table 1.4).

1.2.3 Mechanisms of Resistance to Other Antimicrobials

Regarding other less used antimicrobials, such as tetracyclines and chloramphenicol, a significant decrease in the resistance rates has been observed during the last two decades. Penicillin-resistant pneumococci are also resistant to TMP-SMZ (20–35.9%) and tetracycline (8–16.6%) (Doern, 2001; Hofmann et al., 1995).

Tetracyclines inhibit bacterial protein synthesis by reversible binding on the 30S ribosome that blocks the attachment of transfer RNA (tRNA) to an acceptor site on the messenger RNA (mRNA)-ribosomal complex. Bacteria become resistant to tetracylines by at least two mechanisms: efflux and ribosomal protection. In efflux, a resistance gene encodes a membrane protein that actively pumps the tetracycline out of the cell. This is the mechanism of action of the tetracycline-resistance gene on the artificial plasmid pBR322. In ribosomal protection, the *tetM* gene encodes a protein that binds to the ribosome and prevents tetracyclines from acting on the ribosome. This resistance gene is carried on the same transposon as genes that encode proteins that offer similar protection against TMP-SMZ and chloramphenicol (Burdett, 1991; Jacoby, 1994).

1.2.4 Multidrug-Resistant S. pneumoniae

Multiresistance is defined as resistance to at least three or more classes of antibiotics. In recent years, penicillin-resistant and even multiresistant pneumococci have spread worldwide. Several patient-related factors for acquisition of penicillin/ erythromycin-resistant pneumococcal CAP have been reported. However, factors associated with CAP caused by multidrug-resistant *S. pneumoniae* (MDRSP) have not been extensively studied. Non-β-lactam antibiotic resistance tends to be more common among strains not susceptible to penicillin (Butler et al., 1996; Hofmann et al., 1995). The reasons why a pneumococcus develops simultaneous resistance to various types of antibiotics are not clear, but some resistance determinants are carried together in the same transposon.

Currently, the isolation of MDRSP strains, both in adults and children, has been reported all over the world (Appelbaum, 2002; Butler et al., 1996; Chen et al., 1999; Friedland and McCracken, 1994; Hofmann et al., 1995; McCullers et al., 2000; Verhaegen and Verbist, 1999). Jacobs et al. (1978) reported the emergence of pneumococci resistant to multiple antimicrobial agents in Durban, South Africa, in 1978. After this sentinel report, MDRSP has become a widespread problem. Between 1995 and 1998, the proportion of MDRSP increased from 9 to 14% in the USA (Whitney et al., 2000). As part of the Alexander Project (global surveillance study, 1992–2001) (Mera et al., 2005), it has been recently reported that the combined resistance to penicillin and erythromycin have increased 4.9 times, up to 15.3%, and that three of four penicillin-resistant *S. pneumoniae* isolates are also multiresistant. A report from the SENTRY Antimicrobial Surveillance Program (1997–2003) (Pottumarthy et al., 2005) showed that the multiple resistance rate ranged from 17.6% (penicillin and erythromycin resistance only) to 5.7% (resistance to five drugs). An evaluation of 6,362 *S. pneumoniae* isolates collected during the 2000–2001 community-acquired respiratory tract infection season showed that 88.1 and 80.2% of the 1,077 isolates with high-level penicillin-resistance were also resistant to trimethoprim–sulfamethoxazole and azithromycin, respectively (Kelly, 2001). The Prospective Resistant Organism Tracking and Epidemiology for the Ketolide Telithromicyn (PROTECKT) study also showed a steady rise in pneumococcal resistance among common antibiotics, as well as an increase in MDRSP (Karchmer, 2004). Doern et al. (2005), in 1,817 *S. pneumoniae* isolates obtained from patients with community-acquired respiratory tract infections in 44 US medical centers (2002–2003) reported that 22.2% of isolates were multidrug resistant. In Spain, Fenoll et al. (1998) have also described multiple drug resistance increasing, with rates ranging from 1.1% in 1979–1984 to 7.7 in 1985–1989 and 12.5 in 1990–1996 (9.5% in 1990 to 16.6% in 1996).

1.3 Streptococcal and Enterococcal Resistance to Streptogramin, Oxazolidinones, Vancomycin—Emerging Issues

In recent years, there has been a dramatic increase in the number of infections caused by gram-positive bacteria. This is compounded by the emergence of resistance in enterococci, staphylococci, and pneumococci. For many years, both streptococcal and enterococcal bacteria have uniformly maintained their sensitivity to vancomycin. However, in 1988, this situation began to change when the first isolation of *Enterococcus faecium* resistant to vancomycin was reported (Leclerq et al, 1988). Shortly after, the first isolations of *Enterococcus faecalis* resistant to vancomycin were also documented (Uttley et al., 1988) and, at present, enterococci are best known as antibiotic-resistant opportunistic pathogens that are commonly recovered from patients who have received multiple courses of antibiotics and have been hospitalized for prolonged periods.

The emergence of enterocci with resistance to vancomycin (VRE) has been followed by an increase in the frequency with which these species are recovered. For many patients infected with these resistant organisms, there may not be effective antimicrobial therapy. In recent years, only the combination of streptogramins (quinupristin/dalfopristin), an oxazolidonone (linezolid), and a lipopeptide (daptomycin) have been introduced into clinical practice (Hancock, 2005).

1.3.1 Glycopeptides: Vancomycin

Vancomycin is a glycopeptide antibiotic that exerts its main bactericidal effect by inhibiting the biosynthesis of the major structural polymer of the bacterial cell wall, peptidoglycan. Vancomycin complexes with the D-alanyl-D-alanine (D-Ala-D-Ala) portion of peptide precursor units, which fits into a "pocket" in the vancomycin molecule, thereby inhibiting vital peptidoglycan polymerase and transpeptidation reactions. It inhibits the second stage of synthesis of peptidoglycan at a site earlier than the site of action of penicillin; thus, no cross-resistance occurs. In addition, vancomycin may impair RNA synthesis (Jordan and Inniss, 1959) and alter the permeability of cytoplasmic membranes (Jordan and Mallory, 1964). Its large size effectively prevents it from crossing the outer cell membrane of gram-negative bacteria; thus, no useful activity is conferred against these organisms. In spite of its toxicity, vancomycin is the last line of defense against organisms such as *Enterococcus*, some strains of which are resistant to most other antibiotics. Teicoplanin is chemically similar to vancomycin, but it has important differences that are responsible for its unique physical and chemical properties. Overall, significant adverse effects have seldom been encountered, although further studies are needed to evaluate toxicity of teicoplanin in the current conditions of use.

1.3.1.1 Mechanisms of Resistance

From a theoretical point of view, four types of vancomycin resistance may be considered: enzymatic inactivation of the antibiotic (not yet observed); increased synthesis of precursors to which vancomycin binds the terminal dipeptide D-Ala-D-Ala (staphylococci); sequestration of the antibiotic in the cell wall (staphylococci); and target modification, the most important mechanism of resistance in enterococci. Target is altered by enzymes that make cell-wall precursors ending in D-alanyl-D-lactate (D-Ala-D-Lac) or D-alanyl-D-serine (D-Ala-D-Ser), to which vancomycin binds with very low affinity. Other enzymes prevent synthesis of or modify endogenous cell-wall precursors ending in D-Ala-D-Ala, to which vancomycin binds with high affinity.

Thus far, six glycopeptide resistance phenotypes have been described in enterococci (VanA, VanB, VanC, VanD, VanE, and VanG), which differ by the genetic

Table 1.5 Characteristics of different phenotypes of glycopeptide-resistant enterococci

Characteristics	VANA	VANB	VANC	VAND	VANE	VANG
Vancomycin MIC (mg/l)	64–1024	4–1024	2–32	16–64	16–32	16–32
Teicoplanin MIC (mg/l)	16–512	<1	<1	2–16	<0.5	<0.5
Genetic determinant	Acquired	Acquired	Intrinsec	Acquired	Acquired	Acquired
Tranferable	Yes	Yes	No	No	No	No
New precursor of ligase	D-ala-D-lac	D-ala-D-lac	D-ala-D-ser	D-ala-D-lac	D-ala-D-ser	D-ala-D-ser
Expression of resistance	Inducible	Inducible	Constitutive	Constitutive	Constitutive or Inducible	Constitutive or Inducible

base, the regulation of expression, and the level of resistance conferred (Table 1.5). A seventh phenotype of glycopeptide resistance (VanF) has been described in *Paenibacillus* species (Fraimow et al., 2005; Patel et al., 2000). These phenotypes can be distinguished by the sequence of the structural gene for the resistance ligase (*vanA, vanB, vanC, vanD, vanE,* and *vanG*). The genes for types A, B, D, E, and G are acquired. In contrast, the genes encoding the VanC type of vancomycin resistance are constitutively present in *Enterococcus gallinarum* and *Enterococcus casseliflavus/Enterococcus flavescens*.

Vancomycin resistance is not due to the acquisition of only one gene. Each resistance phenotype is associated with a complex cluster of genes. These genes are physically grouped in operons and are located on plasmids or in the chromosome, and they can be easily transferred, even between species. The mechanism of resistance has been best characterized for the *vanA* cluster of nine genes carried on transposon Tn*1546f*. This mobile genetic element is 10,851 base pairs long and encodes two genes responsible for transposition functions (*orf1* and *orf2*); five genes responsible for the regulation and expression of resistance (*vanR, vanS, vanH, vanA,* and *vanX*); and two genes with auxiliary roles (*vanY* and *vanZ*). Similar gene clusters are found in the remaining resistance phenotypes. Phenotypes VanA, VanB, and VanD, found in *E. faecalis, E. faecium* and, to a much lesser extent, in other enterococci are associated with high-level resistance. The terminal dipeptide D-Ala-D-Ala, to which vancomycin binds, is replaced with the terminal dipeptide D-Ala-D-Lac, which has a notably lower affinity for the antibiotic. The synthesis of D-Ala-D-Lac requires the presence of a ligase (VanA, VanB, or VanD) and a dehydrogenase (VanH, VanH$_B$ or VanH$_D$), which converts pyruvate into lactate. Moreover, two enzymes are responsible for eliminating the terminal precursors in D-Ala: a dipeptidase (VanX, VanX$_B$ or VanX$_D$) that hydrolyzes the normal dipeptide D-Ala-D-Ala, and a carboxypeptidase (VanY, VanY$_B$, or VanY$_D$). VanC, VanE, and VanG isolates exhibit only low-level resistance to vancomycin. These phenotypic changes can be explained by the replacement of D-Ala by D-Ser, which causes a conformational change and reduces vancomycin affinity, although not as markedly as with D-Lac. As we have said, to be phenotypically detectable and significant, resistance requires the coordinated action of several enzymes. Thus, the bacteria need to synthesize D-Lac (VanH) or D-Ser (VanT) and D-Ala-D-Lac or D-Ala-D-Ser (VanA, B, D, or C, E, G), and to degrade D-Ala-D-Ala (VanX) or to remove D-Ala from growing precursors (VanY) or both (VanXY). Moreover, a two-component regulatory system (VanS–VanR) resulting in induction by either vancomycin (VanB, C, E, G phenotype) or vancomycin and teicoplanin (VanA phenotype) plays a critical role.

1.3.1.2 Clinical Implications of Glycopeptide Resistance

Glycopeptide resistance in *S. aureus* first emerged in the form of strains with elevated MIC values towards vancomycin (or both vancomycin and teicoplanin), which have been named vancomycin-intermediate *S. aureus* (VISA) or glycopeptide-intermediate *S. aureus* (GISA). Originally described in Japan in 1996 (Hiramatsu et al., 1997) in methicillin-resistant *S. aureus* (MRSA), VISA and GISA strains have now been isolated in numerous countries, particularly from patients having received prolonged vancomycin therapy (Hamilton-Miller, 2002). This lowering of susceptibility can be explained by the production of an altered peptidoglycan with an increased proportion of free D-Ala-D-Ala termini, which can trap vancomycin molecules and prevent their access to the target at the cytosolic membrane. VISA and GISA need to import a larger amount of precursors than normal strains, which compromises their fitness in an antibiotic-free environment. Thus VISA and GISA tend to lose their resistance when relieved from vancomycin pressure, giving rise to the so-called hetero-VISA phenotype (Hiramatsu et al., 2002). Two MRSA strains with high levels of resistance to vancomycin and teicoplanin have been reported in two different hospitals in the US (Bozdogan et al., 2003; Tenover et al., 2004). Both of these strains harbor the *vanA* gene cluster, indicating that they had acquired the corresponding set of genes from enterococci. Should such strains spread in hospitals, treatment options would rapidly become very limited (Van Bambeke et al., 2004).

Resistance develops in enterococci almost in parallel with vancomycin use. Moreover, carriage of resistant enterococci in animals is frequent in Europe and is thought to be due to the massive use of avoparcin, a glycopeptide antibiotic used as a growth promoter in animal feed. Many hospitals have reported the presence of predominant VRE. Their presence suggests the occurrence of intrahospital spread, which also appears to be an important factor in establishing endemicity. This has led health authorities to implement recommendations to reduce cross-contamination by these resistant strains so the activity of glycopeptides can be maintained. The introduction of new agents, such as quinupristin–dalfopristin and linezolid in the late 1990s has not really changed this situation, since both drugs are expensive, have rare but potentially worrying adverse effects, and are already facing emergence of resistance (Raad et al., 2004).

Recent case reports describe nosocomial infections caused by enterococci with VanA and VanB types that require vancomycin for growth (Dever et al., 1995). Vancomycin dependence may derive from the loss of a D-Ala-D-Ala ligase in a VRE strain, which is then unable to survive unless vancomycin induces the production of

D-Ala-D-Lac ligase. Infecting VRE strains convert to vancomycin dependence only after prolonged exposure to vancomycin.

Vancomycin resistance is rare among streptococci. Only one strain of *Streptococcus bovis,* which has the VanB phenotype, has been described in humans (Woodford, 2005). Resistance to glycopeptides has not yet been reported in pneumococci, although vancomycin tolerance—the ability of bacteria to survive but not grow in the presence of the antibiotic—has been detected. Antibiotic tolerance is particularly insidious because it cannot be detected using conventional in vitro tests. Tolerant strains seem to be sensitive to antibiotics, but they are precursor phenotypes to resistance (Novak et al., 1999).

1.3.2 Lipopeptides: Daptomycin

The emergence of multidrug-resistant pathogens (e.g., MRSA, VRE, and VISA) is one reason for the development of newer antimicrobial agents with enhanced gram-positive activity. Daptomycin is the first in a new and promising class of antibiotics, the cyclic lipopeptides, with a unique bactericidal activity against a range of gram-positive pathogens (Thorne and Alder, 2002). Food and Drug Administration (FDA) approval of daptomycin for the treatment of complicated skin and soft-tissue infections caused by gram-positive bacteria was gained in 2003, with European approval granted in 2006 (Kern, 2006).

Daptomycin has a novel mode of action and is rapidly bactericidal in vitro. It is known to bind to the cytoplasmic membrane of gram-positive bacteria via calcium-dependent binding. Once bound, the lipopeptide tail of the molecule is inserted into the bacterial cell membrane. This tail serves as an ion conduction structure through which potassium and, potentially other ions, can pass, thereby causing rapid depolarization of the cell membrane. Depolarization results in disruption of DNA, RNA and protein synthesis, ultimately causing bacterial cell death (Hancock, 2005). An important attribute of daptomycin is that it remains active throughout all phases of bacterial growth, including stationary phase. After exposure to daptomycin, bacteria are killed but not lysed. Therefore, bacterial cell contents are contained. As a result, there should be minimal activation of the inflammatory cascade in response to bacterial cell wall components, providing a potential advantage of daptomycin over other gram-positive agents. The activity of daptomycin is dependent on the presence of calcium ions. While clinically irrelevant, it does impact the conditions required for in vitro testing. Daptomycin is unable to penetrate the outer membrane of gram-negative organisms, but most gram-positive clinical isolates tested have proved susceptible to this agent, including penicillin-resistant *S. pneumoniae,* MRSA and VRE strains.

1.3.2.1 Mechanisms of Resistance

Daptomycin has a low potential for the development of resistance. Spontaneously resistant mutants were not found when bacteria were exposed to concentrations of eight times the MIC (Thorne and Alder, 2002). Resistance has been induced in vitro in *S. aureus* isolates after serial passage. However, these daptomycin-resistant strains appear to be less virulent that wild-type cells. The exact mechanism by which this resistance is conferred is not known; one proposed mechanism is reduced binding to the cytoplasmic membrane (Silverman et al., 2001). Development of resistance is unlikely to occur when therapeutically effective serum concentrations are maintained (Hancock, 2005). The analysis of a series of unrelated VRE from different areas of the US did not detect isolates with a daptomycin MIC greater than 8 mg/l (Jorgensen et al., 2003). The first clinical isolate of daptomycin-resistant *E. faecium* (MIC > 32 mg/l) has recently been published. High-level resistance appeared during treatment with daptomycin (Lewis et al., 2005).

1.3.3 Oxazolidinones: Linezolid

Linezolid is the first marketed antibiotic in a new class of synthetic antimicrobial agents known as oxazolidinones (Bozdogan and Appelbaum, 2004). FDA approved linezolid to treat infections associated with VRE, including cases with bloodstream infection. Linezolid has a wide spectrum of activity against gram-positive organisms including MRSA, penicillin-resistant streptococci, and enterococal isolates regardless of vancomycin resistance pattern (Cercenado et al., 2001). It lacks significant activity against most gram-negative pathogens.

The oxazolidinones disrupt bacterial growth by inhibiting the initiation of protein synthesis. They bind to the 50S subunit near the interface with the 30S subunit, thereby preventing the formation of the 70S initiation complex (Colca et al., 2003). This site of inhibition occurs earlier in the initiation process than other protein synthesis inhibitors (e.g., macrolides, chloramphenicol, clindamycin, and aminoglycosides) that interfere with the elongation process (Stevens et al., 2004). Because the site of inhibition is unique to linezolid, cross-resistance to other protein synthesis inhibitors has not yet been reported. Linezolid may also inhibit virulence factor expression and decrease toxin production in gram-positive pathogens.

1.3.3.1 Mechanisms of Resistance

Linezolid has been used to treat a number of resistant strains of bacteria. However, clinical resistance to linezolid emerged in a mere seven months after its introduction into the clinic (Fung et al., 2001). Linezolid-resistant strains were initially reported in *Enterococcus* spp. and, more recently, in MRSA, *E. coli*, and other bacteria. Target site modification is virtually the only mechanism of resistance that has, so far, been described. Mutations identified in various species are associated with G–U substitution in the peptidyl transferase center of 23S rRNA at position 2576 and result in reduced affinity of linezolid for the 50S subunit (Prystowsky et al., 2001). Further evaluation of oxazolidinone-resistant *S. pneumoniae* strains demonstrated resistance to be associated with additional mutations (C2610U, C2571U, and A2160G) (Bozdogan and Appelbaum, 2004). In clinical isolates of linezolid-resistant enterococci, resistance usually occurs by means of a G2576T mutation in the 23S rRNA gene. For linezolid-resistant *Enterococcus* spp. strains, increasing MICs seem to be associated with an increase in the number of mutant 23S rRNA genes (Sinclair et al., 2003). Therefore, high levels of resistance to linezolid may be reached through recombination of sensitive and resistant alleles within a single organism. The G2576T mutation has also been observed in one *Streptococcus oralis* isolate (Woodford, 2005).

In gram-positive bacteria, in vitro resistance to linezolid is not easily induced because of its low spontaneous mutation frequency ($<1 \times 10^{-9}$ for methicillin-susceptible and -resistant *S. aureus* and *Staphylococcus epidermidis*). However, it is possible to induce linezolid-resistant *E. faecalis* mutants by using serial passage on spiral gradient plates (Moellering, 2003). Linezolid resistance develops during treatment in patients receiving prolonged courses of antibiotics. The most commonly involved bacterium is *E. faecium* (Moellering, 2003). An outbreak of linezolid-resistant *E. faecium* in six neoplastic patients has recently been reported (Bozdogan and Appelbaum, 2004). Other vancomycin and linezolid-resistant *E. faecium* strains have been isolated in a transplant unit (Sinclair et al., 2003). Clinical isolates of linezolid-resistant *S oralis*, *E. faecalis*, and *E. faecium* have also been reported (Mutnick et al., 2003). Finally, three linezolid-resistant *E. faecalis* strains were isolated from patients who had taken linezolid for more than 30 days. All these strains contained the G2576T mutation in the 23S rRNA gene (Burleson et al., 2004).

1.3.4 Streptogramins: Quinupristin–Dalfopristin

The streptogramins are a family of compounds isolated from *Streptomyces pristinaespiralis*. Streptogramin family comprises several classes of antibiotics, including the mikamycins, pristinamycins, oestreomycins, and virginiamycins. Oral and topical pristinamycins have been used in humans in France for many years, primarily in the management of staphylococcal infections. Streptogramins belong to the macrolide–lincosamide–streptogramin group of antibiotics. All members of the class consist of two components: type A and type B. The type A streptogramins are cyclic polyunsaturated macrolactones, and the type B compounds are cyclic hexa- or heptadepsipeptides. Quinupristin and dalfopristin are derived from the streptogramins pristinamycin IA and IIB. Quinupristin–dalfopristin (Synercid) is composed of chemically modified, water-soluble, injectable derivatives of type B streptogramin (quinupristin) and type A streptogramin (dalfopristin) in a 30:70 ratio. Independently, each class of streptogramins has a bacteriostatic effect; however, in combination they act synergistically and are bactericidal. In 1999, the FDA labeled quinupristin–dalfopristin for use in the treatment of serious or life-threatening infections associated with vancomycin-resistant *E. faecium* bacteremia and complicated skin and skin structure infections caused by methicillin-susceptible *S. aureus* and *Streptococcus pyogenes* (group A streptococcus) (Hancock, 2005). Quinupristin–dalfopristin exhibits bactericidal activity against a wide variety of gram-positive bacteria. However, it is primarily bacteriostatic for *E. faecium* and has poor activity against *E. faecalis*.

The main target of Synercid is the bacterial 50S ribosome. Dalfopristin and quinupristin are thought to bind sequentially to different sites on the 50S subunit. It has been proposed that the binding of dalfopristin alters the conformation of the ribosome such that its affinity for quinupristin is increased. This results in a stable ternary drug–ribosome complex, and the newly synthesized peptide chains cannot be extruded from the ribosome of that complex. Consequently, protein synthesis is interrupted, thus leading to cell death. Dalfopristin directly interferes with peptidyl transferase and blocks an early step in protein synthesis by forming a bond with a ribosome that prevents elongation of the peptide chain. Quinupristin blocks a later step by preventing the extension of peptide chains and causing incomplete chains to be released (Vannufel and Cocito, 1996).

1.3.4.1 Mechanisms of Resistance

As quinupristin and dalfopristin inhibit bacterial protein synthesis by irreversibly binding to different sites on the 50S bacterial ribosomal subunit, cross-resistance between the two components of this association does not exist. Moreover, the potential for selecting resistant strains is limited because of their synergistic activity. When resistance to only one of the components of quinupristin–dalfopristin occurs, the microorganism may continue to be inhibited but not killed. Resistance to mixtures of streptogramin A and B compounds was first reported among staphylococci and requires only resistance to the A component, although resistance to both the A and B components may give a higher level of resistance. Virginiamycin, another streptogramin A and B combination, has been used as a growth promoter in animal feed for many years. It selects for virginiamycin-resistant strains of *E. faecium*, which are cross-resistant to quinupristin–dalfopristin and which may pose a risk to public health. As a consequence, the use of virginiamycin has been banned in the European Union. Resistance to streptogramins can occur by at lest three mechanisms: modification of the target binding site; enzymatic inactivation; and active efflux.

Altered Target Site

The most important mechanism of resistance is a plasmid-borne target modification that confers resistance to macrolides, lincosamides, and quinupristin by methylation of their common binding site. This mechanism is known as phenotype MLS_B and does not affect susceptibility to dalfopristin. The *erm* gene encodes a methylase that dimethylates an adenine residue in the 23S ribosomal RNA, which results in decreased binding of macrolides, lincosamides, and streptogramins of group B. A recent study found two resistant pneumococci strains (MIC 4 mg/l) through mutation in the *rplV* gene that encodes ribosomal protein L22 (Jones et al., 2003).

Enzymatic Inactivation

Enzymatic inactivation of dalfopristin can occur through plasmid-mediated dissemination of several acetyltransferases genes. These enzymes use the acetyl-coenzyme A to acetylate the hydroxyl group of dalfopristin (Kehoe et al., 2003). The plasmid-mediated acetyltranferases confer resistance to streptogramins and limit the clinical usefulness of the combination. Genes that encode these acetyltransferases are referred to as *vat* (virginiamycin acetyltransferase). At least one streptogramin A resistance gene (*vat*) is necessary for an enterococcus to be resistant to Synercid. Two resistance genes that encode acetyltransferases, *vatD* y *vatE*, have been found in *E. faecium*. The *vatD* gene has been transferred between isogenic strains of *E. faecium*, both in vivo and in vitro (Hershberger et al., 2004). The *vatE* gene is more frequently found in *E. faecium* than *vatD*. The *vatE* gene has also been detected in *E. faecalis* isolates from poultry. The *vat* genes are found in transposable genetic elements where they are sometimes associated with *erm* genes that confer resistance to streptogramins B. In staphylococci, other genes responsible for resistance to streptogramin B antibiotics include *vgb*, which is rare, plasmid-mediated, and found only in staphylococci; it encodes a streptogramin-inactivating enzyme, previously thought to play a role as hydrolase. However, careful analysis of the products of the reaction catalyzed by one of these resistance enzymes has identified a new resistance mechanism in which the peptide is linearized by an elimination reaction rather than by hydrolysis (Mukhtar et al., 2001). This has important implications in any attempts to design antimicrobials that are not susceptible to this mechanism. These genes have been shown to be more important as a cause of resistance to quinupristin–dalfopristin in staphylococci when a combination of *vat*, *erm*, and *vgb* was present in the same strain.

Active Efflux

Synercid is not active against *E. faecalis* because of the expression of the *lsa* gene in these organisms. The mechanism of this resistance was recently associated with the predicted ABC (ATP-binding-cassette) transporter Lsa that acts as an efflux pump on dalfopristin and also on clindamycin (Singh et al., 2002). This mechanism of resistance has been described in five of 338 vancomycin-resistant *E. faecium* infected patients (Dowzicky et al., 2000) and during a bacteremic episode caused by this microorganism (Chow et al., 1997). The *msrC* gene that confers resistance to streptogramin B through active transport is commonly found in *E. faecium* (Portillo et al., 2000).

1.4 Future Directions

The remarkable success of antimicrobial drugs generated a misconception in the late 1960s that the battle against bacterial infections was won, at least in the developed world. However, 50 years later, the control of infectious diseases is seriously threatened by the steady increase in the number of microorganisms that are resistant to antimicrobial agents.

The prevalence of resistance to β-lactams among clinical isolates of *S. pneumoniae* has increased during the last two decades. A similar trend has been observed for multiple non-β-lactam antimicrobial classes. Currently, the rates of resistance to β-lactams, macrolides, tetracyclines and TMP-SMZ have either plateaued or have begun to decrease (Aspa et al., 2004; Perez-Trallero et al., 2001; Perez-Trallero et al., 2005). However, as a consequence of increased fluoroquinolone use in the management of respiratory tract infections, it now appears that we have entered a period of substantial change with respect to emerging fluoroquinolone resistance in *S. pneumoniae* (Doern et al., 2005). VRE also continue to be increasingly prevalent in clinical medicine on a global scale. Even more significantly, glycopeptide resistance genes from enterococci have been transferred into *S. aureus*.

International organisations and groups of experts have recently made several attempts to reduce the antibiotic resistances shown by the leading human disease-provoking bacteria. Thus, the consensus group on resistance and prescribing in respiratory tract infections (Ball et al., 2002) made the following recommendations: unnecessary prescribing was recognized as the major factor in influencing resistance and costs; antibiotic therapy must be limited to syndromes in which bacterial infection is the predominant cause and should attempt maximal reduction in bacterial load, with the ultimate aim of bacterial eradication; it should be appropriate in type and context of local resistance prevalence, and optimal in dosage for the pathogen(s) involved; prescribing should be based on pharmacodynamic principles that predict efficacy, bacterial eradication and prevention of resistance emergence.

Pharmacoeconomic analyses confirm that bacteriologically more effective antibiotics can reduce overall management costs, particularly with respect to consequential morbidity and hospital admission. Recently the ESCMID Study Group for Antimicrobial resistance Surveillance (ESGARS) (Cornaglia et al., 2004) has announced the European Recommendations for Antimicrobial Resistance Surveillance. These recommendations focus on the detection of bacterial resistance and its reporting to clinicians, public health officers, and a wider-and ever-increasing audience. The leading concept is that the basis for resistance monitoring is microbial diagnosis. To generate relevant indicators, bacterial resistance data should be reported using adequate denominators and stratification. Reporting of antimicrobial resistance data is necessary for selection of empirical therapy at the local level, for assessing the scale of the resistance problem at the local, national, or international levels, for monitoring changes in resistance rates, and for detecting the emergence and spread of new resistance types. Vaccination with pneumococcal conjugate vaccine is also a novel approach to reduction of the burden of antibiotic resistance in the pneumococcus (Klugman, 2004).

The earliest uses of antimicrobials had a dramatic impact by altering mortality in serious, life-threatening bacterial diseases. Today, antimicrobials are the third most profitable class of drugs, but some large pharmaceutical companies have decided to exit the area of antimicrobial development, especially antibacterial drug development, for several reasons (Powers, 2004). However, there is a need for new antimicrobials to combat disease due to resistant pathogens. A recent analysis showed that there are five new antibacterial drugs in development by large pharmaceutical companies (Spellberg et al., 2004). For instance, oritavancin, and to a lesser extent, dalbavancin, offer considerable possibilities for improvement over vancomycin and teicoplanin. Should these molecules soon become approved for clinical use, it is vital that they are used rationally and prudently to protect against the rapid emergence of resistance and extend their usual life (Van Bambeke et al., 2004).

In conclusion, the rise of antibiotic resistance in the clinic continues unabated and microorganisms can deploy a myriad of mechanisms in the face of use of these toxic agents. Alteration of drug target, changes in target accessibility, degradation by enzymes and efflux, can all be utilized by microorganisms to evade antibiotics. There is much promise, however, in the new strategies described in the literature, new agents in development, and new antibiotics recently introduced to the marketplace that resistance can be met and overcome. The knowledge of the mechanisms of action and resistance, may contribute to the discovery and development of new antimicrobials (Wright 2003).

References

Acar, J.F. and Buu-Hoi, A.Y., (1988), Resistance patterns of important gram-positive pathogens. *J Antimicrob Chemother*, **21**, (Suppl. C), 41–47.

Alfageme, I.; Aspa, J.; Bello, S.; Blanquer, J.; Blanquer, R.; Borderias, L.; Bravo, C.; de Celis, R.; de Gracia, X.; Dorca, J.; Gallardo, J.; Gallego, M.; Menendez, R.; Molinos, L.; Paredes, C.; Rajas, O.; Rello, J.; Rodriguez de Castro, F.; Roig, J.; Sanchez-Gascon, F.; Torres, A. and Zalacain, R., (2005), [Guidelines for the Diagnosis and Management of Community-Acquired Pneumonia. Spanish Society of Pulmonology and Thoracic Surgery (SEPAR).]. *Arch Bronconeumol*, **41**, 272–289.

Allen, G.P.; Kaatz, G.W. and Rybak, M.K., (2003), Activities of mutant prevention concentration-targeted moxifloxacin and levofloxacin against *Streptococcus pneumoniae* in an in vitro pharmacodynamic model. *Antimicrob Agents Chemother*, **47**, 2606–2614.

Ambrose, P.G.; Bhavnani, S.M. and Owes, R.C. Jr., (2003), Clinical pharmacodynamics of quinolones. *Infect Dis Clin North Am*, **17**, 529–543.

Appelbaum, P.C., (1992), Antimicrobial resistance in *Streptococcus pneumoniae*: an overview. *Clin Infect Dis*, **15**, 77–83.

Appelbaum, P.C., (2002), Resistance among *Streptococcus pneumoniae*: Implications for drug selection. *Clin Infect Dis*, **34**, 1613–1620. a. (aldfkald)

Appelbaum, P.C.; Bhamjee, A.; Scragg, J.N.; Hallett, A.F.; Bowen, A.J. and Cooper, R.C., (1977), *Streptococcus pneumoniae* resistant to penicillin and chloramphenicol. *Lancet*, **2**, 995–997.

Aspa, J.; Rajas, O.; Rodriguez de Castro, F.; Blanquer, J.; Zalacain, R.; Fenoll, A.; de Celis, R.; Vargas, A.; Rodriguez Salvanes, F.; Espana, P.P.; Rello, J. and Torres, A., (2004), Drug-resistant pneumococcal pneumonia: clinical relevance and related factors. *Clin Infect Dis*, **38**, 787–798.

Aspa, J., Rajas, O., Rodriguez de Castro, F., Huertas, C., Borderías, L., F.J., C., Tábara, J., Hernández-Flix, S., Martínez-Sanchis, A. and Torres, A., (2006), Impact of Initial antibiotic choice on mortality from pneumococcal pneumonia. *Eur Respir J*, **27**, 1–10.

Austrian, R. and Gold, J., (1964), Pneumococcal bacteremia with special reference to bacteremic pneumococcal pneumonia. *Ann Intern Med*, **60**, 759–776.

Baddour, L.M., Yu, V.L., Klugman, K.P., Feldman, C., Ortqvist, A., Rello, J., Morris, A.J., Luna, C., Snydman, D.R., Ko, W.C., Chedid, M.B., Hui, D.S., Andremont, A., Chiou, C.C., International Pneumococcal Study Group, (2004), Combination antibiotic therapy lowers mortality among severely ill patients with pneumococcal bacteremia. *Am J Respir Crit Care Med*, **170**, 440–444.

Ball, P., Baquero, F., Cars, O., File, T., Garau, J., Klugman, K., Low, D.E., Rubinstein, E. and Wise, R., (2002), Antibiotic therapy of community respiratory tract infections: strategies for optimal outcomes and minimized resistance emergence. *J Antimicrob Chemother*, **49**, 31–40.

Baquero, F., (2001), Low-level antibacterial resistance: a gateway to clinical resistance. *Drug Resist Updat*, **4**, 93–105.

Baquero, F., Barrett, J.F., Courvalin, P., Morrissey, I., Piddock, L. and Novick, W.J., (1998a), Epidemiology and mechanisms of resistance among respiratory tract pathogens. *Clin Microbiol Infect*, **4** (Suppl. 2), S19–S26.

Baquero, F., Blazquez, J., Loza, E. and Canton, R., (1998b), Molecular basis of resistance to beta-lactams in infections by *Streptococcus pneumoniae*. *Med Clin (Barc)*, **110** (Suppl. 1), 8–11.

Baquero, F., García-Rodriguez, J.A., Garcia de Lomas, J. and Aguilar, L., (1999), Antimicrobial resistance of 1,113 *Streptococcus pneumoniae* isolates from patients with respiratory tract infections in Spain: results of a 1-year (1996–1997) multicenter surveillance study. The Spanish Surveillance Group for Respiratory Pathogens. *Antimicrob Agents Chemother*, **43**, 357–359.

Barnett, E.D. and Klein, J.O., (1995), The problem of resistant bacteria for the management of acute otitis media. *Pediatr Clin North Am*, **42**, 509–517.

Bast, D., Low, D., Duncan, C., Kilburn, L., Mandell, L., Davidson, R. and de Azavedo, J., (2000), Fluoroquinolone resistance in clinical isolates of *Streptococcus pneumoniae*: contributions of type II topoisomerase mutations and efflux to levels of resistance. *Antimicrob Agents Chemother*, **44**, 3049–3054.

Bauer, T., Ewig, S., Marcos, M.A., Schultze-Werninghaus, G. and Torres, A., (2001), *Streptococcus pneumoniae* in community-acquired pneumonia. How important is drug resistance? *Med Clin North Am*, **85**, 1367–1379.

Bedos, J.P., Chevret, S., Chastang, C., Geslin, P. and Regnier, B., (1996), Epidemiological features of and risk factors for infection by *Streptococcus pneumoniae* strains with diminished susceptibility to penicillin: findings of a French survey. *Clin Infect Dis*, **22**, 63–72.

Berry, V., Page, R., Satterfield, J., Singley, C., Straub, R. and Woodnutt, G., (2000), Comparative in vivo activity of gemifloxacin in a rat model of respiratory tract infections. *J Antimicrob Chem*, **45 (S1)**, 79–85.

Bhavnani, S.M., Hammel, J., Jones, R., Ambrose, P., (2005). Relationship between increased levofloxacin use and decreased susceptibility of *Streptococcus pneumoniae* in the United States. *Diag Microbiol Infect Dis*, **51**, 31–37.

Blondeau, J.M., Zhao, X., Hansen G., Drlica, K., (2001), Mutant prevention concentrations of fluoroquinolones for clinical isolates of *Streptococcus pneumoniae*. *Antimicrob Agents Chemother*, **45**, 433–438.

Bonnefoy, A., Girard, A.M., Agouridas, C. and Chantot, J.F., (1997), Ketolides lack inducibility properties of MLS(B) resistance phenotype. *J Antimicrob Chemother*, **40**, 85–90.

Boswell, F.J., Andrews, J.M., Ashby, J.P., Fogarty, C., Brenwald, N.P. and Wise, R., (1998), The in-vitro activity of HMR 3647, a new ketolide antimicrobial agent. *J Antimicrob Chemother*, **42**, 703–709.

Bozdogan, B. and Appelbaum, P.C., (2004), Oxazolidinones: activity, mode of action, and mechanism of resistance. *Int J Antimicrob Agents*, **23**, 113–119.

Bozdogan, B., Esel, D., Whitener, C., Browne, F.A. and Appelbaum, P.C., (2003), Antibacterial susceptibility of a vancomycin-resistant *Staphylococcus aureus* strain isolated at the Hershey Medical Center. *J Antimicrob Chemother*, **52**, 864–868

Breiman, R.F., Butler, J.C., Tenover, F.C., Elliott, J.A. and Facklam, R.R., (1994), Emergence of drug-resistant pneumococcal infections in the United States. *JAMA*, **271**, 1831–1835.

Brown, R.B., Iannini, P., Gross, P. and Kunkel, M., (2003), Impact of initial antibiotic choice on clinical outcomes in community-acquired pneumonia: analysis of a hospital claims-made database. *Chest*, **123**, 1503–1511.

Brueggemann, A.B., Coffman, S.L., Rhomberg, P., Huynh, H., Almer, L., Nilius, A., Flamm, R. and Doern, G.V., (2002), Fluoroquinolone resistance in *Streptococcus pneumoniae* in United States since 1994–1995. *Antimicrob Agents Chemother*, **46**, 680–688.

Buckingham, S.C., Brown, S.P. and Joaquin, V.H., (1998), Breakthrough bacteremia and meningitis during treatment with cephalosporins parenterally for pneumococcal pneumonia. *J Pediatr*, **132**, 174–176.

Burdett, V., (1991), Purification and characterization of Tet(M), a protein that renders ribosomes resistant to tetracycline. *J Biol Chem*, **266**, 2872–2877.

Burgess, D.S., (1999), Pharmacodynamic principles of antimicrobial therapy in the prevention of resistance. *Chest*, **115**, 19S-23S.

Burgess, D.S., Lewis, J.S., 2nd., (2000), Effect of macrolides as part of initial empiric therapy on medical outcomes for hospitalized patients with community-acquired pneumonia. *Clin Ther*, **22**, 872–878.

Burleson, B.S., Ritchie, D.J., Micek, S.T. and Dunne, W.M., (2004), *Enterococcus faecalis* resistant to linezolid: case series and review of the literature. *Pharmacotherapy*, **24**, 1225–1231.

Butler, J.C., Hofmann, J., Cetron, M.S., Elliott, J.A., Facklam, R.R. and Breiman, R.F., (1996), The continued emergence of drug-resistant *Streptococcus pneumoniae* in the United States: an update from the Centers for Disease Control and Prevention's Pneumococcal Sentinel Surveillance System. *J Infect Dis*, **174**, 986–993.

Campbell, G.D., Jr and Silberman, R., (1998), Drug-resistant *Streptococcus pneumoniae*. *Clin Infect Dis*, **26**, 1188–1195.

Catalan, M.J., Fernandez, J.M., Vazquez, A., Varela de Seijas, E., Suarez, A. and Bernaldo de Quiros, J.C., (1994), Failure of cefotaxime in the treatment of meningitis due to relatively resistant *Streptococcus pneumoniae*. *Clin Infect Dis*, **18**, 766–769.

Cercenado, E., Garcia-Garrote, F. and Bouza, E., (2001), In vitro activity of linezolid against multiply resistant Gram-positive clinical isolates. *J Antimicrob Chemother*, **47**, 77–81.

Cerda Zolezzi, P., Laplana, L. M., Calvo, C.R., Cepero, P.G., Erazo, M.C. and Gómez-Lus, R., (2004), Molecular basis of resistance to macrolides and other antibiotics in commensal *viridans* group streptococci and *Gemela spp.* and transfer of resistance genes to *Streptococcus pneumoniae*. Antimicrob Agents Chemother, **48**, 3462–3467.

Chabbert, Y.A., (1956), Antagonisme in vitro entre l'erythromycine et la spiramycine. *Ann Inst Pasteur (Paris)*, **90**, 787–790.

Chabbert, Y.A. and Courvalin, P., (1971), Synergism of antibiotic components of the streptogramin group. *Pathol Biol (Paris)*, **19**, 613–619.

Chen, D.K., McGeer, A., de Azavedo, J.C. and Low, D.E., (1999), Decreased susceptibility of *Streptococcus pneumoniae* to fluoroquinolones in Canada. Canadian Bacterial Surveillance Network. *N Engl J Med*, **341**, 233–239.

Chenoweth, C.E., Saint, S., Martinez, F., Lynch, J.P., 3rd and Fendrick, A.M., (2000), Antimicrobial resistance in *Streptococcus pneumoniae*: implications for patients with community-acquired pneumonia. *Mayo Clin Proc*, **75**, 1161–1168.

Choi, E.H. and Lee, H.J., (1998), Clinical outcome of invasive infections by penicillin-resistant *Streptococcus pneumoniae* in Korean children. *Clin Infect Dis*, **26**, 1346–1354.

Chow, J.W., Donahedian, S.M. and Zervos, M.J., (1997), Emergence of increased resistance to quinupristin/dalfopristin during therapy for *Enterococcus faecium* bacteremia. *Clin Infect Dis*, **24**, 90–91.

Clavo-Sanchez, A.J., Giron-Gonzalez, J.A., Lopez-Prieto, D., Canueto-Quintero, J., Sanchez-Porto, A., Vergara-Campos, A., Marin-Casanova, P. and Cordoba-Dona, J.A., (1997), Multivariate analysis of risk factors for infection due to penicillin-resistant and multidrug-resistant *Streptococcus pneumoniae*: a multicenter study. *Clin Infect Dis*, **24**, 1052–1059.

Clay, K.D., Hanson, J.S., Pope, S.D., Rissmiller, R.W., Purdum, P.P., 3rd & Banks, P.M., (2006), Brief communication: severe hepatotoxicity of telithromycin: three case reports and literature review. *Ann Intern Med*, **144**, 415–420.

Coburn, A.F., (1949), *The epidemiology of haemolytic Streptococcus during World War in the United States Navy*. Williams & Williams, Baltimore.

Colca, J.R., McDonald, W.G., Waldon, D.J., Thomasco, L.M., Gadwood, R.C., Lund, E.T., Cavey, G.S., Mathews, W.R., Adams, L.D., Cecil, E.T., Pearson, J.D., Bock, J.H., Mott, J.E., Shinabarger, D.L., Xiong, L. and Mankin, A.S., (2003), Cross-linking in the living cell locates the site of action of oxazolidinone antibiotics. *J Biol Chem*, **278**, 21972–21979.

Cornaglia, G., Hryniewicz, W., Jarlier, V., Kahlmeter, G., Mittermayer, H., Stratchounski, L. and
    Baquero, F., (2004), European recommendations for antimicrobial resistance surveillance. *Clin
    Microbiol Infect*, **10**, 349–383.
Craig, W.A., (1998a), Pharmacokinetic/pharmacodynamic parameters: rationale for antibacterial
    dosing of mice and men. *Clin Infect Dis*, **26**, 1–10; quiz 11–12.
Craig, W.A., (1998b), Choosing an antibiotic on the basis of pharmacodynamics. *Ear Nose Throat
    J*, **77**, 7–12.
Critchley, I.A., Blosser-Middleton, R.S., Jones, M.E., Karlowsky, J.A., Karginova, E.A.,
    Thornsberry, C. and Sahm, D.F., (2002), Phenotypic and genotypic analysis of levofloxacin-
    resistant clinical isolates of *Streptococcus pneumoniae* collected in 13 countries during
    1999–2000. *Intern J Antimicrob Agents*, **20**, 100–107.
Dalhoff, A., Krasemann, C., Begener, S. and Tillotson, G., (2001), Penicillin-resistant *Streptococ-
    cus pneumoniae*: review of moxifloxacin activity. *Clin Infect Dis*, **32**, S22–S29.
Davidson, R., Cavalcanti, R., Brunton, J.L., Bast, D.J., de Azavedo, J.C., Kibsey, P., Fleming, C.
    and Low, D.E., (2002), Resistance to levofloxacin and failure of treatment of pneumococcal
    pneumonia. *N Engl J Med*, **346**, 747–750.
Davies, T., Goering, R.V., Lovgren, M., Talbot, J.A., Jacobs, M.R. and Appelbaum, P.C., (1999),
    Molecular epidemiological survey of penicillin-resistant *Streptococcus pneumoniae* from Asia,
    Europe, and North America. *Diagn Microbiol Infect Dis*, **34**, 7–12.
Davies, T.A., Evangelista, A., Pfleger, S., Bush, K., Sahm, D.F. and Goldschmidt, R., (2002),
    Prevalence of single mutations in topoisomerase type II genes among levofloxacin-susceptible
    clinical strains of *Streptococcus pneumoniae* isolated in the United States in 1992 to 1996 and
    1999 to 2000. *Antimicrob Agents Chemother*, **46**, 119–124.
De la Campa, A.G., Balsalobre, I., Ardanuy, C., Fenoll, A. Perez-Trallero, E. and Liñares, J.,
    (2004), Fluoroquinolone resistance in penicillin-resistant *Streptococcus pneumoniae* clones,
    Spain. *Emerg Infect Dis*, **10**, 1751–1759.
Deeks, S.L., Palacio, R., Ruvinsky, R., Kertesz, D.A., Hortal, M., Rossi, A., Spika, J.S.
    and Di Fabio, J.L., (1999), Risk factors and course of illness among children with inva-
    sive penicillin-resistant *Streptococcus pneumoniae*. The *Streptococcus pneumoniae* Working
    Group. *Pediatrics*, **103**, 409–413.
Dever, L.L., Smith, S.M., Handwerger, S. and Eng, R.H., (1995), Vancomycin-dependent *Entero-
    coccus faecium* isolated from stool following oral vancomycin therapy. *J Clin Microbiol*, **33**,
    2770–2773.
Doern, G.V., (2001), Antimicrobial resistance with *Streptococcus pneumoniae*: much ado about
    nothing? *Semin Respir Infect*, **16**, 177–185.
Doern, G.V., Pfaller, M.A., Kugler, K., Freeman, J. and Jones, R.N., (1998), Prevalence of antimi-
    crobial resistance among respiratory tract isolates of *Streptococcus pneumoniae* in North Amer-
    ica: 1997 results from the SENTRY antimicrobial surveillance program. *Clin Infect Dis*, **27**,
    764–770.
Doern, G.V., Richter, S.S., Miller, A., Miller, N., Rice, C., Heilmann, K. and Beekmann, S.,
    (2005), Antimicrobial resistance among *Streptococcus pneumoniae* in the United States: have
    we begun to turn the corner on resistance to certain antimicrobial classes? *Clin Infect Dis*, **41**,
    139–148.
Dong, Y., Zhao, X., Domagala, J., and Drlica, K., (1999), Effect of fluoroquinolone concentration
    on selection of resistant mutants of *Mycobacterium bovis* BCG and *Staphylococcus aureus*.
    Antimicrob. Agents Chemother. **43**: 1756–1758.
Dowell, S.F., Smith, T., Leversedge, K. and Snitzer, J., (1999), Failure of treatment of pneumonia
    associated with highly resistant pneumococci in a child. *Clin Infect Dis*, **29**, 462–463.
Dowzicky, M., Talbot, G.H., Feger, C., Prokocimer, P., Etienne, J. and Leclercq, R., (2000), Charac-
    terization of isolates associated with emerging resistance to quinupristin/dalfopristin (Synercid)
    during a worldwide clinical program. *Diagn Microbiol Infect Dis*, **37**, 57–62.
Drlica, K. and Zhao, X., (1997), DNA gyrase, topoisomerase IV, and the 4-quinolones. *Microbiol
    Mol Biol Rev*, **61**, 377–392.
Ekdahl, K., Ahlinder, I., Hansson, H.B., Melander, E., Molstad, S., Soderstrom, M. and
    Perssons, K., (1997), Duration of nasopharyngeal carriage of penicillin-resistant *Streptococcus*

pneumoniae: experiences from the South Swedish Pneumococcal Intervention Project. *Clin Infect Dis*, **25**, 1113–1117.

Endimiani, A., Brigante, G., Bettaccini, A., Luzzaro, F., Grossi, P., Toniolo, A.Q., (2005), Failure of levofloxacin treatment in community-acquired pneumococcal pneumonia. *BMC Infect Dis*, **5**, 106–109.

Empey, P.E., Jennings, H.R., Thornton, A.C., Rapp, R.P. and Evans, M.E., (2001), Levofloxacin failure in a patient with pneumococcal pneumonia. *Ann Pharmacother*, **35**, 687–690.

Eriksen, K., (1945), Studies on induced resistance to penicillin in a Pneumococcus type 1. *Act Pathol Microbiol Scan*, **22**, 398–405.

Ewig, S., Ruiz, M., Torres, A., Marco, F., Martinez, J.A., Sanchez, M. and Mensa, J., (1999), Pneumonia acquired in the community through drug-resistant *Streptococcus pneumoniae*. *Am J Respir Crit Care Med*, **159**, 1835–1842.

Feikin, D.R., Dowell, S.F., Nwanyanwu, O.C., Klugman, K.P., Kazembe, P.N., Barat, L.M., Graf, C., Bloland, P.B., Ziba, C., Huebner, R.E. and Schwartz, B., (2000a), Increased carriage of trimethoprim/sulfamethoxazole-resistant *Streptococcus pneumoniae* in Malawian children after treatment for malaria with sulfadoxine/pyrimethamine. *J Infect Dis*, **181**, 1501–1505.

Feikin, D.R., Schuchat, A., Kolczak, M., Barrett, N.L., Harrison, L.H., Lefkowitz, L., McGeer, A., Farley, M.M., Vugia, D.J., Lexau, C., Stefonek, K.R., Patterson, J.E. and Jorgensen, J.H., (2000b), Mortality from invasive pneumococcal pneumonia in the era of antibiotic resistance, 1995–1997. *Am J Public Health*, **90**, 223–229.

Feldman, C., (2004), Clinical relevance of antimicrobial resistance in the management of pneumococcal community-acquired pneumonia. *J Lab Clin Med*, **143**, 269–283.

Felmingham, D. and Gruneberg, R.N., (2000), The Alexander Project 1996–1997: latest susceptibility data from this international study of bacterial pathogens from community-acquired lower respiratory tract infections. *J Antimicrob Chemother*, **45**, 191–203.

Fenoll, A., Asensio, G., Jado, I., Berron, S., Camacho, M.T., Ortega, M. and Casal, J., (2002), Antimicrobial susceptibility and pneumococcal serotypes. *J Antimicrob Chemother*, **50** (Suppl. 2), 13–19.

Fenoll, A., Jado, I., Vicioso, D., Perez, A. and Casal, J., (1998), Evolution of *Streptococcus pneumoniae* serotypes and antibiotic resistance in Spain: update (1990 to 1996). *J Clin Microbiol*, **36**, 3447–3454.

Fenoll, A., Martin Bourgon, C., Munoz, R., Vicioso, D. and Casal, J., (1991), Serotype distribution and antimicrobial resistance of *Streptococcus pneumoniae* isolates causing systemic infections in Spain, 1979–1989. *Rev Infect Dis*, **13**, 56–60.

Fernandez-Moreira, E., Balas, D., Gonzalez, I. and De la Campa, A.G., (2000), Fluoroquinolones inhibit preferentially *Streptococcus pneumoniae* DNA topoisomerase IV than DNA gyrase native proteins. *Microb Drug Resist*, **6**, 259–267.

File, T.M. and Tillotson, G.S., (2004), Gemifloxacin: a new, potent fluoroquinolone for the therapy of lower respiratory tract infections. *Expert Ref Anti Infect Ther*, **2**(6), 831–843.

Fine, M.J., Auble, T.E., Yealy, D.M., Hanusa, B.H., Weissfeld, L.A., Singer, D.E., Coley, C.M., Marrie, T.J. and Kapoor, W.N., (1997), A prediction rule to identify low-risk patients with community-acquired pneumonia. *N Engl J Med*, **336**, 243–250.

Finland M., (1955), Changing patterns of resistance of certain common pathogenic bacteria to antimicrobial agents. *N Engl J Med*, **252**:570–80.

Finland, M., (1971), Increased resistance in the pneumococcus. *N Engl J Med*, **284**, 212–214.

Finland, M., Jones, W.F., Jr. and Nichols, R.L., (1956), Development of resistance and cross-resistance in vitro to erythromycin, carbomycin, spiramycin, oleandomycin and streptogramin. *Proc Soc Exp Biol Med*, **93**, 388–393.

Fleming, A., (1929), On the antibacterial action of culture of a Penicillium, with special reference to their use in isolation of H. influenzae. *Br J Exp Pathol*, **10**, 226–236.

Florey, M.E., (1943), General and local administration of penicillin. *Lancet*, **1**, 387–397.

Fogarty, C., Goldschmidt, R. and Bush, K., (2000), Bacteremic pneumonia due to multidrug-resistant pneumococci in 3 patients treated unsuccessfully with azithromycin and successfully with levofloxacin. *Clin Infect Dis*, **31**, 613–615.

Fraimow, H., Knob, C., Herrero, I. and, Patel, R., (2005), Putative VanRS-like two component regulatory system associated with the inducible glycopeptide resistance cluster of *Paenibacillus popilliae*. *Antimicrob Agents Chemother*, **49**, 2625–2633.

Friedland, I.R., (1995), Comparison of the response to antimicrobial therapy of penicillin-resistant and penicillin-susceptible pneumococcal disease. *Pediatr Infect Dis J*, **14**, 885–890.

Friedland, I.R. and McCracken, G.H., Jr., (1994), Management of infections caused by antibiotic-resistant *Streptococcus pneumoniae*. *N Engl J Med*, **331**, 377–382.

Fukuda, H. and Hiramatsu, K., (1999),,, Primary Targets of fluoroquinolones in *Streptococcus pneumoniae*. *Antimicrob Agents Chemother*, **43**, 410–412.

Fuller, J.D. and Low, D.E., (2005), A review of *Streptococcus pneumoniae* infection treatment failures associated with fluoroquinolones-resistance. *Clin Infect Dis*, **41**, 118–121.

Fung, H.B., Kirschenbaum, H.L. and Ojofeitimi, B.O., (2001), Linezolid: an oxazolidinone antimicrobial agent. *Clin Ther*, **23**, 356–391.

Garau, J., (2002), Treatment of drug-resistant pneumococcal pneumonia. *Lancet Infect Dis*, **2**, 404–415.

Garcia-Rey, C., Aquilar, L. and Baquero, F., (2000), Influences of different factors on prevalence of ciprofloxacin resistance in *Streptococcus pneumoniae* in Spain. *Antimicrob Agents Chemother*, **44**, 3481–3482.

Garrod, L.P., (1957), The erythromycin group of antibiotics. *Br Med J*, **13**, 57–63.

Gay, K., Baughman, W., Miller, Y., Jackson, D., Whitney, C., Schuchat, A., Farley, M., Tenover, F. and Stephens, D., (2000), The emergence of *Streptococcus pneumoniae* resistant to macrolide antimicrobial agents: a 6-year population-based assessment. *J Infect Dis*, **182**, 1417–1424.

Gillespie, S.H., Voelker, L.L., Ambler, J.E., Traini. C. and Dickens, A., (2003), Fluoroquinolone resistance in *Streptococcus pneumoniae*: evidence that gyrA mutations arise at a lower rate and that mutation in *gyrA* or *parC* predisposes to further mutation. *Microb Drug Resist*, **9**, 17–24.

Gilmore, M.S., Behnke, D. and Ferreti, J.J., (1982), Evolutionary relatedness of MLS resistance and replication function sequences on streptococcal antibiotic resistance plasmids. In D. Schlessinger, editor, *Microbiology-1992*. Washington DC, pp. 174–176.

Graham, D.J., (2006), Telithromycin and acute liver failure. *N Engl J Med*, **355**, 2260–2261.

Hamilton-Miller, J.M., (2002), Vancomycin-resistant *Staphylococcus aureus*: a real and present danger? *Infection*, **30**, 118–124.

Hancock, R.E., (2005), Mechanisms of action of newer antibiotics for Gram-positive pathogens. *Lancet Infect Dis*, **5**, 209–218.

Hansman, D. and Bullen, M.M., (1967), A resistant pneumococcus (letter). *Lancet*, **2**, 264–265.

Hansman, D., Glasgow, H.N., Sturt, J., Devitt, L. and Douglas, R., (1971a), Increased resistance to penicillin of pneumococci isolated from man. *N Engl J Med*, **284**, 175–177.

Hansman, D., Glasgow, H.N., Sturt, J., Devitt, L. and Douglas, R.M., (1971b), Pneumococci insensitive to penicillin. *Nature*, **230**, 407–408.

Heffelfinger, J.D., Dowell, S.F., Jorgensen, J.H., Klugman, K.P., Mabry, L.R., Musher, D.M., Plouffe, J.F., Rakowsky, A., Schuchat, A. and Whitney, C.G., (2000), Management of community-acquired pneumonia in the era of pneumococcal resistance: a report from the Drug-Resistant *Streptococcus pneumoniae* Therapeutic Working Group. *Arch Intern Med*, **160**, 1399–1408.

Hershberger, E., Donabedian, S., Konstantinou, K. and Zervos, M.J., (2004), Quinupristin-dalfopristin resistance in gram-positive bacteria: mechanism of resistance and epidemiology. *Clin Infect Dis*, **38**, 92–98.

Hiramatsu, K., Hanaki, H., Ino, T., Yabuta, K., Oguri, T. and Tenover, F.C., (1997), Methicillin-resistant *Staphylococcus aureus* clinical strain with reduced vancomycin susceptibility. *J Antimicrob Chemother*, **40**, 135–136.

Hiramatsu, K., Okuma, K., Ma, X.X., Yamamoto, M., Hori, S. and Kapi, M., (2002), New trends in *Staphylococcus aureus* infections: gycopeptide resistance in hospital and methicillin resistance in the community. *Curr Opin Infect Dis*, **15**, 407–413.

Ho, P.L., Tse, W.S., Tsang, K.W., Kwok, T.K., Ng, T.K., Cheng, V.C. and Chan, R.M., (2001), Risk factors for acquisition of levofloxacin-resistant *Streptococcus pneumoniae*: a case-control study. *Clin Infect Dis*, **32**, 701–707.

Hoban, D.J., Doern, G.V., Fluit, A.C., Roussel-Delvallez, M. and Jones, R.N., (2001), Worldwide prevalence of antimicrobial resistance in *Streptococcus pneumoniae*, *Haemophilus influenzae*, and *Moraxella catarrhalis* in the SENTRY Antimicrobial Surveillance Program, 1997–1999. *Clin Infect Dis*, **32** (Suppl. 2), S81–S93.

Hofmann, J., Cetron, M.S., Farley, M.M., Baughman, W.S., Facklam, R.R., Elliott, J.A., Deaver, K.A. and Breiman, R.F., (1995), The prevalence of drug-resistant *Streptococcus pneumoniae* in Atlanta. *N Engl J Med*, **333**, 481–486.

Hooper, D.C. and Wolfson, J.S., (1993), *Quinolone Antimicrobial Agents*. American Society for Microbiology, Washington.

Hyde, T.B., Gay, K., Stephens, D.S., Vugia, D.J., Pass, M., Johnson, S., Barrett, N.L., Schaffner, W., Cieslak, P.R., Maupin, P.S., Zell, E.R., Jorgensen, J.H., Facklam, R.R. and Whitney, C.G., (2001), Macrolide resistance among invasive *Streptococcus pneumoniae* isolates. *JAMA*, **286**, 1857–1862.

Jacobs, M.R., (1996), Increasing importance of antibiotic-resistant *Streptococcus pneumoniae* in acute otitis media. *Pediatr Infect Dis J*, **15**, 940–943.

Jacobs, M.R., Koornhof, H.J., Robins-Browne, R.M., Stevenson, C.M., Vermaak, Z.A., Freiman, I., Miller, G.B., Witcomb, M.A., Isaacson, M., Ward, J.I. and Austrian, R., (1978), Emergence of multiply resistant pneumococci. *N Engl J Med*, **299**, 735–740.

Jacoby, G.A., (1994), Prevalence and resistance mechanisms of common bacterial respiratory pathogens. *Clin Infect Dis*, **18**, 951–957.

Janoir, C., Zeller, V., Kitzis, M.D., Moreau, N.J. and Gutmann, L., (1996), High-level fluoroquinolone resistance in *Streptococcus pneumoniae* requires mutations in parC and gyrA. *Antimicrob Agents Chemother*, **40**, 2760–2764.

Johansen, H.K., Jensen, T.G., Dessau, R.B., Lundgren, B. and Frimodt-Moller, N., (2000), Antagonism between penicillin and erythromycin against *Streptococcus pneumoniae* in vitro and in vivo. *J Antimicrob Chemother*, **46**, 973–980.

Johnston, N.J., De Azavedo, J.C., Kellner, J.D. and Low, D.E., (1998), Prevalence and characterization of the mechanisms of macrolide, lincosamide, and streptogramin resistance in isolates of *Streptococcus pneumoniae*. *Antimicrob Agents Chemother*, **42**, 2425–2426.

Jones, R.N., (1999), The impact of antimicrobial resistance: changing epidemiology of community-acquired respiratory-tract infections. *Am J Health Syst Pharm*, **56**, S4–S11.

Jones, R.N., Farrell, D.J. and Morrissey, I., (2003), Quinupristin-dalfopristin resistance in *Streptococcus pneumoniae*: novel L22 ribosomal protein mutation in two clinical isolates from the SENTRY antimicrobial surveillance program. *Antimicrob Agents Chemother*, **47**, 2696–2698.

Jordan, D.C. and Inniss, W.E., (1959) Selective inhibition of ribonucleic acid synthesis in *Staphylococcus aureus* by vancomycin. *Nature*, **184**, 1984–1985.

Jordan, D.C. and Mallory H.D., (1964), Site of action of vancomycin on *Staphylococcus aureus*. *Antimicrob Agents Chemother*, **10**, 489–494.

Jorgensen, J.H., Crawford, S.A., Kelly, C.C. and Patterson, J.E., (2003), In vitro activity of daptomycin against vancomycin-resistant enterococci of various Van types and comparison of susceptibility testing methods. *Antimicrob Agents Chemother*, **47**, 3760–3763.

Kaplan, S.L. and Mason, E.O., Jr., (1998), Management of infections due to antibiotic-resistant *Streptococcus pneumoniae*. *Clin Microbiol Rev*, **11**, 628–644.

Karchmer, A.W., (2004), Increased antibiotic resistance in respiratory tract pathogens: PROTEKT US-an update. *Clin Infect Dis*, **39** (Suppl. 3), S142–S150.

Karlowsky, J.A., Thornsberry, C., Jones, M.E., Evangelista, A.T., Critchley, I.A. and Sahm, D.F., (2003), Factors associated with relative rates of antimicrobial resistance among *Streptococcus pneumoniae* in the United States: results from the TRUST Surveillance Program (1998–2002). *Clin Infect Dis*, **36**, 963–970.

Kehoe, L.E., Snidwongse, J., Courvalin, P., Rafferty, J.B. and Murray, I.A., (2003), Structural basis of Synercid (quinupristin-dalfopristin) resistance in Gram-positive bacterial pathogens. *J Biol Chem*, **278**, 29963–29970.

Kelley, M.A., Weber, D.J., Gilligan, P. & Cohen, M.S., (2000), Breakthrough pneumococcal bacteremia in patients being treated with azithromycin and clarithromycin. *Clin Infect Dis*, **31**, 1008–1011.

Kelly, L.J., (2001), Multidrug-resistant pneumococci isolated in the US: 1997–2001 TRUST surveillance (abstract). In: Program and abstrats of the 41st Interscience Conference on Antimicrobial Agents and Chemotherapy (Chicago) (ASM), p.142,Washington DC.

Kern, W.V., (2006), Daptomycin: first in a new class of antibiotics for complicated skin and soft-tissue infections. *Int J Clin Pract*, **60**, 370–378

Kislak, J.W., Razavi, L.M., Daly, A.K. and Finland, M., (1965), Susceptibility of pneumococci to nine antibiotics. *Am J Med Sci*, **250**, 261–268.

Klugman, K.P., (1990), Pneumococcal resistance to antibiotics. *Clin Microbiol Rev*, **3**, 171–196.

Klugman, K.P., (2002), Bacteriological evidence of antibiotic failure in pneumococcal lower respiratory tract infections. *Eur Respir J*, **20** (Suppl. 36), 3S–8S.

Klugman, K.P., (2004), Vaccination: a novel approach to reduce antibiotic resistance. *Clin Infect Dis*, **39**, 649–651.

Kramer, M.R., Rudensky, B., Hadas-Halperin, I., Isacsohn, M. and Melzer, E., (1987), Pneumococcal bacteremia–no change in mortality in 30 years: analysis of 104 cases and review of the literature. *Isr J Med Sci*, **23**, 174–180.

Leclercq, R. and Courvalin, P., (1991), Bacterial resistance to macrolide, lincosamide, and streptogramin antibiotics by target modification. *Antimicrob Agents Chemother*, **35**, 1267–1272.

Leclercq, R., Derlot, E., Duval, J. and Courvalin, P., (1988), Plasmid-mediated resistance to vancomycin and teicoplanin in Enterococcus faecium. *N Engl J Med*, **319**, 157–161.

Lewis, J.S., 2nd, Owens, A., Cadena, J., Sabol, K., Patterson, J.E. and Jorgensen, J.H., (2005), Emergence of daptomycin resistance in *Enterococcus faecium* during daptomycin therapy. *Antimicrob Agents Chemother*, **49**, 1664–1665.

Li, X, Zhao, X., and Drlica, K., (2002), Selection of Streptococcus pneumoniae mutants having reduced susceptibility to levofloxacin and moxifloxacin. Antimicrob. Agents Chemother. **46**: 522–524.

Lim, S., Bast, D., McGeer, A., de Azavedo, J., Low, D.E., (2003), Antimicrobial susceptibility breakpoints and first-step *parC* putations in *Streptococcus pneumoniae*: redefining fluoroquinolone resistance. *Emerg Infect Dis*, **9**, 833–837.

Linares, J., De la Campa, A.G. and Pallares, R., (1999), Fluoroquinolone resistance in *Streptococcus pneumoniae*. *N Engl J Med*, **341**, 1546–1547; author reply 1547–1548.

Linares, J., Pallares, R., Alonso, T., Perez, J.L., Ayats, J., Gudiol, F., Viladrich, P.F. and Martin, R., (1992), Trends in antimicrobial resistance of clinical isolates of *Streptococcus pneumoniae* in Bellvitge Hospital, Barcelona, Spain (1979–1990). *Clin Infect Dis*, **15**, 99–105.

Lonks, J.R., Garau, J., Gomez, L., Xercavins, M., Ochoa de Echaguen, A., Gareen, I.F., Reiss, P.T. and Medeiros, A.A., (2002), Failure of macrolide antibiotic treatment in patients with bacteremia due to erythromycin-resistant *Streptococcus pneumoniae*. *Clin Infect Dis*, **35**, 556–564.

Low, D.E., (2004), Quinolone resistance among pneumococci therapeutic and diagnostic implications. *Clin Infect Dis*, **38**, S357–S362.

Low, D.E., (2005), Fluoroquinolone-resistant pneumococci: maybe resistance isn´t futile?. *Clin Infect Dis*, **40**, 236–238.

Low, D.E., de Azavedo, J., Weiss, K., Mazzulli, T., Kuhn, M., Church, D., Forward, K., Zhanel, G., Simor, A. and McGeer, A., (2002), Antimicrobial resistance among clinical isolates of *Streptococcus pneumoniae* in Canada during 2000. *Antimicrob Agents Chemother*, **46**, 1295–1301.

Lujan, M., Gallego, M., Fontanals, D., Mariscal, D. and Rello, J., (2004), Prospective observational study of bacteremic pneumococcal pneumonia: Effect of discordant therapy on mortality. *Crit Care Med*, **32**, 625–631.

Lynch, I.J. and Martinez, F.J., (2002), Clinical relevance of macrolide-resistant *Streptococcus pneumoniae* for community-acquired pneumonia. *Clin Infect Dis*, **34** (Suppl. 1), S27–S46.

Lynch, III. J. and Zhanel, G., (2005), Escalation of antimicrobial resistance among *Streptococcus pneumoniae*: Implications for therapy. *Sem Respir Crit Care Med*, **26**, 575–616.

Maclean, I.H. and Fleming, A., (1939), M & B 639 and pneumococci. *Lancet*, **1**, 562–568.

Mandell, L.A., Wunderink, R.G., Anzueto, A., Bartlett, J.G., Campbell, G.D., Dean, N.C., Dowell, S.F., File, T.M., Jr., Musher, D.M., Niederman, M.S., Torres, A. and Whitney, C.G., (2007), Infectious Diseases Society of America/American Thoracic Society consensus guidelines on the management of community-acquired pneumonia in adults. Clin Infect Dis, *44 Suppl. (2)*, S27–72.

Markiewicz, Z. and Tomasz, A., (1989), Variation in penicillin-binding protein patterns of penicillin-resistant clinical isolates of pneumococci. *J Clin Microbiol*, **27**, 405–410.

Marston, B.J., Plouffe, J.F., File, T.M., Jr, Hackman, B.A., Salstrom, S.J., Lipman, H.B., Kolczak, M.S. and Breiman, R.F., (1997), Incidence of community-acquired pneumonia requiring hospitalization. Results of a population-based active surveillance Study in Ohio. The Community-Based Pneumonia Incidence Study Group. *Arch Intern Med*, **157**, 1709–1718.

Martínez-Beltrán, J.C., (1994), Mecanismos de resistencia a los antimicrobianos en Grampositivos. *Rev Clin Esp*, 803–813.

Martinez, J.A., Horcajada, J.P., Almela, M., Marco, F., Soriano, A., Garcia, E., Marco, M.A., Torres, A. and Mensa, J., (2003), Addition of a macrolide to a beta-lactam-based empirical antibiotic regimen is associated with lower in-hospital mortality for patients with bacteremic pneumococcal pneumonia. *Clin Infect Dis*, **36**, 389–395.

McCullers, J.A., English, B.K. and Novak, R., (2000), Isolation and characterization of vancomycin-tolerant *Streptococcus pneumoniae* from the cerebrospinal fluid of a patient who developed recrudescent meningitis. *J Infect Dis*, **181**, 369–373.

McDougal, L.K., Facklam, R., Reeves, M., Hunter, S., Swenson, J.M., Hill, B.C. and Tenover, F.C., (1992), Analysis of multiply antimicrobial-resistant isolates of *Streptococcus pneumoniae* from the United States. *Antimicrob Agents Chemother*, **36**, 2176–2184.

Menendez, R., Cordoba, J., Cuadra, P., Cremades, M., Lopez-Hontagas, J., Salavert, M., and Gobernado, M., (1999), Value of the polymerase chain reaction assay in noninvasive respiratory samples for diagnosis of community-acquired pneumonia. *Am J Respir Crit Care Med*, **159**, 1868–1873.

Mera, R.M., Miller, L.A., Daniels, J.J., Weil, J.G. and White, A.R., (2005), Increasing prevalence of multidrug-resistant *Streptococcus pneumoniae* in the United States over a 10-year period: Alexander Project. *Diagn Microbiol Infect Dis*, **51**, 195–200.

Metlay, J.P., Hofmann, J., Cetron, M.S., Fine, M.J., Farley, M.M., Whitney, C. and Breiman, R.F., (2000), Impact of penicillin susceptibility on medical outcomes for adult patients with bacteremic pneumococcal pneumonia. *Clin Infect Dis*, **30**, 520–528.

Mills, G.D., Oehley, M.R. and Arrol, B., (2005), Effectiveness of beta-lactam antibiotics compared with antibiotics active against atypical pathogens in non-severe community acquired pneumonia: meta-analysis. *BMJ*, **330**, 456.

Moellering, R.C., (2003), Linezolid: the first oxazolidinone antimicrobial. *Ann Intern Med*, **138**, 135–142.

Montanari, M.P., Mingoia, M., Giovanetti, E. and Varaldo, P.E., (2001), Differentiation of resistance phenotypes among erythromycin-resistant pneumococci. *J Clin Microbiol*, **39**, 9 1311–1315.

Moore, H.F. and Chesney, A.M., (1917), A study of ethylhydrocupreine (optochin) in the treatment of acute lobar pneumonia. *Arch Intern Med*, **19**, 611–613.

Moreno, S., Garcia-Leoni, M.E., Cercenado, E., Diaz, M.D., Bernaldo de Quiros, J.C. and Bouza, E., (1995), Infections caused by erythromycin-resistant *Streptococcus pneumoniae*: incidence, risk factors, and response to therapy in a prospective study. *Clin Infect Dis*, **20**, 1195–1200.

Mukhtar, T.A., Koteva, K.P., Hughes, D.W. and Wright, G.D., (2001), Vgb from *Staphylococcus aureus* inactivates streptogramin B antibiotics by an elimination mechanism not hydrolysis. *Biochemestry* , **40**, 8877–8886.

Muñoz, R. and De La Campa, A.G., (1996), ParC subunit of DNA topoisomerase IV of *Streptococcus pneumoniae* is a primary target of fluoroquinolones and cooperates with DNA gyrase A subunit in forming resistance phenotype. *Antimicrob Agents Chemother*, **40**, 2252–2257.

Musher, D.M., (2000), *Streptococcus pneumoniae*. In L. Mandell, J.E. Bennet and R. Dolin, editors, *Principles and Practice of Infections Diseases* , 5th edition. Philadelphia, Churchill Livingstone, pp. 2128–2142.

Musher, D.M., Bartlett, J.G. and Doern, G.V., (2001), A fresh look at the definition of susceptibility of *Streptococcus pneumoniae* to beta-lactam antibiotics. *Arch Intern Med*, **161**, 2538–2544.

Musher, D.M., Dowell, M.E., Shortridge, V.D., Flamm, R.K., Jorgensen, J.H., Le Magueres, P. and Krause, K.L., (2002), Emergence of macrolide resistance during treatment of pneumococcal pneumonia. *N Engl J Med*, **346**, 630–631.

Mutnick, A.H., Enne, V. and Jones, R.N., (2003), Linezolid resistance since 2001: SENTRY Antimicrobial Surveillance Program. *Ann Pharmacother*, **37**, 769–774.

National Committee for Clinical Laboratory Standards (1999) Performance Standards for Antimicrobial Susceptibility Testing: Ninth Informational Supplement, M100-S9. National Committee for Clinical Laboratory Standards, Wayne, PA, 2000.

National Committee for Clinical Laboratory Standards (2000) Performance Standards for Antimicrobial Susceptibility Testing: Tenth Informational Supplement, M100-S10. National Committee for Clinical Laboratory Standards, Wayne, PA, 2000.

National Committee for Clinical Laboratory Standards (2002) Performance Standards for Antimicrobial Susceptibility Testing: Twelfth Informational Supplement. Document M100-S12. National Committee for Clinical Laboratory Standards, Wayne, PA, 2002.

National Committee for Clinical Laboratory Standards. (1997). *Methods for dilution antimicrobial susceptibility test for bacteria that grow aerobically*, 4th edition. Approved standard M7-A4. National Committee for Clinical Laboratory Standards, Wayne, PA.

Nava, J.M., Bella, F., Garau, J., Lite, J., Morera, M.A., Marti, C., Fontanals, D., Font, B., Pineda, V., Uriz, S., (1994), Predictive factors for invasive disease due to penicillin-resistant *Streptococcus pneumoniae*: a population-based study. *Clin Infect Dis*, **19**, 884–890.

Niederman, M.S., (2001), Impact of antibiotic resistance on clinical outcomes and the cost of care. *Crit Care Med*, **29**, N114—N120.

Novak, R., Henriques, B., Charpentier, E., Normark, S. and Tuomanen, E., (1999), Emergence of vancomycin tolerance in *Streptococcus pneumoniae*. *Nature*, **399**, 590–593.

Nuorti, J.P., Butler, J.C., Crutcher, J.M., Guevara, R., Welch, D., Holder, P. and Elliott, J.A., (1998), An outbreak of multidrug-resistant pneumococcal pneumonia and bacteremia among unvaccinated nursing home residents. *N Engl J Med*, **338**, 1861–1868.

Pallares, R., Fenoll, A. and Linares, J., (2003), The epidemiology of antibiotic resistance in *Streptococcus pneumoniae* and the clinical relevance of resistance to cephalosporins, macrolides and quinolones. *Int J Antimicrob Agents*, **22** (Suppl. 1), S15–S24; discussion S25–S16.

Pallares, R., Gudiol, F., Linares, J., Ariza, J., Rufi, G., Murgui, L., Dorca, J. and Viladrich, P.F., (1987), Risk factors and response to antibiotic therapy in adults with bacteremic pneumonia caused by penicillin-resistant pneumococci. *N Engl J Med*, **317**, 18–22.

Pallares, R., Linares, J., Vadillo, M., Cabellos, C., Manresa, F., Viladrich, P.F., Martin, R. and Gudiol, F., (1995), Resistance to penicillin and cephalosporin and mortality from severe pneumococcal pneumonia in Barcelona, Spain. *N Engl J Med*, **333**, 474–480.

Pan, X.S., Ambler, J., Mehtar, S. and Fisher, L.M., (1996), Involvement of topoisomerase IV and DNA gyrase as ciprofloxacin targets in *Streptococcus pneumoniae*. *Antimicrob Agents Chemother*, **40**, 2321–2326.

Pasteur, L., (1881), Note sur la maladie nouvelle provoquéé par la salive d'un enfant mort de la rage. *Bul Acad Méd (Paris)*, [Series 2].

Patel, R., Piper, K., Cockerill, F.R., 3rd, Steckelberg, J.M. and Yousten, A.A., (2000), The biopesticide *Paenibacillus popilliae* has a vancomycin resistance gene cluster homologous to the enterococcal VanA vancomycin resistance gene cluster. *Antimicrob Agents Chemother*, **44**, 705–709.

Perez-Trallero, E., Fernandez-Mazarrasa, C., Garcia-Rey, C., Bouza, E., Aguilar, L., Garcia-de-Lomas, J. and Baquero, F., (2001), Antimicrobial susceptibilities of 1,684 *Streptococcus pneumoniae* and 2,039 *Streptococcus pyogenes* isolates and their ecological relationships: results of a 1-year (1998–1999) multicenter surveillance study in Spain. *Antimicrob Agents Chemother*, **45**, 3334–3340.

Perez-Trallero, E., Garcia-de-la-Fuente, C., Garcia-Rey, C., Baquero, F., Aguilar, L., Dal-Re, R. and Garcia-de-Lomas, J., (2005), Geographical and ecological analysis of resistance, coresistance, and coupled resistance to antimicrobials in respiratory pathogenic bacteria in Spain. *Antimicrob Agents Chemother*, **49**, 1965–1972.

Pestova, E., Beyer, R., Cianciotto, N.P., Noskin, G.A. and Peterson, L.R., (1999), Contribution of topoisomerase IV and DNA gyrase mutations in *Streptococcus pneumoniae* to resistance to novel fluoroquinolones. *Antimicrob Agents Chemother*, **43**, 2000–2004.

Piddock, L.J., (1999), Mechanisms of fluoroquinolone resistance: an update 1994–1998. *Drugs*, **58** (Suppl. 2), 11–18.

Pons, J.L., Mandement, M.N., Martin, E., Lemort, C., Nouvellon, M., Mallet, E., Lemelnad, J.F., (1996), Clonal and temporal patterns of nasopharyngeal penicillin-susceptible and penicillin-resistant *Streptococcus pneumoniae* strains in children attending a day care center. *J Clin Microbiol*, **34**, 3218–3222.

Poole, M.D., (1995), Otitis media complications and treatment failures: implications of pneumococcal resistance. *Pediatr Infect Dis J*, **14**, S23–S26.

Portillo, A., Ruiz-Larrea, F., Zarazaga, M., Alonso, A., Martinez, J.L. and Torres, C., (2000), Macrolide resistance genes in *Enterococcus spp. Antimicrob Agents Chemother*, **44**, 967–971.

Pottumarthy, S., Fritsche, T.R. and Jones, R.N., (2005), Comparative activity of oral and parenteral cephalosporins tested against multidrug-resistant *Streptococcus pneumoniae*: report from the SENTRY Antimicrobial Surveillance Program (1997–2003). *Diagn Microbiol Infect Dis*, **51**, 147–150.

Powers JH., (2004), Antimicrobial drug development – the past, the present, and the future. *Clin Microbiol Infect* , **10** (Suppl. 4), 23–31.

Prystowsky, J., Siddiqui, F., Chosay, J., Shinabarger, D.L., Millichap, J., Peterson, L.R. and Noskin, G.A., (2001), Resistance to linezolid: characterization of mutations in rRNA and comparison of their occurrences in vancomycin-resistant enterococci. *Antimicrob Agents Chemother*, **45**, 2154–2156.

Raad, II., Hanna, H.A., Hachem, R.Y., Dvorak, T., Arbuckle, R.B., Chaiban, G. and Rice, L.B., (2004), Clinical-use-associated decrease in susceptibility of vancomycin-resistant *Enterococcus faecium* to linezolid: a comparison with quinupristin-dalfopristin. *Antimicrob Agents Chemother*, **48**, 3583–3585.

Rosato, A., Vicarini, H., Bonnefoy, A., Chantot, J.F. and Leclercq, R., (1998), A new ketolide, HMR 3004, active against streptococci inducibly resistant to erythromycin. *Antimicrob Agents Chemother*, **42**, 1392–1396.

Ross, J.J., Worthington, M.G. and Gorbach, S.L., (2002), Resistance to levofloxacin and failure of treatment of pneumococcal pneumonia. *N Engl J Med*, **347**, 65–67; author reply 65–67.

Ruiz, M., Ewig, S., Torres, A., Arancibia, F., Marco, F., Mensa, J., Sanchez, M. and Martinez, J.A., (1999), Severe community-acquired pneumonia. Risk factors and follow-up epidemiology. *Am J Respir Crit Care Med*, **160**, 923–929.

Sahm, D.F., Karlowsky, J.K. and Kelly, L.J., (2001), Need for annual surveillance of antimicrobial resistance in the United States: 2 year longitudinal analysis. *Antimicrob Agents Chemother*, **45**, 1037–1042.

Sanchez, F., Mensa, J., Martinez, J.A., Garcia, E., Marco, F., Gonzalez, J., Marcos, M.A., Soriano, A. and Torres, A., (2003), Is azithromycin the first-choice macrolide for treatment of community-acquired pneumonia? *Clin Infect Dis*, **36**, 1239–1245.

Schmidt, L.H., (1943), Development of resistance to penicillin by pneumococci. *Proc Soc Exp Biol Med*, **53**, 353–357.

Schmitz, F.J., Verhoef, J., Milatovic, D. and Fluit, A.C., (2001), Treatment options for *Streptococcus pneumoniae* strains resistant to macrolides, tetracycline, quinolones, or trimethoprim/sulfamethoxazole. *Eur J Clin Microbiol Infect Dis*, **20**, 827–829.

Schrag, S.J., Beall, B. and Dowell, S.F., (2000), Limiting the spread of resistant pneumococci: biological and epidemiologic evidence for the effectiveness of alternative interventions. *Clin Microbiol Rev*, **13**, 588–601.

Schrag, S.J., McGee, I., Whitney, C.G.,Beall, B., Craig, A.S., Choate, M.E., Jorgensen, J.H., Facklam, R.R., Klugman, K.P., and The Active Bacterial Core Surveillance Team, (2004), Emergence of *Streptococcus pneumoniae* with very-high-level resistance to penicillin. *Antimicrob Agents Chemother*, **48**, 3016–3023.

Schuchat, A., Robinson, K., Wenger, J.D., Harrison, L.H., Farley, M., Reingold, A.L., Lefkowitz, L. and Perkins, B.A., (1997), Bacterial meningitis in the United States in 1995. Active Surveillance Team. *N Engl J Med*, **337**, 970–976.

Shefet, D., Robenshtok, E., Paul, M. and Leibovici, L., (2005), Empirical atypical coverage for inpatients with community-acquired pneumonia: systematic review of randomized controlled trials. *Arch Intern Med*, **165**, 1992–2000.

Silverman, J.A., Oliver, N., Andrew, T. and Tongchuani, L., (2001), Resistance studies with daptomycin. *Antimicrob Agents Chemother* , **45**, 1799–1802.

Sinclair, A., Arnold, C. and Woodford, N., (2003), Rapid detection and estimation by pyrosequencing of 23S rRNA genes with a single nucleotide polymorphism conferring linezolid resistance in Enterococci. *Antimicrob Agents Chemother*, **47**, 3620–3622.

Singh, K.V., Weinstock, G.M. and Murray, B.E., (2002), An Enterococcus faecalis ABC homologue (Lsa) is required for the resistance of this species to clindamycin and quinupristindalfopristin. *Antimicrob Agents Chemother*, **46**, 1845–1850.

Sloas, M.M., Barrett, F.F., Chesney, P.J., English, B.K., Hill, B.C., Tenover, F.C. and Leggiadro, R.J., (1992), Cephalosporin treatment failure in penicillin- and cephalosporin-resistant *Streptococcus pneumoniae* meningitis. *Pediatr Infect Dis J*, **11**, 662–666.

Smith, H.J., Walters, M., Hisanaga, T., Zhanel, G.G. and Hoban, D.J., (2004a), Mutant prevention concentrations for single-step fluoroquinolone-resistant mutants of wild-type, efflux-positive, or ParC or GyrA mutation-containing *Streptococcus pneumoniae* isolates. *Antimicrob Agents Chemother*, **48**, 3954–3958.

Smith, H.J., Noreddin, A.M., Siemens, C.G., Schurek, H., Greisman, J., Hoban, C., Homan, D. and Zhanel, G., (2004b), Designing fluoroquinolone breakpoints for *Streptococcus pneumoniae* by using genetics instead of pharmacokinetics-pharmacodynamics. *Antimicrob Agents Chemother*, **48**, 3630–3635.

Spellberg, B., Powers, J.H., Brass, E.P., Miller, L.G. and Edwards, J.E., Jr., (2004), Trends in antimicrobial drug development: implications for the future. *Clin Infect Dis*, **38**, 1279–1286.

Stahl, J.E., Barza, M., DesJardin, J., Martin, R. and Eckman, M.H., (1999), Effect of macrolides as part of initial empiric therapy on length of stay in patients hospitalized with community-acquired pneumonia. *Arch Intern Med*, **159**, 2576–2580.

Sternberg, G., (1881), A fatal form of septicaemia in the rabbit, produced by the subcutaneous injection of human saliva. *Ann resp Nath Board Health*, **3**, 87–108.

Stevens, D.L., Dotter, B. and Madaras-Kelly, K., (2004), A review of linezolid: the first oxazolidinone antibiotic. *Expert Rev Anti Infect Ther*, **2**, 51–59.

Sutcliffe, J., Tait-Kamradt, A. and Wondrack, L., (1996), *Streptococcus pneumoniae* and *Streptococcus pyogenes* resistant to macrolides but sensitive to clindamycin: a common resistance pattern mediated by an efflux system. *Antimicrob Agents Chemother*, **40**, 1817–1824.

Tankovic, J., Lascols, C., Sculo, Q., Petit, J.C. and Soussy, C.J., (2003), Single and double mutations in *gyrA* but not in *gyrB* are associated with low- and high-level fluoroquinolone resistance in Helicobacter pylori. *Antimicrob Agents Chemother*, **47**, 3942–3944.

Tenover, F.C., Weigel, L.M., Appelbaum, P.C., McDougal, L.K., Chaitram, J., McAllister, S., Clark, N., Killgore, G., O'Hara, C.M., Jevitt, L., Patel, J.B. and Bozdogan, B., (2004), Vancomycin-resistant *Staphylococcus aureus* isolate from a patient in Pennsylvania. *Antimicrob Agents Chemother*, **48**, 275–280.

Thorne, G.M. & Alder, J., (2002), Daptomycin: a novel lipopeptide antibiotic. *Clin Microbiol News L*, **24**, 33–40.

Tillotson, G., Zhao, X. & Drlica, K., (2001), Fluoroquinolones as pneumococcal therapy: closing the barn door before the horse escapes. *Lancet Infect Dis*, **1**, 145–146.

Tillotson, G. and Watson, S., (2001), Antimicrobial resistence mechanisms: what's hot and what's not in respiratory pathogens. *Semin Respir Infect*, **16**, 155–168.

Tomasz A., (1999), New faces of an old pathogen: emergence and spread of multidrug-resistant *Streptococcus pneumoniae*. *Am J Med*, **107** (Suppl. 1A), 55S–62S.

Turett, G.S., Blum, S., Fazal, B.A., Justman, J.E. and Telzak, E.E., (1999), Penicillin resistance and other predictors of mortality in pneumococcal bacteremia in a population with high human immunodeficiency virus seroprevalence. *Clin Infect Dis*, **29**, 321–327.

Urban, C., Rahman, N., Zhao, X., Mariano, N., Segal-Maurer, S., Drlica, K. and Rahal, J.J., (2001), Fluoroquinolone-resistant *Streptococcus pneumoniae* associated with levofloxacin therapy. *J Infect Dis*, **184**, 794–798.

Uttley, A.H., Collins, C.H., Naidoo, J. and George, R.C., (1988), Vancomycin-resistant entero-cocci. *Lancet*, **1**, 57–58.

Van Bambeke, F., Van Laethem, Y., Courvalin P. and Tulkens PM., (2004), Glycopeptide antibiotics from conventional molecules to new derivatives. *Drugs*, **64**, 913–936.

Vanderkooi, O.G., Low D.E., Green, K. Powis, J.P. and McGeer, A., (2005), Predicting antimicro-bial resistance in invasive pneumococcal infections. *Clin Infect Dis*, **40**, 1288–97

Vannufel, P. and Cocito, C., (1996), Mechanism of action of streptogramins and macrolides. *Drugs*, **51** (Suppl. 1), 20–30.

Verhaegen, J. and Verbist, L., (1999), In-vitro activities of 16 non-beta-lactam antibiotics against penicillin-susceptible and penicillin-resistant *Streptococcus pneumoniae*. *J Antimicrob Chemother*, **43**, 563–567.

Waterer, G.W., Somes, G.W. and Wunderink, R.G., (2001), Monotherapy may be suboptimal for severe bacteremic pneumococcal pneumonia. *Arch Intern Med*, **161**, 1837–1842.

Waterer, G.W., Wunderink, R.G. and Jones, C.B., (2000), Fatal pneumococcal pneumonia attributed to macrolide resistance and azithromycin monotherapy. *Chest*, **118**, 1839–1840.

Weiss, K. and Tillotson, G., (2005), The controversy of combination vs monotherapy in the treat-ment of hospitalized community-acquired pneumonia. *Chest*, **128**, 940–946.

Whitney, C.G., Farley, M.M., Hadler, J., Harrison, L.H., Lexau, C., Reingold, A., Lefkowitz, L., Cieslak, P.R., Cetron, M., Zell, E.R., Jorgensen, J.H. and Schuchat, A., (2000), Increasing prevalence of multidrug-resistant *Streptococcus pneumoniae* in the United States. *N Engl J Med*, **343**, 1917–1924.

Woodford, N., (2005), Biological counterstrike: antibiotic resistance mechanisms of Gram-positive cocci. *Clin Microbiol Infect*, **11** (Suppl. 3), 2–21.

Woods, D., (1962), The biochemical mode of sulphonamide drug. *J Gen Microbiol*, **29**, 687–702.

Wright, G. D., (2003), Mechanisms of resistance to antibiotics. *Curr Opin Chem Biol*, **7**, 563–569.

Yu, V.L., Chiou, C.C., Feldman, C., Ortqvist, A., Rello, J., Morris, A.J., Baddour, L.M., Luna, C.M., D.R., S., Ip, M., Ko, W.C., M.B.F., C., Andremont, A. and Klugman, H.P., (2003), An international prospective study of pneumococcal bacteremia: Correlation with in vitro resis-tance, antibiotics administered, and clinical outcome. *Clin Infect Dis*, **37**, 230–237.

Zhanel, G.G., Hoban, D.J. and Chan, C.K., (2002), Resistance to levofloxacin and failure of treat-ment of pneumococcal pneumonia. *N Engl J Med*, **347**, 65–67; author reply 65–67.

Zhanel, G.G., Palatnich, L., Nichol, K.A., Bellyou, T., Low, D.E. and Hoban, D.J., (2003), Antimicrobial resistance in respiratory tract *Streptococcus pneumoniae* isolates: results of the Canadian Respiratory Organism Susceptibility Study, 1997 to 2002, *Antimicrob Agents Chemother*, **47**, 1867–1874.

Zhao, X., and Drlica, K., (2001), Restricting the selection of antibiotic-resistant mutants: a general strategy derived from fluoroquinolone studies. *Clin Infect Dis*, **33**, S147–S156.

Zhao, X. and Drlica, K., (2002), Restricting the selection of antibiotic-resistant mutants: measure-ment and potenial uses of the mutant selection window. *J Inf Dis*, **185**, 561–565.

Chapter 2
Emergence of MRSA in the Community

Adam L. Cohen, Rachel Gorwitz, and Daniel B. Jernigan

2.1 Introduction

Staphylococcus aureus has been recognized as a cause of human infection for over a hundred years, and its role in causing diseases such as sepsis and abscesses was first described by Ogston in the late nineteenth century (Ogston, 1882). *S. aureus* can colonize human hosts without causing disease, and infections with *S. aureus*, especially antimicrobial-resistant strains such as methicillin-resistant *S. aureus* (MRSA), now pose a large and growing health burden.

 S. aureus has emerged as one of the major pathogens of public health significance in the new millennium, and MRSA is on the rise. Over the last 20 years, the percentage of *S. aureus* infections in hospitalized US patients that are resistant to methicillin has continually increased from 2% in 1975 to 34% in 1992 and 64% in 2003 (Klevens et al., 2006; Panlilio et al., 1992). Skin and soft tissue infections, which annually account for an estimated 11.6 million visits to medical providers in the United States (four visits for every 100 persons) (McCaig et al., 2006), are most commonly caused by *S. aureus;* the majority of abscesses cultured in emergency departments in the United States are now MRSA (Cohen et al., in press; Moran et al., 2005, 2006). The percentage of skin infections caused by MRSA in a emergency department in Los Angeles has increased from 29 to 64% from 2001 to 2004 (Moran et al., 2005).

 Penicillin was first introduced to treat patients with bacterial infections in 1941, and resistance to penicillin was first reported in *S. aureus* within 1–2 years (Kirby, 1944). These resistant strains were first found in hospitals after the Second World War, where patients were exposed to this new antimicrobial agent (Barber and Rozwadowska-Dowzenko, 1948). *S. aureus* had quickly acquired the ability to produce penicillinase, an enzyme that inactivates penicillin. An "epidemic strain" of antimicrobial-resistant *S. aureus*, which was characteristically lysed by bacteriophage 80 and 81, was noted to cause hospital outbreaks in Australia, Canada, and the United States in the 1950s. This strain was reported to more commonly cause disease in hospitalized children and otherwise healthy young adults (Fekety and Bennett, 1959).

 In Copenhagen in the late 1960s, the first large-scale study of penicillin-resistant *S. aureus* discovered that not only were the majority of *S. aureus* found

I. W. Fong and K. Drlica (eds.), *Antimicrobial Resistance and Implications*
for the Twenty-First Century. © Springer 2008

in hospitals resistant to penicillin but also the resistance gene had spread to a majority of *S. aureus* strains in community settings (Jessen et al., 1969). Within a decade, a majority of community *S. aureus* strains in the United States were penicillin-resistant (Ross et al., 1974). New drug development found a solution with semisynthetic penicillins (such as methicillin, oxacillin, and nafcillin) that resisted penicillinases produced in the majority of *S. aureus* strains.

The first *S. aureus* isolates resistant to methicillin were obtained from patients in England within months of methicillin introduction in 1959 (Jevons, 1961). Reports of MRSA in the United States soon followed (Barrett et al., 1968). As with penicillin-resistant *S. aureus*, MRSA strains were first seen in hospitals, prompting concerns that MRSA would soon spread outside the hospital. Nearly 50 years later, we are now seeing the emergence of MRSA in the community; however, the source of the resistant strains does not appear to be the hospital. It appears that both the absolute number of MRSA infections and the proportion of MRSA infections that are community-associated are increasing. At three urban children's hospitals in the United States, the percentage of MRSA infections associated with the community have increased as much as 60% in the 3 or 4 years after 2000 (Buckingham et al., 2004; Mishaan et al., 2005; Ochoa et al., 2005).

Community-associated MRSA (CA-MRSA) has unique characteristics that separates it from methicillin-susceptible *S. aureus* (MSSA) and MRSA in healthcare settings. CA-MRSA typically has different epidemiological risk factors, clinical manifestations, and microbiological characteristics than healthcare-associated MRSA (HA-MRSA). In this chapter we discuss the epidemiology and mechanisms of resistance of CA-MRSA, an approach to management of CA-MRSA infections, recommendations for prevention of MRSA in the community, and future directions for research.

2.2 Epidemiology and Mechanisms of Resistance in MRSA in the Community

2.2.1 Epidemiology and Clinical Presentation of MRSA

In 1999, the Centers for Disease Control and Prevention (CDC) published a report concerning four children from 12 months to 13 years of age who had died from MRSA infections (Centers for Disease Control and Prevention, 1999). None of the children had risk factors for HA-MRSA, which included recent hospitalization or surgery, residence in a long-term care facility, or a history of injecting drug use. Although CA-MRSA had been recently reported in children (Adcock et al., 1998; Herold et al., 1998), most documented US MRSA infections up to that point were associated with healthcare settings or in adult injecting drug users (Levine et al., 1982; Saravolatz et al., 1982). These four cases highlighted the fact that not only could patients develop MRSA disease outside the hospital, but also that CA-MRSA disease could be severe and fatal. In addition to the four pediatric deaths, CA-MRSA infections were reported in other populations, such as prisoners (Centers for

Disease Control and Prevention, 2001). In response to these reports, CDC began active surveillance of all culture-confirmed MRSA infections in the United States to study the epidemiology and drug resistance patterns of MRSA isolates in the community (Fridkin et al., 2005).

Numerous reports have noted that CA-MRSA can present with different clinical manifestations than HA-MRSA and, typically, causes less severe disease for the majority of infections. Most *S. aureus* infections are of the skin and soft tissue and often start out as a small papule or pustule. MRSA and other *S. aureus* skin and soft tissue infections may be misdiagnosed as "spider bites" due to their characteristic presentation. Possible reasons for this are that patients may detect a spontaneous, red bump and assume that it is a spider bite and that clinicians may note necrotic features of the lesions and associate this with a brown recluse spider bite. Patients and medical providers may misdiagnose MRSA skin infections as spider bites even in the absence of having seen a spider or having brown recluses endemic to the local community. In addition, brown recluse spiders have been implicated as the cause of skin lesions across the United States and Canada, despite the fact that these spiders are found only in the southern and central Midwestern United States (Vetter and Bush, 2002).

CA-MRSA infection is defined as an MRSA infection with onset in the community in an individual lacking established HA-MRSA risk factors. For MRSA active surveillance conducted through the CDC-sponsored Active Bacterial Core Surveillance program, an MRSA infection is considered to be community-associated if *S. aureus* isolates are cultured in an outpatient or less than 2 days after hospitalization and if the patient has had no hospitalization, surgery, dialysis, or residence in a long-term care facility within the previous year; no permanent indwelling catheter or percutaneous medical device; and no previous history of MRSA (Fridkin et al., 2005). When case definitions based on risk factors are extended to newborns, admission of a neonate to the hospital at birth has been disregarded as a risk factor for HA-MRSA since nearly all neonates are hospitalized at birth (Buckingham et al., 2004). From studies using these definitions, several demographic characteristics of CA-MRSA have been identified.

2.2.1.1 Geographic Characteristics

Cases of CA-MRSA infection have been reported worldwide (Boyce et al., 2005) and regional differences exist within countries. The percentage of MRSA that was community-associated varied from 8 to 20% in three states in the United States (Fridkin et al., 2005); however, prevalence of MRSA in some populations may be higher. MRSA strains present in the community vary by continent, region, and country, but the strains seen worldwide have many microbiological characteristics in common, such as presence of genes for Panton–Valentine leukocidin toxin (Harbarth et al., 2005; Ma et al., 2005; Mulvey et al., 2005; Vandenesch et al., 2003). Although there are microbiological similarities in MRSA strains from different countries, these similarities, such as the development of resistance to methicillin, may have arisen independently in each country or region (Ma et al., 2005).

2.2.1.2 Age Characteristics

As discussed earlier, some of the first reports of MRSA infection without traditional healthcare-associated risk factors were in children (Centers for Disease Control and Prevention, 1999; Herold et al., 1998). In two urban areas in the United States where MRSA surveillance was in place from 2001 through 2002, CA-MRSA was more common among children less than 2 years of age than in other age groups (Fridkin et al., 2005). Age differences exist in nasal colonization rates as well. Prevalence of *S. aureus* nasal colonization is highest in children 6–11 years of age, and MRSA nasal carriage is highest in individuals 60 years or older (Kuehnert et al., 2006).

2.2.1.3 Racial, Ethnic, and Socioeconomic Characteristics

Rates of CA-MRSA vary between racial and ethnic groups. Compared with Australians of European descent, high rates of CA-MRSA have been noted in Maori and Pacific Islanders in Australia (Hill et al., 2001). Similarly, Pacific Islanders in Hawaii have much higher rates of CA-MRSA infection than Hawaiians of Asian descent. In one investigation of MRSA in Hawaii, Pacific Islanders comprised 76% of patients with CA-MRSA but only 35% of patients who received care in the facility under investigation (Estivariz et al., 2007). In some urban centers in the United States, African Americans have incidence rates higher than whites (Fridkin et al., 2005). Pediatric patients hospitalized with CA-MRSA infection in an urban center in the United States were more likely to be African American than patients hospitalized with community-associated MSSA infections (Ochoa et al., 2005; Sattler et al., 2002).

The racial and ethnic differences that are seen in CA-MRSA may be due to intrinsic genetic factors associated with different races and ethnicities; however, further study is needed (Fine et al., 2005). Geographic isolation of some groups, such as Pacific Islanders in Hawaii, may also contribute to increased rates of MRSA infection (Estivariz et al., 2007). Concurrent diseases such as diabetes may also be a factor (Estivariz et al., 2007). A more likely explanation for the differences may be socioeconomic factors, such as limited access to healthcare and crowded housing conditions, which are more common in some racial groups (Estivariz et al., 2007). Surveillance of CA-MRSA infections shows that the disease occurs across all income categories (Fridkin et al., 2005); however, a study in Minnesota found that the median household income for patients with MRSA infection was lower than the median income for state residents overall (Naimi et al., 2003).

2.2.1.4 Sex Characteristics

Although the trends are not consistent across age groups, men and women have different rates of MRSA infection and nasal colonization. Male infants are more likely than female infants both to be colonized by *S. aureus* and to have *S. aureus* skin infections (Enzenauer et al., 1984, 1985; Thompson et al., 1963, 1966). The cause of this difference is unknown and has not been demonstrated in adults. MRSA nasal carriage is more common among females compared to males, but among non-Hispanic whites and Mexican Americans, males are more likely than females to have nasal colonization of *S. aureus* (Kuehnert et al., 2006). Outbreaks involving men have been repeatedly reported; however, it is not clear if this represents a reporting bias or a higher incidence of disease among men (Centers for Disease Control and Prevention, 2003a,b).

2.2.1.5 Urban, Suburban, and Rural Characteristics

CA-MRSA is found in both urban and rural settings. The four pediatric deaths in 1999 from MRSA were reported in rural areas of North Dakota and Minnesota, and different strain types of CA-MRSA continue to cause disease in rural western United States (Stevenson et al., 2005) and among rural native populations in Alaska (Baggett et al., 2003), the Midwestern United States (Groom et al., 2001), and the Northern Territory in Australia (Maguire et al., 1996). A study in Minnesota found that the percentage of MRSA infections that were community-associated were higher outside urban areas (Naimi et al., 2003). Recent investigations have shown increases in prevalence of CA-MRSA in rural settings, possibly associated with increases in methamphetamine use (Cohen et al., in press).

2.2.2 Transmission of CA-MRSA

As late as the 1960s, transmission of MRSA was not well understood. Early studies of transmission of *S. aureus* among newborns in a hospital nursery suggested that *S. aureus* was transmitted from patient to patient on the hands of nursery staff and not by airborne particles (Mortimer et al., 1962; Wolinksky et al., 1960). In the hospital, healthcare workers can become transiently colonized after contact with colonized or infected patients and subsequently transmit the MRSA to other patients (Mulligan et al., 1993). While staphylococci can be spread through contaminated fomites, environmental spread does not appear to be the most common method of transmission (Mortimer et al., 1962). As in earlier studies of *S. aureus* in the hospital, CA-MRSA is now considered to be transmitted primarily person-to-person and less commonly through environmental contamination.

When considering transmission of MRSA in the community, it may be helpful to categorize individuals into one of three states: (1) infected with MRSA, (2) colonized with MRSA, or (3) susceptible to MRSA. An individual may move in and out of each of these states, but individuals colonized with MRSA are more likely than non-colonized individuals to develop infection (Ellis et al., 2004, von Eiff et al., 2001). Colonized and infected individuals can also contaminate the environment, allowing environmental surfaces, shared items, and hands to serve as transient vehicles for transmission.

2.2.2.1 The Colonized State

Colonization with *S. aureus* is common in humans. Although colonization may occur in many parts of the body (including the axillae, perineum, groin, rectum, skin, and umbilical stump in neonates), the anterior nares are the most consistent site of colonization. Colonization of the nares can lead to contamination of the hands and other parts of the body, and likelihood of colonization can increase after contact with an infected individual (Adcock et al., 1998).

The majority of individuals are either transiently or persistently colonized by *S. aureus* at some point during their life. Staphylococcal carriage studies have found that 16–36% of individuals are persistently colonized, 15–70% are intermittently or occasionally colonized, and 6–47% are never colonized (VandenBergh et al., 1999; Williams, 1963). This has been colloquially referred to as "The Rule of Thirds": one third of individuals are persistently colonized, one third occasionally, and one third never. However, the Rule of Thirds likely underestimates the number of individuals who are persistently colonized.

In a representative sample survey, nearly a third (an estimated 32.4%) of the United States population carried *S. aureus* in their nares in 2001 and 2002, but less than 1% were carriers of MRSA at that time (Kuehnert et al., 2006). Rates of MRSA colonization based solely on nasal carriage have been as low as less than 1% in certain populations (Hussain et al., 2001; Kuehnert et al., 2006; Sa-Leao et al., 2001), but this may underestimate the carriage on other body parts or intermittent colonization. MRSA nasal carriage can be much higher in certain populations, and children may have higher rates of colonization than adults (Shopsin et al., 2000). In 2001, MRSA nasal colonization in a healthy pediatric population was less than 1%, but the prevalence had increased to 9.2% by 2004 (Creech et al., 2005). The MRSA nasal carriage rate for patients admitted to a public urban hospital in the United States was 7.3% in 1996–1998 (Hidron et al., 2005).

2.2.2.2 The Infected State

S. aureus causes a broad spectrum of disease. MRSA also causes infections in many parts of the body and has manifestations similar to MSSA (Herold et al., 1998). Those that are infected can be treated but become susceptible again since there is no natural lasting immunity to *S. aureus*. Severity of disease can vary by site of infection, and CA-MRSA may also cause recurrent infections. In one study, approximately one quarter of patients with CA-MRSA infection were hospitalized for their infection (Fridkin et al., 2005), but this percentage may be higher among those with severe or invasive disease (Naimi et al., 2001).

Multi-state surveillance for CA-MRSA infections in the United States shows that skin and soft tissue are the most common sites of infection, causing between 77 and 84% of disease (Fridkin et al., 2005; Naimi et al., 2001) (Figure 2.1). The majority of skin and soft tissue infections are abscesses or cellulitis, but up to one quarter are superficial infections such as impetigo (Naimi et al., 2001). In addition to skin and soft tissue infections, MRSA causes wound infections, sinus infections, urinary tract infections, and pneumonia. Invasive syndromes, such as bacteremia, meningitis,

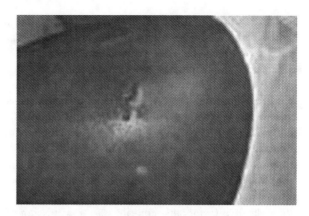

Fig. 2.1 MRSA skin and soft tissue infection in prison inmate (CDC, 2005; photo credit Bruno Coignard, MD and Jeff Hageman, MHS).

osteomyelitis, bursitis, and arthritis, are uncommon manifestations (approximately 5%) of MRSA infection (Fridkin et al., 2005). MRSA pneumonia has been reported as a bacterial super-infection after influenza infection in both adults and children (Bhat et al., 2005; Podewils et al., 2005). MRSA can be a cause of otitis media in children (Santos et al., 2000) and accounted for as much as 12% of cases of otorrhea in one series (Hwang et al., 2002). MRSA has also been a reported cause of pyomyositis (Kaplan, 2005a) and Waterhouse–Friderichsen syndrome in children (Adem et al., 2005). Recently, cases of CA-MRSA necrotizing fasciitis have been reported (Miller et al., 2005).

2.2.2.3 The Susceptible State

The development of MRSA infection in an individual is related to both the virulence of the bacteria and the susceptibility of the host. Host factors that increase susceptibility to MRSA include a weakened immune response, a non-intact skin barrier, and a genetic predisposition to infection. Specifically, defects in chemotaxis (such as in Wiskott–Aldrich and Chediak–Higashi syndromes), phagocytosis, and humoral immunity can cause an increased risk for staphylococcal infection. MRSA skin infections may be more common in persons with HIV, but this is more likely due to behavioral risk factors than immune suppression by HIV (Lee et al., 2005). Other factors leading to increased transmission of MRSA in susceptible hosts have been elucidated by evaluating outbreaks of MRSA in the community.

2.2.2.4 Outbreaks of MRSA in the Community

Since the early 1980s when MRSA was recognized as a pathogen that can cause outbreaks in the community in groups such as intravenous drug users, outbreaks of CA-MRSA have been reported in a number of diverse groups: Native American, Alaskan Native, and Pacific Islander communities (Baggett et al., 2003, 2004; Estivariz et al., 2007; Groom et al., 2001; Hill et al., 2001); prisoners (Centers for Disease Control and Prevention, 2001, 2003b); amateur and professional sports participants, such as football players, wrestlers, rugby players, fencers, and divers (Begier et al., 2004; Centers for Disease Control and Prevention, 2003a; Kazakova et al., 2005; Stacey et al., 1998; Wang et al., 2003); child care attendees (Adcock et al., 1998); military personnel (Campbell et al., 2004; Zinderman et al., 2004); men who have sex with men (Lee et al., 2005); methamphetamine and injecting drug users (Cohen et al., in press; Fleisch et al., 2001); survivors of natural disasters (Centers for Disease Control and Prevention, 2005); recipients of tattoos (Centers for Disease Control and Prevention, 2006); and isolated religious communities (Coronado et al., 2006). MRSA can cause infections in animals and pets and has been reported to cause infections in humans who have contact with infected animals (Weese et al., 2005).

Although these groups are diverse, they have common factors that may underlie the transmission of MRSA in the community. Based on investigations of community outbreaks, five factors that contribute to transmission of MRSA in the community have been characterized as the "Five Cs":

1. *Crowding*. Outbreaks have occurred in populations living in crowded quarters such as prisons and military barracks. Living in a house with more than one person per bedroom has been independently associated with developing a CA-MRSA skin or soft tissue infection (Cohen et al., in press).

2. *Contact, skin-to-skin*. Participants in contact sports may have frequent skin-to-skin contact, which may act as a method of transmitting MRSA skin and soft tissue infection. Outbreaks among professional and college football teams have been attributed to frequent skin-to-skin contact (Begier et al., 2004; Kazakova et al., 2005). Similarly, wrestlers who have significant skin-to-skin contact have experienced outbreaks of MRSA infection (Centers for Disease Control and Prevention, 2003a). High-risk sexual behavior (Lee et al., 2005) and sexual contact with someone with a skin infection (Cohen et al., in press; Lee et al., 2005) have both been associated with CA-MRSA skin and soft tissue infections. These factors have been described both in rural and urban communities.

3. *Cut or compromised skin*. Breaks in the skin can be a portal for MRSA bacteria to enter the body. For example, in an outbreak of MRSA infections among a college football team, MRSA infections were associated with abrasions from artificial grass ("turf burns") and cosmetic body shaving (Begier et al., 2004). In an outbreak among military recruits, most of the MRSA skin and soft tissue infections were on exposed skin of the arms, legs, and knees, where abrasions are common during field training (Zinderman et al., 2004). Skin picking behavior has also been associated with MRSA skin and soft tissue infections (Cohen et al., in press). Injection drug use, where the skin is compromised by insertion of contaminated needles, has been associated with MRSA infections (Miller et al., 2005; Young et al., 2004), but injection may not be the only method by which MRSA is transmitted among drug users (Cohen et al., in press).

4. *Contaminated surfaces and shared items*. Although environmental transmission of MRSA may not be the most common mode of transmission, the environment may have played a role in some outbreaks of MRSA in the community. Outbreaks have been associated with whirlpools (Begier et al., 2004) MRSA-contaminated sauna benches (Baggett et al., 2004) (Figure 2.2). An outbreak

among fencers was unusual because there is typically little skin-to-skin contact in that sport; however, investigators surmised that the cluster of cases was due to shared fencing equipment (Centers for Disease Control and Prevention, 2003a). In a correctional facility in Mississippi, sharing personal items such as linens was associated with infection (Centers for Disease Control and Prevention, 2001), while sharing bars of soap was implicated in an outbreak among members of a college football team (Nguyen et al., 2005).

5. *Cleanliness.* Cleanliness includes both personal bathing and laundering of clothing, linens, and towels, all of which has been noted as potential contributing factors to CA-MRSA infection among prison inmates (Centers for Disease Control and Prevention, 2001). Investigations of MRSA transmission in prisons suggest that lack of access to basic hygiene is a contributing factor (Centers for Disease Control and Prevention, 2003b). Homelessness has also been associated with MRSA skin and soft tissue infections (Young et al., 2004).

In addition to the "Five Cs," previous use of antimicrobial agents has also been shown to be a factor in the development of CA-MRSA (Baggett et al. 2003, Kazakova et al., 2005).

6. *Antimicrobial agent use.* The prior use of antimicrobial agents has been associated with MRSA infections in the healthcare setting and may also be a contributing factor in the community. Use of antimicrobial agents may predispose individuals to the acquisition of resistant organisms, such as MRSA strains. An outbreak of CA-MRSA skin infections in southwestern Alaska found that patients with the skin infections received significantly more antimicrobial agents in the year before the outbreak compared to community members without skin infections (Baggett et al., 2004). Nasal MRSA colonization, which can precede MRSA infection, was more common in new military recruits who had received antimicrobial agents within 6 months of arriving at training camp than those who had received no recent antibiotics (Ellis et al., 2004). In an outbreak of MRSA

Fig. 2.2 Scanning electron micrograph (magnification 1719×) of *Staphylococcus*-like organisms on samples of wood collected from steam baths responsible for a furunculosis outbreak in rural Alaska (CDC, 2004; photo taken by Janice Carr, Division of Healthcare Quality Promotion).

in a closed religious community in the United States, investigators found use of antimicrobial agents was associated with infection (Coronado et al., 2006).

It is important to note that in a study at an urban emergency department in the United States, most patients with MRSA skin infections had none of these characteristic risk factors associated with recent reports of CA-MRSA outbreaks (Moran et al., 2005).

2.2.3 Microbiology and Mechanisms of Resistance

S. aureus is a non-motile, non-spore-forming Gram-positive coccus that appears as characteristic grape-like clusters (Figure 2.3). The bacteria are catalase-positive, facultative aerobes that are usually unencapsulated. They can survive on fomites in the environment for months. Antimicrobial resistance and virulence factors in *S. aureus* help explain the clinical presentations and the transmission of CA-MRSA. Microbiologic differences of MRSA isolated from patients with CA- and HA-MRSA infections have been identified based on molecular typing, antimicrobial susceptibility testing, and identification of methicillin resistance and toxin genes (Table 2.1). However, these differences are becoming less distinct as MRSA strains that emerged in the community develop resistance to additional classes of antimicrobial agents and enter healthcare settings.

2.2.3.1 Mechanisms of Resistance

Penicillin resistance in *S. aureus* is conferred by a plasmid-associated gene (*blaZ*) that codes for a β-lactamase. Methicillin resistance is usually conferred by an altered chromosomally encoded penicillin-binding protein (PBP-2a) that causes resistance to all β-lactam antimicrobial agents (including penicillin) and cephalosporins. The staphylococcal chromosomal cassette contains the gene for this altered penicillin-

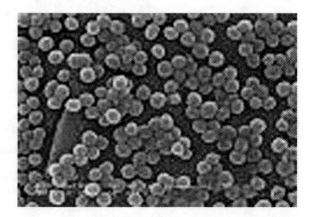

Fig. 2.3 Scanning electron micrograph of *S. aureus* (CDC, 2005; Photo taken by Janice Carr, Division of Healthcare Quality Promotion).

Table 2.1 Microbiological differences initially noted between community- and healthcare-associated MRSA in the United States.

	Community-associated	Healthcare-associated
Pulsed-field gel electrophoresis (PFGE) type	USA300, USA400 commonly; USA1000, USA1100 less commonly	USA100, USA200
Staphylococcal chromosomal cassette or *mec* type	*SCCmec* Type IV	*SCCmec Types* I, II, III (most commonly Type II)
Accessory gene regular (agr) allele	agr3	agr2
Panton–Valentine leukocidin (PVL) toxin production	Common	Rare
Antimicrobial susceptibility [a]	Generally susceptible to antimicrobials other than β-lactams and erythromycin	Generally resistant to multiple agents
Chloramphenicol	Usually susceptible	Often resistant
Clindamycin	Usually susceptible	Often resistant
Erythromycin	Usually resistant	Usually resistant
Fluoroquinolone	Variable	Usually resistant
trimpethoprim /sulfamethoxazole (TMP-SFZ)	Usually susceptible	Usually susceptible

[a] Antimicrobial susceptibility patterns may change.
References: Naimi et al. (2003), Baba et al. (2002), Fey et al. (2003) and Weber (2005).

binding protein (*SCCmec*). *SCCmec* can be mobilized for transfer between organisms in vitro (Katayama et al., 2000), although this is thought to be a rare occurrence (Chambers, 2001). *SCCmec* contains two genes [(cassette chromosome recombinase A and B (*ccrA* and *ccrB*)] that encode recombinases that integrate the cassette into its chromosomal locus.

Five types of *SCCmec* have been described; Types II and IV are the primary types seen in the United States. Genotypes associated with community transmission almost exclusively contain *SCCmec* Type IV; it is the smallest of the five types. *SCCmec* Type IV has been identified in MRSA strains from CA-MRSA infections in the United States and worldwide. MRSA strains associated with healthcare transmission in the United States most commonly contain *SCCmec* Type II and less commonly Types I and III. *SCCmec* Type IV is also typical in some healthcare-associated strains, such as USA800. Types II and III often carry genes conferring resistance to other antimicrobial agents (such as aminoglycosides, tetracyclines, erythromycin, and clindamycin), whereas Type IV typically does not. This difference in community- and healthcare-associated *SCCmec* types leads to different antimicrobial agent susceptibility patterns between healthcare- and community-associated MRSA infections. However, these patterns are likely to change over time.

2.2.3.2 Mechanisms of Virulence

Virulence factors enhance the ability of bacteria to cause infection by evading the host's defenses, increasing adherence to tissues, or spreading through tissues. Examples of virulence factors in *S. aureus* include production of coagulase, toxins, and proteins intrinsic to the cell wall. *S. aureus* produces coagulase, which interacts with fibrinogen causing plasma to clot. This clumping creates a loose polysaccharide capsule that can interfere with phagocytosis. The combination of these virulence factors may cause localization of an infection, such as in an abscess, a common clinical manifestation of CA-MRSA infection.

Panton–Valentine Leukocidin (PVL) is a cytotoxin (coded by the *lukS-PV* and *lukF-PV* genes) first identified in methicillin-susceptible *S. aureus* (Panton and Valentine, 1932). PVL kills leukocytes by creating pores in the cell membrane of affected cells or by activating apoptosis pathways. Pore formation leads to increased cell wall permeability and leakage of protein from the cell causing cell death and tissue necrosis. PVL genes have been associated with severe abscesses, necrotizing pneumonia, and increased complications in osteomyelitis (Lina et al., 1999, Martinez-Aguilar et al., 2004). Genes encoding PVL are more commonly found among strains of MRSA originating in the community and not strains traditionally associated with health care settings (Naimi et al., 2003). The presence of PVL may explain the clinical manifestations of abscesses and skin and soft tissue infections in CA-MRSA infections.

In addition to PVL, other toxins may be produced by *S. aureus*: α-toxin, which causes tissue necrosis and acts on cell membranes; exfoliatin A and B, which cause skin separation in diseases such as bullous impetigo and staphylococcal scalded skin syndrome; enterotoxins A, B, C_1, C_2, D, and E, which can cause vomiting and diarrhea associated with food poisoning; and toxic shock syndrome toxin 1 (TSST-1), which induces production of interleukin-1 and tumor necrosis factor leading to shock. Peptidoglycans, which comprise 50% by weight of the cell wall of staphylococci, can have endotoxin properties as well. Other cell wall polymers, such as teichoic acid, and cell surface proteins, such as protein A, fibronectin-, and collagen-binding proteins, may also be virulence factors for *S. aureus*. Recent sequencing of the most common molecular type of CA-MRSA (USA300) suggests that encoded gene products might enhance the ability of the strain to live on the host's skin (Diep et al., 2006).

2.2.3.3 Antimicrobial Susceptibility Testing

Antimicrobial agent susceptibility testing is commonly used in clinical laboratories to guide the clinical treatment of *S. aureus* infection. The most standardized and accurate methods of antimicrobial agent susceptibility testing are disk diffusion tests and broth microdilution tests. In disk diffusion tests, a disk impregnated with an antimicrobial agent is placed on an agar plate containing a lawn of bacteria to test whether the antimicrobial agent inhibits the growth of bacteria. (However, the vancomycin disk diffusion test does not detect vancomycin intermediate *S. aureus* isolates.) One variation of the disk diffusion test is the E-test, a plastic strip with a gradient of antimicrobial agent concentrations used to determine the minimal inhibitory concentration (MIC) to specific antimicrobial agents. Broth microdilution tests determine the lowest concentration of antimicrobial agents that inhibit bacterial growth in a broth medium using a standard inoculum size. In an agar screen test, a standardized suspension of the microorganism is inoculated directly onto an agar plate impregnated with an antimicrobial agent. Rapid automated instrumentation, such as with devices offered by Vitek^{TM}, Microscan^{TM}, and others, are most commonly used in laboratories to determine the susceptibility pattern of *S. aureus.*

Clindamycin resistance may be constitutive or inducible, and testing for this resistance can impact clinical treatment decisions. Resistance to clindamycin is closely tied to resistance to erythromycin, the latter of which is encoded by two different genes: *msrA* and *erm* (Siberry et al., 2003). The *msrA* gene encodes an adenosine triphosphate (ATP)-dependent efflux pump that confers resistance to erythromycin but not clindamycin. The *erm* (or erythromycin ribosomal methylase) gene confers constitutive resistance to erythromycin and either constitutive or inducible clindamycin resistance. MRSA isolates with inducible clindamycin resistance are resistant to erythromycin and sensitive to clindamycin on routine testing, but can be induced to express resistance to clindamycin in vitro.

Rates of inducible clindamycin resistance among strains of MRSA vary widely across the United States, from less than 10% (Sattler et al., 2002) to greater than 90% (Frank et al., 2002). Inducible clindamycin resistance has become less common among MRSA isolates from an urban pediatric population in the United States (Chavez-Bueno et al., 2005). Inducible clindamycin resistance can be identified with the D-zone test, a double disk diffusion test in which the zone of inhibition is measured around both erythromycin and clindamycin disks (Fiebelkorn et al., 2003). The "D" is formed when the zone of inhibition around the clindamycin disk is blunted on the side adjacent to the erythromycin disk. A positive D-zone test indicates inducible clindamycin resistance. The Clinical and Laboratory Standards Institute (CLSI), formerly the National Committee for Clinical Laboratory Standards (NCCLS), recommends performing a D-zone test on all erythromycin-resistant, clindamycin-susceptible *S. aureus* isolates before reporting clindamycin susceptibility results (Clinical and Laboratory Standards Institute, 2006).

2.2.3.4 Molecular Typing of MRSA

Typing MRSA strains has been used in the past to link cases in a cluster and to locate sources of specific outbreaks. Using the antimicrobial agent susceptibility profile to determine genetic relatedness of strains of *S. aureus* is unreliable. Since this typing method has low discriminatory power, pulsed-field gel electrophoresis (PFGE) is one of the most frequently used methods of typing strains; pulsed-field types are commonly recognized in the United States and around the world. Spa (*Staphylococcus* protein A) typing, multi-locus sequence typing (MLST), multi-locus variable-number tandem-repeat analysis (MLVA), or a combination of typing methods can also be used to differentiate among isolates.

In the United States, a limited number of MRSA strains have been implicated in most community outbreaks. Most CA-MRSA disease in the United States is now caused by pulsed-field type USA300, but other genotypes (USA400, USA1000, and USA1100) also cause disease (McDougal et al., 2003). A highly conserved USA300 strain (USA300-0114) has been implicated in multiple outbreaks across the United States in diverse populations that are not epidemiologically related, such as athletes, prisoners, and children (Kazakova et al., 2005).

2.2.3.5 Molecular Origins of MRSA

The microbiological differences between strains of MRSA isolated from CA-MRSA and HA-MRSA infections give us clues as to the origins of MRSA in the community. Some experts postulate that CA-MRSA strains are descendants of healthcare-associated strains that leaked from hospitals into the community. However, CA-MRSA and HA-MRSA are so microbiologically different that it is unlikely that CA-MRSA strains descended directly or recently from healthcare-associated strains. Instead, the *mecA* gene was likely acquired by MSSA strains established in the community, possibly from coagulase-negative staphylococci or HA-MRSA strains (Chambers, 2001). This theory is bolstered by the finding that MSSA and MRSA strains in a series of community-associated *S. aureus* infections in an urban city in the United States had indistinguishable PFGE patterns and differed only by *SCCmec* (Mongkolrattanothai et al., 2003).

2.3 Considerations in Management of *S. aureus* Infection in the Community

Given that most skin and soft tissue infections are treated empirically, a majority of CA-MRSA infections may be treated with antimicrobial agents to which the bacteria are not susceptible in vitro (Naimi et al., 2001). Treatment should be based on the susceptibility profile of the organism and may differ with severity of the infection. Even though wound, skin, and soft tissue MRSA infections are generally less severe than invasive MRSA infections, they account for nearly 90% of all MRSA infections (Fridkin et al., 2005) and comprise the majority of the burden of MRSA disease.

2.3.1 Management of Severe or Invasive MRSA Infections

Severe or invasive MRSA infections include sepsis, pneumonia, endocarditis, osteomyelitis, and the progression of localized infections such as of the skin or soft tissue. MRSA should be considered in cases of community-acquired pneumonia, particularly those following influenza-like illness (Hageman et al., 2006). Many patients with severe or invasive MRSA infections require parenteral antimicrobial therapy; vancomycin remains a first-line treatment. Prompt empiric treatment with antimicrobial agents can be essential in treating severe or invasive disease; however, a culture should be obtained before starting antimicrobial therapy if possible. In some cases, other antimicrobial agents such as ceftriaxone, gentamicin, or rifampin may be added to ensure optimal coverage of other potential pathogens. In addition to antimicrobial agent therapy, incision and drainage is mandatory for drainable skin and soft tissue infections and should be considered for severe or deep infections, such as septic joints.

2.3.2 Management of MRSA Skin or Soft Tissue Infections

Few MRSA skin and soft tissue infections will progress to severe or invasive disease, but early diagnosis and treatment is important to prevent progression and limit further spread. MRSA should now be considered a potential causative organism in all community-associated skin and soft tissue infections. MRSA may be the most common bacterial cause of certain skin and soft tissue infections in some areas (Cohen et al., in press; King et al., 2006; Moran et al., 2006). Local data on prevalence of MRSA and susceptibilities to alternative antimicrobial agents should be used to guide treatment decisions. Although data from controlled trials are limited, steps that may be considered part of management of CA-MRSA skin and soft tissue infections are as follows (Gorwitz et al., 2006):

1. *Incision and drainage should be routine for skin lesions that are able to be drained.* Incision and drainage has been the primary mode of treatment for skin and soft tissue abscesses for many centuries, and the introduction of antimicrobial agents should not change that practice. Incision and drainage may become more important as MRSA and other antimicrobial agent-resistant organisms cause more diseases. Studies that assess whether antimicrobial treatment with incision and drainage is more effective than incision and drainage alone have not shown a clear benefit to adding antimicrobial agents for uncomplicated disease (Gorwitz et al., 2006). Clinicians may consider prescribing antimicrobial agents if the skin and soft tissue infection (1) is severe, rapidly progressing, or greater than 5 cm (Lee et al., 2004); (2) has spread systemically; (3) is in a location that may be difficult to drain completely; (4) does not respond to incision and drainage alone; or (5) is in a patient with extremes of age or other co-morbidities (e.g., diabetes mellitus, neoplastic disease, or HIV infection).

2. *Collect diagnostic specimens for culture.* Incision and drainage should be performed for all drainable skin lesions; cultures should be obtained from patients with both draining and non-draining, purulent lesions. Obtaining isolates not only helps guide treatment of individual patients but also monitors the antimicrobial agent susceptibility patterns in the community. Cultures are also important when clinicians or public health officials suspect that the patient may be part of a cluster or outbreak of cases.

3. *Provide careful and thorough wound care.* MRSA skin and soft tissue infections are transmissible through contact with draining lesions. After incision and drainage, wounds should be adequately covered, bandages should be appropriately disposed of, and hand hygiene should be continued to prevent further spread of MRSA.

4. *Ensure adequate follow-up.* Clinicians should ensure that all culture and antimicrobial susceptibility testing results are reviewed to identify cases of MRSA. Patients should be monitored for response to treatment and educated about symptoms that suggest treatment failure, i.e., worsening of local symptoms, development of systemic symptoms, or lack of improvement of symptoms in 48 h.

5. *Use appropriate antimicrobial agents.* Although practice patterns are changing as medical providers learn about CA-MRSA, a large percentage of pediatric and adult patients with MRSA skin infections still receive an antimicrobial agent to which the bacteria are not susceptible in vitro (Lee et al., 2004; Moran et al., 2006). Local antimicrobial susceptibility data for outpatient *S. aureus* skin and soft tissue infections should guide empiric treatment decisions. Ideally, antimicrobial susceptibility testing results from the individual patient should be used to select definitive therapy of infections that are yet to resolve or have not resolved by the time culture results are available. Empiric therapy for MRSA may be warranted in areas where prevalence of MRSA in the community is high. A prevalence of 10–15% of methicillin resistance in community *S. aureus* strains has been suggested by some experts (Kaplan, 2005b).

The choice of antimicrobial therapy for MRSA infections depends on a number of factors: Is the therapy empiric or directed? Is the MRSA disease severe or invasive versus non-invasive? Options for therapy with commonly recommended antimicrobial agents are listed below:

1. *Clindamycin.* Clindamycin is bacteriostatic and, by binding to the bacterial ribosomal 50S subunit, inhibits protein synthesis. Clindamycin is available in both oral and parenteral formulations. Resistance to clindamycin in *S. aureus* can be either constitutive or inducible. Inducible clindamycin resistance, as discussed earlier, can be detected with the D-zone test. If using clindamycin, it is important to consider whether a *S. aureus* isolate possesses inducible clindamycin resistance. Although some *S. aureus* infections in which the isolate possesses inducible clindamycin resistance improve on clindamycin therapy (Frank et al., 2002), clindamycin is not recommended for any infection in which the infecting isolate is known to possess inducible resistance. If inducible resistance is detected after empiric clindamycin therapy has been initiated, the decision to

modify therapy should be based on the patient's clinical response. A theoretical benefit of clindamycin treatment of *S. aureus* infections is inhibition of protein synthesis in toxin-mediated bacterial syndromes; however, no data have shown this benefit for *S. aureus* strains producing the PVL cytotoxin.

2. *Daptomycin.* Daptomycin is a cyclic lipopeptide that binds to bacterial membranes, causing a loss of membrane potential and inhibiting protein, DNA, and RNA synthesis. It was approved in 2003 in the United States for treatment of complicated MRSA skin and soft tissue infections in adults. Daptomycin is only available as an intravenous medication. Daptomycin is inactivated by surfactant and is not recommended for treating *S. aureus* pneumonia (Carpenter and Chambers, 2004).

3. *Linezolid.* Linezolid is an oxazolidinone with both parenteral and oral formulations. Its antibacterial effects are from inhibition of ribosomal protein synthesis. This mechanism of action is unique, so there is no cross-resistance with other known antimicrobial agents. Linezolid was approved in 2000 in the United States for treatment of complicated MRSA skin and soft tissue infections and MRSA pneumonia in both children and adults. Linezolid is expensive compared to other antimicrobial agents and side effects such as reversible myelosuppression have been reported. In order to limit development of resistance, clinicians may choose to reserve linezolid for more severe infections and infections that do not respond to other antimicrobial therapy. Consultation with an infectious disease specialist may be helpful.

4. *Mupirocin.* Mupirocin is a topical antimicrobial agent that works by inhibiting bacterial protein and RNA synthesis. Mupirocin is recommended for the topical therapy of impetigo, but not for other *S. aureus* or MRSA skin and soft tissue infections. Mupirocin is also available in an intra-nasal formulation for nasal decolonization.

5. *Quinupristin-dalfopristin.* Quinupristin-dalfopristin is an intravenous-only streptogramin antibacterial agent that inhibits peptide bond formation in the 50S ribosome. It is approved by the United States Food and Drug Administration for the treatment of complicated MRSA skin and soft tissue infections.

6. *Rifampin.* Rifampin is an antimicrobial agent thought to inhibit RNA polymerase. Rifampin should not be used as a single agent because *S. aureus* can develop resistance to rifampin rapidly when used alone (Strausbaugh et al., 1992). Rifampin has been used in combination with other antimicrobial agents (Iyer and Jones, 2004) and may be useful in decreasing nasal carriage because it reaches high levels in mucosal surfaces.

7. *Tetracyclines.* Tetracyclines are bacteriostatic antimicrobial agents. Doxycycline is approved for treatment of *S. aureus* infection in the United States, and minocycline is also used. There is little information in the literature about treatment of MRSA or serious staphylococcal infection with tetracycline. Tetracyclines should be avoided in pregnant women and children less than 8 years of age due to impaired tooth and bone growth.

8. *Tigecycline.* In 2005, the United States Food and Drug Administration approved intravenous tigecycline for the treatment of MRSA. Tigecycline inhibits protein translation by binding to the 30S ribosomal subunit. Tigecycline has activity

against a wide range of bacteria, but it is too early to tell how this antimicrobial agent should be used in the treatment of MRSA.

9. *Trimethoprim/sulfamethoxazole (TMP/SFX)*. TMP/SFX is available in both oral and intravenous formulations and blocks the production of folic acid. TMP/SFX is a common treatment for *S. aureus* skin and soft tissue infections, because MRSA isolates are often susceptible to TMP/SFX. However, other bacteria that cause skin infections, such as β-hemolytic streptococci (groups A and B), are typically resistant to this agent. For clinical syndromes in which streptococcal infection is likely, such as cellulitis without abscess, substitution or addition of an agent with streptococcal coverage should be considered. TMP/SFX is not recommended for women in their third trimester of pregnancy or for infants less than 2 months of age.

10. *Vancomycin*. Vancomycin is commonly used as first-line therapy for severe and invasive disease and as a second-line therapy when other treatment fails. Vancomycin is a glycopeptide that inhibits cell wall biosynthesis. It is only used in the intravenous form because it is not well absorbed from the gastrointestinal tract. Another glycopeptide, teichoplanin, is available in some countries outside the United States but is not routinely recommended as a first- or second-line agent in the treatment of MRSA. A number of isolates of vancomycin-resistant and vancomycin-intermediate *S. aureus* have been reported and are discussed later.

Prevalence of resistance to other antimicrobial agents, such as fluoroquinolones and macrolides, is relatively high or can rapidly develop among *S. aureus* strains. These agents are not optimal choices for empiric treatment of community-associated skin and soft tissue infections:

1. *Fluoroquinolones*. MRSA strains can easily develop fluoroquinolone resistance. Newer fluoroquinolones, such as moxifloxacin and levofloxacin, have increased activity against *S. aureus*, but increasing resistance suggests that fluoroquinolones may not be an optimal choice for empiric treatment of community-associated skin infections.

2. *Macrolides and azalides*. Resistance to macrolides (e.g., erythromycin and clarithromycin) and azalides (e.g., azithromycin) is common among MRSA isolates (Fridkin et al., 2005), so they are not recommended for empiric treatment of infections possibly caused by MRSA.

2.3.3 Strategies to Eliminate S. aureus Colonization

Decolonization has been suggested as a treatment for recurrent or persistent MRSA infections but has not been proven an effective intervention for the management of MRSA infection. Decolonization may also be useful to halt ongoing transmission in a closely associated, well-defined cohort. Decolonization regimens have been effective in reducing colonization in the short term, but recolonization is common (Laupland and Conly, 2003; Perl et al., 2002). Carriage at sites other than the nares and reports of resistance to mupirocin may also limit the effectiveness of nasal decolonization (Loeb et al., 2003). If decolonization is used, the medical provider should confirm that the patient is colonized with MRSA. Proposed regimens for decolonization include intranasal mupirocin twice a day for 5–10 days and antiseptic body wash, such as with chlorhexidine, for the same duration (Stevens et al., 2005). Decolonization with systemic oral antimicrobial agents such as TMP/SFX, tetracyclines, and clindamycin has also been suggested.

2.4 Prevention of MRSA in the Community

Effective prevention strategies for CA-MRSA need to include public health officials, medical providers and infection control practitioners, and patients and community members.

2.4.1 Public Health Officials

1. *Enhance surveillance.* Both prospective and retrospective surveillance may be important to identify cases of MRSA in the community. Surveillance can also be used as an opportunity to collect isolates and to educate both patients and providers. Contacts of patients with MRSA infection should be notified to identify new cases and to ensure that they are receiving proper treatment.
2. *Initiate public health investigations when appropriate.* When MRSA is detected in a group of individuals in the community who are linked epidemiologically, public health officials should consider whether to investigate the causes of the cluster. When considering whether to investigate, public health officials should weigh the number and clustering of time and space of cases, the setting of the cluster, the severity of illness, the presence of ongoing transmission, and the likelihood that an intervention could be successfully implemented.

2.4.2 Medical Providers and Infection Control Practitioners

1. *Use appropriate treatment.* As discussed earlier, abscesses should be drained and antimicrobial treatment should be chosen based on local antimicrobial susceptibility patterns. Since CA-MRSA infections have been associated with previous antimicrobial use, antimicrobials should be used appropriately and judiciously to prevent development of resistance. Since skin and soft tissue infections may start out innocuously, patients may attempt to care for their wounds without medical treatment. Inadequate self-care has been associated with developing skin infections and may contribute to transmission in an outbreak (Centers for Disease Control and Prevention, 2001).

2. *Educate patients and providers on diagnosis, treatment, and prevention.* Clinicians should ask patients with MRSA infections if other contacts and household members also have suspicious lesions or infections.
3. *Consider decolonization.* Testing for MRSA colonization may be useful in public health investigations. Decolonization may be useful along with other measures, when there is ongoing transmission in members of a closely associated, well-defined cohort. If decolonization is being considered for the prevention of MRSA transmission in a discrete population, then all members of the population should be decolonized simultaneously.
4. *Continue effective healthcare infection control measures.* The control of MRSA in the community needs to include infection control measures in healthcare settings, since many of these patients interface with the medical system. Recommendations for preventing MRSA infections in the healthcare setting include (1) isolating or cohorting patients with MRSA, (2) wearing gloves and gowns during contact with MRSA patients, (3) using appropriate antimicrobials for treatment, and (4) ensuring appropriate hand hygiene (Siegel et al., 2007). Hospitals may also educate staff on MRSA transmission and treatment, increase the number of infection control personnel, and ensure environmental cleaning (Nijssen et al., 2005). Contaminated surfaces in healthcare locations should be cleaned with an approved hospital detergent/disinfectant or a 1:100 solution of diluted bleach (1 tablespoon bleach in a 1 quart water) (Sehulster and Chinn, 2003). Targeted interventions to prevent MRSA infection can be successful; however, many interventions use multiple approaches and it is often unclear which activities led to the successful prevention of disease (Wootton et al., 2004). The proportion of *S. aureus* infections caused by MRSA has remained very low in the Netherlands, which has been attributed to vigilant active surveillance and strict patient isolation (Vandenbroucke-Grauls, 1996). Preventive measures in the hospital vary from institution to institution, and surveillance cultures may not effectively measure the success of infection control policies (Nijssen et al., 2005). Passive surveillance of hospitalized patients with MRSA may fail to identify asymptomatic patients who are colonized with MRSA. No study has conclusively shown that patient isolation or cohorting of patients with similar disease is alone sufficient to reduce nosocomial spread of MRSA (Cepeda et al., 2005). However, isolation and cohorting should still be used since they have been part of successful campaigns to stop transmission (Cooper et al., 2004).

2.4.3 Patients and Members of the Community

1. *Educate patients on treatment and prevention.* Patients should be encouraged to keep wounds covered, to maintain good personal and hand hygiene, and to avoid sharing potentially contaminated items. In addition to patients, individuals in high-risk groups, such as incarcerated individuals, those involved in contact sports, and injecting drug users, should also be educated on prevention measures. Prevention recommendations may need to be tailored for each individual group.

This education is critical to ensure that patients get correct treatment and stop the spread of infection in the community.

2. *Care for and contain wounds.* Wounds should be covered with clean, dry dressings. Patients with open skin wounds, such as draining skin and soft tissue infections that cannot be covered, may need to be excluded from activities that could lead to transmission. Specifically, sports participants with draining lesions that are unable to be adequately contained during sports play may need to be excluded until the lesion can be adequately contained.

3. *Encourage personal hygiene*, especially hand hygiene. Use soap and water or alcohol-based hand gels to clean hands, and encourage regular bathing. Do not share personal items that may transmit infection. Launder contaminated clothes and linens with detergent, soap, or bleach, and dry thoroughly (Sehulster and Chinn, 2003).

4. *Maintain a clean environment.* Facilities where patrons and staff have close contact (e.g., homeless shelters) or shared equipment or surfaces (e.g., gyms) should consider environmental interventions to prevent transmission of MRSA. Since surfaces contaminated with *S. aureus* have been implicated in outbreaks (Baggett et al., 2004), targeted cleaning may be warranted on areas and equipment where known cases had recent contact. A barrier such as clothes or a towel should be used when in contact with shared equipment or surfaces, such as at a gymnasium.

2.5 Future Directions

Why has MRSA emerged in the community? An increase in use of antimicrobial agents, more virulent strains, and transmission of methicillin-resistance genes may be contributing to the spread of MRSA in the community (Weber, 2005). MRSA will continue to evolve, and we will need to adhere to known methods for the control of *S. aureus* while we identify other treatment and prevention strategies. Research will continue to elucidate why some people are more susceptible to MRSA than others and what host factors are linked to MRSA infection.

Vancomycin is a primary treatment for severe, invasive MRSA infections that fail to respond to other antimicrobial agents. *S. aureus* developed resistance first to penicillin and then to semisynthetic penicillins, such as methicillin, oxacillin, and nafcillin; similarly, recently identified isolates of *S. aureus* have developed resistance to vancomycin. Clinical isolates of *S. aureus* with intermediate resistance to vancomycin (MICs of 8–16 mug/ml) were first reported in Japan in the late 1990s (Hiramatsu et al., 1997). Intermediate resistance to vancomycin in strains of *S. aureus* may be due to the development of thicker cell walls in the bacteria.

Resistance to vancomycin is conferred by the presence of a *vanA* operon, which is thought to be transferred from vancomycin-resistant enterococci (Weigel et al., 2003). In 2002, reports of *S. aureus* resistant to vancomycin (MICs $\geq$ 32mug/ml) came from two states (Michigan and Pennsylvania) in the United States (Centers for Disease Control and Prevention, 2002a,b); four subsequent cases have occurred in Michigan and New York (Centers for Disease Control and Prevention, 2004). All these vancomycin-resistant *S. aureus* (VRSA) patients have risk factors for healthcare-associated disease.

New mechanisms of preventing *S. aureus* infection are also on the horizon. Passive immunizations for high-risk patients (namely premature infants and adults with blood stream infections) are being evaluated. Active vaccination may target staphylococcal toxins or the bacterial cell wall through peptidoglycans, lipids, or associated cell wall proteins (Gotz, 2004; Shinefield and Black, 2005). Other treatments and antimicrobial agents are being developed. Intravenous immunoglobulin can neutralize toxins such as PVL, but this has not been sufficiently tested as a treatment for PVL-positive MRSA infections.

The epidemiology of the MRSA infection continues to evolve. Many described strains of CA-MRSA are now the cause of disease and outbreaks in healthcare settings, such as in maternal and neonatal wards (Bratu et al., 2005; Healy et al., 2004; Saiman et al., 2003), in patients receiving prosthetic joints (Kourbatova et al., 2005), and in patients with healthcare-associated bloodstream infections (Seybold et al., 2006). We will need to be vigilant in our identification and treatment of MRSA infections in the future to prevent further spread and development of new resistant strains.

References

Adcock, P.M., Pastor, P., Medley, F., Patterson, J.E. and Murphy, T.V., (1998), Methicillin-resistant *Staphylococcus aureus* in two child care centers. *J Infect Dis*, **178**, 577–580.

Adem, P.V., Montgomery, C.P., Husain, A.N., Koogler, T.K., Arangelovich, V., Humilier, M., Boyle-Vavra, S. and Daum, R.S., (2005), *Staphylococcus aureus* sepsis and the Waterhouse–Friderichsen syndrome in children. *N Engl J Med*, **353**, 1245–1251.

Baba, T., Takeuchi, F., Kuroda, M., Yuzawa, H., Aoki, K., Oguchi, A., Nagai, Y., Iwama, N., Asano, K., Naimi, T., Kuroda, H., Cui, L., Yamamoto, K. and Hiramatsu, K., (2002), Genome and virulence determinants of high virulence community-acquired MRSA. *Lancet*, **359**, 1819–1827.

Baggett, H.C., Hennessy, T.W., Leman, R., Hamlin, C., Bruden, D., Reasonover, A., Martinez, P. and Butler, J.C., (2003), An outbreak of community-onset methicillin-resistant *Staphylococcus aureus* skin infections in southwestern Alaska. *Infect Control Hosp Epidemiol*, **24**, 397–402.

Baggett, H.C., Hennessy, T.W., Rudolph, K., Bruden, D., Reasonover, A., Parkinson, A., Sparks, R., Donlan, R.M., Martinez, P., Mongkolrattanothai, K. and Butler, J.C., (2004), Community-onset methicillin-resistant *Staphylococcus aureus* associated with antibiotic use and the cytotoxin Panton–Valentine leukocidin during a furunculosis outbreak in rural Alaska. *J Infect Dis*, **189**, 1565–1573.

Barber, M. and Rozwadowski-Dowzenko, M., (1948), Infection by penicillin-resistant staphylococci. *Lancet*, **1**, 641–1644.

Barrett, F.F., McGehee, R.F., Jr and Finland, M., (1968), Methicillin-resistant *Staphylococcus aureus* at Boston City Hospital. Bacteriologic and epidemiologic observations. *N Engl J Med*, **279**, 441–448.

Begier, E.M., Frenette, K., Barrett, N.L., Mshar, P., Petit, S., Boxrud, D.J., Watkins-Colwell, K., Wheeler, S., Cebelinski, E.A., Glennen, A., Nguyen, D. and Hadler, J.L., (2004), A high-morbidity outbreak of methicillin-resistant *Staphylococcus aureus* among players on a college football team, facilitated by cosmetic body shaving and turf burns. *Clin Infect Dis*, **39**, 1446–1453.

Bhat, N., Wright, J. G., Broder, K.R., Murray, E.L., Greenberg, M.E., Glover, M.J., Likos, A.M., Posey, D.L., Klimov, A., Lindstrom, S.E., Balish, A., Medina, M.J., Wallis, T.R., Guarner, J., Paddock, C.D., Shieh, W.J., Zaki, S.R., Sejvar, J.J., Shay, D.K., Harper, S.A., Cox, N.J., Fukuda, K. and Uyeki, T.M., (2005), Influenza-associated deaths among children in the United States, 2003–2004. *N Engl J Med*, **353**, 2559–2567.

Boyce, J.M., Cookson, B., Christiansen, K., Hori, S., Vuopio-Varkila, J., Kocagoz, S., Oztop, A.Y., Vandenbroucke-Grauls, C.M., Harbarth, S. and Pittet, D., (2005), Meticillin-resistant *Staphylococcus aureus*. *Lancet Infect Dis*, **5**, 653–663.

Bratu, S., Eramo, A., Kopec, R., Coughlin, E., Ghitan, M., Yost, R., Chapnick, E.K., Landman, D. and Quale, J., (2005), Community-associated methicillin-resistant *Staphylococcus aureus* in hospital nursery and maternity units. *Emerg Infect Dis*, **11**, 808–813.

Buckingham, S.C., McDougal, L.K., Cathey, L.D., Comeaux, K., Craig, A.S., Fridkin, S.K. and Tenover, F.C., (2004), Emergence of community-associated methicillin-resistant *Staphylococcus aureus* at a Memphis, Tennessee Children's Hospital. *Pediatr Infect Dis J*, **23**, 619–624.

Campbell, K.M., Vaughn, A.F., Russell, K.L., Smith, B., Jimenez, D.L., Barrozo, C.P., Minarcik, J.R., Crum, N.F. and Ryan, M.A. (2004) Risk factors for community-associated methicillin-resistant *Staphylococcus aureus* infections in an outbreak of disease among military trainees in San Diego, California, in 2002. *J Clin Microbiol*, **42**, 4050–4043.

Carpenter, C.F. and Chambers, H.F., (2004), Daptomycin: another novel agent for treating infections due to drug-resistant gram-positive pathogens. *Clin Infect Dis*, **38**, 994–1000.

Centers for Disease Control and Prevention, (1999), Four pediatric deaths from community-acquired methicillin-resistant *Staphylococcus aureus*—Minnesota and North Dakota, 1997–1999. *MMWR Morb Mortal Wkly Rep*, **48**, 707–710.

Centers for Disease Control and Prevention, (2001), Methicillin-resistant *Staphylococcus aureus* skin or soft tissue infections in a state prison—Mississippi, 2000. *MMWR Morb Mortal Wkly Rep*, **50**, 919–922.

Centers for Disease Control and Prevention, (2002a), *Staphylococcus aureus* resistant to vancomycin—United States, 2002. *MMWR Morb Mortal Wkly Rep*, **51**, 565–567.

Centers for Disease Control and Prevention, (2002b), Vancomycin-resistant *Staphylococcus aureus*—Pennsylvania, 2002. *MMWR Morb Mortal Wkly Rep*, **51**, 902.

Centers for Disease Control and Prevention, (2003a), Methicillin-resistant *Staphylococcus aureus* infections among competitive sports participants—Colorado, Indiana, Pennsylvania, and Los Angeles County, 2000–2003. *MMWR Morb Mortal Wkly Rep*, **52**, 793.

Centers for Disease Control and Prevention, (2003b), Methicillin-resistant *Staphylococcus aureus* infections in correctional facilities—Georgia, California, and Texas, 2001–2003. *MMWR Morb Mortal Wkly Rep*, **52**, 992–996.

Clinical and Laboratory Standards Institute, (2006), Performance standards for antimicrobial susceptibility testing; 16th informational supplement. M100-S16. Wayne (PA): The Institute.

Centers for Disease Control and Prevention, (2004), Vancomycin-resistant *Staphylococcus aureus*—New York, 2004. *MMWR Morb Mortal Wkly Rep*, **53**, 322–323.

Centers for Disease Control and Prevention, (2005), Infectious disease and dermatologic conditions in evacuees and rescue workers after Hurricane Katrina—multiple states, August–September, 2005. *MMWR Morb Mortal Wkly Rep*, **54**, 961–964.

Centers for Disease Control and Prevention, (2006), Methicillin-resistant *Staphylococcus aureus* skin infections among tattoo recipients—Ohio, Kentucky, and Vermont, 2004–2005. *MMWR Morb Mortal Wkly Rep*, **55**, 677–679.

Cepeda, J.A., Whitehouse, T., Cooper, B., Hails, J., Jones, K., Kwaku, F., Taylor, L., Hayman, S., Cookson, B., Shaw, S., Kibbler, C., Singer, M., Bellingan, G. and Wilson, A.P., (2005), Isolation of patients in single rooms or cohorts to reduce spread of MRSA in intensive-care units: prospective two-centre study. *Lancet*, **365**, 295–304.

Chambers, H.F., (2001), The changing epidemiology of *Staphylococcus aureus*? *Emerg Infect Dis*, **7**, 178–182.

Chavez-Bueno, S., Bozdogan, B., Katz, K., Bowlware, K.L., Cushion, N., Cavuoti, D., Ahmad, N., McCracken, G.H., Jr and Appelbaum, P.C., (2005), Inducible clindamycin resistance and molecular epidemiologic trends of pediatric community-acquired methicillin-resistant *Staphylococcus aureus* in Dallas, Texas. *Antimicrob Agents Chemother*, **49**, 2283–2288.

Cohen, A.L., Shuler, C., McAllister, S., Fosheim, G.E., Brown, M.G., Abercrombie, D., Anderson, K., McDougal, L.K., Drenzek, C., Arnold, K., Jernigan, D., Gorwitz, R. (In press) Is Methamphetamine Use a Risk Factor for Methicillin-resistant *Staphylococcus aureus* Skin Infections in the Community? *Emerg infect Dis*.

Cooper, B.S., Stone, S.P., Kibbler, C.C., Cookson, B.D., Roberts, J.A., Medley, G.F., Duckworth, G., Lai, R. and Ebrahim, S., (2004), Isolation measures in the hospital management of methicillin resistant *Staphylococcus aureus* (MRSA): systematic review of the literature. *BMJ*, **329**, 533.

Coronado, F., Nicholas, J.A., Wallace, B.J., Kohlerschmidt, D.J., Musser, K., Schoonmaker-Bopp, D.J., Zimmerman, S.M., Boller, A.R., Jernigan, D.B. and Kacica, M.A., (2006), Community-associated methicillin-resistant *Staphylococcus aureus* skin infections in a religious community. *Epidemiol Infect*, 1–10.

Creech, C.B., 2nd, Kernodle, D.S., Alsentzer, A., Wilson, C. and Edwards, K.M., (2005), Increasing rates of nasal carriage of methicillin-resistant *Staphylococcus aureus* in healthy children. *Pediatr Infect Dis J*, **24**, 617–621.

Diep, B.A., Gill, S.R., Chang, R.F., Phan, T.H., Chen, J.H., Davidson, M.G., Lin, F., Lin, J., Carleton, H.A., Mongodin, E.F., Sensabaugh, G.F. and Perdreau-Remington, F., (2006), Complete genome sequence of USA300, an epidemic clone of community-acquired meticillin-resistant *Staphylococcus aureus*. *Lancet*, **367**, 731–739.

Ellis, M.W., Hospenthal, D.R., Dooley, D.P., Gray, P.J. and Murray, C.K., (2004), Natural history of community-acquired methicillin-resistant *Staphylococcus aureus* colonization and infection in soldiers. *Clin Infect Dis*, **39**, 971–979.

Enzenauer, R.W., Dotson, C.R., Leonard, T., Jr, Brown, J., 3rd, Pettett, P.G. and Holton, M.E., (1984), Increased incidence of neonatal staphylococcal pyoderma in males. *Mil Med*, **149**, 408–410.

Enzenauer, R.W., Dotson, C.R., Leonard, T., Reuben, L., Bass, J.W. and Brown, J., 3rd (1985) Male predominance in persistent staphylococcal colonization and infection of the newborn. *Hawaii Med J*, **44**, 389–390, 392, 394–396.

Estivariz, C.F., Park, S.Y., Hageman, J.C., Dvorin, J., Melish, M.M., Arpon, R., Coon, P., Slavish, S., Kim, M., McDougal, L.K., Jensen, B., McAllister, S., Lonsway, D., Killgore, G., Effler, P.E., Jernigan, D.B., (2007), Emergence of community-associated methicillin resistant Staphylococcus aureus in Hawaii, 2001–2003. J Infect, 54, 349–57.

Fekety, F.R. and Bennett, I.L., Jr (1959) The epidemiological virulence of staphylococci. *Yale J Biol Med*, **32**, 23–32.

Fey, P.D., Said-Salim, B., Rupp, M.E., Hinrichs, S.H., Boxrud, D.J., Davis, C.C., Kreiswirth, B.N. and Schlievert, P.M., (2003), Comparative molecular analysis of community- or hospital-acquired methicillin-resistant *Staphylococcus aureus*. *Antimicrob Agents Chemother*, **47**, 196–203.

Fiebelkorn, K.R., Crawford, S.A., McElmeel, M.L. and Jorgensen, J.H., (2003), Practical disk diffusion method for detection of inducible clindamycin resistance in *Staphylococcus aureus* and coagulase-negative staphylococci. *J Clin Microbiol*, **41**, 4740–4744.

Fine, M.J., Ibrahim, S.A. and Thomas, S.B., (2005), The role of race and genetics in health disparities research. *Am J Public Health*, **95**, 2125–2128.

Fleisch, F., Zbinden, R., Vanoli, C. and Ruef, C., (2001), Epidemic spread of a single clone of methicillin-resistant *Staphylococcus aureus* among injection drug users in Zurich, Switzerland. *Clin Infect Dis*, **32**, 581–586.

Frank, A.L., Marcinak, J.F., Mangat, P.D., Tjhio, J.T., Kelkar, S., Schreckenberger, P.C. and Quinn, J.P., (2002), Clindamycin treatment of methicillin-resistant *Staphylococcus aureus* infections in children. *Pediatr Infect Dis J*, **21**, 530–534.

Fridkin, S.K., Hageman, J.C., Morrison, M., Sanza, L.T., Como-Sabetti, K., Jernigan, J.A., Harriman, K., Harrison, L.H., Lynfield, R. and Farley, M.M., (2005), Methicillin-resistant *Staphylococcus aureus* disease in three communities. *N Engl J Med*, **352**, 1436–1444.

Gorwitz, R.J., Jernigan, D.B., Powers, J.H., Jernigan, J.A., and Participants in the Centers for Disease Control and Prevention-Convened Experts Meeting on Management of MRSA in the Community, (2006), Strategies for Clinical Management of MRSA in the Community. Available at http://0-www.cdc.gov.mill1.sjlibrary.org/ncidod/dhqp/pdf/ar/CAMRSA_ExpMtgStrategies.pdf

Gotz, F., (2004), Staphylococci in colonization and disease: prospective targets for drugs and vaccines. *Curr Opin Microbiol*, **7**, 477–487.

Groom, A.V., Wolsey, D.H., Naimi, T.S., Smith, K., Johnson, S., Boxrud, D., Moore, K.A. and Cheek, J.E., (2001), Community-acquired methicillin-resistant *Staphylococcus aureus* in a rural American Indian community. *JAMA*, **286**, 1201–1205.

Hageman, J.C., Uyeki, T.M., Francis, J.S., Jernigan, D.B., Wheeler, J.G., Bridges, C.B., Barenkamp, S.J., Sievert, D.M., Srinivasan, A., Doherty, M.C., McDougal, L.K., Killgore, G.E., Lopatin, U.A., Coffman, R., MacDonald, J.K., McAllister, S.K., Fosheim, G.E., Patel, J.B. and McDonald, C., (2006), Severe community-acquired pneumonia caused by an emerging strain of *Staphylococcus aureus* during the 2003–04 influenza season. *Emerg Infect Dis*, 12, 894–899.

Harbarth, S., Francois, P., Shrenzel, J., Fankhauser-Rodriguez, C., Hugonnet, S., Koessler, T., Huyghe, A. and Pittet, D., (2005), Community-associated methicillin-resistant *Staphylococcus aureus*, Switzerland. *Emerg Infect Dis*, **11**, 962–965.

Healy, C.M., Hulten, K.G., Palazzi, D.L., Campbell, J.R. and Baker, C.J., (2004), Emergence of new strains of methicillin-resistant *Staphylococcus aureus* in a neonatal intensive care unit. *Clin Infect Dis*, **39**, 1460–1466.

Herold, B.C., Immergluck, L.C., Maranan, M.C., Lauderdale, D.S., Gaskin, R.E., Boyle-Vavra, S., Leitch, C.D. and Daum, R.S., (1998), Community-acquired methicillin-resistant *Staphylococcus aureus* in children with no identified predisposing risk. *JAMA*, **279**, 593–598.

Hidron, A.I., Kourbatova, E.V., Halvosa, J.S., Terrell, B.J., McDougal, L.K., Tenover, F.C., Blumberg, H.M. and King, M.D. (2005) Risk factors for colonization with methicillin-resistant *Staphylococcus aureus* (MRSA) in patients admitted to an urban hospital: emergence of community-associated MRSA nasal carriage. *Clin Infect Dis*, **41**, 159–166.

Hill, P.C., Birch, M., Chambers, S., Drinkovic, D., Ellis-Pegler, R.B., Everts, R., Murdoch, D., Pottumarthy, S., Roberts, S.A., Swager, C., Taylor, S.L., Thomas, M.G., Wong, C.G. and Morris, A.J., (2001), Prospective study of 424 cases of *Staphylococcus aureus* bacteraemia: determination of factors affecting incidence and mortality. *Intern Med J*, **31**, 97–103.

Hiramatsu, K., Hanaki, H., Ino, T., Yabuta, K., Oguri, T. and Tenover, F.C., (1997), Methicillin-resistant *Staphylococcus aureus* clinical strain with reduced vancomycin susceptibility. *J Antimicrob Chemother*, **40**, 135–136.

Hussain, F.M., Boyle-Vavra, S. and Daum, R.S., (2001), Community-acquired methicillin-resistant *Staphylococcus aureus* colonization in healthy children attending an outpatient pediatric clinic. *Pediatr Infect Dis J*, **20**, 763–767.

Hwang, J.H., Tsai, H.Y. and Liu, T.C., (2002), Community-acquired methicillin-resistant *Staphylococcus aureus* infections in discharging ears. *Acta Otolaryngol*, **122**, 827–830.

Iyer, S. and Jones, D.H., (2004), Community-acquired methicillin-resistant *Staphylococcus aureus* skin infection: a retrospective analysis of clinical presentation and treatment of a local outbreak. *J Am Acad Dermatol*, **50**, 854–858.

Jessen, O., Rosendal, K., Bulow, P., Faber, V. and Eriksen, K.R. (1969) Changing staphylococci and staphylococcal infections. A ten-year study of bacteria and cases of bacteremia. *N Engl J Med*, **281**, 627–635.

Jevons, M.P., (1961), "Celbenin"-resistance staphylococci. *BMJ*, **1**, 124.

Kaplan, S.L. (2005a) Implications of methicillin-resistant *Staphylococcus aureus* as a community-acquired pathogen in pediatric patients. *Infect Dis Clin North Am*, **19**, 747–757.

Kaplan, S.L. (2005b) Treatment of community-associated methicillin-resistant *Staphylococcus aureus* infections. *Pediatr Infect Dis J*, **24**, 457–458.

Katayama, Y., Ito, T. and Hiramatsu, K., (2000), A new class of genetic element, staphylococcus cassette chromosome *mec*, encodes methicillin resistance in *Staphylococcus aureus*. *Antimicrob Agents Chemother*, **44**, 1549–1555.

Kazakova, S.V., Hageman, J.C., Matava, M., Srinivasan, A., Phelan, L., Garfinkel, B., Boo, T., McAllister, S., Anderson, J., Jensen, B., Dodson, D., Lonsway, D., McDougal, L.K., Arduino, M., Fraser, V.J., Killgore, G., Tenover, F.C., Cody, S. and Jernigan, D.B. (2005) A clone of methicillin-resistant *Staphylococcus aureus* among professional football players. *N Engl J Med*, **352**, 468–475.

King, M.D., Humphrey, B.J., Wang, Y.F., Kourbatova, E.V., Ray, S.M. and Blumberg, H.M., (2006), Emergence of community-acquired methicillin-resistant *Staphylococcus aureus* USA300 clone as the predominant cause of skin and soft-tissue infections. *Ann Intern Med*, **144**, 309–317.

Kirby, W.M.M., (1944), Extraction of a highly potent penicillin inactivator from penicillin resistant staphylococci. *Science*, **99**, 452–453.

Klevens, R.M., Edwards, J.R., Tenover, F.C., McDonald, L.C., Horan, T. and Gaynes, R., (2006), Changes in the epidemiology of methicillin-resistant *Staphylococcus aureus* in intensive care units in US hospitals, 1992–2003. *Clin Infect Dis*, **42**, 389–391.

Kourbatova, E.V., Halvosa, J.S., King, M.D., Ray, S.M., White, N. and Blumberg, H.M., (2005), Emergence of community-associated methicillin-resistant *Staphylococcus aureus* USA300 clone as a cause of health care-associated infections among patients with prosthetic joint infections. *Am J Infect Control*, **33**, 385–391.

Kuehnert, M.J., Kruszon-Moran, D., Hill, H.A., McQuillan, G., McAllister, S.K., Fosheim, G., McDougal, L.K., Chaitram, J., Jensen, B., Fridkin, S.K., Killgore, G. and Tenover, F.C., (2006), Prevalence of *Staphylococcus aureus* nasal colonization in the United States, 2001–2002. *J Infect Dis*, **193**, 172–179.

Laupland, K.B. and Conly, J.M., (2003), Treatment of *Staphylococcus aureus* colonization and prophylaxis for infection with topical intranasal mupirocin: an evidence-based review. *Clin Infect Dis*, **37**, 933–938.

Lee, M.C., Rios, A.M., Aten, M.F., Mejias, A., Cavuoti, D., McCracken, G.H., Jr and Hardy, R.D., (2004), Management and outcome of children with skin and soft tissue abscesses caused by community-acquired methicillin-resistant *Staphylococcus aureus*. *Pediatr Infect Dis J*, **23**, 123–127.

Lee, N.E., Taylor, M.M., Bancroft, E., Ruane, P.J., Morgan, M., McCoy, L. and Simon, P.A., (2005), Risk factors for community-associated methicillin-resistant *Staphylococcus aureus* skin infections among HIV-positive men who have sex with men. *Clin Infect Dis*, **40**, 1529–1534.

Levine, D.P., Cushing, R.D., Jui, J. and Brown, W.J., (1982), Community-acquired methicillin-resistant *Staphylococcus aureus* endocarditis in the Detroit Medical Center. *Ann Intern Med*, **97**, 330–338.

Lina, G., Piemont, Y., Godail-Gamot, F., Bes, M., Peter, M.O., Gauduchon, V., Vandenesch, F. and Etienne, J., (1999), Involvement of Panton–Valentine leukocidin-producing *Staphylococcus aureus* in primary skin infections and pneumonia. *Clin Infect Dis*, **29**, 1128–1132.

Loeb, M., Main, C., Walker-Dilks, C. and Eady, A., (2003), Antimicrobial drugs for treating methicillin-resistant *Staphylococcus aureus* colonization. *Cochrane Database Syst Rev*, **4**, CD003340.

Ma, X.X., Galiana, A., Pedreira, W., Mowszowicz, M., Christophersen, I., Machiavello, S., Lope, L., Benaderet, S., Buela, F., Vincentino, W., Albini, M., Bertaux, O., Constenla, I., Bagnulo, H., Llosa, L., Ito, T. and Hiramatsu, K., (2005), Community-acquired methicillin-resistant *Staphylococcus aureus*, Uruguay. *Emerg Infect Dis*, **11**, 973–976.

Maguire, G.P., Arthur, A.D., Boustead, P.J., Dwyer, B. and Currie, B.J., (1996), Emerging epidemic of community-acquired methicillin-resistant *Staphylococcus aureus* infection in the Northern Territory. *Med J Aust*, **164**, 721–723.

Martinez-Aguilar, G., Avalos-Mishaan, A., Hulten, K., Hammerman, W., Mason, E.O., Jr and Kaplan, S.L., (2004), Community-acquired, methicillin-resistant and methicillin-susceptible *Staphylococcus aureus* musculoskeletal infections in children. *Pediatr Infect Dis J*, **23**, 701–706.

McCaig LF, McDonald LC, Mandal S, Jernigan DB., (2006), Staphylococcus aureus-associated skin and soft tissue infections in ambulatory care. *Emerg Infect Dis* [serial on the Internet], **112**, Available from http://www.cdc.gov/ncidod/EID/vol12no1/06-0190.htm

McDougal, L.K., Steward, C.D., Killgore, G.E., Chaitram, J.M., McAllister, S.K. and Tenover, F.C., (2003), Pulsed-field gel electrophoresis typing of oxacillin-resistant *Staphylococcus aureus* isolates from the United States: establishing a national database. *J Clin Microbiol*, **41**, 5113–5120.

Miller, L.G., Perdreau-Remington, F., Rieg, G., Mehdi, S., Perlroth, J., Bayer, A.S., Tang, A.W., Phung, T.O. and Spellberg, B., (2005), Necrotizing fasciitis caused by community-associated methicillin-resistant *Staphylococcus aureus* in Los Angeles. *N Engl J Med*, **352**, 1445–1453.

Mishaan, A.M., Mason, E.O., Jr, Martinez-Aguilar, G., Hammerman, W., Propst, J.J., Lupski, J.R., Stankiewicz, P., Kaplan, S.L. and Hulten, K., (2005), Emergence of a predominant clone of community-acquired *Staphylococcus aureus* among children in Houston, Texas. *Pediatr Infect Dis J*, **24**, 201–206.

Mongkolrattanothai, K., Boyle, S., Kahana, M.D. and Daum, R.S. (2003) Severe *Staphylococcus aureus* infections caused by clonally related community-acquired methicillin-susceptible and methicillin-resistant isolates. *Clin Infect Dis*, **37**, 1050–1058.

Moran, G.J., Amii, R.N., Abrahamian, F.M. and Talan, D.A., (2005), Methicillin-resistant *Staphylococcus aureus* in community-acquired skin infections. *Emerg Infect Dis*, **11**, 928–930.

Moran, G.J., Krishnadasan, A., Gorwitz, R.J., Fosheim, G.E., McDougal, L.K., Carey, R.B. and Talan, D.A., (2006), Methicillin-resistant *Staphylococcus aureus* infections among emergency department patients in 11 U.S. cities. *N Engl J Med*, 355, 666–674.

Mortimer, E.A., Lipsitz, P.J., Wolinksky, E., Gonzaga, A.J. and Rammelkamp, C.H., (1962), Transmission of *Staphylococci* between newborns: importance of the hands of personnel. *Am J Dis Child*, **104**, 289–295.

Mulligan, M.E., Murray-Leisure, K.A., Ribner, B.S., Standiford, H.C., John, J.F., Korvick, J.A., Kauffman, C.A. and Yu, V.L. (1993) Methicillin-resistant *Staphylococcus aureus*: a consensus review of the microbiology, pathogenesis, and epidemiology with implications for prevention and management. *Am J Med*, **94**, 313–328.

Mulvey, M.R., MacDougall, L., Cholin, B., Horsman, G., Fidyk, M. and Woods, S., (2005), Community-associated methicillin-resistant *Staphylococcus aureus*, Canada. *Emerg Infect Dis*, **11**, 844–850.

Naimi, T.S., Ledell, K.H., Boxrud, D.J., Groom, A.V., Steward, C.D., Johnson, S.K., Besser, J.M., O'Boyle, C., Danila, R.N., Cheek, J.E., Osterholm, M.T., Moore, K.A. and Smith, K.E., (2001), Epidemiology and clonality of community-acquired methicillin-resistant *Staphylococcus aureus* in Minnesota, 1996–1998. *Clin Infect Dis*, **33**, 990–996.

Naimi, T.S., Ledell, K.H., Como-Sabetti, K., Borchardt, S.M., Boxrud, D.J., Etienne, J., Johnson, S.K., Vandenesch, F., Fridkin, S., O'Boyle, C., Danila, R.N. and Lynfield, R., (2003), Comparison of community- and health care-associated methicillin-resistant *Staphylococcus aureus* infection. *JAMA*, **290**, 2976–2984.

Nguyen, D.M., Mascola, L. and Brancoft, E., (2005), Recurring methicillin-resistant *Staphylococcus aureus* infections in a football team. *Emerg Infect Dis*, **11**, 526–532.

Nijssen, S., Bonten, M.J. and Weinstein, R.A., (2005), Are active microbiological surveillance and subsequent isolation needed to prevent the spread of methicillin-resistant *Staphylococcus aureus*? *Clin Infect Dis*, **40**, 405–409.

Ochoa, T.J., Mohr, J., Wanger, A., Murphy, J.R. and Heresi, G.P. (2005) Community-associated methicillin-resistant *Staphylococcus aureus* in pediatric patients. *Emerg Infect Dis*, **11**, 966–968.

Ogston, A., (1882), Micrococcus poisoning. *J Anat*, **17**, 24–58.

Panlilio, A.L., Culver, D.H., Gaynes, R.P., Banerjee, S., Henderson, T.S., Tolson, J.S. and Martone, W.J. (1992) Methicillin-resistant *Staphylococcus aureus* in U.S. hospitals, 1975–1991. *Infect Control Hosp Epidemiol*, **13**, 582–586.

Panton, P.N. and Valentine, F.C., (1932), Staphylococcal toxin. *Lancet*, **222**, 506–508.

Perl, T.M., Cullen, J.J., Wenzel, R.P., Zimmerman, M.B., Pfaller, M.A., Sheppard, D., Twombley, J., French, P.P. and Herwaldt, L.A. (2002) Intranasal mupirocin to prevent post-operative *Staphylococcus aureus* infections. *N Engl J Med*, **346**, 1871–1877.

Podewils, L.J., Liedtke, L.A., McDonald, L.C., Hageman, J.C., Strausbaugh, L.J., Fischer, T.K., Jernigan, D.B., Uyeki, T.M. and Kuehnert, M.J., (2005), A national survey of severe influenza-associated complications among children and adults, 2003–2004. *Clin Infect Dis*, **40**, 1693–1696.

Ross, S., Rodriguez, W., Controni, G. and Khan, W., (1974), Staphylococcal susceptibility to penicillin G. The changing pattern among community strains. *JAMA*, **229**, 1075–1077.

Sa-Leao, R., Sanches, I.S., Couto, I., Alves, C.R. and de Lencastre, H., (2001), Low prevalence of methicillin-resistant strains among *Staphylococcus aureus* colonizing young and healthy members of the community in Portugal. *Microb Drug Resist*, **7**, 237–245.

Saiman, L., O'Keefe, M., Graham, P.L., 3rd, Wu, F., Said-Salim, B., Kreiswirth, B., Lasala, A., Schlievert, P.M. and Della-Latta, P., (2003), Hospital transmission of community-acquired methicillin-resistant *Staphylococcus aureus* among postpartum women. *Clin Infect Dis*, **37**, 1313–1319.

Santos, F., Mankarious, L.A. and Eavey, R.D., (2000), Methicillin-resistant *Staphylococcus aureus*: pediatric otitis. *Arch Otolaryngol Head Neck Surg*, **126**, 1383–1385.

Saravolatz, L.D., Markowitz, N., Arking, L., Pohlod, D. and Fisher, E., (1982), Methicillin-resistant *Staphylococcus aureus*. Epidemiologic observations during a community-acquired outbreak. *Ann Intern Med*, **96**, 11–16.

Sattler, C.A., Mason, E.O., Jr and Kaplan, S.L., (2002), Prospective comparison of risk factors and demographic and clinical characteristics of community-acquired, methicillin-resistant versus methicillin-susceptible *Staphylococcus aureus* infection in children. *Pediatr Infect Dis J*, **21**, 910–917.

Sehulster, L. and Chinn, R.Y., (2003), Guidelines for environmental infection control in health-care facilities. Recommendations of CDC and the Healthcare Infection Control Practices Advisory Committee (HICPAC). *MMWR Recomm Rep*, **52**, 1–42.

Seybold, U., Kourbatova, E.V., Johnson, J.G., Halvosa, S.J., Wang, Y.F., King, M.D., Ray, S.M. and Blumberg, H.M., (2006), Emergence of community-associated methicillin-resistant *Staphylococcus aureus* USA300 genotype as a major cause of health care-associated blood stream infections. *Clin Infect Dis*, **42**, 647–656.

Shinefield, H.R. and Black, S., (2005), Prevention of *Staphylococcus aureus* infections: advances in vaccine development. *Expert Rev Vaccines*, **4**, 669–676.

Shopsin, B., Mathema, B., Martinez, J., Ha, E., Campo, M.L., Fierman, A., Krasinski, K., Kornblum, J., Alcabes, P., Waddington, M., Riehman, M. and Kreiswirth, B.N., (2000), Prevalence of methicillin-resistant and methicillin-susceptible *Staphylococcus aureus* in the community. *J Infect Dis*, **182**, 359–362.

Siberry, G.K., Tekle, T., Carroll, K. and Dick, J., (2003), Failure of clindamycin treatment of methicillin-resistant *Staphylococcus aureus* expressing inducible clindamycin resistance in vitro. *Clin Infect Dis*, **37**, 1257–1260.

Siegel, J.D., Rhinehart, E., Jackson, M., Chiarello, L., Healthcare Infection Control Practices Advisory Committee, (2007), Guideline for Isolation Precautions: Preventing Transmission of Infectious Agents in Healthcare Settings 2007. Available at http://www.cdc.gov/ncidod/dhqp/pdf/guidelines/Isolation2007.pdf

Stacey, A.R., Endersby, K.E., Chan, P.C. and Marples, R.R., (1998), An outbreak of methicillin resistant *Staphylococcus aureus* infection in a rugby football team. *Br J Sports Med*, **32**, 153–154.

Stevens, D.L., Bisno, A.L., Chambers, H.F., Everett, E.D., Dellinger, P., Goldstein, E.J., Gorbach, S.L., Hirschmann, J.V., Kaplan, E.L., Montoya, J.G. and Wade, J.C., (2005), Practice guidelines for the diagnosis and management of skin and soft-tissue infections. *Clin Infect Dis*, **41**, 1373–1406.

Stevenson, K.B., Searle, K., Stoddard, G.J. and Samore, M., (2005), Methicillin-resistant *Staphylococcus aureus* and vancomycin-resistant *Enterococci* in rural communities, western United States. *Emerg Infect Dis*, **11**, 895–903.

Strausbaugh, L.J., Jacobson, C., Sewell, D.L., Potter, S. and Ward, T.T., (1992), Antimicrobial therapy for methicillin-resistant *Staphylococcus aureus* colonization in residents and staff of a Veterans Affairs nursing home care unit. *Infect Control Hosp Epidemiol*, **13**, 151–159.

Thompson, D.J., Gezon, H.M., Hatch, T.F., Rycheck, R.R. and Rogers, K.D., (1963), Sex distribution of *Staphylococcus aureus* colonization and disease in newborn infant. *N Engl J Med*, **269**, 337–341.

Thompson, D.J., Gezon, H.M., Rogers, K.D., Yee, R.B. and Hatch, T.F., (1966), Excess risk of staphylococcal infection and disease in newborn males. *Am J Epidemiol*, **84**, 314–328.

VandenBergh, M.F., Yzerman, E.P., Van Belkum, A., Boelens, H.A., Sijmons, M. and Verbrugh, H.A. (1999) Follow-up of *Staphylococcus aureus* nasal carriage after 8 years: redefining the persistent carrier state. *J Clin Microbiol*, **37**, 3133–3140.

Vandenbroucke-Grauls, C.M., (1996), Methicillin-resistant *Staphylococcus aureus* control in hospitals: the Dutch experience. *Infect Control Hosp Epidemiol*, **17**, 512–513.

Vandenesch, F., Naimi, T., Enright, M.C., Lina, G., Nimmo, G.R., Heffernan, H., Liassine, N., Bes, M., Greenland, T., Reverdy, M.E. and Etienne, J., (2003), Community-acquired methicillin-resistant *Staphylococcus aureus* carrying Panton–Valentine leukocidin genes: worldwide emergence. *Emerg Infect Dis*, **9**, 978–984.

Vetter, R.S. and Bush, S.P., (2002), Reports of presumptive brown recluse spider bites reinforce improbable diagnosis in regions of North America where the spider is not endemic. *Clin Infect Dis*, **35**, 442–445.

von Eiff, C., Becker, K., Machka, K., Stammer, H. and Peters, G. (2001) Nasal carriage as a source of *Staphylococcus aureus* bacteremia. Study Group. *N Engl J Med*, **344**, 11–16.

Wang, J., Barth, S., Richardson, M., Corson, K. and Mader, J. (2003) An outbreak of methicillin-resistant *Staphylococcus aureus* cutaneous infection in a saturation diving facility. *Undersea Hyperb Med*, **30**, 277–284.

Weber, J.T., (2005), Community-associated methicillin-resistant *Staphylococcus aureus*. *Clin Infect Dis*, **41** (Suppl 4), S269–S272.

Weese, J.S., Archambault, M., Willey, B.M., Hearn, P., Kreiswirth, B.N., Said-Salim, B., McGeer, A., Likhoshvay, Y., Prescott, J.F. and Low, D.E., (2005), Methicillin-resistant *Staphylococcus aureus* in horses and horse personnel, 2000–2002. *Emerg Infect Dis*, **11**, 430–435.

Weigel, L.M., Clewell, D.B., Gill, S.R., Clark, N.C., McDougal, L.K., Flannagan, S.E., Kolonay, J.F., Shetty, J., Killgore, G.E. and Tenover, F.C., (2003), Genetic analysis of a high-level vancomycin-resistant isolate of *Staphylococcus aureus*. *Science*, **302**, 1569–1571.

Williams, R.E.O., (1963), Healthy carriage of *Staphylococcus aureus*: its prevalence and importance. *Bacteriol Rev*, **27**, 56–71.

Wolinksky, E., Lipsitz, P.J., Mortimer, E.A. and Rammelkamp, C.H. (1960) Acquisition of *Staphylococci* by newborns: direct versus indirect transmission. *Lancet*, **2**, 620–622.

Wootton, S.H., Arnold, K., Hill, H.A., McAllister, S., Ray, M., Kellum, M., Lamarre, M., Lane, M.E., Chaitram, J., Lance-Parker, S. and Kuehnert, M.J., (2004), Intervention to reduce the incidence of methicillin-resistant *Staphylococcus aureus* skin infections in a correctional facility in Georgia. *Infect Control Hosp Epidemiol*, **25**, 402–407.

Young, D.M., Harris, H.W., Charlebois, E.D., Chambers, H., Campbell, A., Perdreau-Remington, F., Lee, C., Mankani, M., Mackersie, R. and Schecter, W.P., (2004), An epidemic of methicillin-resistant *Staphylococcus aureus* soft tissue infections among medically underserved patients. *Arch Surg*, **139**, 947–951; discussion 951–953.

Zinderman, C.E., Conner, B., Malakooti, M.A., Lamar, J.E., Armstrong, A. and Bohnker, B.K., (2004), Community-acquired methicillin-resistant *Staphylococcus aureus* among military recruits. *Emerg Infect Dis*, **10**, 941–944.

Chapter 3
Antimicrobial Resistance to Sexually Transmitted Infections

Hillard Weinstock, David Trees, and John Papp

3.1 Introduction

According to the World Health Organization (WHO), "Sexually transmitted infections are a major global cause of acute illness, infertility, disability and death with severe medical and psychological economic consequences for millions of men, women, and infants." (World Health Organization, 2001). In 1999 WHO estimated that 340 million curable sexually transmitted infections (STIs) occurred throughout the world in men and women aged 15–49 years: new cases of syphilis, gonorrhea, chlamydia, and trichomonas. This estimate did not account for the millions of new cases of viral sexually transmitted infections such as human papillomavirus and herpes. In the United States, sexually transmitted infections are among the most common notifiable diseases (Centers for Disease Control and Prevention, 2006d). The Centers for Disease Control and Prevention (CDC) estimated that in 2000, 18.9 million new sexually transmitted infections occurred in the United States in that year alone (Weinstock et al., 2004). A report by the Institute of Medicine indicated that the scope of the sexually transmitted diseases (STD) epidemic is largely hidden from the American public despite its tremendous cost in health and economic terms (Institute of Medicine, 1997).

Major advances have been made in the clinical recognition, diagnosis, and treatment of many STI. Sensitive and specific assays have become available for noninvasive diagnosis of specific infections. New agents have advanced the treatment of these infections. However, resistance remains a continual challenge, especially, for the treatment and prevention of some STIs. In this chapter, we review the development of antimicrobial resistance among several STIs and discuss the prevalence and mechanisms of resistance along with recommendations for treatment.

3.2 Genital Herpes

Genital herpes is caused by herpes simplex virus type 1 (HSV-1) or herpes simplex virus type 2 (HSV-2). It is a chronic life-long infection. The primary route of acquisition of HSV-2 infections is through genital sexual contact with an infected

I. W. Fong and K. Drlica (eds.), *Antimicrobial Resistance and Implications for the Twenty-First Century.* © Springer 2008

partner who is shedding virus symptomatically or asymptomatically. Risk of infection correlates with number of lifetime sexual partners (Fleming et al., 1997). Genital HSV-1 infection is usually acquired through oral–genital contact. Reports from a variety of clinical settings in the United States show that genital herpes accounts for a large majority of patient requests for treatment of symptomatic genital ulcers. Only a minority of infections with herpes simplex virus type 2 (HSV-2), the main virus that causes genital herpes, are recognized as symptomatic by those who are infected (Fleming et al., 1997). Thus, most infections are transmitted by persons who are unaware that they have the infection or who are asymptomatic when transmission occurs.

The acyclic nucleoside analogues acyclovir, valacyclovir, and famciclovir are recommended for the management of genital herpes (Centers for Disease Control and Prevention, 2006c). All are effective for the treatment of a first episode of genital herpes, for treatment of recurrent genital herpes, and, when taken daily, for the prevention of a clinical recurrence. HSV infections that are resistant to acyclovir, valacyclovir, or famciclovir are rare; when they do occur, they are usually in immunocompromised patients. Acyclovir-resistant strains also are resistant to valacyclovir, and most are resistant to famciclovir. Foscarnet is often effective for treatment of acyclovir-resistant genital herpes. Because of its toxicity and cost, use of this drug is usually limited to patients with extensive mucocutaneous infections. Topical cidofovir gel applied to the genital lesions may be effective (Centers for Disease Control and Prevention, 2006c).

Acyclovir, valacyclovir, and famciclovir all require activation by phosphorylation with viral thymidine kinase; then, following two successive phosphorylation steps, the active triphosphate form of these drugs acts to inhibit viral DNA polymerase. Little effect on cellular DNA polymerase is observed. Mutant viruses may be thymidine kinase deficient, have a thymidine kinase with altered substrate specificity, or have a DNA polymerase with altered substrate specificity. Thymidine kinase deficiency is the most common mechanism of resistance to acyclovir (Tyring et al., 2002). In animal studies, resistant strains are less virulent and less able to establish neural latency. However, they are capable of causing extensive mucocutaneous disease in immunocompromised hosts. Thymidine kinase-altered or DNA polymerase-altered phenotypes have variable pathogenicity.

Despite widespread use of acyclovir, resistance to acyclovir in immunocompetent persons is rare. Estimates of in vitro HSV resistance in immunocompetent patients show a prevalence of about 0.1–0.4% (Reyes et al., 2003; Tyring et al., 2002). This prevalence does not appear to increase in patients receiving several years of suppressive therapy. The presence of in vitro acyclovir resistance in immunocompetent patients generally has not been associated with clinical treatment failure.

In immunocompromised patients, prevalence of resistance is about 5–6% (Tyring et al., 2002), and almost all clinically significant acyclovir resistance occurs in this population (Corey, 2000; Englund et al., 1990; Erlich et al., 1989). While the vast majority of resistant isolates are from immunocompromised patients undergoing multiple courses of acyclovir for established infection, risk factors associated with resistance have not been well defined. Routine in vitro testing of HSV isolates for acyclovir sensitivity is not recommended. Isolates from patients with persistent

HSV infections unresponsive to acyclovir, especially from those individuals with advanced HIV disease, should be tested for acyclovir resistance.

Transmission of resistant strains from person to person has not been documented, but the potential exists. The development of HSV-1 resistance is also a concern due to the wide availability of topical antiviral preparations for herpes labialis, though such resistance has been uncommon (Bacon et al., 2002).

3.3 *Trichomonas vaginalis*

Trichomoniasis is caused by the parasitic protozoan *Trichomonas vaginalis*. It is primarily sexually transmitted. Many men who are infected do not have symptoms; others have nongonococcal urethritis. Infected women may have symptoms characterized by a diffuse, malodorous, yellow-green discharge with vulvar irritation, though many also have minimal or no symptoms. It is believed that more than 170 million people are infected with this organism worldwide (World Health Organization, 1995) and between 5 and 8 million people in North America annually (Upcroft and Upcroft, 2001; World Health Organization, 1995). Prevalence rates have ranged from 3–10% of women in the general population to 60% in certain populations (Krieger and Alderete, 1999; Lossick, 1990; Sutton et al., 2006). Trichomoniasis has been associated with upper reproductive tract post-surgical infections and premature rupture of membranes, premature labor, and low-birth weight infants in pregnant women. As with other sexually transmitted infections, it has also been implicated in facilitating the transmission of HIV (Krieger and Alderete, 1999; Wasserheit, 1992).

The treatment of trichomoniasis was revolutionized in 1959 (Dunne et al., 2003) when metronidazole was found to be highly effective. Prior to that time, topical vaginal preparations provided some symptomatic relief but were ineffective as cures. The nitroimidazoles remain the only class of drugs useful for the oral or parenteral therapy of trichomoniasis; of these, only metronidazole and tinidazole are available in the United States and approved by the FDA for the treatment of trichomoniasis. In randomized clinical trials, the recommended metronidazole regimens have resulted in cure rates of approximately 90–95% when sex partners were also treated, and the recommended tinidazole regimen has resulted in cure rates of approximately 86–100% (Centers for Disease Control and Prevention, 2006c). Treatment of patients and sex partners results in relief of symptoms, microbiologic cure, and reduction of transmission.

Follow-up is unnecessary for men and women who become asymptomatic after treatment or who are initially asymptomatic. Certain strains of *T. vaginalis* can have diminished susceptibility to metronidazole; however, infections caused by most of these organisms respond to tinidazole or higher doses of metronidazole. Clinical metronidazole resistance has been reported since 1962 (Robinson, 1962). Resistance does not strictly correlate with aerobic minimal lethal concentrations (MLC) to metronidazole, but clinically resistant cases tend to have MLCs greater than or equal to 200 µg/ml (Lossick et al., 1986; Nanda et al., 2006). Clinicians make a diagnosis of metronidazole resistance clinically in patients who fail to respond to therapy and have no risk factors for reinfection (Lossick et al., 1986; Nanda et al., 2006). Low-level in vitro metronidazole resistance has been identified in 2–10% of cases of vaginal trichomoniasis (Perez et al., 2001; Schmid et al., 2001; Schwebke and Barrientes, 2006). While cross-resistance among different nitroimidazoles has been reported, tinidazole appears more active in vitro than metranidazole against resistant *T. vaginalis* strains (Crowell et al., 2003; Schwebke and Barrientes, 2006).

If treatment failure occurs with a metronidazole 2 g single dose, the patient can be treated with metronidazole 500 or 375 mg orally twice daily for 7 days or tinidazole 2 g single dose. Patients failing a 7-day course of metronidazole should be treated with tinidazole 2 g single dose. If failure occurs again, patients should be treated with metronidazole or tinidazole 2 g orally for 3–5 days. Patients with laboratory documented infection who do not respond to the 3 to 5-day treatment regimen and who have not been reinfected should be managed in consultation with a specialist; evaluation of such cases should include determination of the susceptibility of *T. vaginalis* to metronidazole and tinidazole (Centers for Disease Control and Prevention, 2006c).

The mechanism of metronidazole resistance in *T. vaginalis* associated with treatment failures is not as well understood as resistance which develops under laboratory conditions of increasing drug pressures. Resistance of *T. vaginalis* to metronidazole is classified as either aerobic or anaerobic. In aerobic resistance, oxygen-scavenging pathways and possibly ferredoxin are involved. In anaerobic resistance these pathways are not involved; resistance is due instead to a reduction or cessation of activity of the enzyme pyruvate:ferredoxin oxidoreductase (PFOR), which reduces metronidazole in the hydrogenosomes of *T. vaginalis* and eliminates activation of metronidazole by the parasite.

Because there are no ongoing, systematically collected surveillance data of vaginal trichomonas and clinical and microbiologic response to treatment, few data exist regarding temporal trends in incidence of metronidazole or tinidazole resistance despite the widespread use of those drugs. While cases have been reported throughout the United States and in many foreign countries, no clusters have been reported, suggesting that patients with resistant isolates may not be highly sexually active or that resistant organisms are unstable.

3.4 Syphilis

Syphilis is a systemic disease caused by *Treponema pallidum*. Penicillin G administered parenterally is the preferred treatment of all stages of syphilis. The preparations used (i.e., benzathine, aqueous procaine, or aqueous crystalline), the dosage, and length of treatment depend on the stage and clinical manifestations of disease. No evidence of penicillin resistance has developed over the years, although follow-up may be required to ensure the effectiveness of treatment (Centers for Disease Control and Prevention, 2006c).

Azithromycin, which has a long tissue half-life and can be administered orally, has been found effective in the treatment of syphilis in some studies (Hook et al.,

2002; Riedner et al., 2005). Because of its convenience and apparent efficacy, some disease control programs have used azithromycin as a single oral regimen for their patients with early syphilis and their sexual partners (Centers for Disease Control and Prevention, 2004a). Azithromycin was also used in a targeted mass treatment program in Vancouver, BC (Rekart et al. 2000). In Vancouver, after the program ended, there was an initial decline in rates, but rates then rebounded to levels higher than expected. In San Francisco, patient-delivered partner therapy with azithromycin was discontinued after azithromycin treatment failures were documented in 2002 and 2003.

Macrolide resistance in *T. pallidum* has been identified as resulting from a point mutation within the 23S ribosomal RNA gene, at a ribosomal site targeted by macrolides (Stamm et al., 1988). This mutation has been found in 22% of lesion samples from San Francisco, 11% from Baltimore, 13% from Seattle, and 88% from Dublin, Ireland. All but one of the strains tested with the point mutation were from men who have sex with men (MSM) (Lukehart et al., 2004).

The prevalence of macrolide resistance in the United States and elsewhere is not known, and it is unknown how selective pressures from macrolide use for syphilis or other conditions contribute to the emergence of macrolide-resistant *T. pallidum*. Due to its emergence and the continued effectiveness of penicillin G benzathine, azithromycin has not been recommended for the treatment of syphilis. Patients who are treated with azithromycin for early syphilis require close follow-up (Centers for Disease Control and Prevention, 2006c; Holmes, 2005).

3.5 Chlamydia

Chlamydiae are obligate intracellular Gram-negative bacterial pathogens of eukaryotic cells that have been associated with a variety of insidious diseases in a wide range of animals. Among humans, *Chlamydia trachomatis* is the most prevalent sexually transmitted bacterial infection in the United States. Women bear the burden of disease with tubal infertility, ectopic pregnancy, and pelvic inflammatory disease being well-recognized sequelae following *C. trachomatis* infections of the reproductive tract. Infection is amenable to treatment with macrolides, tetracyclines, and some fluoroquinolones (Centers for Disease Control and Prevention, 2006c). There have been a few reports of clinical treatment failures among patients who complied with therapy (Jones et al., 1990; Somani et al., 2000), but a stable, antibiotic-resistant *C. trachomatis* isolate has not been recovered from human infection. The inability to select for antibiotic-resistant *C. trachomatis* during a mass treatment campaign with azithromycin was recently demonstrated in the global effort to eliminate blinding trachoma (Solomon et al., 2005).

The chlamydial life cycle is slowed or completely stopped by some endogenous molecules such interferon γ (Beatty et al., 1995). Antibiotics used to treat chlamydiae are solely active against replicating organisms and would not be effective during stages of induced quiescence. An intriguing, yet unanswered question is the impact these molecules may have on the clinical management of *C. trachomatis* infection. Resistance to treatment at either the microbial or clinical level remains a controversial topic (Wang et al., 2005).

3.6 Gonorrhea

With 339,593 gonorrhea cases reported in 2005, gonorrhea is the second most frequently reported disease in the United States (Centers for Disease Control and Prevention, 2006b). Gonorrhea rates in the United States declined 74.3% from 1975 through 1997 following the implementation of national gonorrhea control programs in the mid-1970s. After 1997, gonorrhea rates appeared to plateau. The rate in 2005 was 115.6 per 100,000 persons. Overall, in 2005, gonorrhea rates continued to remain high in the South, among African-Americans, and among adolescents and young adults of all racial and ethnic groups. The health impact of gonorrhea is largely related to its role as a major cause of pelvic inflammatory disease, which frequently leads to infertility or ectopic pregnancy. In addition, data suggest that gonorrhea facilitates HIV transmission (Wasserheit, 1992).

The treatment and control of gonorrhea has been complicated by the ability of *Neisseria gonorrhoeae* to develop resistance to antimicrobial agents. Antimicrobial resistance in *N. gonorrhoeae* is mediated through either the cumulative effect of chromosomal mutations (to penicillin, tetracycline, spectinomycin, and the fluoroquinolones) or by the single-step acquisition of plasmids that encode for high-level resistance (to penicillin and tetracycline). Sentinel surveillance to monitor antimicrobial resistance in *N. gonorrhoeae* in the United States has been conducted since 1986 by the Gonococcal Isolate Surveillance Project (GISP), sponsored by the CDC in collaboration with local and state health departments (Centers for Disease Control and Prevention, 2005, 2006b; Schwarcz et al., 1990). In GISP, antimicrobial susceptibilities of *N. gonorrhoeae* isolated from men in approximately 26–30 cities in the United States are determined.

By the end of World War II, penicillin became the therapy of choice for gonorrheal infections as reports indicated an efficacy of close to 100% (Hook and Handsfield, 1999). Since that time, recommendations for gonorrhea therapy have been repeatedly revised with the continuing development of resistance. From the 1950s until the mid-1970s, gradually increasing chromosomal penicillin resistance led to periodic increases in the amount of penicillin required. Similar trends were seen for the macrolide and tetracycline antibiotics. In 1975–1976, strains of *N. gonorrhoeae* with high-level penicillin resistance due to plasmid-mediated β-lactamase production were reported in the United States, Western Europe, the Philippines, and western Africa (Perine et al., 1979). Due to single-step plasmid acquisition, penicillin at any dose was no longer effective against gonorrhea. By the early 1980s, penicillinase-producing strains accounted for up to 50% of *N. gonorrhoeae* isolates in some parts of the developing world and up to 11% of isolates in the United States. By the mid-1980s, these strains were common throughout the world (Sparling and Handsfield, 2000). In 1985, plasmid-mediated high-level resistance to tetracycline was first described in the United States (Centers for Disease

Control and Prevention, 1985). In 2004, chromosomal and plasmid-mediated resistance to tetracycline was found in 10.4 and 3.9% *N. gonorrhoeae* isolates, respectively, in the United States (Centers for Disease Control and Prevention, 2005).

Spectinomycin has long been recognized as a safe and effective option for treating gonococcal infections. It is particularly useful for the treatment of patients who cannot tolerate cephalosporins and for whom quinolones are not appropriate therapy. Spectinomycin-resistant gonorrhea has been rare in the United States, perhaps due to its low use in this country because it is expensive, must be given as an intramuscular injection, and has poor efficacy against pharyngeal infection (Newman et al., 2007). Sporadic cases of high-level resistance have been reported elsewhere (Ison et al., 1983; World Health Organization, 2006). In the 1980s, as spectinomycin was adopted as a drug of choice among US servicemen in Korea, increases in chromosomally mediated spectinomycin resistance in *N. gonorrhoeae* were noted (Boslego et al., 1987). Currently, spectinomycin is not manufactured in the United States or elsewhere in the world (Centers for Disease Control and Prevention, 2006a).

Increases in penicillin and tetracycline-resistant *N. gonorrhoeae* over the last 25 years eventually led to the abandonment of these drugs as therapies for gonorrhea. Because of the spread of gonococcal isolates resistant to penicillin and tetracycline, extended-spectrum cephalosporins (ceftriaxone and cefixime, which is not currently available in the United States (Centers for Disease Control and Prevention, 2002)) and the fluoroquinolones (ciprofloxacin, ofloxacin, levofloxacin) have been recommended as primary therapies to treat uncomplicated gonorrhea (Centers for Disease Control and Prevention, 2006c). At this time, while small numbers of isolates with decreased susceptibility to ceftriaxone have been identified, no failures of gonococcal infections to respond to ceftriaxone have been confirmed. However, failures of gonococcal infections to respond to fluoroquinolones have been seen. Fluoroquinolone-resistant (QRNG) isolates are very prevalent in Asia and have been isolated with increasing frequency in other parts of the world, including western Europe, Canada, and the United States (Australian Gonococcal Surveillance Programme, 2005; Centers for Disease Control and Prevention, 2005; GRASP Steering Group, 2006; Martin et al., 2006; World Health Organization, 2006) (Figure 3.1). QRNG increases in MSM and in some regions of the United States, led the CDC to recommend in 2004 that quinolones not be used for infections in MSM, as well as in those with a history of recent foreign travel or partners' travel, or for infections acquired in other areas with increased QRNG prevalence (Centers for Disease Control and Prevention, 2004b). In 2007, continued increases in QRNG in MSM and heterosexuals throughout the United States led the CDC to recommend that fluoroquinolones no longer be used for treating gonorrhea in that country (Newman et al., 2007) (Figure 3.2).

The determination of antimicrobial resistances in *N. gonorrhoeae* can be performed using a number of methods. These include disk diffusion, E-test, and agar plate dilution assays. Agar plate dilution susceptibility testing is considered the most accurate, but it is also the most costly and time consuming. Once the specific assay has been performed, results are interpreted by National Committee for Clinical Laboratory Standards (NCCLS) recommended methods, as described

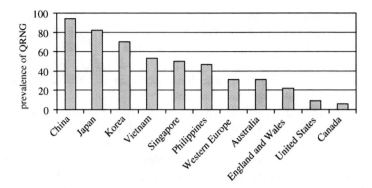

Fig. 3.1 Areas around the world with increased prevalence of fluoroquinolone resistance, 2004–2005. Reference: Centers for Disease Control and Prevention (2006) Areas Around the World with Increased Prevalence of Fluoroquinolone Resistance, 2004–2005. Available at http://www.cdc.gov/std/Gonorrhea/arg/world-prev.htm (31 March 2007).

elsewhere (National Committee for Clinical Laboratory Standards, 1993, 1998). These NCCLS standards allow for the assignment of a specific *N. gonorrhoeae* isolate as susceptible, intermediate resistant (if applicable), or resistant to the antimicrobial being examined. However, the three levels of classifications are not defined for all antimicrobials (Table 3.1).

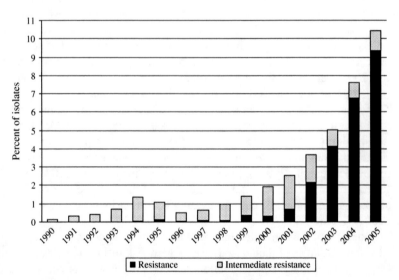

Fig. 3.2 Percentage of isolates with intermediate resistance or resistance to ciprofloxacin, United States, Gonococcal Isolate Surveillance Project (GISP), 1990–2005. Reference: Centers for Disease Control and Prevention (2007).

Table 3.1 Antimicrobial resistance criteria

Antimicrobial resistance in *N. gonorrhoeae* is defined by the criteria recommended by the National Committee on Clinical Laboratory Standards (NCCLS) (1993, 1998)

Penicillin, MIC $\geq$ 2.0 μg/ml
Tetracycline, MIC $\geq$ 2.0 μg/ml
Spectinomycin, MIC $\geq$ 128.0 μg/ml
Ciprofloxacin, MIC 0.125–0.5 μg/ml (intermediate resistance)
Ciprofloxacin, MIC $\geq$ 1.0 μg/ml (resistance)
Ceftriaxone, MIC $\geq$ 0.5 μg/ml (decreased susceptibility)
Cefixime, MIC $\geq$ 0.5 μg/ml (decreased susceptibility)

NCCLS criteria for resistance to ceftriaxone, cefixime, erythromycin, and azithromycin and for susceptibility to erythromycin and azithromycin have not been established for *N. gonorrhoeae*.

3.6.1 Mechanisms of Resistance

3.6.1.1 Azithromycin

Azithromycin and erythromycin belong to the macrolide family of antibiotics. While neither antimicrobial is currently recommended by CDC for the treatment of gonorrhea, a 2-g dose of azithromycin is FDA approved for the treatment of uncomplicated gonorrhea. The 2-g dose of azithromycin causes a high level of gastrointestinal distress and is expensive; additionally, this dose produces low serum drug levels, thought to be favorable for the rapid development of resistance (Ehret et al., 1996; Handsfield, 1997; Young et al., 1997). A 1-g dose of azithromycin is better tolerated but is insufficiently effective with several studies documenting treatment failures at that dose. It is thought that this dose raises even greater concerns about the emergence of antimicrobial resistance than the 2-g dose (McLean et al., 2004; Tapsall, 2002; Tapsall et al., 1998; Waters et al., 2005; Young et al., 1997).

One of the mechanisms of resistance to azithromycin and erythromycin in *N. gonorrhoeae* is the multiple transferable resistance (mtr) efflux system, which confers resistance to structurally diverse hydrophobic agents such as detergents, dyes, and macrolide antibiotics (Hagman and Shafer, 1995; Hagman et al., 1995; Lucas et al., 1997; Pan and Spratt, 1994; Shafer et al., 1995). Recent studies have demonstrated that mutations in the promotor regions of the *mtr*R and *mtr*C genes, which are part of the *mtr* efflux operon, produce increased minimal inhibitory concentrations (MICs) of azithromycin ranging from 0.25 to 0.5 μg/ml of azithromycin (Xia et al., 2000; Zarantonelli et al., 1999).

Reduced susceptibility of *N. gonorrhoeae* to macrolide antibiotics, specifically azithromycin and erythromycin, has been demonstrated to be a function of the expression of an efflux system regulated by the *mtr*R product (Hagman et al., 1995; Pan and Spratt, 1994). While earlier isolates exhibited only slightly increased MICs of azithromycin, isolates obtained in Kansas City over a 5-month period (McLean et al., 2005) demonstrated significantly elevated values that ranged from 1.0 to

4.0 µg/ml of azithromycin. This collection of isolates could be divided into two groups distinguishable from one another on the basis of their azithromycin MICs (Zarantonelli et al., 1999), but were identical based on auxotype, serovar, other MIC values, and Lip type.

The high MICs of these isolates were presumed to result from mutations in *mtr*R that derepressed the efflux system. They were initially investigated by sequencing *mtr*R and flanking sequences from 12 with decreased susceptibility and four susceptible strains from Kansas City and from an azithromycin/erythromycin-susceptible laboratory strain. These sequences were compared with the published sequence of FA19 (Hagman et al., 1995; Short et al., 1977). For the strains with decreased susceptibility, a number of changes from the *mtr* sequence in FA19 were noted. These alterations included the insertion of a 153 bp sequence at a site identical for all the 12 strains with decreased susceptibility. This sequence was highly similar to the published sequences of the transposon-like Correia elements (Correia et al., 1988). Further, 14 identical, single base substitutions, six of which altered the mtr protein, were also detected. Substitutions at bp 1178 and 1296 (Pan and Spratt coordinates) (Hagman and Shafer, 1995) were within the helix–turn–helix motif of *mtr*R where substitutions have been shown to alter the function of the gene product and may be involved in decreased susceptibility for some of the isolates (Shafer et al., 1995). Single base substitutions were also identified at four additional positions within the *mtr*R structural gene, but the consequences of these substitutions, which also altered the amino acid sequence of the mtrR protein, were not readily apparent. Also of potential significance were three substitutions within the overlapping promoter regions of *mtr*R and *mtr*C. One of these at bp 1008 replaces A with C, and the others at bp 1002 and 1005 alter the –35 sequences for both *mtr*R and *mtr*C (Hagman et al., 1995). In general, the conservation of the 153 bp insertion and single base substitutions confirmed the relatedness of the isolates as demonstrated by the other tests used to characterize the isolates. The sequence data also offered explanations for some of the phenotypic differences noted among the isolates. The five isolates that exhibited greater decreased susceptibility contained an additional T residue inserted at bp 1120 that created an early frame shift, and a single isolate with similarly decreased susceptibility possessed a single base change at bp 1606 that created a translational stop codon. Thus, functional elimination of the *mtr*R gene product appeared to confer higher levels of decreased susceptibility in these clinical isolates.

3.6.1.2 Tetracycline

Tetracycline resistance in *N. gonorrhoeae* is mediated in one of the two ways. Chromosomal resistance is mediated by the chromosomal loci *tet* and *mtr* and results in low-level resistance with minimum inhibitory concentrations (MICs) ranging from 1.0 to 8.0 µg/ml of tetracycline (Sparling, 1977). High-level resistance (usually MICs of ≥ 16.0 µg/ml of tetracycline) is due to the presence of a 25.2-MDa conjugative plasmid that contains the *tet*M gene (Knapp et al., 1987, 1988). *Tet*M is present in a number of bacteria and has been shown to have a highly mosaic structure depending on the host species (Oggioni et al., 1996). Two distinct *tet*M genes have been identified in *N. gonorrhoeae* and have been designated as either "Dutch-type" (Netherlands) or "American-type" (United States) based on the location where they were originally isolated (Gascoyne et al., 1991; Gascoyne-Binzi et al., 1993). A polymerase chain reaction (PCR)/restriction endonuclease analysis has been described which can determine the *tet*M type present in a gonococcal isolate (Ison et al., 1993).

3.6.1.3 Penicillin

As with tetracycline, resistance to penicillin can be either chromosomally or plasmid mediated. Chromosomal resistance can occur through the modification of various genetic loci that can result in resistance by a number of mechanisms, such as decreased cell entry of the antibiotic, increased efflux of the antibiotic, or alteration of the target site for penicillin.

Plasmid-mediated resistance is due to strains acquiring one of the β-lactamase-producing plasmids which leads to the isolate being designated penicillinase-producing *N. gonorrhoeae* (PPNG). PPNG isolates were first identified in 1976 and contain a TEM-1-type β-lactamase plasmid. The four plasmids that have been described are the 3.05 MDa "Toronto", 3.2 MDa "African", 4.0 MDa "Nimes", and 4.4 MDa "Asian" forms.

3.6.1.4 Ciprofloxacin (Fluoroquinolones)

Studies of the genetic mutations resulting in resistance of gonococcal isolates to fluoroquinolones have shown that mutations within the quinolone resistance-determining regions (QRDRs) of *gyr*A and *par*C are responsible for resistance to these agents (Belland et al., 1994; Deguchi et al., 1995, 1996a). These mechanisms are analogous to those observed in *Escherichia coli* and other bacteria (Ferrero et al., 1994; Yoshida et al., 1990). However, unlike *gyr*B mutations in *E. coli*, mutations in the *N. gonorrhoeae gyr*B gene do not appear to have a clinically significant impact on fluoroquinolone resistance (Deguchi et al., 1996b; Ferrero et al., 1994). The order in which mutations and the resulting amino acid alterations occur and result in increased fluoroquinolone MICs in gonococcal strains follows a fairly predictable pattern: mutations which confer alterations at Ser-91 and/or Asp-95 of GyrA result in decreased susceptibility to ciprofloxacin (CipI, MICs 0.125–0.5 µg/ml of ciprofloxacin). On the other hand, clinically significant resistance to CDC-recommended fluoroquinolone regimens (CipR, MIC ≥ 1.0 µg/ml of ciprofloxacin) appears to require one or more mutations in *par*C in addition to both *gyr*A mutations. However, this order of mutations is not absolute, and a number of GyrA/ParC protein "alteration patterns" have been described.

3.6.1.5 Cephalosporins

Currently, few gonococcal isolates are exhibiting decreased susceptibility to the cephalosporins. Reports from Japan suggest that MICs of some cephalosporins, particularly, cefixime, may be increasing (Akasaka et al., 2001; Ito et al., 2004; Muratani et al., 2001). Ameyama et al. (2002) have described alterations in the *pen*A gene that they attribute to reduced susceptibility to cefixime. There have also been reports in the United States of a small number of isolates with decreased susceptibility to cephalosporins such as cefixime and ceftriaxone (Centers for Disease Control and Prevention, 2005; Wang et al., 2003). Studies to determine the mechanism(s) responsible for the elevated MICs are currently underway.

3.7 *Haemophilus ducreyi* (Chancroid)

Chancroid is a sexually transmitted infection caused by *H. ducreyi*. It is distributed worldwide, though it is particularly common in parts of Africa, Asia, and Latin America where it may exceed syphilis as a cause of genital ulceration (Ballard and Morse, 2003). It is most often seen in uncircumcised men; only 10% of reported cases are in women. Chancroid is a co-factor for HIV transmission, as are other genital ulcerative diseases (Jessamine et al., 1990). While it is not considered a common sexually transmitted infection in Europe, Canada, and the United States, where only 17 cases were reported in 2005 (Centers for Disease Control and Prevention, 2006b), it is probably underdiagnosed (Centers for Disease Control and Prevention, 2006b; Mertz et al., 1998; Schulte et al., 1992). Major outbreaks of infection have been reported in the United States and Canada (Blackmore et al., 1985; Flood et al., 1993; Hammond et al., 1980; Ronald and Plummer, 1985; Schmid et al., 1987).

A definitive diagnosis of chancroid requires the identification of *H. ducreyi* on special culture media that is not widely available commercially; even when these media are used, sensitivity is less than 80% (Trees and Morse, 1995). PCR assays for detection of *H. ducreyi* have been evaluated in clinical studies and appear to be sensitive and specific. The diagnosis of chancroid on clinical grounds is difficult and inaccurate because genital ulcerations caused by herpes simplex virus, syphilis, and lymphogranuloma venereum may be similar and may be seen in association with chancroid.

Methodology and the results of susceptibility studies have been reviewed in detail elsewhere (Dangor et al., 1990a,b; Morse, 1989). Most studies have used agar dilution; there has been little experience with disk diffusion susceptibility testing of *H. ducreyi*. Geographical and temporal differences in antimicrobial susceptibilities have been observed. Many of these differences are due to the presence or absence of resistance plasmids (Morse, 1989; Motley et al., 1992; Sarafian and Knapp, 1992; Sarafian et al., 1991). Plasmids have been identified in *H. ducreyi*, which encode resistance, either separately or in combination, to sulfonamides (Sarafian and Knapp, 1992; Willson et al., 1989), aminoglycosides (Sanson-Le Pors et al.,

1985; Willson et al., 1989), tetracyclines (Marshall et al., 1984; Roberts, 1989, 1990), chloramphenicol (Roberts et al., 1985), and β-lactam antibiotics (Brunton et al., 1982; Maclean et al., 1980, 1992). These plasmids have been reviewed in detail elsewhere (Morse, 1989). A single *H. ducreyi* isolate may contain multiple resistance plasmids (Motley et al., 1992; Sarafian and Knapp, 1992; Sarafian et al., 1991), including more than one plasmid that confers resistance to β-lactam antibiotics. Strains containing both a 7.0-MDa TEM-1 β-lactamase plasmid and a 3.5-MDa ROB-1 β-lactamase plasmid have been isolated frequently in Thailand (Motley et al., 1992).

The appearance of Tet M in *H. ducreyi* (Roberts, 1989) is consistent with its presence in other genitourinary tract pathogens (Roberts, 1990). The Tet M determinant is located on a 34-MDa conjugative plasmid. Unlike the Tet M-containing conjugative plasmid in *N. gonorrhoeae* (Roberts and Knapp, 1988), this plasmid is unable to mobilize Amp^r plasmids (Totten et al., 1982).

Very little information is available concerning the development of chromosomally mediated antimicrobial resistance in *H. ducreyi*. Approximately 30% of strains isolated in the United States between 1982 and 1990 were shown to lack resistance plasmids (Morse, 1989; Sarafian and Knapp, 1992). Some of these strains exhibited decreased susceptibilities to trimethoprim, ciprofloxacin, and ofloxacin, suggesting the presence of chromosomally mediated resistance to these antimicrobials. MICs to penicillin G of 0.06–0.12 μg/ml have been reported for some strains of *H. ducreyi* (Bilgeri et al., 1982) indicating in vitro activity by this class of antibiotics. However, increased MICs to penicillin G (0.25–1.0 μg/ml) have been observed in β-lactamase-negative strains of *H. ducreyi* (Morse, 1989), which are consistent with the development of chromosomally mediated resistance.

Regimens of azithromycin (once), ceftriaxone (once), erythromycin (over 7 days) or ciprofloxacin (over 3 days) remain effective for the treatment of chancroid in the United States (Centers for Disease Control and Prevention, 2006c). Trimethoprim–sulfamethoxazole combination has been used previously in the treatment of chancroid, but resistance to the sulfonamides has been reported worldwide, and many strains are now resistant to trimethoprim. Clinical failures and in vitro resistance have been demonstrated with this drug combination (Schulte and Schmid, 1995).

The emergence of β-lactamase-producing strains of *H. ducreyi* has resulted in the abandonment of penicillin and other β-lactamase-susceptible antibiotics for treatment of chancroid, such as ampicillin and amoxicillin. Many strains of *H. ducreyi* are resistant to tetracycline, which should not be used for the treatment of chancroid any longer (Ballard et al., 1989; Bilgeri et al., 1982; Rutanarugsa et al., 1990).

No cephalosporin other than ceftriaxone has been clinically tested for treatment of chancroid since 1993. Before that, cefotaxime and ceftazidime were also tested and, as with ceftriaxone, no resistance had been observed (Schmid, 1999). While ceftriaxone may be given as a single dose, it must be given intramuscularly, and failures have been reported in those with HIV co-infection (Martin et al., 1995; Taylor et al., 1985; Tyndall et al., 1993). Ciprofloxacin and other quinolones as a single dose have had limited success, but 3-day regimens have been shown to be more efficacious; treatment failures have not been due to resistance (Bodhidatta et al., 1988; Naamara et al., 1987; Schmid, 1999). In vitro studies of the use of quinolones have also shown high susceptibility rates. Similarly, little clinical or in vitro resistance has been found to the macrolides, erythromycin, and azithromycin (Schmid, 1999).

References

Akasaka, S., Muratani, T., Yamada, Y., Inatomi, H., Takahashi, K. and Matsumoto, T., (2001), Emergence of cephem- and aztreonam-high-resistant *Neisseria gonorrhoeae* that does not produce beta-lactamase. *J Infect Chemother*, **7**, 49–50.

Ameyama, S., Onodera, S, Takahata, M, Minami, S, Maki, N., Endo, K, Goto, H., Suzuki, H. and Oishi, Y., (2002), Mosaic-like structure of penicillin-binding protein 2 Gene (penA) in clinical isolates of *Neisseria gonorrhoeae* with reduced susceptibility to cefixime. *Antimicrob Agents Chemother*, **46**, 3744–3749.

Australian Gonococcal Surveillance Programme (2005) Annual report of the Australian Gonococcal Surveillance Programme, 2004. *Commun Dis Intell*, **29**, 137–142.

Bacon, T.H., Boon, R.J., Schultz, M. and Hodges-Savola, C., (2002), Surveillance for antiviral-agent-resistant herpes simplex virus in the general population with recurrent herpes labialis. *Antimicrob Agents Chemother*, **46**, 3042–3044.

Ballard, R. and Morse, S., (2003), Chancroid. In S. Morse, R. Ballard, K.K. Holmes and A. Moreland, editors, *Atlas of Sexually Transmitted Diseases and AIDS*, 3rd Edition, Edinburgh, Mosby, pp. 53–71.

Ballard, R.C., Duncan, M.O., Fehler, H.G., Dangor, Y., Exposto, F.L. and Latif, A.S., (1989), Treating chancroid: summary of studies in southern Africa. *Genitourin Med*, **65**, 54–57.

Beatty, W.L., Morrison, R.P. and Byrne, G.I., (1995), Reactivation of persistent *Chlamydia trachomatis* infection in cell culture. *Infect Immun*, **63**, 199–205.

Belland, R.J., Morrison, S.G., Ison, C. and Huang, W.M., (1994), *Neisseria gonorrhoeae* acquires mutations in analogous regions of gyrA and parC in fluoroquinolone-resistant isolates. *Mol Microbiol*, **14**, 371–380.

Bilgeri, Y.R., Ballard, R.C., Duncan, M.O., Mauff, A.C. and Koornhof, H.J., (1982), Antimicrobial susceptibility of 103 strains of *Haemophilus ducreyi* isolated in Johannesburg. *Antimicrob Agents Chemother*, **22**, 686–688.

Blackmore, C.A., Limpakarnjanarat, K., Rigau-Perez, J.G., Albritton, W.L. and Greenwood, J.R., (1985), An outbreak of chancroid in Orange County, California: descriptive epidemiology and disease-control measures. *J Infect Dis*, **151**, 840–844.

Bodhidatta, L., Taylor, D.N., Chitwarakorn, A., Kuvanont, K. and Echeverria, P., (1988), Evaluation of 500- and 1,000-mg doses of ciprofloxacin for the treatment of chancroid. *Antimicrob Agents Chemother*, **32**, 723–725.

Boslego, J.W., Tramont, E.C., Takafuji, E.T., Diniega, B.M., Mitchell, B.S., Small, J.W., Khan, W.N. and Stein, D.C., (1987), Effect of spectinomycin use on the prevalence of spectinomycin-resistant and of penicillinase-producing *Neisseria gonorrhoeae*. *N Engl J Med*, **317**, 272–278.

Brunton, J., Meier, M., Ehrman, N., MacLean, I., Slaney, L. and Albritton, W.L., (1982), Molecular epidemiology of beta-lactamase-specifying plasmids of *Haemophilus ducreyi*. *Antimicrob Agents Chemother*, **21**, 857–863.

Centers for Disease Control and Prevention (1985) Tetracycline-resistant *Neisseria gonorrhoeae*—Georgia, Pennsylvania, New Hampshire. *MMWR Morb Mortal Wkly Rep*, **34**, 563.

Centers for Disease Control and Prevention (2002) Notice to readers: discontinuation of cefixime tablets—United States. *MMWR Morb Mortal Wkly Rep*, **51**, 1056.

Centers for Disease Control and Prevention (2004a) Azithromycin treatment failures in syphilis infections—San Francisco, California, 2002–2003. *MMWR Morb Mortal Wkly Rep*, **53**, 197–198.

Centers for Disease Control and Prevention (2004b) Increases in fluoroquinolone-resistant *Neisseria gonorrhoeae* among men who have sex with men—United States, 2003, and revised recommendations for gonorrhea treatment, 2004. *MMWR Morb Mortal Wkly Rep*, **53**, 335–338.

Centers for Disease Control and Prevention (2005) Sexually Transmitted Disease Surveillance 2004 Supplement: Gonococcal Isolate Surveillance Project (GISP) Annual Report. Atlanta, GA, US Department of Health and Human Services.

Centers for Disease Control and Prevention (2006a) Notice to readers: discontinuation of spectinomycin. *MMWR Morb Mortal Wkly Rep*, **55**, 370.

Centers for Disease Control and Prevention (2006b) Sexually Transmitted Disease Surveillance, 2005. Atlanta, GA, US Department of Health and Human Services.

Centers for Disease Control and Prevention (2006c) Sexually Transmitted Diseases Treatment Guidelines, 2006. Atlanta, GA, US Department of Health and Human Services.

Centers for Disease Control and Prevention (2006d) Summary of notifiable diseases—United States, 2004. *MMWR Morb Mortal Wkly Rep*, **53**.

Centers for Disease Control and Prevention (2007) Sexually Transmitted Disease Surveillance 2005 Supplement, Gonococcal Isolate Surveillance Project (GISP) Annual Report. Atlanta, GA, US Department of Health and Human Services.

Corey, L., (2000), Herpes simplex virus. In G. Mandell, J. Bennett and R. Dolin, editors, *Principles and Practice of Infectious Diseases*, 5th Edition. New York, Churchill Livingstone, pp. 1564–1580.

Correia, F.F., Inouye, S. and Inouye, M., (1988), A family of small repeated elements with some transposon-like properties in the genome of *Neisseria gonorrhoeae*. *J Biol Chem*, **263**, 12194–12198.

Crowell, A.L., Sanders-Lewis, K.A. and Secor, W.E., (2003), In vitro metronidazole and tinidazole activities against metronidazole-resistant strains of *Trichomonas vaginalis*. *Antimicrob Agents Chemother*, **47**, 1407–1409.

Dangor, Y., Ballard, R.C., da L Exposto, F., Fehler, H.G., Miller, S.D. and Koornhof, H.J., (1990a), Accuracy of clinical diagnosis of genital ulcer disease. *Sex Transm Dis*, **17**, 184–189.

Dangor, Y., Ballard, R.C., Miller, S.D. and Koornhof, H.J., (1990b), Antimicrobial susceptibility of *Haemophilus ducreyi*. *Antimicrob Agents Chemother*, **34**, 1303–1307.

Deguchi, T., Yasuda, M., Asano, M., Tada, K, Iwata, H., Komeda, H., Ezaki, T, Saito, I. and Kawada, Y., (1995), DNA gyrase mutations in quinolone-resistant clinical isolates of *Neisseria gonorrhoeae*. *Antimicrob Agents Chemother*, **39**, 561–563.

Deguchi, T., Yasuda, M., Nakano, M., Ozeki, S., Ezaki, T., Saito, I. and Kawada, Y., (1996a), Quinolone-resistant *Neisseria gonorrhoeae*: correlation of alterations in the GyrA subunit of DNA gyrase and the ParC subunit of topoisomerase IV with antimicrobial susceptibility profiles. *Antimicrob Agents Chemother*, **40**, 1020–1023.

Deguchi, T., Yasuda, M., Nakano, M., Ozeki, S., Kanematsu, E, Kawada, Y., Ezaki, T. and Saito, I., (1996b), Uncommon occurrence of mutations in the gyrB gene associated with quinolone resistance in clinical isolates of *Neisseria gonorrhoeae*. *Antimicrob Agents Chemother*, **40**, 2437–2438.

Dunne, R.L., Dunn, L.A., Upcroft, P., O'Donoghue, P.J. and Upcroft, J.A., (2003), Drug resistance in the sexually transmitted protozoan *Trichomonas vaginalis*. *Cell Res*, **13**, 239–249.

Ehret, J.M., Nims, L.J. and Judson, F.N., (1996), A clinical isolate of *Neisseria gonorrhoeae* with in vitro resistance to erythromycin and decreased susceptibility to azithromycin. *Sex Transm Dis*, **23**, 270–272.

Englund, J.A., Zimmerman, M.E., Swierkosz, E.M., Goodman, J.L., Scholl, D.R. and Balfour, Jr, H.H., (1990), Herpes simplex virus resistant to acyclovir. A study in a tertiary care center. *Ann Intern Med*, **112**, 416–422.

Erlich, K.S., Mills, J., Chatis, P. Mertz, G.J., Busch, D.F., Follansbee, S.E., Grant, R.M. and Crumpacker, C.S., (1989), Acyclovir-resistant herpes simplex virus infections in patients with the acquired immunodeficiency syndrome. *N Engl J Med*, **320**, 293–296.

Ferrero, L., Cameron, B., Manse, B., Lagneaux, D., Crouzet, J., Famechon, A. and Blanche, F., (1994), Cloning and primary structure of *Staphylococcus aureus* DNA topoisomerase IV: a primary target of fluoroquinolones. *Mol Microbiol*, **13**, 641–653.

Fleming, D.T., McQuillan, G.M., Johnson, R.E., Nahmias, A.J., Aral, S.O., Lee, F.K. and St Louis, M.E., (1997), Herpes simplex virus type 2 in the United States, 1976 to 1994. *N Engl J Med*, **337**, 1105–1111.

Flood, J.M., Sarafian, S.K., Bolan, G.A., Lammel, C, Engelman, J., Greenblatt, R.M., Brooks, G.F., Back, A. and Morse, S.A., (1993), Multistrain outbreak of chancroid in San Francisco, 1989–1991. *J Infect Dis*, **167**, 1106–1111.

Gascoyne, D.M., Heritage, J., Hawkey, P.M., Turner, A. and van Klingeren, B., (1991), Molecular evolution of tetracycline-resistance plasmids carrying TetM found in *Neisseria gonorrhoeae* from different countries. *J Antimicrob Chemother*, **28**, 173–183.

Gascoyne-Binzi, D.M., Heritage, J. and Hawkey, P.M (1993) Nucleotide sequences of the tet(M) genes from the American and Dutch type tetracycline resistance plasmids of *Neisseria gonorrhoeae*. *J Antimicrob Chemother*, **32**, 667–676.

GRASP Steering Group (2006) The Gonococcal Resistance to Antimicrobials Surveillance Programme (GRASP) Year 2005 Report. London, Health Protection Agency.

Hagman, K.E., Pan, W., Spratt, B.G., Balthazar, J.T., Judd, R.C. and Shafer, W.M., (1995), Resistance of *Neisseria gonorrhoeae* to antimicrobial hydrophobic agents is modulated by the mtr-RCDE efflux system. *Microbiology*, **141**, 611–622.

Hagman, K.E. and Shafer, W.M., (1995), Transcriptional control of the mtr efflux system of *Neisseria gonorrhoeae*. *J Bacteriol*, **177**, 4162–4165.

Hammond, G.W., Slutchuk, M., Scatliff, J., Sherman, E., Wilt, J.C. and Ronald, A.R., (1980), Epidemiologic, clinical, laboratory, and therapeutic features of an urban outbreak of chancroid in North America. *Rev Infect Dis*, **2**, 867–879.

Handsfield, H.H., (1997), Azithromycin in gonorrhoea. *Int J STD AIDS*, **8**, 472–473.

Holmes, K.K., (2005), Azithromycin versus penicillin G benzathine for early syphilis. *N Engl J Med*, **353**, 1291–1293.

Hook, E.W., III and Handsfield, H.H., (1999), Gonococcal infections in the adult. In K.K. Holmes, P.F. Sparling, P.A. Mardh, S. Lemon, W.E. Stamm, P. Piot and J.N. Wasserheit, editors, *Sexually Transmitted Diseases*, 3rd Edition. New York, McGraw-Hill, pp. 451–466.

Hook, E.W., III, Martin, D.H., Stephens, J., Smith, B.S. and Smith, K., (2002), A randomized, comparative pilot study of azithromycin versus benzathine penicillin G for treatment of early syphilis. *Sex Transm Dis*, **29**, 486–490.

Institute of Medicine (1997) *The Hidden Epidemic, Confronting Sexually Transmitted Diseases*. Eng, TR and Butler, WT. Washington, DC, National Academy Press, pp. 196–197.

Ison, C.A., Littleton, K., Shannon, K.P., Easmon, C.S. and Phillips, I., (1983), Spectinomycin resistant gonococci. *Br Med J (Clin Res Ed)*, **287**, 1827–1829.

Ison, C.A., Tekki, N. and Gill, M.J., (1993), Detection of the tetM determinant in *Neisseria gonorrhoeae*. *Sex Transm Dis*, **20**, 329–333.

Ito, M., Yasuda, M., Yokoi, S., Ito, S, Takahashi, Y., Ishihara, S., Maeda, S. and Deguchi, T., (2004), Remarkable increase in central Japan in 2001–2002 of *Neisseria gonorrhoeae* isolates with decreased susceptibility to penicillin, tetracycline, oral cephalosporins, and fluoroquinolones. *Antimicrob Agents Chemother*, **48**, 3185–3187.

Jessamine, P.G., Plummer, F.A., Ndinya Achola, J.O., Wainberg, M.A., Wamola, I., D'Costa, L.J., Cameron, D.W., Simonsen, J.N., Plourde, P. and Ronald, AR., (1990), Human immunodeficiency virus, genital ulcers and the male foreskin: synergism in HIV-1 transmission. *Scand J Infect Dis Suppl*, **69**, 181–186.

Jones, R.B., Van der, P.B., Martin, D.H. and Shepard, M.K., (1990), Partial characterization of *Chlamydia trachomatis* isolates resistant to multiple antibiotics. *J Infect Dis*, **162**, 1309–1315.

Knapp, J.S., Johnson, S.R., Zenilman, J.M., Roberts, M.C. and Morse, S.A., (1988), High-level tetracycline resistance resulting from TetM in strains of Neisseria spp., *Kingella denitrificans*, and *Eikenella corrodens*. *Antimicrob Agents Chemother*, **32**, 765–767.

Knapp, J.S., Zenilman, J.M., Biddle, J.W., Perkins, G.H., DeWitt, W.E., Thomas, M.L., Johnson, S.R. and Morse, S.A., (1987), Frequency and distribution in the United States of strains of

Neisseria gonorrhoeae with plasmid-mediated, high-level resistance to tetracycline. *J Infect Dis*, **155**, 819–822.

Krieger, J.N. and Alderete, J., (1999), *Trichomonas vaginalis* and Trichomoniasis. In K.K. Holmes, P. Sparling, P.A. Mardh, S. Lemon, W.E. Stamm, P. Piot and J. Wasserheit, editors, *Sexually Transmitted Diseases,* 3rd Edition. New York, McGraw-Hill.

Lossick, J.G., (1990), Epidemiology of urogenital trichomonas. In B.M. Honigberg, editor, *Trichomonads Parasitic in Humans*. New York, Springer-Verlag, pp. 311–323.

Lossick, J.G., Muller, M. and Gorrell, T.E., (1986), In vitro drug susceptibility and doses of metronidazole required for cure in cases of refractory vaginal trichomoniasis. *J Infect Dis*, **153**, 948–955.

Lucas, C.E., Balthazar, J.T., Hagman, K.E. and Shafer, W.M., (1997), The MtrR repressor binds the DNA sequence between the mtrR and mtrC genes of *Neisseria gonorrhoeae*. *J Bacteriol*, **179**, 4123–4128.

Lukehart, S.A., Godornes, C., Molini, B.J., Sonnett, P., Hopkins, S., Mulcahy, F., Engelman, J., Mitchell, S.J., Rompalo, A.M., Marra, C.M. and Klausner, J.D., (2004), Macrolide resistance in *Treponema pallidum* in the United States and Ireland. *N Engl J Med*, **351**, 154–158.

Maclean, I.W., Bowden, G.H. and Albritton, W.L., (1980), TEM-type beta-lactamase production in *Haemophilus ducreyi*. *Antimicrob Agents Chemother*, **17**, 897–900.

Maclean, I.W., Slaney, L., Juteau, J.M., Levesque, R.C., Albritton, W.L. and Ronald, A.R., (1992), Identification of a ROB-1 beta-lactamase in *Haemophilus ducreyi*. *Antimicrob Agents Chemother*, **36**, 467–469.

Marshall, B., Roberts, M., Smith, A. and Levy, S.B., (1984), Homogeneity of transferable tetracycline-resistance determinants in Haemophilus species. *J Infect Dis*, **149**, 1028–1029.

Martin, D.H., Sargent, S.J., Wendel, Jr, G.D., McCormack, W.M., Spier, N.A. and Johnson, R.B., (1995), Comparison of azithromycin and ceftriaxone for the treatment of chancroid. *Clin Infect Dis*, **21**, 409–414.

Martin, I.M., Hoffmann, S. and Ison, C.A., (2006), European Surveillance of Sexually Transmitted Infections (ESSTI): the first combined antimicrobial susceptibility data for Neisseria gonorrhoeae in Western Europe. *J Antimicrob Chemother*, **58**, 587–593.

McLean, C.A., Wang, S.A., Hoff, G.L., Dennis, L.Y., Trees, D.L., Knapp, J.S., Markowitz, L.E. and Levine, W.C., (2004), The emergence of *Neisseria gonorrhoeae* with decreased susceptibility to azithromycin in Kansas City, Missouri, 1999 to 2000. *Sex Transm Dis*, **31**, 73–78.

Mertz, K.J., Weiss, J.B., Webb, R.M., Levine, W.C., Lewis, J.S., Orle, K.A., Totten, P.A., Overbaugh, J., Morse, S.A., Currier, M.M., Fishbein, M. and St Louis, M.E., (1998), An investigation of genital ulcers in Jackson, Mississippi, with use of a multiplex polymerase chain reaction assay: high prevalence of chancroid and human immunodeficiency virus infection. *J Infect Dis*, **178**, 1060–1066.

Morse, S.A., (1989), Chancroid and *Haemophilus ducreyi*. *Clin Microbiol Rev*, **2**, 137–157.

Motley, M., Sarafian, S.K., Knapp, J.S., Zaidi, A.A. and Schmid, G., (1992), Correlation between in vitro antimicrobial susceptibilities and beta-lactamase plasmid contents of isolates of *Haemophilus ducreyi* from the United States. *Antimicrob Agents Chemother*, **36**, 1639–1643.

Muratani, T., Akasaka, S., Kobayashi, T., Yamada, Y., Inatomi, H., Takahashi, K. and Matsumoto, T., (2001), Outbreak of cefozopran (penicillin, oral cephems, and aztreonam)-resistant *Neisseria gonorrhoeae* in Japan. *Antimicrob Agents Chemother*, **45**, 3603–3606.

Naamara, W., Plummer, F.A., Greenblatt, R.M., D'Costa, L.J., Ndinya-Achola, J.O. and Ronald, A.R., (1987), Treatment of chancroid with ciprofloxacin. A prospective, randomized clinical trial. *Am J Med*, **82**, 317–320.

Nanda, N., Michel, R.G., Kurdgelashvili, G. and Wendel, K.A., (2006), Trichomoniasis and its treatment. *Expert Rev Anti Infect Ther*, **4**, 125–135.

National Committee for Clinical Laboratory Standards (1993) Approved standard M7-A3. Methods for Dilution Antimicrobial Susceptibility Tests for Bacteria That Grow Aerobically. Villanova, PA, National Committee for Clinical Laboratory Standards.

National Committee for Clinical Laboratory Standards (1998) Approved Standard M100-38. Performance Standards for Antimicrobial Susceptibility Testing. Wayne, PA, National Committee for Clinical Laboratory Standards.

Newman, L., Moran, J.S. and Workowski, K.A., (2007), Update on the management of gonorrhea in adults in the United States. *Clin Infect Dis*, 44, S84–S101.

Oggioni, M.R., Dowson, C.G., Smith, J.M., Provvedi, R. and Pozzi, G., (1996), The tetracycline resistance gene tet(M) exhibits mosaic structure. *Plasmid*, 35, 156–163.

Pan, W. and Spratt, B.G., (1994), Regulation of the permeability of the gonococcal cell envelope by the mtr system. *Mol Microbiol*, 11, 769–775.

Perez, S., Fernandez-Verdugo, A., Perez, F. and Vazquez, F., (2001), Prevalence of 5-nitroimidazole-resistant *Trichomonas vaginalis* in Oviedo, Spain. *Sex Transm Dis*, 28, 115–116.

Perine, P., Morton, R., Piot, P., Siegel, M.S. and Antal, G.M., (1979), Epidemiology and treatment of penicillinase-producing *Neisseria gonorrhoeae*. *Sex Transm Dis*, 6(Suppl), S152–S158.

Rekart, M., Patrick, D., Jolly, A., Wong, T., Morshed, M., Jones, H., Montgomery, C., Knowles, L., Chakraborty, N. and Maginley, J., (2000), Mass treatment/prophylaxis during an outbreak of infectious syphilis in Vancouver, British Columbia. *Can Commun Dis Rep*, 26, 101–105.

Reyes, M., Shaik, N.S., Graber, J.M., Nisenbaum, R., Wetherall, N.T., Fukuda, K. and Reeves, W.C., (2003), Acyclovir-resistant genital herpes among persons attending sexually transmitted disease and human immunodeficiency virus clinics. *Arch Intern Med*, 163, 76–80.

Riedner, G., Rusizoka, M., Todd, J., Maboko, L., Hoelscher, M., Mmbando, D., Samky, E., Lyamuya, E., Mabey, D., Grosskurth, H. and Hayes, R., (2005), Single-dose azithromycin versus penicillin G benzathine for the treatment of early syphilis. *N Engl J Med*, 353, 1236–1244.

Roberts, M.C., (1989), Plasmid-mediated Tet M in *Haemophilus ducreyi*. *Antimicrob Agents Chemother*, 33, 1611–1613.

Roberts, M.C., (1990), Characterization of the Tet M determinants in urogenital and respiratory bacteria. *Antimicrob Agents Chemother*, 34, 476–478.

Roberts, M.C., Actis, L.A. and Crosa, J.H., (1985), Molecular characterization of chloramphenicol-resistant *Haemophilus parainfluenzae* and *Haemophilus ducreyi*. *Antimicrob Agents Chemother*, 28, 176–180.

Roberts, M.C. and Knapp, J.S., (1988), Transfer of beta-lactamase plasmids from *Neisseria gonorrhoeae* to *Neisseria meningitidis* and commensal Neisseria species by the 25.2-megadalton conjugative plasmid. *Antimicrob Agents Chemother*, 32, 1430–1432.

Robinson, S.C., (1962), *Trichomonas vaginitis* resistant to metronidazole. *Can Med Assoc J*, 86, 665.

Ronald, A.R. and Plummer, F.A., (1985), Chancroid and *Haemophilus ducreyi*. *Ann Intern Med*, 102, 705–707.

Rutanarugsa, A., Vorachit, M., Polnikorn, N. and Jayanetra, P., (1990), Drug resistance of *Haemophilus ducreyi*. *Southeast Asian J Trop Med Public Health*, 21, 185–193.

Sanson-Le Pors, M.J., Casin, I.M. and Collatz, E., (1985), Plasmid-mediated aminoglycoside phosphotransferases in *Haemophilus ducreyi*. *Antimicrob Agents Chemother*, 28, 315–319.

Sarafian, S.K., Johnson, S.R., Thomas, M.L. and Knapp, J.S., (1991), Novel plasmid combinations in *Haemophilus ducreyi* isolates from Thailand. *J Clin Microbiol*, 29, 2333–2334.

Sarafian, S.K. and Knapp, J.S., (1992), Molecular epidemiology, based on plasmid profiles, of *Haemophilus ducreyi* infections in the United States. Results of surveillance, 1981–1990. *Sex Transm Dis*, 19, 35–38.

Schmid, G., Narcisi, E., Mosure, D., Secor, W.E., Higgins, J. and Moreno, H., (2001), Prevalence of metronidazole-resistant *Trichomonas vaginalis* in a gynecology clinic. *J Reprod Med*, 46, 545–549.

Schmid, G.P., (1999), Treatment of chancroid, 1997. *Clin Infect Dis*, 28(Suppl 1), S14–S20.

Schmid, G.P., Sanders, Jr, L.L., Blount, J.H. and Alexander, E.R., (1987), Chancroid in the United States. Reestablishment of an old disease. *JAMA*, 258, 3265–3268.

Schulte, J.M., Martich, F.A. and Schmid, G.P., (1992), Chancroid in the United States, 1981–1990: evidence for underreporting of cases. *MMWR CDC Surveill Summ*, 41, 57–61.

Schulte, J.M. and Schmid, G.P., (1995), Recommendations for treatment of chancroid, 1993. *Clin Infect Dis*, 20(Suppl 1), S39–S46.

Schwarcz, S.K., Zenilman, J.M., Schnell, D., Knapp, J.S., Hook, III, E.W., Thompson, S., Judson, F.N. and Holmes, K.K., (1990), National surveillance of antimicrobial resistance in *Neisseria gonorrhoeae*. The Gonococcal Isolate Surveillance Project. *JAMA*, 264, 1413–1417.

Schwebke, J.R. and Barrientes, F., (2006), Prevalence of *Trichomonas vaginalis* isolates with resistance to metronidazole and tinidazole. *Antivir Chem Chemother*, 50, 4209–4210.

Shafer, W.M., Balthazar, J.T., Hagman, K.E. and Morse, S.A., (1995), Missense mutations that alter the DNA-binding domain of the MtrR protein occur frequently in rectal isolates of *Neisseria gonorrhoeae* that are resistant to faecal lipids. *Microbiology*, 141, 907–911.

Short, H.B., Ploscowe, V.B., Weiss, J.A. and Young, F.E., (1977), Rapid method for auxotyping multiple strains of *Neisseria gonorrhoeae*. *J Clin Microbiol*, 6, 244–248.

Solomon, A.W., Mohammed, Z., Massae, P.A., Shao, J.F., Foster, A., Mabey, D.C. and Peeling, R.W., (2005), Impact of mass distribution of azithromycin on the antibiotic susceptibilities of ocular *Chlamydia trachomatis*. *Antimicrob Agents Chemother*, 49, 4804–4806.

Somani, J., Bhullar, V.B., Workowski, K.A., Farshy, C.E. and Black, C.M., (2000), Multiple drug-resistant *Chlamydia trachomatis* associated with clinical treatment failure. *J Infect Dis*, 181, 1421–1427.

Sparling, P., (1977), Antibiotic resistance in the gonococcus. In R. Roberts, editor, *The gonococcus*. New York, Wiley, pp. 111–136.

Sparling, P.F. and Handsfield, H.H., (2000), *Neisseria gonorrhoeae*. In G. Mandell and J. Bennett and R. Dalin, editors, *Principles and Practice of Infectious Diseases*, 5th Edition, New York, Churchill Livingstone, pp. 2242–2258.

Stamm, L.V., Stapleton, J.T. and Bassford, Jr, P.J., (1988), In vitro assay to demonstrate high-level erythromycin resistance of a clinical isolate of *Treponema pallidum*. *Antimicrob Agents Chemother*, 32, 164–169.

Sutton, M.Y., Sternberg, M.R., Kouman, E., McQuillan, G., Berman, S. and Markowitz, L.E., (2006), Prevalence of *Trichomonas vaginalis* in the United States, 2001–2002. Program and Abstracts, 2006 National STD Prevention Conference, Jacksonville, FL.

Tapsall, J.W., (2002), Current concepts in the management of gonorrhoea. *Expert Opin Pharmacother*, 3, 147–157.

Tapsall, J.W., Shultz, T.R., Limnios, E.A., Donovan, B., Lum, G. and Mulhall, B.P., (1998), Failure of azithromycin therapy in gonorrhea and discorrelation with laboratory test parameters. *Sex Transm Dis*, 25, 505–508.

Taylor, D.N., Pitarangsi, C., Echeverria, P., Panikabutra, K. and Suvongse, C., (1985), Comparative study of ceftriaxone and trimethoprim–sulfamethoxazole for the treatment of chancroid in Thailand. *J Infect Dis*, 152, 1002–1006.

Totten, P.A., Handsfield, H.H., Peters, D., Holmes, K.K. and Falkow, S., (1982), Characterization of ampicillin resistance plasmids from *Haemophilus ducreyi*. *Antimicrob Agents Chemother*, 21, 622–627.

Trees, D.L. and Morse, S.A., (1995), Chancroid and *Haemophilus ducreyi*: an update. *Clin Microbiol Rev*, 8, 357–375.

Tyndall, M., Malisa, M., Plummer, F.A., Ombetti, J., Ndinya-Achola, J.O. and Ronald, A.R., (1993), Ceftriaxone no longer predictably cures chancroid in Kenya. *J Infect Dis*, 167, 469–471.

Tyring, S.K., Baker, D. and Snowden, W., (2002), Valacyclovir for herpes simplex virus infection: long-term safety and sustained efficacy after 20 years' experience with acyclovir. *J Infect Dis*, 186(Suppl 1), S40–S46.

Upcroft, P. and Upcroft, J.A., (2001), Drug targets and mechanisms of resistance in the anaerobic protozoa. *Clin Microbiol Rev*, 14, 150–164.

Wang, S.A., Lee, M.V., O'Connor, N., Iverson, C.J., Ohye, R.G., Whiticar, P.M., Hale, J.A., Trees, D.L., Knapp, J.S., Effler, P.V. and Weinstock, H.S., (2003), Multidrug-resistant *Neisseria gonorrhoeae* with decreased susceptibility to cefixime—Hawaii, 2001. *Clin Infect Dis*, 37, 849–852.

Wang, S.A., Papp, J.R., Stamm, W.E., Peeling, R.W., Martin, D.H. and Holmes, K.K., (2005), Evaluation of antimicrobial resistance and treatment failures for *Chlamydia trachomatis*: a meeting report. *J Infect Dis*, 191, 917–923.

Wasserheit, J.N., (1992), Epidemiological synergy. Interrelationships between human immunodeficiency virus infection and other sexually transmitted diseases. *Sex Transm Dis*, 19, 61–77.

Waters, L.J., Boag, F.C. and Betournay, R., (2005), Efficacy of azithromycin 1 g single dose in the management of uncomplicated gonorrhoea. *Int J STD AIDS*, 16, 84.

Weinstock, H., Berman, S. and Cates, Jr, W., (2004), Sexually transmitted diseases among American youth: incidence and prevalence estimates, 2000. *Perspect Sex Reprod Health,* **36**(1), 6–10.

Willson, P.J., Albritton, W.L., Slaney, L. and Setlow, J.K., (1989), Characterization of a multiple antibiotic resistance plasmid from *Haemophilus ducreyi. Antimicrob Agents Chemother,* **33**, 1627–1630.

World Health Organization (1995) An Overview of Selected Curable Sexually Transmitted Diseases. Geneva, World Health Organization.

World Health Organization (2001) Global Prevalence and Incidence of Selected Curable Sexually Transmitted Infections: Overview and Estimates. Geneva, World Health Organization.

World Health Organization (2006) Gonococcal antimicrobial surveillance programme. Surveillance of anitbiotic resistance in *Neisseria gonorrhoeae* in the WHO Western Pacific Region, 2004. *Commun Dis Intell,* **30**, 129–132.

Xia, M., Whittington, W.L., Shafer, W.M. and Holmes, K.K., (2000), Gonorrhea among men who have sex with men: outbreak caused by a single genotype of erythromycin-resistant *Neisseria gonorrhoeae* with a single-base pair deletion in the mtrR promoter region. *J Infect Dis,* **181**, 2080–2082.

Yoshida, H., Bogaki, M., Nakamura, M. and Nakamura, S., (1990), Quinolone resistance-determining region in the DNA gyrase gyrA gene of *Escherichia coli. Antimicrob Agents Chemother,* **34**, 1271–1272.

Young, H., Moyes, A. and McMillan, A., (1997), Azithromycin and erythromycin resistant *Neisseria gonorrhoeae* following treatment with azithromycin. *Int J STD AIDS,* **8**, 299–302.

Zarantonelli, L., Borthagaray, G., Lee, E.H. and Shafer, W.M., (1999), Decreased azithromycin susceptibility of *Neisseria gonorrhoeae* due to mtrR mutations. *Antimicrob Agents Chemother,* **43**, 2468–2472.

Chapter 4
Resistance of Gram-Negative Bacilli to Antimicrobials

Patricia A. Bradford and Charles R. Dean

4.1 Background and Epidemiology of the Emerging Problem World Wide with Multiresistant Gram-Negative Bacilli

At the beginning of the twenty-first century, we now find ourselves experiencing a taste of what life was like prior to the advent of the antibiotic age in the twentieth century. We are again faced with life-threatening infections for which there are very few antibiotic options for treatment. However, unlike the previous century, these infections are often caused by gram-negative pathogens (Hujer et al., 2004).

An important factor in the upswing in antibiotic resistance are intensive care units (ICU), which have been deemed "factories for creating, disseminating and amplifying resistance to antibiotics (Carlet et al., 2004)." A majority of patients in the ICU receive antibiotics during their stay, often in combination to attempt to circumvent the development of resistance. For example in the latest report from the National Nosocomial Infections Surveillance (NNIS) System, there was a 47% increase in *Klebsiella pneumoniae* resistant to third-generation cephalosporins isolated in the ICUs in the USA in 2003 compared to the previous four years (CDC, 2004). Furthermore, this study revealed a 20% increase in quinolone resistance in *Pseudomonas aeruginosa*.

Another factor contributing to the increase in antibiotic resistance involves the increasing age of the general population of Western countries. With increasing frequency, geriatric patients reside in long-term care facilities (LTCF). Among the residents of LTCFs, one of the foremost causes of morbidity and mortality is infection (Hujer et al., 2004). The most frequently prescribed antibiotics in LTCFs are oral and parenteral cephalosporins. The excessive use of these antibiotics has caused the emergence of gram-negative pathogens that are resistant to third-generation cephalosporins to become endemic in LTCFs. Many of these patients have an indwelling bladder catheter that can become colonized with a resistant organism (Hujer et al., 2004). A number of hospital outbreaks involving ESBL-producing Enterobacteriaceae have been attributed to patients coming to the hospital out of nursing homes (Bradford et al., 1995; Rice et al., 1990, 1996; Schiappa et al., 1996). An epidemiological survey in Chicago showed that 35 out of 55 hospitalized patients infected or colonized with ceftazidime-resistant Enterobacteriaceae had been admitted to the hospital from a LTCF (Wiener et al., 1999). These studies suggest that

I. W. Fong and K. Drlica (eds.), *Antimicrobial Resistance and Implications for the Twenty-First Century.* © Springer 2008

patients in LCTFs serve as a reservoir for antibiotic resistance that then gets transferred into the hospital setting along with the patient. Several multidrug resistant pathogens are of specific concern. Whereas prior to the 1990s *Acinetobacter baumannii* were almost universally susceptible to broad spectrum antibiotics, during this decade they became increasingly resistant to penicillins, cephalosporins, fluoroquinolones and aminoglycosides (Bergogne-Berezin and Towner, 1996). Thus in recent years, many of these antimicrobials are no longer reliable for treatment of infections caused by this organism. Most notable is the increase in resistance to the carbapenems which are caused by a variety of β-lactamases and changes in penicillin-binding proteins (PBPs) (Nordmann and Poirel, 2002). There are now reports of multidrug resistant *A. baumannii* strains that are susceptible only to polymixin B and colistin (Levin et al., 1998).

The problem of antibiotic resistance has gained recognition in the mainstream media as well as in scientific publications. Before the problem can be tackled, the mechanisms must be understood. The present threats of resistance to currently available therapies for infections with gram-negative bacteria are outlined in this chapter.

4.2 Mechanisms of Resistance

4.2.1 New β-Lactamases

Production of β-lactamase is the most common resistance mechanism against β-lactam antibiotics in gram-negative bacteria. These enzymes hydrolyze the β-lactam ring of all classes of β-lactam antibiotics, thus inactivating the drug. β-lactamases had been in observed in bacterial strains long before penicillin was available for use in the treatment of bacterial infections (Abraham and Chain, 1940). Because most β-lactamases share an active site Ser-XX-Lys motif with PBPs, it is thought that serine β-lactamases evolved from PBPs as protection against β-lactam antibiotics produced by molds in the environment (Ghuysen, 1991).

Over the last 20 years many new β-lactam antibiotics have been developed that were specifically designed to be resistant to hydrolysis by β-lactamases. However, with each new class of β-lactam antibiotics that has been used to treat patients, new β-lactamases have emerged that caused resistance to that drug. The first plasmid-mediated β-lactamase, TEM-1 was initially described in the early 1960s in a single *Escherichia coli* isolate from a patient in Greece (Datta and Kontomichalou, 1965). Since that time, TEM-1 has spread worldwide and is now found in many different species of Enterobacteriaceae, *P. aeruginosa*, *Haemophilus influenzae*, and *Neisseria gonorrhoeae*. Following the increased use of the oxyimino-cephalosporins in the early 1980s, extended-spectrum β-lactamases (ESBLs) emerged. Following the increased use of β-lactam/β-lactamase inhibitor combinations containing clavulanic acid in the late 1980s, inhibitor-resistant TEM enzymes emerged. Following the increased use of cephamycins in the early 1990s, plasmid-mediated AmpC-type enzymes emerged. Finally, following the increased use of carbapenems in the late 1990s, carbapenemases have emerged.

4.2.1.1 ESBLS

Resistance to expanded spectrum β-lactam antibiotics in Enterobacteriaceae is often due to the presence of an ESBL. ESBLs are most often derivatives of TEM or SHV enzymes, and are molecular Class A and belong to Functional Group 2be (Table 4.1) (Bush et al., 1995; Jacoby and Medeiros, 1991; Jacoby and Bush, 2005). With both of these groups of enzymes a few point mutations at selected loci within the structural gene encoding the enzyme give rise to the extended-spectrum phenotype. TEM and SHV-type ESBLs are most often found in *E. coli* and *K. pneumoniae*, however they are also found in *Proteus* spp., *Enterobacter* spp. and other members of the Enterobacteriaceae.

ESBLs continue to be a problem in hospitalized patients worldwide. The incidence of ESBLs in strains of *E. coli* and *K. pneumoniae* varies from 0–40%, depending on the country (Shah et al., 2004). In the USA, the overall incidence is around 3%; however, it can be as high as 40% in some institutions (Jacoby, 2005). Many of the patients infected with ESBL-producing pathogens are in ICUs, although they can occur in surgical wards as well. Organisms with ESBLs are also being isolated with increasing frequency from patients in extended care facilities (Bradford et al., 1995; Wiener et al., 1999). Hospitals have reported outbreaks of ESBL-producing organisms, some of which have been successfully managed using infection control methods and through the restriction of the use of oxyimino-cephalosporins (Lucet et al., 1999; Rahal et al., 1998).

The newest family of plasmid-mediated ESBLs that has arisen preferentially hydrolyzes cefotaxime. Members include the CTX-M type enzymes, CTX-M-1 through CTX-M-40 (Bonnet et al., 2000; Bush et al., 1995; Gazouli et al., 1998b; Jacoby and Bush, 2005; Sabaté et al., 2000) as well as Toho enzymes 1 and 2 (Tzouvelekis et al., 2000). These enzymes are not closely related to TEM or SHV β-lactamases, but instead share a high homology to the Class A chromosomally encoded β-lactamase of *Klebsiella oxytoca*. In addition to the preferential hydrolysis of cefotaxime, another unique feature of these enzymes is that they are inhibited better by tazobactam than the other β-lactamase inhibitors sulbactam and clavulanate (Bradford et al., 1998; Ma et al., 1998). There have been outbreaks of strains expressing these enzymes in Eastern Europe, Argentina, and Japan. In Argentina, CTX-M-2 is the most prevalent ESBL, whereas the TEM and SHV-type ESBLs remain more common in North America, Asia, and Europe (Tzouvelekis et al., 2000). More recently, isolates expressing these enzymes have been identified in Greece, Spain, and Brazil, however the isolates from Greece came from immigrants from the outbreak areas in Eastern Europe (Bonnet et al., 2000; Sabaté et al., 2000). Interestingly, a number of these enzymes have been found among isolates of *Salmonella enterica* ser typhimurium (Bauernfeind et al., 1992, 1996a; Bradford et al., 1998). In a 1994 outbreak of *S.* typhimurium in Morocco, numerous different ESBLs including SHV-type, TEM-type, CTX-M-type enzymes were described (Mhand et al., 1999). Large outbreaks of *S.* typhimurium strains resistant to cefotaxime have been reported in Eastern Europe (Bradford et al., 1998; Gazouli et al., 1998a,b). These strains have all been found to express plasmid-mediated variants of the CTX-M-type β-lactamase. Several cases of *S.* typhimurium strains harboring these enzymes have also been reported in Greece (Tzouvelekis et al., 1998).

Table 4.1 Classification of β-lactamases[a]

Functional Group	Molecular Class	Descriptor	Preferred substrates	Inhibited by		Representative Enzymes
				CA	EDTA	
1	C	AmpC-type β-lactamases	Cephalosporins	–	–	Inducible enzymes from *Enterobacter, Citrobacter, Serratia,* and *Pseudomonas*; plasmid mediated enzymes ACT-1, CMY-type, FOX-type
2a	A	gram-positive penicillinases	Penicillins	+		*S. aureus* PC1 penicillinase
2b	A	Broad spectrum enzymes	Penicillin and narrow-spectrum cephalosporins	+	–	TEM-1, TEM-2, SHV-1, OHIO-1
2be	A	Extended-spectrum β-lactamases	Penicillins, all cephalosporins and monobactams	+	–	TEM-3, TEM-10, TEM-26 etc., SHV-2, SHV-5, SHV-7 etc.; CTX-M-1 to CTX-M-9, *K. oxytoca* K1
2br	A	Inhibitor-resistant β-lactamases	Penicillin and narrow-spectrum cephalosporins	–		TEM-30 to TEM-36 etc., SHV-10
2c	A	Carbenicillin hydrolyzing β-lactamases	Penicillins including carbenicillin	+	–	PSE -1, PSE-3, and PSE-4; *Aeromonas* AER-1; CARB-3 and CARB-4
2d	D	Oxacillin-hydrolyzing β-lactamases	Penicillins including cloxacillin	±	–	OXA-1 to OXA-30
2e	A	Cephalosporinases inhibited by clavulanic acid	Cephalosporins	+	–	*B. fragilis* CepA, *P. vulgaris* FPM-1, *S. maltophilia* L2
2f	A	Non-metallo carbapenemases	Penicillins, cephalosporins, and carbapenems	+	–	IMI-1, NMC-A, Sme-1
3a	B	Metallo-β-lactamases	Penicillins, cephalosporins and carbapenems	–	+	*B. cereus* II, CcrA, IMP-1, L1, VIM-1, VIM-2
3b	B	Metallo-β-lactamases	carbapenems	–	+	CphA, AsbM1, ASA-1, ImiS
3c	B	Metallo-β-lactamases	Cephalosporins, carbapenems	–	+	β-lactamase from *Legionella gormanii*.
4	Not determined	Penicillinase not inhibited by clavulanate	Penicillins	–	–	β-lactamase from *Burkholderia cepacia*, SAR-2

[a] adapted from Ambler (1980), Bush et al. (1995) and Rasmussen and Bush (1997).

Another growing family of ESBLs is the OXA-type enzymes. These enzymes differ from the TEM and SHV enzymes in that they belong to molecular class D and Functional Group 2d (Table 4.1) (Bush et al., 1995). To date, at least 45 OXA variants have been identified, although not all of these possess the ESBL phenotype (Jacoby and Bush, 2005; Siu et al., 2000). Several of the OXA enzymes with extended-spectrum phenotype have been found in *P. aeruginosa* isolated in Turkey (Hall et al., 1993). OXA-11, -14, -16, and -17 are all derivatives of OXA-10 (PSE-2), whereas OXA-15 is a derivative of OXA-2. This group of ESBL provides weak resistance to oxyimino-cephalosporins.

As the prevalence of Enterobacteriaceae-producing ESBLs increases, the greater the need for laboratory testing methods that accurately detect these enzymes in clinical isolates. In the years since ESBLs were first described, a number of different testing methods have been suggested. These include the double-disk approximation test (Jarlier et al., 1988), the three-dimensional test (Thomson and Sanders, 1992), Etest ESBL strips (Cormican et al., 1996), and the Vitek ESBL card (Sanders et al., 1996). These tests employ the β-lactamase inhibitor clavulanate in combination with an oxyimino-cephalosporin such as ceftazidime or cefotaxime. With each of these methods, the clavulanate inhibits the ESBL, thereby reducing the level of resistance to the cephalosporin. Of the tests that have been developed to date, the double-disk approximation test recommended by Jarlier et al. (1988) and the broth-dilution MIC reduction method (CLSI confirmatory test, (CLSI, 2005)) are the easiest and most cost-effective methods for use by many clinical laboratories. While each of these tests has their merit, none accurately detect all strains that produce ESBLs. In a recent survey of detection of ESBLs in clinical isolates, Tenover et al. found that only 18% of laboratories correctly identified challenge organisms as potential ESBL producers and reported the susceptibility of oxyimino-cephalosporins to be resistant (Tenover et al., 2003). Furthermore, false positive results can be obtained with strains expressing a high level of a non-ESBL β-lactamase such as SHV-1 (Miró et al., 1998; Rasheed et al., 1997).

4.2.1.2 Inhibitor-Resistant β-Lactamases

In the early 1990s, β-lactamases resistant to inhibition by clavulanic acid were discovered. Nucleotide sequencing revealed these enzymes to be variants of the TEM-1 or TEM-2 β-lactamases. These enzymes were at first given the designation "IRT" for *I*nhibitor *R*esistant *T*EM β-lactamase; however, all been subsequently renamed with numerical TEM designations. The inhibitor-resistant TEM β-lactamases belong to molecular class A and Functional Group 2br (Table 4.1) (Bush et al., 1995). There are at least 17 distinct inhibitor-resistant TEM β-lactamases (Bermudes et al., 1999). Inhibitor resistant TEMs have been found mainly in clinical isolates of *E. coli*, but also in some strains of *K. pneumoniae*, *K. oxytoca*, *P. mirabilis*, and *Citrobacter freundii* (Bret et al., 1996; Lemozy et al., 1995). Although the inhibitor-resistant TEM variants are resistant to inhibition by clavulanic acid and sulbactam, thereby conferring clinical resistance to the β-lactam/β-lactamase inhibitor combinations of amoxicillin/clavulanate, ticarcillin/clavulanate, and ampicillin/sulbactam, they remain extremely susceptible to inhibition by tazobactam, and subsequently the combination of piperacillin/tazobactam (Bonomo et al., 1997; Chaibi et al., 1999). To date these β-lactamases have primarily been detected in France and a few other locations within Europe; however, there have been a few reports of these enzymes occurring in the USA (Bradford et al., 2004; Chaibi et al., 1999; Kaye et al., 2004).

Point mutations that lead to the inhibitor-resistant phenotype occur at specific amino acid residues within the structural gene of the TEM enzyme, which are Met-69, Arg-244, Arg-275, and Asn-276 (Yang et al., 1999). The sites of these amino acid substitutions are generally distinct from those that lead to the ESBL phenotype, although the TEM-50 enzyme was identified that had mutations common to both ESBLs and inhibitor-resistant TEMs. TEM-50 was weakly resistant to clavulanate and also conferred a slight resistance to the expanded-spectrum cephalosporins (Sirot et al., 1997). In addition to the variants of TEM, inhibitor-resistant variants of SHV-1 and the related enzyme OHIO-1 have been detected (Bonomo et al., 1992; Prinarakis et al., 1997).

4.2.1.3 AmpC

AmpC β-lactamases are naturally occurring, chromosomally encoded enzymes found in *Enterobacter* spp., *Citrobacter* spp., *Serratia* spp., *P. aeruginosa*, and *Hafnei alvei*. In these genera, the β-lactamase is normally expressed at low levels and is induced after exposure to some β-lactam antibiotics. Chromosomally encoded AmpC β-lactamases are under the regulatory control of an activator–repressor gene, *ampR*. Another regulatory gene, *ampD* provides the repressor protein that binds to *ampR*. Certain β-lactam antibiotics can induce the production of increased amounts of β-lactamase via the sensing of recycled cell wall components (Jacobs et al., 1994). AmpC-type β-lactamases, belonging to molecular Class C and Functional group 1, generally confer resistance to oxyimino-cephalosporins such as ceftazidime, cefotaxime and ceftriaxone, and cephamycins such as cefoxitin and cefotetan (Table 4.1) (Bush et al., 1995). As a group, Class C enzymes are not inhibited by the β-lactamase inhibitors clavulanate, sulbactam and tazobactam.

The use of expanded-spectrum cephalosporins has led to resistance caused by the hyperproduction of AmpC in strains where the enzyme is normally produced at low levels. This occurs as a result of derepression of the inducible AmpC in *Enterobacter* spp., *Citrobacter* spp. or *P. aeruginosa* causing the enzyme to be expressed constitutively at high levels. Although they do not possess the AmpR gene to enable induction, hyperproduction of AmpC can also occur in *E. coli* and *Shigella* spp. and cause cephalosporin resistance in these organisms via a promoter mutation (Caroff et al., 1999; Nelson and Elisha, 1999).

In the 1990s AmpC β-lactamases found their way onto mobile elements, especially plasmids, and are being expressed constitutively at high levels in *K. pneumoniae* and *E. coli*. The MIR-1 and ACT-1 β-lactamases appear to have originated from the AmpC enzyme of *Enterobacter cloacae* (Bradford et al., 1997; Jacoby, 1999); BIL-1, CMY-2, -3,-4, -5 and LAT-1, -2, -3, -4 from *C. freundii* (Bush et al., 1995; Gazouli et al., 1998c); CMY-1, FOX-1, -2, -3, -4 and MOX-1 are from *P. aeruginosa* (Bauernfeind et al., 1996b; Bou et al., 2000; Bush et al., 1995); DHA-1 from *Morganella morganii* (Barnaud et al., 1998) and ACC-1 is from *H. alvei* (Bauernfeind et al., 1999; Girlich et al., 2000). The ACC-1 enzyme was recently implicated in an outbreak of *K. pneumoniae* in an ICU in Tunisia (Nadjar et al., 2000). Most of the plasmid-mediated AmpC β-lactamases have lost the *ampR* regulatory gene and are therefore produced constitutively at high levels. Of the plasmid-mediated Amp-C-type β-lactamases described to date, only the plasmid harboring DHA-1 retained the *ampR* region necessary to retain the inducible phenotype (Barnaud et al., 1998). Although most AmpC enzymes are not inhibited by any of the commercially available β-lactamase inhibitors, a unique feature of the CMY-2 enzyme and related β-lactamases is that they are susceptible to inhibition by tazobactam.

Plasmid-mediated AmpC enzymes have also been detected among *Salmonella* spp. isolates. In one case report, it was demonstrated that a strain of *Salmonella* spp. expressing the CMY-2 β-lactamase was apparently transferred to a child from infected cattle (Fey et al., 2000). It has also been recognized that most strains of *S. enterica* ser. Typhimurium phage type DT104 possess a transposon that harbors the PSE-1 β-lactamase (Buerra et al., 2000). Previously ampicillin and more recently, ceftriaxone have been the antibiotics of choice to treat serious infections caused by *Salmonella* spp. The appearance of ESBLs and AmpC-type β-lactamases in this species will now require alternative therapeutic approaches.

In addition to these plasmid-mediated AmpC enzymes conferring general resistance to cephalosporins and cephamycins, they can also contribute to carbapenem resistance when combined with the loss of outer-membrane porin proteins in *K. pneumoniae* (Bradford et al., 1997; Martínez-Martínez et al., 1999). These strains are particularly disturbing because imipenem has been considered to be the drug of choice for treatment of infections caused by multiply resistant Enterobacteriaceae. This has lead to cases where there was no commercially available antibiotic with which to treat these patients (Bradford et al., 1997).

4.2.1.4 Carbapenemases

Reports of carbapenem-hydrolyzing β-lactamases have been increasing over the last few years. This phenotypic grouping of enzymes is comprised of a rather heterogeneous mixture of β-lactamases belonging to either molecular Class A, which has a serine in the active site of the enzyme, or molecular Class B, the metallo-β-lactamases which have zinc in the active site (Table 4.1) (Walsh et al., 2005). The Class A, functional group 2f carbapenem-hydrolyzing β-lactamases have been found in *E. cloacae* (IMI-1 and NMC-A) (Nordmann et al., 1993; Rasmussen et al., 1996), *Serratia marcescens* (Sme-1-3) (Bush, 1998; Rasmussen and Bush, 1997; Yang et al 1990) *K. pneumoniae* (KPC-1-3) (Bradford et al., 2004; Yigit et al., 2001), *P. aeruginosa* (GES-2) (Poirel et al., 2001c) and *A. baumannii* (Afzal-Shah et al., 1999). They are inhibited by clavulanate, but not by EDTA. While these enzymes hydrolyze imipenem well enough to provide resistance, they show much higher hydrolysis rates for ampicillin and cephaloridine and have been found in strains that also possess an AmpC β-lactamase. This combination of enzymes provides resistance to a broad range of β-lactams (Rasmussen and Bush, 1997). For the first ten years after their initial discovery, the class A carbapenemases remained extremely rare. However, in recent years, there have been reports of *Klebsiella* spp., *Enterobacter* spp. and *Salmonella enterica* ser Cubana expressing KPC-type enzymes in cities in the Eastern USA (Bradford et al., 2004; Bratu et al., 2005a; Hossain et al., 2004; Miriagou et al., 2003; Woodford et al., 2004; Yigit et al., 2003). In one study, *K. pneumoniae* expressing the KPC-2 carbapenem-hydrolyzing β-lactamase were recovered from seven hospitals in the New York City boroughs of Brooklyn and Queens (Bradford et al., 2004). Among the isolates was a predominant ribotype that was common to all seven of the hospitals; however, no epidemiologic link among the various institutions was observed. Following this observation, a city-wide surveillance study of carbapenem-resistant *K. pneumoniae* revealed that the KPC-2 enzyme was present in patient isolates in seven hospitals in Brooklyn, NY and this strain was the cause of outbreaks in two of the hospitals (Bratu et al., 2005b).

The Class B metallo-enzymes are inhibited by EDTA, but not by the β-lactamase inhibitors (Bush et al., 1995; Bush, 1998). These β-lactamases have been described as naturally occurring, chromosomally mediated enzymes in diverse genera of gram-positive and gram-negative bacteria such as *Bacillus cereus*, *Burkholderia cepacia*, *Bacteroides fragilis*, *Chryseobacterium* spp., *Legionella gormanii*, *Flavobacterium odoratum*, *Stenotrophomonas maltophilia*, and several different species of *Aeromonas* (Bush, 1998; Rasmussen and Bush, 1997). Metallo-β-lactamases can all hydrolyze imipenem at a measurable rate; however, they differ in their abilities to hydrolyze other β-lactam substrates (Rasmussen and Bush, 1997). Enzymes in Functional subgroup 3a have the ability to hydrolyze penicillins or cephalosporins as well as imipenem and include the *B. cereus* enzyme, CcrA from *B. fragilis*, the L1 enzyme from *S. maltophilia* and the plasmid-mediated enzymes, IMP-type, VIM-type, SPM-1, and GIM-1 (Walsh et al., 2005). These β-lactamases confer resistance to all classes of β-lactam antibiotics except for the monobactams. Enzymes in subgroup 3b preferentially hydrolyze carbapenems and include the enzymes found in *F. odoratum* and *B. cepacia* and the AsbM1 and CphA β-lactamases from *Aeromonas* spp. (Rasmussen and Bush, 1997). In addition to the zinc present in the active site of the enzyme, this subgroup requires additional zinc cations for maximal catalytic activity (Bush, 1998). Although the metallo-β-lactamase in these organisms does not confer resistance to penicillins or cephalosporins, these organisms are found in combination with other chromosomally mediated β-lactamases that do hydrolyze these β-lactam substrates. This combination of metallo- and serine-based β-lactamases confers resistance to all classes of β-lactam antibiotics. Although attempts have been made to develop inhibitors that would inactivate the metallo-β-lactamases, there is currently no β-lactamase inhibitor available that can be used against these enzymes (Payne et al., 1997).

A particularly disturbing event that occurred in Japan in the early 1990s was the discovery of a metallo-β-lactamase, IMP-1, which was encoded on plasmids. These β-lactamase producing plasmids were readily transferable to other strains.

This enzyme was originally found in *S. marcescens* , but has also been found in *K. pneumoniae* and *P. aeruginosa* (Osano et al., 1994; Watanabe et al., 1991). For a number of years, this enzyme was contained within that country. However, in recent years, the IMP-1 β-lactamase and at least 18 related enzymes have been identified, first in other Asian countries, then more recently in Europe, and finally in North America (Jacoby and Bush, 2005; Walsh et al., 2005). Interestingly, some of the more recent reports of the IMP-type of metallo-β-lactamases have been in isolates of *Citrobacter* spp., *E. cloacae*, *K. pneumoniae* , *Shigella flexneri*, *A. baumannii*, *Pseudomonas putida*, and *Achromobacter xylosoxidans* (Walsh et al., 2005). Most of the genes encoding the plasmid-mediated metallo-β-lactamases are also found on gene cassettes in class 1 or class 3 integrons (Walsh et al., 2005).

Another growing family of plasmid-mediated metallo-β-lactamases is the VIM-type enzymes, of which there are now at least 10 derivatives of VIM-1 (Jacoby and Bush, 2005). Unlike the IMP-type enzymes which came from Asia, the VIM-type (*Veronese imi*penemase) enzymes were first found in Verona, Italy in 1997 (Lee et al., 2002). VIM-1 shares 39% amino acid identity with BCII from *B. cereus,* but is not closely related to other metallo-β-lactamases (Lauretti et al., 1999). In the last few years VIM-type enzymes have been reported throughout Europe as well as South East Asia and North America in strains of *P. aeruginosa, K. pneumoniae, E. cloacae* (Walsh et al., 2005). VIM-type enzymes are often harbored in gene cassettes and are also associated with integrons (Giakkoupi et al., 2003; Lauretti et al., 1999).

A few other plasmid-mediated β-lactamases have also been described. In 1997, SPM-1 (*Sao Paulo M*BL) was found in a clinical isolate of *P. aeruginosa* from Brazil (Toleman et al., 2002). SPM-1 is most closely related to IMP-1, but shares only 35.5% identity (Walsh et al., 2005). Also originating from *P. aeruginosa*, in 2002 the GIM-1 (*German imi*penemase) metallo-β-lactamase was found in five patients in a medical center in Germany (Castanheira et al., 2004). GIM-1 shares approximately 43% homology to the IMP-1, -4 and –6 β-lactamases (Walsh et al., 2005). In addition, a new plasmid-mediated metallo-β-lactamase, designated MET-1, was recently discovered in a strain of *S. flexneri* (O'Hara et al., 1998). Transferable resistance to imipenem has also been demonstrated by a plasmid-mediated enzyme similar to CcrA in *B. fragilis* (Bandoh et al., 1992).

Detection of metallo-β-lactamases in clinical isolates can be performed by phenotypic or genetic methods. Unfortunately, many strains carrying metallo-β-lactamases do not result in overt resistance to carbapenems. For example the MIC of imipenem may be only 1–2 µg/ml in isolates of Enterobacteriaceae. Therefore, it is important to be cognizant of the potential for the presence of a metallo-β-lactamase in strains with slightly elevated MICs of imipenem or meropenem. However, discrepancies between various testing methods and automated systems have been noted in detecting carbapenem resistance (Giakkoupi et al., 2005). Several laboratory tests have been proposed to aid in the detection of metallo-β-lactamases. Etest strips (AB Biodisk, Solna Sweden) have been developed that have a gradient of imipenem on one end and imipenem plus EDTA on the other end. A positive test is interpreted as a decrease in the MIC of imipenem by at least three- twofold dilutions in the presence of EDTA. An evaluation of these strips showed that the Etest MBL strips had a sensitivity of 94% and a specificity of 95% in detecting metallo-β-lactamases in clinical isolates (Walsh et al., 2002). However, it was noted in another study that the sensitivity dropped to 36.7% when the isolates bearing metallo-β-lactamases were not resistant to imipenem (Yan et al., 2004a). A double-disk method using ceftazidime or cefepime with or without clavulanate in proximity to a disk containing either EDTA or 2-mercaptopropionic acid (MPA) was also evaluated. Good sensitivity and specificity were noted, especially when the results of both cefepime and ceftazidime were used (Yan et al., 2004a).

The continuing volume of publications on new β-lactamases suggests that bacterial pathogens are continuing to develop new mechanisms of resistance at an alarming rate. The introduction of new classes of β-lactam antibiotics has been likewise met with the emergence of new β-lactamases. Some of these new β-lactamases, like the TEM and SHV-type ESBLs result from simple point mutations in existing β-lactamase genes that lead to a changed substrate profile. Other new β-lactamases such as the plasmid-mediated AmpC-type enzymes have been borrowed from the chromosomally encoded AmpC that naturally occur in several species of Enterobacteriaceae and *P. aeruginosa*. Strains expressing these β-lactamases will all present a host of therapeutic challenges as we head into the twenty-first century.

4.2.2 *Quinolones*

Quinolones and the related fluoroquinolones, introduced in the 1960s and 1970s respectively, display a broad spectrum of antibacterial activity including gram-negative, gram-positive and some atypical coverage (King et al., 2000). The use of first-generation quinolones (e.g. nalidixic acid) was restricted generally to treating urinary tract infections, due largely to sub-optimal systemic distribution and somewhat limited activity. The first fluoroquinolone derivatives (e.g. norfloxacin and ciprofloxacin) showed improved tissue distribution and enhanced gram-negative antibacterial activity, and were approved in the late 1980s for a more broad clinical application. Since then, additional fluoroquinolone derivatives have been developed including ofloxacin, enoxacin, lomefloxacin, levofloxacin, trovafloxacin, gatifloxacin, moxifloxacin, sparfloxacin and grepafloxacin, with improved gram positive and atypical coverage.

Quinolone antibiotics are bactericidal and act by inhibiting the gyrase and topoisomerase IV enzymes that control DNA topology and play essential roles in DNA replication, transcription and recombination (Drlica and Malik, 2003; Zechiedrich et al., 2000). The DNA gyrase holoenzyme tetramer, consisting of two subunits each of the GyrA and GyrB proteins, introduces negative superhelicity into DNA, required for initiation of replication and replication fork movement, and transcription (Champoux, 2001; Levine et al., 1998). Topoisomerase IV, also a tetrameric holoenzyme, consisting of two subunits each of ParC and ParE, functions in relaxing both positive and negative supercoils and mediates decatenation (unlinking) of the replicated chromosome copies to allow partitioning upon cell division (Adams et al., 1992; Kato et al, 1990; Khodursky et al., 1995; Peng and Marians, 1993a, 1993b; Zechiedrich and Cozzarelli, 1995). Both enzymes are type II topoisomerases and function by introducing double stranded breaks in DNA and passing DNA strands/helices through each other. This is accomplished by forming the transient, so-called "cleaved complex" where the enzyme is covalently linked to the DNA and serves as a bridge between the DNA ends, allowing strand breakage, strand passage and resealing (Champoux, 2001). Quinolones have been shown to inhibit gyrase and topoisomerase IV by forming stable ternary complexes with the enzymes bound to DNA (Gellert, et al., 1977; Shen et al., 1989a–c; Shen, 1994, 2001; Sugino et al., 1977). This traps the complex as a bridge between DNA ends, and subsequent loss of the complex without rejoining the ends introduces double stranded DNA breakage (DSB) (Chen et al., 1996; Drlica and Zhao, 1997) leading to cell death. Trapped complexes may also collide with replication forks, halting DNA synthesis (Cox et al., 2000; Hiasa et al., 1996; Marians and Hiasa, 1997; Michel et al., 1997, 2004; Willmott and Maxwell, 1993) and interfere with transcription by blocking RNA polymerase (Willmott et al., 1994) and disrupting the action of DNA helicases (Shea and Hiasa, 1999). These latter activities are likely to be growth inhibitory rather that cidal (Drlica and Malik, 2003), and recent work has shown that quinolones can rapidly kill bacteria even when DNA replication is inhibited, indicating that double strand DNA breakage associated with rapid cell killing is likely not mediated by replication forks colliding with trapped complexes (i.e. it is independent of ongoing DNA replication) (Zhao et al., 2006).

There is homology between the GyrA and ParC, and GyrB and ParE subunits of gyrase and topoisomerase IV, and correspondingly quinolones can inhibit the activities of both enzymes. For many years gyrase was the only known target of quinolones (Drlica, 1984) but the later discovery and characterization of topoiso-merase IV revealed it was a bona fide target as well (Belland et al., 1994; Ferrero et al., 1994; Kato et al., 1990, 1992; Khodursky et al., 1995; Peng and Marians, 1993a). There is however an apparent difference in which of the two enzymes con-stitute the primary and secondary target of quinolones in different bacteria. Overall, DNA gyrase is the more sensitive target in gram negatives, while topoisomerase IV is the more sensitive target in gram positives, although the ratio of sensitivity of these complexes is organism specific (Drlica and Malik, 2003; Hooper, 1999).

By becoming widely available in the late 1980s and through continued develop-ment, fluoroquinolones have become one of the most broadly used classes of antibi-otic (Hooper, 2000; Norrby and Lietman, 1993). This reflects their broad spectrum of activity and tissue penetration, which makes them ideal for empirical therapy. Furthermore, their exposure in the lung makes them useful for treatment of respira-tory tract infections (Croom and Goa, 2003; Grossman et al., 2005; Hooper, 2000), including chronic infections in cystic fibrosis, which currently comprises one of the largest categories of infectious diseases (Gibson et al., 2003). In addition, increasing amounts of quinolones are used in food animals. Reflecting this widespread use, resistance to quinolones is becoming increasingly encountered. The emergence of resistance to quinolones in can be complex due to (a) the presence of two potential intracellular targets (gyrase and topoisomerase IV) with different intrinsic suscepti-bilities to the inhibitory effect of quinolones, (b) the near universal role of efflux in both intrinsic and acquired multidrug resistance and (c) the possibility of plasmid mediated topoisomerase protection, which has been found among members of the

Enterobacteriaceae (Robicsek et al., 2005). Each of these, and the interplay between them, are discussed below.

4.2.2.1 Target Mutations Conferring Quinolone Resistance

The most well-characterized mechanism conferring resistance to quinolones in both gram positives and gram negatives is target alteration resulting from chromosomal mutations in the *gyrA* and/or *parC* genes, with mutations in *gyrB* and *parE* less frequently observed (Hooper, 2001). For all subunits, regions defined as quinolone resistance determining regions (QRDR) have been characterized. In gram negatives, gyrase is usually the primary target and therefore mutations in *gyrA* have been frequently encountered. Mutations in *gyrA* generally cluster in the QRDR corresponding to nucleotides 199–318 of *E. coli gyrA* (amino acids 67–106). This region encompasses the amino-terminal active site tyrosine which interacts covalently with transiently broken DNA and comprises a quinolone binding domain (Morais Cabral et al., 1997; Yoshida et al., 1990). One frequently encountered mutation in this region, Ser-83 to Trp, has little impact on gyrase function but reduces norfloxacin binding substantially (Willmott and Maxwell, 1993). The ParC subunit of topoisomerase IV is highly similar to GyrA and as such possesses a corresponding QRDR (Drlica and Zhao, 1997; Zhao et al., 1997). Although topoisomerase IV is often the secondary target of quinolones, mutations in this QRDR, in conjunction with single or accumulated GyrA mutations, are now understood to play an important role in the development of higher levels of fluoroquinolone resistance in many cases (Belland et al., 1994; Chen et al., 2003; Heisig, 1996; Li et al., 2004a; Shultz et al., 2001). Mutations in the GyrB/ParE subunits are much less common but have been associated with increased resistance to quinolones (Le Thomas et al., 2001; Li et al., 2004b; Rafii et al., 2005) and a QRDR has been described (Yoshida et al., 1991). Mutations generally cluster in a region mediating interaction with their cognate GyrA/ParC subunits.

4.2.2.2 Plasmid-Mediated Quinolone Resistance: Topoisomerase Protection and Quinolone Modification

In 1998, a transmissible, plasmid-borne resistance determinant (*qnrA*) was identified from a quinolone resistant *K. pneumoniae* clinical isolate (Martinez-Martinez et al., 1998) providing confirmation of previously suspected plasmid mediated resistance (Courvalin, 1990). QnrA and recently identified putative homologs (Cheung et al., 2005; Hata et al., 2005; Robicsek et al., 2005) are members of a family of proteins termed the pentapeptide repeat family (Tran and Jacoby, 2002) and apparently confer protection from quinolones (Tran and Jacoby, 2002) by a mechanism involving binding to gyrase (Tran et al., 2005b) and topoisomerase IV (Tran et al., 2005a). Binding is not dependent on DNA or ATP and is presumed to occur prior to establishment of the ternary complex favored by quinolones, thereby blocking or lessening quinolone interaction with the topisomerases. This mechanism has been increasingly identified in members of the Enterobacteriaceae (Cheung et al., 2005; Hata et al., 2005; Mammeri et al., 2005; Robicsek et al., 2005; Rodriguez-Martinez et al., 2003; Wang et al., 2003, 2004). Another member of the pentapeptide repeat family, MfpA, reduces susceptibility to quinolones in *Mycobaterium smegmatis* (Montero et al., 2001). QnrA and related proteins generally increase quinolone resistance only marginally but provide additive resistance with target or efflux mediated mechanisms and likely contribute to the development of clinically relevant resistance.

Intriguingly, a variant of the aminoglycoside modifying enzyme *aac(6')-1b* (see Section 4.2.3.1 on aminoglycoside modifying enzymes), also present on several plasmids harboring the *qnrA* determinant described above, was recently shown to contribute to fluoroquinolone resistance (Robicsek et al., 2006). This variant, designated *aac(6')-1b-cr*, modifies ciprofloxacin and norfloxacin by acetylation of the amino substituent of the piperazinyl ring, thereby reducing their activity. This activity is therefore limited to those fluoroquinolones having the unmodified piperazinyl structure and, correspondingly, no effect on levofloxacin or moxifloxacin was seen. The level of resistance conferred by *aac(6')-1b-cr* alone was modest, however, more significant levels of resistance were observed when *aac(6')-1b-cr* and *qnrA* were found together. Furthermore, the presence of plasmid-borne *aac(6')-1b-cr* in *E. coli* resulted in a greater recovery of resistant mutants during exposure to ciprofloxacin. This again underscores the potential contribution of determinants such as *aac(6')-1-cr* and *qnrA* to the stepwise acquisition of clinically significant resistance. Finally, this altered form of *aac(6')-1b* is significant in that, based on two amino acid substitutions, it has acquired the ability to reduce the activity of a different class of antibiotic, without a substantial loss of activity against the original aminoglycoside substrates. The recent, and apparently wide, geographic penetration of this novel resistance gene suggests that its emergence was selected by the increased use of fluorquinolones during the 1990s. It might therefore have been a significant contributor to resistance development in those populations from where it was isolated, and the dual activity of this resistance determinant might help explain the association often seen between fluoroquinolone resistance and aminoglycoside resistance (Robicsek et al., 2006).

4.2.2.3 Efflux Mediated Quinolone Resistance in Gram-Negative Bacteria

As described in Section 4.2.5, gram-negative RND family efflux pumps contribute significantly to intrinsic and mutationally acquired resistance to many antibiotics and biocides. In the case of certain organisms such as *P. aeruginosa*, the effect of multiple pumps working synergistically with a notably less permeable outer membrane can be an astonishingly effective barrier to the entry of a wide range of noxious compounds including antibacterials (Poole, 2002, 2004). Fluoroquinolones are intriguing and unique in that they have good activity against gram-negative pathogens, including those such as *P. aeruginosa* that exhibit particularly problematic efflux. This indicates that efflux generally provides less intrinsic resistance to these agents. There has been limited analysis suggesting that compound hydrophobicity is correlated with extrusion by RND family pumps (Nikaido et al., 1998) indicating that the hydrophilic (charged) nature of fluoroquinolones (Nikaido and Thanassi, 1993) might underlie the reduced impact of efflux vis a vis intrinsic resistance. Nonetheless, although fluoroquinolones are apparently less subject to the absolute levels of efflux-mediated intrinsic resistance that plagues many antibiotics, they have been shown to be substrates of a wide range of RND family pumps (Poole, 2004). These include the Mex pumps in *P. aeruginosa* (Poole, 2000), the AcrAB–TolC pump of *E. coli* (Jellen-Ritter and Kern, 2001) and *Salmonella* spp. (Baucheron et al., 2002, 2004; Giraud et al., 2003), the AcrEF pumps of *E. coli* (Jellen-Ritter and Kern, 2001) and *Salmonella enterica* (Olliver et al., 2005) and the CmeABC pump of *Campylobacter jejuni* (Ge et al., 2005). Given that quinolones are general pump substrates, it is not surprising that mutations in regulatory genes leading to overexpression of pumps such as MexAB–OprM and AcrAB–TolC can be readily selected by in vitro exposure to quinolones or other pump substrates (Kohler et al., 1997; Poole, 2000), and the resulting increases in pump expression can decrease susceptibility to fluoroquinolones. Efflux is now generally recognized as an important determinant of fluoroquinolone resistance in P. aeruginosa and other gram negatives (Poole, 2000). Efflux pump mutants are routinely found among clinical isolates (Kriengkauykiat et al., 2005; Poole, 2000), and although these can be selected by many different drug classes, there appears to be a correlation in some cases between fluoroquinolone treatment histories and the emergence of pump mutants (Kriengkauykiat et al., 2005; Le Thomas et al., 2001). Since RND pumps usually provide resistance to multiple antibiotics, selection of pump overexpression during previous treatment with other pump substrate antibiotics will result in pump-mediated resistance to fluoroquinolones and vice versa. RND family efflux pumps function in concert with the outer membrane permeability barrier (Li et al., 2000) and therefore any reduction in a compound's ability to cross the outer membrane will have a corresponding enhancing effect on efflux mediated resistance. Fluoroquinolones cross the outer membrane through water filled porin channels, and reductions in porin levels have also been associated with fluoroquinolone resistance (Hooper et al., 1989; Hooper et al., 1992; Miro et al., 2004). Reduced porin levels often occur concomitantly with upregulation of efflux pumps, for example as a result of mutations in the *mar* locus of *E. coli* (Randall and Woodward, 2002). The role of efflux in the emergence of quinolone resistance is discussed further below.

4.2.2.4 Emergence of Quinolone Resistance: Interplay Between Intrinsic and Acquired Resistance and Drug Exposure

If gyrase and topoisomerase IV enzymes are equally sensitive to a given quinolone, mutations in one enzyme should only marginally reduce the antibacterial activity of the quinolone, due to the inherent sensitivity of the other enzyme. Therefore, clinically relevant or high-level resistance in some instances occurs through stepwise increases mediated by the accumulation of mutations in both targets. Alternatively, if one enzyme is significantly more sensitive than the other, mutations in the sensitive enzyme can confer substantial single-step increases in resistance. In other instances, multiple mutations within the primary target are needed to confer clinically relevant resistance. The frequency of resistance based on target mutations can therefore be expected to vary among different organisms. Furthermore, the ability of a target mutation to confer survival in the presence of the quinolone depends on whether the mutation in question sufficiently decreases susceptibility to the level of drug present at the site of infection to avoid killing. The absolute level of resistance resulting from individual or multiple target mutations can be significantly influenced by efflux. Through reductions of intracellular drug levels, pumps ultimately create the same effect as sub-optimal drug exposures, reducing the amount by which the drug concentration exceeds the MIC. Although intrinsic resistance provided by efflux may be comparatively lower for quinolones, the natural resistance they provide can contribute significantly to absolute resistance mediated by a given target mutation. Thus, efflux likely contributes to the survival of stepwise target mutants, particularly in cases where drug levels may not be optimal or in organisms such as *P. aeruginosa* or *K. pneumoniae* where general permeability is reduced. Indeed in some gram negatives, certain target mutations can create clinically significant quinolone resistance only in the presence of a functional efflux pump (Baucheron et al., 2002; Lomovskaya et al., 1999; Oethinger et al., 2000). Perhaps consistent with this interrelationship, target based resistance to quinolones was observed quickly in *P. aeruginosa* following widespread use of fluoroquinolones (Coronado et al., 1995). This may reflect the fact that single point mutations can occur in *gyrA* causing 4- to 16-fold reductions in susceptibility, with the absolute resistance levels conferred by these mutations enhanced by efflux. Mutations resulting in pump overexpression, or expression of silent pumps, can also confer significant resistance in the absence of target mutations (Poole, 2004). For *P. aeruginosa*, in vitro experiments have demonstrated that at lower fluoroquinolone levels, pump mutants may be preferentially selected (Kohler et al., 1997). The combination of accumulated target mutations and pump overexpression can lead to very high levels of resistance (Poole, 2000, 2004). In one recent study, the use of an efflux pump inhibitor to assess the prevalence of pump mediated fluoroquinolone and multidrug resistance among *P. aeruginosa* clinical isolates suggested a correlation between fluoroquinolone treatment and the co-emergence of target and pump mediated multidrug resistance (Kriengkauykiat et al., 2005).

In contrast to *P. aeruginosa*, the development of significant resistance in many gram negatives, in particular Enterobacteriaceae, usually requires the accumulation of mutations in both GyrA and/or ParC. Given that simultaneous mutations in two subunits are expected to be exceedingly rare, this process likely occurs in a stepwise fashion (Drlica and Malik, 2003). The accumulation of mutations thus depends on enrichment of mutants at each step, a process dependent in part on the level of quinolone being within the so-called mutant selection window (Zhao and Drlica, 2002), defined as being between the concentration required to kill 99% of bacteria in culture (MIC_{99}) and the MIC of the least susceptible next step mutant, the mutant prevention concentration (MPC). At concentrations above MPC, two mutations are required and are unlikely to be present in most bacterial populations. Therefore factors such as sub-optimal exposure to drug can contribute to enrichment of earlier stage mutants, and in cases where mutations confer resistance only in conjunction with efflux (Baucheron et al., 2002; Oethinger et al., 2000), the presence of pumps is likely to be very important to this process. Ultimately, efflux will contribute to the development of resistance and to the final absolute level of resistance achieved in most gram-negative bacteria. Therefore, emerging resistance to quinolones in gram negatives is often a stepwise process based on the accumulation of both target mutations and mutations affecting efflux/permeability, and the rapidity of this process will depend on the interplay of these factors and drug exposure level.

Like efflux, topoisomerase protection mediated by QnrA may also play a role in target based resistance development. The impact of QnrA in and of itself is generally less significant, mediating reported 4- to 32-fold decreases in susceptibility in Enterobacteriaceae, which are quite susceptible to fluoroquinolones initially. However, this mechanism is additive with target and efflux/permeability based resistance (Martinez-Martinez et al., 1998, 2003) and can therefore contribute to high-level resistance and to the stepwise accumulation of target mutations. Consistent with this, the presence of this determinant increased the in vitro frequency of quinolone resistance in *E. coli* (Martinez-Martinez et al., 1998; Robicsek et al., 2005) and *E. cloacae* (Robicsek et al., 2005). Of additional concern, *qnrA* appears to be increasingly widespread and is present on the conjugative plasmid pMG252 (Martinez-Martinez et al., 1998) and related plasmids (Robicsek et al., 2005; Wang et al., 2004). Plasmid pMG252 confers resistance to chloramphenicol, streptomycin, sulphonamide, trimethoprim, and mercuric chloride and carries the bla_{FOX-5} β-lactamase gene, while additional *qnrA* plasmids encode the SHV-7 extended spectrum β-lactamase (ESBL) (Wang et al., 2004). Therefore, non-quinolone antibiotic selection pressure may also drive distribution of *qnrA* containing plasmids. Consistent with this, *qnrA* has tracked with plasmid-encoded *ampC* or ESBL's and has been found in *Enterobacter* spp. from a range of geographical regions in the USA (Robicsek et al., 2005). The presence of plasmid-borne *qnrA* in susceptible isolates raises the possibilities of non-quinolone selection for pMG252 and related plasmids priming the ability of Enterobacteriaceae to undergo stepwise development of high-level resistance (Robicsek et al., 2005).

An additional factor that may influence the rate at which resistance to quinolones emerges is the apparent mutagenicity of agents causing DNA damage. Quinolones have been shown to be moderate mutagens as tested by Ames test (Mamber et al., 1993), the likely result of induction of error prone polymerases, via the SOS system, in response to the interruption of DNA synthesis and DNA damage (Cirz et al., 2005; Clerch et al., 1992; Drlica and Zhao, 1997; Gmuender et al., 2001). A recent report showed that in vitro evolution of quinolone resistance in *E. coli* was severely reduced, and eliminated in a mouse thigh model of infection, upon disruption of proper SOS response (Cirz et al., 2005). The significance of this process to the development of resistance clinically remains to be determined, but it is intriguing conceptually that induced mutagenesis may play a role in generating the multiple mutations often required for resistance to quinolones. It has been suggested that blocking or considerably slowing the emergence of resistance may be possible through directly inhibiting the induction of the SOS response (Cirz et al., 2005).

Although the mechanism of killing by quinolones is relatively well character-ized, there may be involvement of additional mechanisms beyond direct target activity. For example a recent report detailing the transcriptomic interrogation of ciprofloxacin treated *P. aeruginosa* implicated induction of the pyocin system in cell killing activity (Brazas and Hancock, 2005). One component encoded within this induced locus resembles a phage lytic enzyme, the overexpression of which has been shown to cause cell lysis (Nakayama et al., 2000). Mutants constructed in vitro lacking this enzyme showed decreased susceptibility to fluoroquinolones (Brazas and Hancock, 2005), and alterations in pyocin production during chronic lung infections has been proposed as a possible factor contributing to the evolution of fluoroquinolone resistance in *P. aeruginosa* chronic lung infections (Brazas and Hancock, 2005).

4.2.2.5 Clinical Emergence of Quinolone Resistance in Gram-Negative Pathogens

The high level of activity of fluoroquinolones against gram-negative bacteria, allow-ing for dosing well above the MIC, and the requirement for multiple mutations necessary for establishing significant resistance might have been expected to curtail the rapid rise of resistance clinically. In general however, resistance among gram negatives has been steadily increasing concomitant with increased use of fluoro-quinolones in the USA (Neuhauser et al., 2003; Zervos et al., 2003). As described above, resistance emerged almost immediately in *P. aeruginosa* (Acar and Gold-stein, 1997; Coronado et al., 1995; Ogle et al., 1988) following widespread introduc-tion of fluoroquinolone use, and is increasing, including a 50% increase in resistance in nosocomial isolates from ICUs between 1994 and 1999 (Bhavnani et al., 2003; NNIS, 1999).

E. coli clinical resistance was rare prior to the 1990s but has since emerged rapidly in several settings, particularly in Europe and China. The incidence of resistance in China appears to have escalated very rapidly to as high as 50–60% in some settings (Wang et al., 2001a,b, 2003). Fluoroquinolone use was linked to steady increases in the rate of resistant strain isolation (0% in 1988 to 7.5% in 1992) from bloodstream infections in Spain (Pena et al., 1995). From 1992 to 1996, the incidence of fluoroquinolone resistance in nosocomial and community acquired *E. coli* infections in Spain increased from 9 to 17% (Garau et al., 1999). Prior exposure to fluoroquinolones and catheterization during complicated urinary tract infection were identified as risk factors (Ena et al., 1995; Garau et al., 1999). Moreover, neutropenic patients with fever carried no resistant isolates as of 1992 but upon implementation of prophylactic fluoroquinolone treatment the incidence of resistant strain isolation escalated to 78% by 1996 (Garau et al., 1999). Intriguingly, there was also a high rate of carriage of resistant bacteria in feces of healthy adults (24%) and children (26 %). Therefore the rapid rise in resistance detected in clinical isolates in Spain may be attributable to fluoroquinolone selection of preexisting resistant bacteria present in the population. Although circumstantial, the high carriage rate was plausibly attributed to the rampant use of fluoroquinolones in veterinary medicine, allowing development of a reservoir of resistant bacteria to occur in the farm environment, which ultimately transferred to humans. Similar reservoirs of *Campylobacter jejuni* resistant to fluoroquinolones, and other antibiotics, were also found (Saenz et al., 2000). In the USA, the emergence of fluoroquinolone resistance in *C. jejuni* was temporally linked to the introduction of fluoroquinolones in veterinary medicine (Endtz et al., 1991; Smith et al., 1999). From 1997–2000 resistance rates as measured by National Antimicrobial Monitoring System (NARMS) held more or less steady from 14–18%, but one study determined an escalation of resistance to 40.5% between 1995 and 2001 within the University of Pennsylvania Health System (Nachamkin et al., 2002). Factors such as veterinary usage seem to have contributed to the relatively rapid emergence of resistance in these cases, where clinical resistance may not be occurring through direct selection of spontaneous mutants during treatment but rather facilitated through creation of resistance reservoirs, which then spread rapidly. An additional setting where resistance might be facilitated is cases of gonococcal infections, facilitated by person-to-person spread. Analysis of resistance in a clinic in Cleveland Ohio revealed the emergence of a single genotype of *N. gonorrhoeae* with decreased susceptibility to fluoroquinolones (Gordon et al., 1996). These isolates did not meet the criteria for resistance but likely have increased potential for development of high-level resistance. In Sydney Australia, there has been a clear progression in the incidence of resistance in *N. gonorrheae* from low level-resistance in 1985 through to high-level resistance and the recommendation to discontinue use of fluoroquinolones for treatment of gonococcal infections in Sydney (Shultz et al., 2001). In other settings, resistance is emerging more slowly. In the Netherlands, the incidence of resistant *E. coli*, as measured with norfloxacin, increased from 1.3% in 1989 to 4% in 1998 (Goettsch et al., 2000). In the USA, large-scale analysis of Enterobacteriaceae clinical isolates showed that susceptibility to levofloxacin and ciprofloxacin remained at 90% or higher while *P. aeruginosa* susceptibility was lower at approximately 70% (Sahm et al., 2001). Resistance rates in *P. aeruginosa* seem to be relatively stable however as measured between 2001 and 2003 (Karlowsky et al., 2005). There was also a high prevalence of multidrug resistance among fluoroquinolone resistant strains. The rate of resistance overall may be increasing at somewhere near 2% per year (Neuhauser et al., 2003).

Despite the general susceptibility of *H. influenzae* to fluoroquinolones, the requirement for multiple mutations for resistance, the possible decreased contribution of efflux in this strain, and a single reservoir being the human respiratory tract, the emergence of resistance has sporadically been observed, albeit at a low frequency (Bastida et al., 2003; Li et al., 2004b). Intriguingly, mutations in *gyrA* in resistant isolates from a clonal outbreak in a long-term care facility (Nazir et al., 2004) may have been acquired by horizontal transfer (Li et al., 2004b). Emergence of stepwise resistance in this potentially lower frequency scenario may be facilitated by occasional dosing lapses or issues related to patient factors such as chronic lung infection (Li et al., 2004a).

Extensive evaluation of mutations conferring resistance in clinical isolates has not been done, but several reports implicate combinations of overexpressed efflux pumps, including MexAB–OprM, MexCD–OprJ, MexXY–OprM and MexEF–OprN, and target mutations in GyrA, GyrB, ParC and ParE in mediating resistance in *P. aeruginosa* clinical isolates (Jalal and Wretlind, 1998; Kriengkauykiat et al., 2005; Le Thomas et al., 2001; Oh et al., 2003). Many fluoroquinolone resistant strains are also multidrug resistant. This is likely due in part to the high prevalence of efflux pump overexpression (Kriengkauykiat et al., 2005) and therefore use of pump substrate antibiotics, whether fluoroquinolones or non-fluoroquinolones has seemingly negatively impacted the usefulness of antipseudomonal agents generally. Analysis of resistant *E. coli* strains from uncomplicated urinary tract infections in several European countries and from Beijing, China, revealed accumulated mutations in the targets GyrA, ParC and ParE, and in the *mar* and *acrR* loci controlling efflux pump expression (Komp Lindgren et al., 2003; Wang et al., 2001b). There was also an apparent correlation between increased mutation rates (intermediate mutator) and the accumulation of multiple mutations leading to high-level resistance in the strains from European centers (Komp Lindgren et al., 2003). In *H. influenzae*, mutations in GyrA, ParC and ParE have been characterized (Bastida et al., 2003; Georgiou et al., 1996; Li et al., 2004b).

4.2.2.6 Newer Fluoroquinolones and Treatment Strategies

When first introduced, the quinolone class of antibiotics was viewed with great optimism due to their potency and broad spectrum activity (Shams and Evans, 2005; Van Bambeke et al., 2005). A mechanism involving two intracellular targets also implied the possibility of lower rates of resistance. Although the rapid emergence of clinical resistance to earlier-generation quinolones and recent observations of resistance to newer quinolones has perhaps tempered this euphoria, the quinolones nonetheless remain one of the most important and effective classes of antibiotic in use today (Van Bambeke et al., 2005). Because of this, and the fact that resistance to newer quinolones has not become overly widespread, there is intense discussion over strategies to preserve the utility of quinolones. Although novel antibiotic development per se has tapered in recent years, quinolones, along with β-lactams, are still the focus of continuous incremental improvement based on the desirable combination of attributes that make these antimicrobials very effective.

The development of newer quinolones focuses on the notions of broadening their spectrum, improving their antibacterial activity and pharmacokinetic characteristics and potentially developing increased activity against both gyrase and topisomerase IV in order to reduce resistance development. Historically, quinolone development has led to four basic "generations": first-generation quinolones had good activity against many gram negatives but were effective only for urinary tract infections due to poor tissue and serum distribution; second-generation quinolones ultimately possessed sufficient tissue distribution and broadened gram-negative spectrum for use in treating systemic infection; third-generation quinolones had further improved pharmacokinetics and increased spectrum to include *Streptococcus pneumoniae*; and fourth-generation quinolones added anaerobes to their spectrum (Shams and Evans, 2005). Another scheme detailing the generations of quinolones takes into account drug structure and in vitro activity (Lu et al., 2001). Accordingly, members of the first generation (e.g. nalidixic acid) do not kill in the presence of chloramphenicol.

Second-generation compounds (e.g. norfloxacin) add a piperizinyl ring and C-6-F, which allows the compound to kill *E. coli* when resuspended in saline. Third generation compounds (e.g. ciprofloxacin) have the addition of *N*-cyclopropyl, which gives the compound the ability to kill in the presence of chloramphenicol. Finally, fourth-generation compounds add a halogen or methoxy group to the eight position. This gives antimutant activity. Attempts to improve both antibacterial activity and pharmacodynamic properties are objectives of antibiotic compound development, generally, for obvious reasons. In the case of quinolones, however, this is intimately connected to the notion of the mutant prevention concentration (MPC) (Zhao and Drlica, 2001). This centers on the idea that when multiple mutations are required to confer clinically significant resistance, maintaining an antibiotic level at which only mutants having acquired simultaneous multiple mutations can survive dramatically reduces the frequency with which resistance can emerge (potentially to statistically insignificant probabilities based on bacterial numbers present) (Zhao and Drlica, 2001). Improving antibacterial activity and tissue distribution both serve to enable more consistent achievement of drug levels above the MPC. Increasing potency against both gyrase and topoisomerase IV should also increase or preserve the necessity for multiple mutations to confer resistance, and this combined with the ability to exceed the mutant selection level is viewed as an important potential strategy to reduce resistance. Supporting this, combinations of a quinolone acting primarily against gyrase with one active primarily against topisomerase IV dramatically reduced resistance frequency in the gram-positive organism *Staphylococcus aureus* (Strahilevitz and Hooper, 2005). Furthermore, compounds requiring multiple mutations within one enzyme (e.g. gyrase) to confer resistance have also been shown to result in lower resistance frequencies (Zhao et al., 1997).

    While these strategies have a good likelihood of reducing the emergence of resistance, the above sections describing resistance indicate that preserving quinolone utility may be a complex endeavor. Even with more favorable drug properties, factors such as the use of quinolones in agriculture, suboptimal treatment monitoring or regimens and the natural variation in intrinsic susceptibility of different bacterial strains can undermine these strategies. Suboptimal exposures can lead to the selection of single resistance mutations (i.e. exposure within the so-called mutant selection window, between the MIC and the MPC), which then increases the likelihood of full resistance emerging during optimal treatment. The impact of quinolone usage to date, i.e. in perhaps selecting first step mutants, will also impact the future success of more refined treatment regimes. This has led to calls for a more thorough understanding of pharmacokinetic and pharmacodynamic properties of quinolones and application of these principles to treatment regimens, thorough monitoring of resistance, decreasing the non-justifiable use of quinolones, and employing dosing strategies to achieve a C_{max}/MIC ratio of at least 10 to increase exposure over the MPC (Shams and Evans, 2005; Van Bambeke et al., 2005). Strategies relying on increased drug exposure levels must, however, take into account the potential for increased toxicity, which is not inconsequential for the quinolone group and is an important factor in determining the appropriate use of this antibotic class (Van Bambeke et al., 2005). The increased possibility of treatment failure with *P. aeruginosa* generally is such that the fluoroquinolone may be used in combination with an appropriate β-lactam (Shams and Evans, 2005). An additional strategy under development for combination therapy with fluoroquinolones in the treatment of *P. aeruginosa* infections is efflux pump inhibitors (Lomovskaya and Watkins that has 2001a and b.).

Defeating efflux in *P. aeruginosa* (through pump deletion or the use of efflux pump inhibitors), in practical terms, results in a significant improvement in the antibacterial potency of fluoroquinolones (Lomovskaya et al., 1999, 2001). Therefore, inhibition of pumps should directly address pump-mediated resistance as well as improve the ability to achieve/maintain drug exposures in excess of the MPC, to minimize target based resistance. Supporting this, a pump deficient strain of *P. aeruginosa* had a resistance frequency to levofloxacin of $<10^{-11}$ (when exposed in vitro to $4\times$ the parental MIC against the pump containing parent) compared to 10^{-6}–10^{-7} for the parent at similar drug levels (Lomovskaya et al., 1999). In the absence of efflux, sufficient resistance for survival apparently required multiple simultaneous target mutations, resulting in essentially undetectable resistance selection in vitro. For the pump-containing parent strain, the exposure levels exceed the MIC by a much narrower margin, and the apparent ability of both pump overexpression and/or target mutations to allow survival results in higher rates of resistance. The use of an efflux pump inhibitor to reduce pump action also caused a similar decrease in resistance frequency (Lomovskaya et al., 2001). MPEX pharmaceuticals has currently begun phase 1b trials of aerosolized MP-601,205, with the eventual desired outcome of using it in combination with fluoroquinolones for the treatment of respiratory infections in cystic fibrosis patients (www.mpexpharma.com/pr_aug1_05.html).

4.2.3 Resistance to Aminoglycosides

The aminoglycosides, including streptomycin, kanamycin, neomycin, gentamicin, amikacin, netilmicin and tobramycin, function by binding the ribosomal 30S subunit and interfering with protein synthesis. The interaction with aminoglycosides involves highly conserved portions of 16S rRNA, in domains essential for controlling codon–anticodon interactions (Moazed and Noller, 1987; Purohit and Stern, 1994; Woodcock et al., 1991; Yoshizawa et al., 1999). Disruption of this function leads to translational errors and also interferes with protein translocation (Davies and Davis, 1968). Aminoglycosides are broad-spectrum, rapidly bactericidal, and show synergy with other antibiotics (particularly β-lactams) and therefore have found widespread use in the treatment of severe nosocomial infections (Bartlett, 2003; Zembower et al., 1998). Indeed, although newer β-lactams and fluoroquinolones are potent against gram negatives, aminoglycosides are still extensively used to treat these infections, often as combination therapy. Aminoglycosides show in vitro activity against *Salmonella* spp. and *Shigella* spp. but are not clinically effective against these organisms and are therefore not recommended for these indications. Combination therapy with an aminoglycoside and β-lactam is a preferred treatment strategy for nosocomial infections with *P. aeruginosa* (Livermore,

2002; Zembower et al., 1998). Unfortunately, resistance has appeared and poses a threat to their continued successful use. Resistance to aminoglycosides is generally mediated by three mechanisms: mutations in the rRNA or ribosomal protein targets of the drug, efflux and permeability mechanisms, and perhaps most importantly, by the action of aminoglycoside modifying enzymes. More recently however, resistance through ribosomal protection has been described. Target mutations generally impact susceptibility to streptomycin (Honore and Cole, 1994; Honore et al., 1995; Kono and O'Hara, 1976; Meier et al., 1994; Springer et al., 2001) and are important mainly in the clinical treatment of *Mycobacteria* infections. Resistance based on target mutation is perhaps more rare in other cases since there are several copies of the 16S rRNAs and there is a minimum number of copies/mutations that must occur in order for resistance to be observed (Kotra et al., 2000; Prammananan et al., 1998) and therefore this mechanism will not be discussed here. Aminoglycoside inactivation, permeability/efflux and ribosomal protection are described below.

4.2.3.1 Inactivating Enzymes

Aminoglycoside inactivating enzymes covalently modify the drug, resulting in a reduced ability to bind to the ribosomal A site target and are perhaps the most significant resistance factor (Miller et al., 1997; Llano-Sotelo et al., 2002). Modifying enzyme-based resistance to aminoglycosides has been known for decades, consistent with aminoglycosides being natural products derivatives (i.e. protection determinants would exist for producing organisms) (Shaw et al., 1993). Because of this, modifying enzymes are generally well understood and significant effort is ongoing to further understand these determinants (Wright, 1999). There are three categories of modification: O phosphorylation (aminoglycoside phosphoryltransferases, AHPs), O adenylation (aminoglycoside nucleotidyltransferase, ANT) and N acetylation (aminoglycoside *N*-acetyltransferases, AACs). Modifying enzymes are further subdivided into classes based on their site of modification of the drug and the spectrum of resistance conferred (Wright, 1999). For example AACs can acetylate aminoglycosides at the 1, 3, 2' and 6' amino groups, and are correspondingly designated AAC(1), AAC(3), AAC(2') and AAC(6') respectively. Individual variants of these classes are further subdivided using roman numerals, for example AAC(3)-I, II and III. The sheer complexity of the naming scheme underscores how many of these enzymes have been characterized over the years and these are reviewed in (Azucena and Mobashery, 2001; Shaw et al., 1993; Smith and Baker, 2002; Wright, 1999). Indeed modifying enzymes have found great utility as genetic tools, with gentamicin acetyltransferase AAC(3)-I in particular being used for genetic manipulation in *P. aeruginosa* (Schweizer, 1993). This underscores a major issue with these determinants, namely that they are mobile, being carried on R factors, transposons and integrons. These elements can therefore transfer readily and are often found on mobile elements with other resistance determinants, providing multidrug resistance to sulfonamides, chloramphenicol, antiseptics and perhaps most worrisome, β-lactams (Poirel et al., 2001a,b; Rubens et al., 1979). Finally, aminoglycoside modifying enzymes of differing specificities can also accumulate to provide pan-aminoglycoside resistance (Chow et al., 2001; Miller et al., 1995, 1997; Rodriguez Esparragon et al., 2000; Shaw et al., 1991). Certain variants of modifying enzymes may also provide resistance by sequestering the drug, through tight drug binding (Magnet et al., 2003; Menard et al., 1993).

4.2.3.2 Impermeability/Efflux

Resistance to aminoglycosides has been repeatedly associated with reduced accumulation of the drug (referred to as impermeability phenotype), mainly in *P. aeruginosa* and other non-fermenting gram-negative bacilli (Mingeot-Leclercq et al., 1999). The impermeability phenotype generally confers low to moderate levels of pan aminoglycoside resistance and is most frequently encountered in *P. aeruginosa*, particularly among isolates from cystic fibrosis patients (Hurley et al., 1995; MacLeod et al., 2000). The mechanisms mediating the impermeability phenotype have been the subject of much investigation. Aminoglycosides are cationic and have a relatively well characterized mode of entry into cells involving their insertion into LPS and self promoted uptake across the outer membrane followed by proton gradient (energy) dependent transport across the cytoplasmic membrane (Mingeot-Leclercq et al., 1999; Peterson et al., 1985; Rivera et al., 1988; Saika et al., 1999). Impermeability resistance to aminoglycosides has frequently been attributed to modifications in LPS structure which are thought to reduce the ability of aminoglycosides to cross the outer membrane (Bryan et al., 1984; Hasegawa et al., 1997; Shearer and Legakis, 1985; Yoneyama et al., 1991). Aminoarabinose substitution of lipid A, potentially controlled by regulatory components such as PhoP/PhoQ, may also play a role in this process in *P. aeruginosa* (McPhee et al., 2003). There are also energy deficient small colony variants in which changes in cytoplasmic membrane proteins or alterations in energy dependent uptake across the inner membrane are implicated in reduced accumulation (Bayer et al., 1987; Parr and Bayer, 1988). In *E. coli*, the oligopeptide transporter OppA has also been specifically implicated in transporting aminoglycosides, and reduced levels of this transporter can confer resistance (Acosta et al., 2000). Adaptive resistance, where the initial rapid accumulation of drug and killing effect in cell populations is followed by a change to slow accumulation and corresponding reduced susceptibility, which is reversible upon removal of the aminoglycoside, has often been observed in vitro and in vivo in *P. aeruginosa* and is a consideration in establishing effective dosing regimens (Barclay et al., 1992, 1996; Barclay and Begg, 2001; Daikos et al., 1990, 1991; Karlowsky et al., 1996; Xiong et al., 1997). Adaptive resistance may be partially the result of induction of anaerobic respiration genes upon exposure to aminoglycosides (Karlowsky et al., 1997). Bacteria grown under anaerobic or low pH conditions exhibit a general aminoglycoside transport defect (Schlessinger, 1988) and the induction of anaerobic respiration by aminoglycosides may therefore contribute to this adaptive resistance.

With recent increased understanding and appreciation of efflux mechanisms in gram negatives, it has become apparent that efflux may be playing a significant role in aminoglycoside resistance in some cases. Most RND family pumps extrude mainly lipophilic compounds; however, efflux pumps of the RND family have now been identified that extrude the cationic aminoglycosides, including AcrD of *E. coli* (Aires and Nikaido, 2005; Rosenberg et al., 2000), AmrAB-OprA and BpeAB-OprB of *Burkholderia pseudomallei* (Chan et al., 2004; Chan and Chua, 2005; Moore et al., 1999), and MexXY–OprM of *P. aeruginosa* (Aires et al., 1999; Mine et al., 1999; Westbrock-Wadman et al., 1999). In addition, the *P. aeruginosa* outer membrane proteins OpmG and OpmI appear to be involved in intrinsic resistance to aminoglycosides, potentially as additional outer membrane channel components of the MexXY pump or another efflux pump (Jo et al., 2003). The MexXY components of the MexXY–OprM efflux pump are unusual among RND pumps in that their expression is strongly induced by agents interfering with protein synthesis (Jeannot et al., 2005). It thus appeared likely that this pump would contribute to both impermeability and adaptive aminoglycoside resistance and this is now known to be the case (Hocquet et al., 2003; Westbrock-Wadman et al., 1999). The exact nature of MexXY induction and its role in conferring aminoglycoside resistance remains to be fully elucidated. Typical of RND family efflux pumps, the *mexXY* operon has an associated putative repressor gene, *mexZ* (Aires et al., 1999; Westbrock-Wadman et al., 1999), and inactivation of this regulator increases expression of MexXY. However, this did not lead to aminoglycoside resistance (Westbrock-Wadman et al., 1999), and the pump is apparently still further inducible by drugs even when *mexZ* is mutated (Jeannot et al., 2005), suggesting that regulation of MexXY expression is not controlled solely by MexZ. Furthermore, MexZ or other regulatory components apparently do not directly interact with inducing compounds, but rather, the impact of drugs at the ribosome may be generating an intracellular signal(s) leading to the regulatory cascade (Jeannot et al., 2005). Although significant evidence shows that RND pump components recognize and transport aminoglycosides directly (Aires and Nikaido, 2005), it stands to reason that the MexXY pump could have a natural function inextruding a toxic molecule related to the interference with protein synthesis. Whether this also indirectly contributes to survival in the presence of these inhibitors, thereby indirectly conferring increased resistance, is currently unknown. Interestingly, recent reports show that despite the inducibility of the MexXY pump, pan-aminoglycoside resistant clinical isolates with constitutively derepressed MexXY production do occur in clinical settings, and these isolates may or may not have mutations in *mexZ* (Sobel et al., 2003; Vogne et al., 2004). Moreover, although inactivating the pump in resistant strains increased susceptibility confirming a contribution, there is not always a clear association between expression level and resistance, and some isolates were further drug inducible for pump expression while others were not (Sobel et al., 2003). It is tempting to speculate that one selective advantage of having the inducible pump expressed constitutively is the elimination of lag time for induction of the pump in cell populations intermittently exposed to aminoglycosides, which may be the case particularly in lung infections or during dosing strategies designed to avoid adaptive resistance. It seems clear that induction of the pump and generating aminoglycoside resistance is a complex process and further research is necessary to fully understand the contributions of the pump and other mechanisms to impermeability and adaptive resistance.

4.2.3.3 Ribosomal Protection

Although resistance in the form of aminoglycoside modifying enzymes and permeability has been studied for decades, a more recent and ominous development is the characterization of resistance mediated by post transcriptional methylation of rRNA. This mechanism was initially characterized as a self defense strategy in antibiotic-producing organisms (Beauclerk and Cundliffe, 1987) but has now been shown to exist in several important gram-negative nosocomial pathogens. The first example of a modifying enzyme in this context, *armA* (aminoglycoside resistance methyltransferase) was initially sequenced on multidrug resistance plasmid pCTX-M-3 in *C. freundii* isolated in Poland and later characterized from *K. pneumoniae* isolated in France (Galimand et al., 2003). The plasmid borne *armA* gene is associated with CTX-M-3 extended spectrum β-lactamase and was shown to be part of a composite transposon also encoding resistance to streptomycin-spectinomycin, sulfonamides, and trimethoprim. Subsequently, the *armA* gene has been shown to be disseminated worldwide in *E. coli*, *C. freundii*, *K. pneumoniae*, *E. cloacae*, *Salmonella enterica*, *S. flexneri* and *A. baumannii*, likely through conjugation and transposition mediated transfer (Galimand et al., 2005; Gonzalez-Zorn et al., 2005; Yamane et al., 2005; Yan et al., 2004b). Additional methylases RmtA (Yokoyama et al., 2003) and RmtB (Doi et al., 2004) have been identified in *P. aeruginosa* and *S. marcescens* respectively.

It is still unclear as to the exact origin of the *armA* or *rmt* genes, although this mechanism is employed by aminoglycoside producers and indeed may have originated from these organisms (Yamane et al., 2005). As such, it would be predicted to confer significant protection and this is indeed the case. Therefore, these genes comprise a rapidly transmissible single-step route to very high resistance, and are generally associated with multidrug resistance genes, most importantly β-lactamases. Furthermore, ribosomal protection shows less specificity than aminoglycoside modifying enzymes, conferring resistance to most of the clinically important aminoglycosides including tobramycin. The high-level pan aminoglycoside resistance conferred by this mechanism and its potential for rapid spread may be a significant concern with respect to the longevity of aminoglycosides.

4.2.3.4 Biofilms

Biofilms are communities of bacterial cells that exist attached to a surface. These communities can establish on abiotic surface such as plastics or metals used for medical devices or biotic surfaces such as human tissue. It is now well appreciated that biofilms are very intrinsically resistant to antibiotics and that biofilms likely contain sub-populations of extremely tolerant cells, termed persisters (Lewis, 2005). Furthermore, many chronic infections, such as *P. aeruginosa* infections in the CF lung, are now known to involve the biofilm mode of growth. Much research is focused on establishing the underlying mechanisms by which biofilms manifest such high levels of resistance. Although this is an issue for many, if not all, antibiotic classes, it is of particular interest recently with respect to aminoglycosides. As mentioned above, aminoglycosides are still one of the more effective agents used for *P. aeruginosa* infections, particularly the use of inhaled tobramycin for CF lung infections (Burns et al., 1999; Ramsey et al., 1999). Intriguingly, aminoglycosides have been shown experimentally to induce biofilm formation by *E. coli* and *P. aeruginosa* (Hoffman et al., 2005). This effect is likely mediated through the regulator Arr, which controls c-di-GMP levels, and biofilms formed by *arr* mutants are 100-fold more susceptible to tobramycin than the parent strain (Hoffman et al., 2005). This locus is not present in all *P. aeruginosa* isolates, however, it has been found in several CF isolates recently examined, and its impact could be clinically relevant for tobramycin treatment. Furthermore, the *ndvB* locus is involved in glucan synthesis during the biofilm mode of growth, and these glucans were shown to bind tobramycin (Mah et al., 2003), potentially mediating resistance by sequestering the drug in the periplasm and preventing entry. Finally, biofilms in the CF lung are likely growing anaerobically and, as discussed above, aminoglycosides are generally inactive against bacteria utilizing anaerobic respiration (Hassett et al., 2002; Yoon et al., 2002). Therefore, biofilm formation, perhaps partially induced by exposure to aminoglycosides, may be an important component of resistance in certain settings.

4.2.3.5 Emergence of Clinical Resistance

Despite the preponderance, and mobility, of aminoglycoside resistance mechanisms employed by gram negatives, this class of drug is still generally effective in some settings. A comprehensive analysis of aminoglycoside resistance data from North America between the years 1999 and 2001 (from the MYSTIC surveillance program) revealed that the prevalence of resistance to gentamicin and tobramycin ranged between 2.7–4.5% and 1.7–3.8 %, respectively, in *E. coli* (Mutnick et al., 2004). As might be expected, resistance was higher in *P. aeruginosa*, ranging between 8.4–18.4% (gentamicin) and 5.7–9.1% (tobramycin) (Mutnick et al., 2004). Rates of susceptibility to gentamicin in *P. aeruginosa* analyzed between 2001–2003 in US isolates decreased from 75.5 to 72.3%. Worldwide, the rates of resistance are greatly influenced by geographic location. For example areas of Europe and Latin America report very high rates of *P. aeruginosa* resistance. These range for gentamicin to approximately 26.8% (blood isolates (Unal et al., 2004)) to 38.9% (ICU isolates (Garcia-Rodriguez and Jones, 2002)) in Europe and hover around the 50% mark for amikacin, gentamicin and tobramycin for urinary (Gales et al., 2002) and skin and soft tissue (Sader et al., 2002) infections in Latin America. This impact of geography may partly reflect differences in prescribing and infection control practices and indicates how quickly severe resistance can emerge. As described above, a major factor in resistance is aminoglycoside-modifying enzymes. Since these enzymes generally affect only certain aminoglycosides, the acquisition of one enzyme would often not preclude the use of other aminoglycosides. However the observation that isolates are emerging clinically with several modifying enzymes of different specificities is disturbing. The recent description of ribosomal modification carried on mobile genetic elements is also reason for concern.

4.2.3.6 Novel Approaches and Treatment Strategies

Despite the plethora of resistance mechanisms that can conspire to decrease their effectiveness, aminoglycosides have remained useful for decades, and are still clinically important for treating serious nosocomial gram-negative infections. Their potential for toxicity and the emergence of resistance are both concerns, however, and this prompts research into novel compound development. The structural knowledge continually emerging about the ribosome may allow for the development of newer, more potent compounds that have reduced recognition by aminoglycoside modifying enzymes. Indeed, isepamycin, amikacin, dibekacin and tobramycin are examples of existing potent compounds that are less susceptible to certain inactivating enzymes (Jana and Deb, 2005). There is structural information available for several aminoglycoside modifying enzymes (Burk et al., 2001; Hon et al., 1997; Sakon et al., 1993; Wolf et al., 1998; Wybenga-Groot et al., 1999), which may contribute to this rational design strategy, and may also be exploited for the development of inhibitors of enzymatic inactivation. Ideally, targeting the aminoglycoside-binding pocket of inactivating enzymes by aminoglycoside-like molecules would provide broadly active inhibitors of inactivating enzymes; however, there has not been significant progress in this area. Derivatives of neamine and kanamycin with poor antibiotic activity but excellent binding to APH (3') have been shown to act as suicide inhibitors of this enzyme (Roestamadji and Mobashery, 1998). Various dimer forms of neamine have been tested for antibiotic activity and avoidance of enzymatic inactivation. Some of these have antibiotic activity and are either poor substrates for inactivating enzymes or, alternatively, interfered with inactivating enzymes (Allen et al., 1982; Greenberg et al., 1999). The crystal structure of the aminoglycoside kinase APH(3')-IIIa revealed significant similarity to eukaryotic protein kinases (Hon et al., 1997), and known kinase inhibitors were shown to inhibit aminoglycoside kinases (Daigle et al., 1997). This raises the possibility of reversing resistance mediated by this class of enzyme with kinase inhibitors. These would only target this class of enzyme and the design of kinase inhibitors with sufficient selectivity vis a vis host kinases may be a significant hurdle in this strategy. Selective inhibition may be achieved by utilizing a strategy of combining a kinase inhibitor and an aminoglycoside molecule that would exhibit dual binding to the cofactor binding site and aminoglycoside binding site. Linking of adenosine and neamine resulted in compounds able to inhibit modifying enzymes (Liu et al., 2000), although the inherent large size of such inhibitors may pose problems in their further development. Although no widely effective molecules have emerged to date providing broad reversal of aminoglycoside resistance, increasing knowledge of the mechanisms and structures of the target and the inactivating enzymes may yield such compounds in the future. The discovery of ribosomal modification as an aminoglycoside resistance mechanism introduces an unfortunate additional threat to the continued effectiveness of these drugs. However there is precedence for overcoming ribosomal modification-based resistance by developing compounds with improved ribosome binding. Indeed tigecycline, a glycyl derivative of minocycline, overcomes ribosomal modification by TetM at least partially through increased binding to its ribosomal target (Bauer et al., 2004). Further work will hopefully result in new aminoglycosides or other molecules extending the usefulness of this class. In the meantime increased monitoring of resistance and judicious use of these drugs is essential.

4.2.4 Resistance to Tetracyclines

Tetracycline antibiotics, first introduced into medicine in 1948, offer a broad spectrum of antibacterial activity that includes gram-positive, gram-negative anaerobic and atypical bacteria. Tetracyclines inhibit protein synthesis from occurring in the bacterial cell by binding to the A-site of the 30S subunit of the bacterial ribosome. This prohibits the elongation phase of protein synthesis by inhibiting the entrance of the aminoacyl-tRNA into the binding site and preventing incorporation of new amino acids into the growing polypeptide chain (Hierowski, 1965; Suarez and Nathans, 1965). Their widespread use in humans, veterinary medicine and agriculture has, in turn led to widespread resistance in most clinically relevant bacterial species (Schnappinger and Hillen, 1996). Resistance to tetracyclines is caused by two diverse mechanisms; tetracycline-specific efflux pumps and ribosomal protection proteins (Table 4.2).

Table 4.2 Common tetracycline-specific resistance determinants found in gram-negative pathogens

Type	Genes	Organisms
Efflux	*tet*(A)	Enterobacteriaceae, *P. aeruginosa*, *Aeromonas* spp., *Vibrio* spp., *Plesiomonas* spp.
	tet(B)	Enterobacteriaceae, *Aeromonas* spp., *Vibrio* spp., *Plesiomonas* spp., *Actinobacillus* spp., *Moraxella* spp., *Pasteurella multocida*
	tet(C)	Enterobacteriaceae, *P. aeruginosa*, *Vibrio* spp.,
	tet(D)	Enterobacteriaceae, *Vibrio* spp., *Yersinia* spp., *Pasteurella multocida*
	tet(E)	*E. coli*, *Providencia* spp., *Proteus* spp, *S. marcescens*, *P. aeruginosa*, *Aeromonas* spp, *Alcaligenes* spp.
	tet(G)	*Salmonella* spp., *P. aeruginosa*, *Vibrio* spp., *Pasteurella multocida*
	tet(I)	*E. coli*, *Providencia* spp.
Ribosomal protection	*tet*(M)	*Neisseria* spp., *H. influenzae*, *Eikenella corrodens*, *Pasteurella multocida*, *Bacteroides spp.*, *Fusobacterium* spp.
	tet(O)	*Campylobacter* spp.
	tet(Q)	*Bacteroides* spp., *Porphrymonas* spp., *Prevotella* spp., *Veillonella* spp.,

4.2.4.1 Tetracycline Specific Efflux Pumps

The efflux pumps that remove tetracyclines from bacterial cells belong to the energy-dependant "Major Facilitator" family of efflux pumps. These pumps are comprised of a single membrane associated protein that has 12 membrane spanning domains (Levy, 2002). The efflux is coupled to a protonmotive force that exchanges a proton for the tetracycline-cation complex (Nikaido and Thanassi, 1993). They function by exporting tetracycline from the cell, thereby reducing the intracellular concentrations of tetracycline to a level that is no longer effective in inhibiting protein synthesis.

In gram-negative pathogens, the tetracycline-specific efflux pumps are encoded by a number of different genes, designated *tet*(A-E), *tet* (G), *tet*(H), *tet*(J), *tet*(Y) *tet*(Z), *tet*(30), *tet*(31) *tet*(33) and *tet*(38) (Table 4.2) (Berens and Hillen, 2003; Butaye et al., 2003; Chopra and Roberts, 2001). The *tet*(B) gene appears to have a wide host range and has been found in diverse genera of gram-negative bacteria including Enterobacteriaceae, non-fermenters and respiratory pathogens (Chopra and Roberts, 2001). Resistance genes *tet*(A), *tet*(B), *tet*(D) and *tet*(H) are often found on transposons, and *tet*(C), *tet*(E) and *tet*(G) are often found on large conjugative plasmids that may also harbor other antibiotic resistance genes (Jones et al., 1992). The *tet*(A-E) genes are found in many clinical isolates of Enterobacteriaceae, *P. aeruginosa* and *Aeromonas* spp.(Chopra and Roberts, 2001).

The proteins encoded by these genes are approximately 46 kDa in size and contain 12 transmembrane domains. Each structural gene is under the regulation of an accompanying tetracycline-responsive repressor gene, *tet*R, which is encoded in the opposite orientation (Schnappinger and Hillen, 1996). In the absence of tetracycline, dimers of the repressor protein TetR bind to the operator in the intergenic region and prevent transcription of both the repressor and the efflux pump protein (Berens and Hillen, 2003). The expression of the tetracycline efflux pump can be induced by very small amounts of the tetracycline antibiotic. Induction occurs when the tetracycline molecule complexes with Mg^{++} and binds to TetR, which causes a conformational change and the dissociation of the protein from the operator (Orth et al., 2000; Roberts, 1996; Takehasi et al., 1986). The functional efflux pump then extrudes tetracycline from the bacterial cell in an energy-dependent manner, exchanging a proton for a tetracycline-cation complex (Yamaguchi et al., 1990). The end result of this action is that the antibiotic fails to accumulate in the cell in sufficient concentrations to inhibit protein syntheses and block growth of the bacterium. With the exception of *tet*(B), the tetracycline efflux pumps recognize and expel tetracycline, chlortetracycline and doxycycline, but not minocycline nor tigecycline, whereas *tet*(B) also confers resistance to minocycline (Chopra and Roberts, 2001).

4.2.4.2 Ribosomal Protection Proteins

Resistance to tetracyclines can also be conferred by the so called ribosomal-protection proteins, most often encoded by *tet*(M) and *tet*(O). The ribosomal protection proteins are more common in gram-positive bacteria, but have been noted in some strains of gram-negatives as well. Tet(O) was first found on a transferable plasmid in a clinical isolate of *Campylobacter jejunii* (Taylor, 1986). Sequence analysis of *tet*(O) revealed a high homology to the *otrA* determinant of *Streptomyces rimosus*, which is the natural producer of oxytetracycline (Doyle et al., 1991; Sougakoff et al., 1987). The presence of *tet*(O) on mobile genetic elements suggests that the gene moved from its role of protecting the producing organism from the antibiotic it produces to a role of causing resistance to that same antibiotic in another host organism. The genes encoding TetM and TetQ have also been found on conjugative plasmids (Chopra and Roberts, 2001). Structurally, the ribosomal protection proteins also bear some similarity to the ribosomal elongation factors, EF-Tu and EF-G (Sanchez-Pescador et al., 1998). Like the elongation factors, the ribosomal protection proteins also function by utilizing a ribosomal-dependent hydrolysis of GTP (Burdett, 1991). In addition, a number of studies by using a variety of techniques such as cryoelectron microscopy and chemical probing have shown that the ribosomal protection proteins bind to a similar spot on the 50S subunit of the bacterial ribosome as the elongation factors (Connell et al., 2002, 2003b; Dantley et al., 1998; Spahn et al., 2001).

The ribosomal protection proteins cause resistance to tetracyclines by releasing the bound tetracycline from the ribosome. Although the ribosomal protection protein binds to a site on the 50S subunit of the bacterial ribosome at a site distant from that of the tetracycline binding site on the 30S subunit, the conformational change on the entire ribosome is sufficient to dislodge tetracycline from the A-site (Connell et al., 2003a). Concurrent with the removal of tetracycline from the ribosome, Tet(M) and Tet(O) also promote the binding of the tRNA to the A-site, thereby protecting the active site from further inhibition by a tetracycline molecule (Burdett, 1991).

In addition to the common mechanisms of tetracycline-specific efflux pumps and the ribosomal protection protein, a gene was identified in *B. fragilis* (*tet*(X)) that enzymatically degrades tetracycline (Speer and Salyers, 1989). However, Tet(X) functions only aerobically and does not confer resistance to the anaerobe *B. fragilis*. Resistance to tetracyclines can also be mediated by mutations to the genes encoding the active site rRNA in *Propionibacterium acnes* (Ross et al., 1998) and *Helicobacter pylori* (Gerrits et al., 2002; Trieber and Taylor, 2002). Furthermore, tetracyclines are also affected by non-specific multidrug efflux pumps, such as AcrAB, that confer resistance to a number of structurally related antibiotics, dyes, and biocides (Nikaido, 1998).

4.2.4.3 New Agents to Combat Tetracycline Resistance

Several attempts have been made to overcome tetracycline resistance through rational drug design. The glycylcyclines were discovered as a result of chemical modification of existing tetracycline molecules by the addition of a glycylamido side chain at the nine position (Sum and Petersen, 1999). Several molecules were synthesized that extended the spectrum of minocycline and DMDOT to recapture activity against strains resistant to classical tetracyclines by means of both efflux and ribosomal protection (Sum and Petersen, 1999; Testa et al., 1993). The glycylcyclines bind to a similar site on the bacterial ribosome as the tetracyclines, but with a much higher affinity (Bauer et al., 2004; Bergeron et al., 1996). The first glycylcycline to enter development is tigecycline (formerly GAR-936), the 9-*t*-butylglycylamido derivative of minocycline. Tigecycline is active against many strains of clinically relevant gram-positive and gram-negative pathogens, including those that express tetracycline-specific efflux pumps or ribosomal protection proteins (Petersen et al., 1999). Tigecycline has recently been approved in the USA for use in complicated skin and skin structure infections and complicated intraabdominal infections caused by a wide range of gram-positive, some Enterobacteriaceae and anaerobic pathogens (Tygacil, 2005).

In recent years there has also been a considerable effort to discover inhibitors of efflux pumps that confer resistance to tetracyclines. Several pharmaceutical companies have developed screens to identify inhibitors of the Tet efflux proteins (Nelson and Levy, 1999; Rothstein et al., 1993). Although compounds were identified that showed inhibition in in vitro assays, none of the compounds were put into development. One reason for the difficulty in finding a suitable inhibitor for the various tetracycline efflux pumps is that there is very little structural similarity between the efflux pumps found in gram-positive and gram-negative bacteria, thus it would be difficult to find an inhibitor for all of the various efflux proteins (Lomovskaya and Watkins, 2001a).

4.2.5 Efflux Mediated Resistance

Efflux refers to the active extrusion (pumping) of molecules from the cell and, in the case of antibiotics, the ability of efflux pumps to recognize and extrude these compounds serves to decrease the intracellular accumulation of the drug, thereby reducing or preventing access to the intracellular target. Historically, outer membrane impermeability was thought to be largely responsible for intrinsic resistance to toxic compounds associated with gram negatives. It is now recognized that the outer membrane permeability barrier is insufficient alone to provide resistance in many cases, but works in conjunction with efflux pumps, by reducing the rate of compound influx. Correspondingly, the level of impermeability of the outer membrane, which varies among gram negatives, is thought to determine to a significant extent the level of resistance conferred by efflux pumps (Li et al., 2000; Sanchez et al., 1997). There are five broad categories of efflux pumps: the ATP binding cassette family (ABC); major facilitator superfamily (MFS); the small multidrug resistance family (SMR); the multidrug and toxic compound extrusion family (MATE); and the resistance-nodulation-division family (RND) (Reviewed in (Kumar and Schweizer, 2005)). The ABC family derives energy from ATP hydrolysis whereas the other efflux pump families are compound-ion antiporters. In various contexts, each of these families can foster resistance to antimicrobials, but in gram-negative bacteria, the RND family is very widely distributed (examples summarized in Table 4.3)

Table 4.3 Example RND efflux pumps in gram negative bacteria[a]

Organism	Pump component[b]			Antibiotic substrates[c]
	MFP	RND	OMF	
Acinetobacter baummani	AdeA	AdeB	AdeC	AG, CM, FQ, TC
Agrobacterium tumefaciens	AmeA	AmeB	AmeC	CB, NO
Burkholderia cepacia	CeoA	CeoB	OpcM	CM, FQ, TM
Campylobacter jejuni	CmeA	CmeB	CmeC	AP, CM, EM, FQ, TC
E. coli	AcrA	AcrB	TolC	BL, CM, FQ, ML, NO, RF
	AcrA	AcrD	TolC	AG, FU, NO
	AcrE	AcrF	TolC	FQ
H. influenzae	AcrA	AcrB	TolC	EM, NO
K. pnuemoniae	AcrA	AcrB	?	FQ
P. mirabilis	AcrA	AcrB	?	NO, TG
P. aeruginosa	MexA	MexB	OprM	AG, BL, CM, ML, NO, TC, TG, TMCM, CP, FQ, TC
	MexC	MexD	OprJ	CM, FQ
	MexE	MexF	OprN	EM, TC
	MexJ	MexK	OprM/H	CM, EM, FQ, TC
	MexV	MexW	OprM	AG, ML, TC, TG
	MexX	MexY	OprM	AG, ML, TC, TG
Salmonella enterica serovar Typhimurium	AcrA	AcrB	TolC	BL, CM, EM, FQ, NO, RF, TC
S. maltophilia	SmeA	SmeB	SmeC	AG, BL, FQ
	SmeD	SmeE	SmeF	EM, FQ, TC

[a] Table summarized from Kumar and Schweizer (2005) and Poole (2004). For additional pumps and detail regarding substrate ranges and regulation, and corresponding references, please consult these publications.
[b] MFP, membrane fusion protein; RND, resistance nodulation division pump component; OMF, outer membrane factor.
[c] AG, aminoglycosides; AP, ampicillin; BL, -lacatamases; CB, carbenicillin; CM, chloramphenicol; CP, cephalosporins; EM, erythromycin; FQ, fluoroquinolones; FU, fusidic acid; ML, macrolides; NO, novobiocin; RF, rifampicin; TC, tetracycline; TG, tigecycline; TM, trimethoprim

and is by far the most relevant clinically, and therefore this family is the subject of intense interest. RND pumps in gram negatives function through a tripartite architecture: the inner-membrane-located RND pump component interacts with an outer membrane channel component (outer membrane factor; OMF) and a periplasmic membrane fusion protein (MFP) (Figure 4.1) (Kumar and Schweizer, 2005; Poole, 2004). This organization serves to span the double membrane envelope of the gram-negative cell to allow efficient compound transport across two membranes (Zgurskaya and Nikaido, 2000). Evidence has been accumulating, however, that RND family pumps may often extrude compounds from the periplasmic space or the outer leaflet of the inner membrane. This was initially suspected since RND pumps could confer resistance to β-lactam antibiotics, which act on periplasmically accessible targets (Li et al., 1994; Nikaido et al., 1998). Structural studies have revealed significant information on the prototypical AcrB inner membrane pump component, strongly suggesting that compounds are recognized in the periplasm and could enter into the pump through vestibules, accessed via the inner membrane

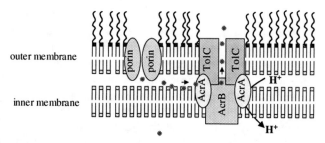

Fig. 4.1 Tripartite architecture of the resistance-nodulation-cell division (RND) family of efflux pumps in the Gram-negative cell envelope.

The AcrAB-To lC pump is shown (AcrB; RND pump component; AcrA, membrane fusion protein (MFP); TolC, outer membrane factor (OMF). Compounds are extruded from the periplasmic space (outer leaflet of inner membrane) or from inside the cell to the external environment. RND pumps are driven by H^+ antiport. Exchange of H^+ from periplasm to cytoplasm linked to extrusion of compounds across two membranes has led to the term ion-periporter for this class of pumps (Lomovskaya and Totrov, 2005).

(Elkins and Nikaido, 2003; Murakami et al., 2002; Yu et al., 2003, 2005).* This is consistent with mounting evidence that amino acid residues in periplasmic loops of RND inner membrane pump components are involved in determining substrate specificity (Eda et al., 2003; Elkins and Nikaido, 2002; Mao et al., 2002; Middlemiss and Poole, 2004). Furthermore, there is now experimental evidence using the *E. coli* AcrD pump showing that aminoglycosides can be transported from either the cytoplasm or the periplasm (Aires and Nikaido, 2005). Taken together, this evidence has prompted the recent likening of RND pumps to periplasmic vacuum cleaners, with the outer membrane channel component forming an exhaust pipe expelling compounds directly into the surrounding medium (Lomovskaya and Totrov, 2005). The capacity of multi-subunit RND family pumps to extrude compounds through the outer membrane to the external environment is a fundamental difference from single component pumps, for example from the MFS family, which can only pump compounds into the periplasm. In general, single component pumps are less clinically relevant in gram negatives, although there are a small number that may be important (Kumar and Schweizer, 2005). In cases where an RND pump and single component pump have overlapping substrate specificity, the ability of the RND pump to capture the substrate from the periplasm likely means that the two pumps can work together, in series, to provide an overall higher level of resistance (Lee et al., 2000). In addition to being very widely distributed (Table 4.3), RND family pumps can have a wide substrate range, allowing the extrusion of structurally dissimilar compounds. These include most classes of antibiotics, biocides, dyes, organic solvents, detergents, bile salts, β-lactamase inhibitors and several others (Kumar and Schweizer, 2005; Poole, 2004; Thanassi et al., 1997). The broad

---

* Two recent publications provide critical new structural insight into the assembly and function of RND family efflux pumps (Murakami et al., 2006, Seeger et al., 2006) and the reader is directed to these publications for more details.

specificity may result from the nature of the large amorphous binding cavity of the RND transporters (Yu et al., 2003). Moreover, some bacteria possess many different RND pumps with overlapping substrate specificities, providing a potential extreme and redundant ability to cope with toxic compounds. Presumably RND pumps exist in part to provide environmental adaptability and allow bacteria to survive in different harsh environments. Perhaps reflecting this, an environmental organism such as *P. aeruginosa*, which has a large and highly regulated genome, also has approximately 12 different putative RND family efflux pumps (Kumar and Schweizer, 2005; Poole, 2001; Stover et al., 2000). In contrast, *H. influenzae*, having a very limited environmental niche (i.e. the human respiratory tract) has only one RND pump (Fleischmann et al., 1995; Sanchez et al., 1997). The *E. coli* AcrAB–TolC pump is likely involved in resisting bile salts in the enteric tract (Thanassi et al., 1997) while the *N. gonorrhoeae* MtrCDE pump provides resistance to fecal lipids (Shafer et al., 1995). RND pumps are also thought to control exposure to toxic metabolites (Helling et al., 2002). Although the role of pumps in providing drug resistance is of primary concern, recent efforts have begun to investigate other natural roles for efflux pumps. Clearly some of these pumps serve other physiological roles besides drug resistance; but little is known about potential cellular substrates of RND pumps. The MexAB–OprM and MexEF–OprN pumps of *P. aeruginosa* may influence virulence or other functions through extrusion of quorum sensing molecules important for cell density dependent gene regulation (Evans et al., 1998; Kohler et al., 2001; Pearson et al., 1999) and MexAB–OprM itself may play a role in invasiveness (Hirakata et al., 2002). The IefABC pump of *Agrobacterium tumefaciens* pumps isoflavanoid compounds that are involved in root colonization (Palumbo et al., 1998), further supporting non-resistance roles for some efflux pumps.

4.2.5.1 Intrinsic and Mutationally Acquired Efflux Mediated Resistance

Whether as general toxic compound resistance mechanisms or as a consequence of other physiological functions, efflux pumps are one of the more intriguing examples of the notion that bacteria exist in harsh environments, and the evolutionary response to this contributes to the difficulty in developing new antibacterial agents. Unlike other target based or enzymatic resistance mechanisms that may exist and spread or be selected directly by drug exposure, efflux pumps are typical features of gram negatives and therefore constitute a pre-existing determinant of intrinsic resistance. This hampers efforts to achieve broad-spectrum activity in novel antibacterials. Recent examples include tigecycline, a glycyl derivative of minocycline, which circumvents both ribosomal protection and efflux by tetracycline-specific efflux pumps (e.g. TetA), but which is still subject to RND mediated efflux in *P. aeruginosa* (Dean et al., 2003), *Proteus mirabilis* (Visalli et al., 2003) and *K. pneumoniae* (Ruzin et al., 2005). Another novel compound, the peptide deformylase inhibitor LBM415, is subject to AcrAB–TolC mediated efflux in *H. influenzae*, a target organism for this class of antimicrobial (Dean et al., 2005). The ability of RND family pumps to recognize even novel compound derivatives or entirely new entities and reduce susceptibility is a major impediment to target based drug discovery. Indeed, a recent report detailing target identification efforts for a large series of antimicrobial compounds identified from direct antibacterial screening in *E. coli* revealed that the vast majority of active compounds were also AcrAB–TolC pump substrates (Li et al., 2004c). Despite the general impact of these pumps on antimicrobial effectiveness, there are nonetheless compounds that are effective and currently useful. These include, for example, the fluoroquinolones, many β-lactams and aminoglyocosides which, while being substrates for various pumps (Table 4.3), often have sufficient potency and are perhaps more modestly impacted by intrinsic levels of pump expression. In the case of currently useful antimicrobials, however, mutations causing increased pump expression can decrease susceptibility substantially. In bacteria with several pumps, there are often those that are constitutively expressed (e.g. MexAB–OprM of *P. aeruginosa* and AcrAB–TolC of *E. coli*) and other pumps that are not appreciably expressed, at least under laboratory conditions (e.g. MexCD-OprJ of *P. aeruginosa* and AcrEF–TolC of *E. coli*). The levels of pumps such as MexAB–OprM and AcrAB–TolC can be increased by mutations in their cognate repressors (i.e. MexR (Adewoye et al., 2002) and AcrR (Wang et al., 2001b)) or in other genes (e.g. *nalC* (Cao et al., 2004) and *nalD* (Sobel et al., 2005) increasing MexAB–OprM expression). Additionally, complex regulatory circuits such as the multiple antibiotic resistance (MAR) locus control expression of pumps such as AcrAB–TolC (Alekshun and Levy, 1997) and also the expression of outer membrane porins through the involvement of the global regulators MarA, SoxS and Rob (Delihas and Forst, 2001; Li and Nikaido, 2004; Randall and Woodward, 2002). This circuit can therefore coordinately decrease porin expression, reducing antibiotic influx, and increase AcrAB–TolC expression, increasing efflux. Mutations in regulatory genes for pumps such as MexCD-OprJ (*nfxB*), which are not appreciably expressed under laboratory conditions, will result in these pumps being turned on to generally high levels with corresponding increases in resistance to their substrate antibiotics. The natural regulation of pump expression and mutational events leading to pump overexpression in vivo are not well understood and are likely very complex and as such exceed the scope of this discussion (for a recent review see (Kumar and Schweizer, 2005)). However, loss of function mutations in repressors or other regulatory mutations can be selected in vitro at relatively high frequencies. Since these pumps can accommodate a wide range of compounds, pump upregulation selected by exposure to one antibiotic can result in resistance to many antibiotics, and the extent of multidrug resistance conferred can be quite broad in bacteria such as *P. aeruginosa*. Pump overexpressing mutants are increasingly being identified among clinical isolates (Bert and Lambert-Zechovsky, 1996; Kriengkauykiat et al., 2005; Wang et al., 2001b; Williams et al., 1984). Furthermore, the ability of RND pumps to extrude common biocides such as triclosan (Chuanchuen et al., 2003) and the inducibility of certain pumps by biocides (Morita et al., 2003) have prompted speculation regarding the potential of residual environmental biocides selecting for pump mutants contributing to the development of multidrug resistance (Poole, 2004; Randall et al., 2004). Whether this is occurring to an appreciable extent is not known, however, the emergence of pump overexpressors generally is compromising the utility of currently available antimicrobials and efforts to develop novel antimicrobials. A major focus of devising antimicrobials with new mechanisms of action is to avoid cross resistance conferred by pre-existing target-based resistance mechanisms (or selection of cross resistance by the new agent). The presence of pump overexpressors clinically, presumably selected by previous drug exposures, serves to defeat this strategy in cases where the novel compound is a pump substrate. Finally, the reduced accumulation of antimicrobial compounds conferred by efflux pumps also enhances the emergence and impact of target based resistance mechanisms.

4.2.5.2 Circumventing Efflux and Pump Inhibitors

Given the general negative impact of efflux on the usefulness of existing antibiotics as well as on the development of novel compounds, inhibitors of efflux pumps could find utility in improving the effectiveness of pump substrate antibiotics. The most advanced effort in this area has been focused on the efflux pumps of *P. aeruginosa*, through the development of efflux pump inhibitors (EPIs) based on the initial pump inhibitor MC 207,110 (*phe-arg-β*-napthylamide), identified through an EPI screening program (Lomovskaya et al., 2001). MC 207,110 and similar compounds have shown effective reversal of pump mediated resistance in *P. aeruginosa* (Renau et al., 1999, 2001, 2002, 2003), Enterobacteriaceae and *H. influenzae* (Lomovskaya and Watkins, 2001b), *Enterobacter aerogenes* (Mallea et al., 2002) and *Campylobacter jejuni* (Mamelli et al., 2003). Inhibitors such as MC 207,110 are pump substrates and are thought to act through competitive binding and interference with substrate recognition (Lomovskaya et al., 2001). Competitive inhibitors may be somewhat pump and substrate specific, necessitating the pairing of the inhibitor with a particular antibiotic to facilitate optimal development. As mentioned in Section 4.2.2.6, an inhaled formulation of a pump inhibitor is undergoing Phase 1b testing for use in cystic fibrosis patients suffering from *P. aeruginosa* infections (www.mpexpharma.com/pr_aug1_05.html), with the ultimate goal to extend the utility of fluoroquinolones. Recent reports detail additional compounds with pump inhibitory activity against pumps in *E. aerogenes* and *K. pneumoniae* (Chevalier et al., 2004; Mallea et al., 2002, 2003). Another strategy for combating efflux is the development of antimicrobials that avoid efflux, either by escaping recognition or potentially by rapidly influxing and binding tightly to the intracellular target. However, the ability of RND pumps to recognize such a wide range of substrates, and a relatively sparse knowledge of substrate recognition and cell penetration makes this difficult. To date only gross chemical descriptors such as hydrophobicity have been generally correlated to efflux (Li et al., 2004c; Nikaido et al., 1998). Effective gram-negative antibiotics generally need to be amphiphilic in order to traverse both membranes, and this property also seems to determine efflux recognition (Lomovskaya and Watkins, 2001a). Nonetheless, some antibiotics are less impacted by efflux, which suggests that there is a rationale for developing improved compounds. The recent solving of crystal structures for pump proteins, including AcrB bound to pump substrates (Elkins and Nikaido, 2003; Yu et al., 2003, 2005), the MFP (Akama et al., 2004b) and OMF (Akama et al., 2004a; Koronakis et al., 2000) components and the evolving understanding of pump assembly and function should facilitate the development of novel pump inhibitors and/or antimicrobial compounds avoiding efflux.*

4.3 Future Directions

4.3.1 Preventive Measures

Most intensive care physicians recognize the continuing problem of antibiotic resistance among the severely ill in the hospital. Many hospitals choose to restrict the use of new antibiotics, to reserve their use for only those cases where older antibiotics have failed (Carlet et al., 2004). In addition, several other methods have been suggested to reduce the problem of antibiotic resistance.

4.3.1.1 Antibiotic Cycling

Antibiotic cycling or rotation has been suggested as a method for preventing the development of antibiotic resistance in the hospital setting. The practice of antibiotic cycling involves the scheduled switching of empiric antibiotic therapy regimens within a given hospital from one drug class to another after a pre-determined period of time. The theory behind antibiotic cycling is that as the antibiotics are switched from one class to the next, it will result in lower prevalence of antibiotic resistance to any one antibiotic. A number of studies have been performed to test this hypothesis.

One of the early studies was conducted over an 18-month period of time in a hematology-oncology unit (Dominguez et al., 2000). Patients were randomized into four treatment regimens: (1) ceftazidime + vancomycin, (2) imipenem, (3) aztreonam + cefazolin and (4) ciprofloxacin + clindamycin. Antibiograms revealed no increase in bacterial resistance over the course of the study. They did, however, note an increase in the incidence of enterococcal infections (Dominguez et al., 2000). In another study examining ventilator associated pneumonia (VAP) patients in the ICU, an empiric regimen of ceftazidime + ciprofloxacin was replaced by a monthly cycle of a β-lactam (cefepime, piperacillin-tazobactam, imipenem or ticarcillin-clavulanic acid) plus an aminoglycoside (amikacin, tobramycin, netilmycin or isepamycin) for a two-year period (Gruson et al., 2000). Compared to the two-year period prior to the study start, susceptibility of a number of the test antibiotics increased, especially for *P. aeruginosa* and *B. cepacia*. In addition the rates of *S. aureus* susceptible to methicillin increased from 40 to 60%. Furthermore, there was an overall reduction in the incidence of VAP in the ICU patients (Gruson et al., 2003). The authors concluded that the strategy of antibiotic cycling could be an effective method for reducing resistance in the pathogens that cause VAP (Gruson et al., 2000, 2003).

It has also been noted that a rotation of antibiotics within the ICU can also have a positive effect on the non-ICU wards in that hospital-wide resistance rates dropped during the time of antibiotic rotation studies (Hughes et al., 2004).

Although the rates of resistance may decrease with antibiotic cycling, this technique may have no effect on the rates of colonization in hospitalized patients. Toltzis et al. conducted a study in the neonatal intensive care unit (NICU) involving proven or suspected gram-negative infections (Toltzis et al., 2002). Patients were given gentamicin, piperacillin-tazobactam, or ceftazidime on a rotating monthly basis. The patients in a control cohort were prescribed antibiotics of the physician's choice. At the end of the study, no differences in colonization with antibiotic resistant gram-negative bacilli were noted between control and rotation groups (Toltzis et al., 2002).

Despite the promising results that have been published from these studies, this practice has not been made commonplace. Some of the skepticism of cycling antibiotics may be due to some noted shortcomings in the studies including the indication that as many as 10–50% of patients get exposed to antibiotics that are off cycle, and the lack of compliance with treating physicians (Fridkin, 2003). Additional studies of antibiotic cycling will have to be conducted before widespread implementation of this practice is accepted.

4.3.1.2 Infection Control

The control of the increase in antibiotic appears to require several strategies including the ongoing surveillance of antibiotic resistance and implementing strict hygiene controls to limit the spread of resistance (Weinstein, 2001). Simple hand washing techniques remain the primary means by which the spread of antibiotic resistance is kept to a minimum within the healthcare setting. However, surveys of hand washing compliance reveal that hand washing actually takes place only after 25–50% of patient contacts. In addition, healthcare workers appear to have an inflated estimate of their compliance with hand washing as 85% of hospital personnel reported that they wash their hands after patient contact, however the observed rate was only 28% (Weinstein, 2001).

A systematic screening of patients for resistant bacteria upon admission has been suggested (Carlet et al., 2004). For example in some institutions, every patient is cultured for nasal carriage of MRSA (Lucet et al., 2003). These results are used to place patients in strict isolation until the infection is cleared. In others, every patient coming from a LCTF is treated as if they are colonized with MRSA until the cultures prove otherwise (Lucet et al., 2003). Mupericin has been used topically to eradicate MRSA from the nasal passage, however the results from this intervention has been mixed (Carlet et al., 2004). In general, good basic isolation and infection control measures remain the best way to limit the spread of antibiotic resistance.

4.3.2 New Agents

The always-present problem of bacterial resistance to existing therapies for infections caused by gram-negative bacteria highlights the need for new antibacterial agents. Unfortunately, there have been few new antibiotics developed in recent years that include coverage for gram-negative pathogens. However, there are a few new agents that are in various stages of development. Two of these new agents are β-lactams that add to the spectrum of an existing class of antibiotics. Ceftobiprole (Ro 63-9141, BAL9141) is a pyrrolidinone cephalosporin that has increased affinity for essential PBPs. This includes PBP2' which is responsible for methicillin-resistance in *S. aureus* (Hebeisen et al., 2001). For gram-negative bacteria, ceftobiprole showed similar in vitro activity to cefotaxime and ceftriaxone for non-ESBL producing strains of *E. coli, K. pneumoniae* and had better activity against *E. cloacae* and *C. freundii* (Hebeisen et al., 2001; Issa et al., 2004). In addition, it has good to moderate activity against ceftazidime susceptible *P. aeruginosa* (Hebeisen et al., 2001; Zbindien et al., 2002). Ceftobiprole is currently undergoing phase 3 clinical trials. Doripenem (S-4661) is a new parenteral 1-β-methyl carbapenem that has a broad-spectrum of activity. It has similar activity to imipenem against gram-positive bacteria (Brown and Traczewski, 2005; Ge et al., 2004). Against Enterobacteriaceae, the activity of doripenem is slightly better than that of imipenem, and similar to the activity of meropenem (Brown and Traczewski, 2005; Ge et al., 2004). Doripenem is also active against most strains of *P. aeruginosa*, however results in a wide range of MICs for other non-fermentors (Brown and Traczewski, 2005). Like the other carbapenems, resistance to doripenem in *P. aeruginosa* results from the loss of the OprD porin protein (Mushtaq et al., 2004).

A novel class of antimicrobial agents is peptide deformylase inhibitors. One of these, LBM415, is active against *M. catharralis, H. influenzae, N. meningitidis, L. pneumophila, Bacteroides* and *Fusobcterium* spp., but has no appreciable activity against Enterobacteriaceae or non-fermentative gram-negatives (Fritsche and Jones, 2004; Fritsche et al., 2004; Ryder et al., 2004).

Glycylcyclines are a unique chemical class of antibiotics that act to inhibit protein synthesis at the level of the bacterial ribosome. Two glycylcyclines have been chosen for development. The first of these, tigecycline was recently approved for the treatment of complicated skin and skin structure infections and complicated intraabdominal infections in the USA (Wyeth Pharmaceuticals, 2005). It has a broad spectrum of antibacterial activity, including inhibition of gram-positive, gram-negative, atypical and anaerobic bacteria. It is active against multiply resistant gram-positive bacteria, including methicillin-resistant *S. aureus* (MRSA), penicillin-resistant *S. pneumoniae* (PRSP), vancomycin-resistant *Enterococcus* spp. (VRE), and extended-spectrum β-lactamase (ESBL), producing *E. coli* and *K. pneumoniae* (Bradford et al., 2005; Petersen et al., 1999; Petersen and Bradford, 2005). In addition, tigecycline is active against strains that carry any of the two major types of tetracycline resistance genes for ribosomal protection and efflux-mediated resistance (Fluit et al., 2005; Petersen et al., 1999). In clinical trials, tigecycline was generally well tolerated, safe, and effective in treating skin and soft tissue, as well as complicated intraabdominal infections (Babinchak et al., 2005; Ellis-Grosse et al., 2005). In preclinical studies PTK 0796 (BAY 73-6944) has shown a similar potency and spectrum of activity to tigecycline, although somewhat less activity against gram-negative pathogens (Macone et al., 2003; Traczewski and Brown, 2003). However, clinical trial data has yet to generated for this compound.

4.4 Concluding Remarks

As long as antibacterial agents are used for treatment, resistance to those therapies will continue to develop and disseminate among bacterial pathogens. For most resistant strains, alternative therapies are still available; however, some truly "pan-resistant" strains have been emerging for which there are no therapeutic options left (Ahmad et al., 1999). This comes at a time when many large pharmaceutical companies are ceasing research activities for new antimicrobial agents (Projan, 2003). There is a great worry among the infectious disease community that there is a serious risk that a growing proportion of infections will be untreatable (Livermore, 2004). The need for continued surveillance of antibiotic resistance is apparent.

References

Abraham, E.P. and Chain, E., (1940), An enzyme from bacteria able to destroy penicillin. *Nature*, **146**, 837.

Acar, J.F. and Goldstein, F.W., (1997), Trends in bacterial resistance to fluoroquinolones. *Clin Infect Dis*, **24** (Suppl. 1), S67–73.

Acosta, M.B., Ferreira, R.C., Padilla, G., Ferreira, L.C. and Costa, S.O., (2000), Altered expression of oligopeptide-binding protein (OppA) and aminoglycoside resistance in laboratory and clinical *Escherichia coli* strains. *J Med Microbiol*, **49**, 409–413.

Adams, D.E., Shekhtman, E.M., Zechiedrich, E.L., Schmid, M.B. and Cozzarelli, N.R., (1992), The role of topoisomerase IV in partitioning bacterial replicons and the structure of catenated intermediates in DNA replication. *Cell*, **71**, 277–288.

Adewoye, L., Sutherland, A., Srikumar, R. and Poole, K., (2002), The *mexR* repressor of the *mexAB-oprM* multidrug efflux operon in *Pseudomonas aeruginosa*: characterization of mutations compromising activity. *J Bacteriol*, **184**, 4308–4312.

Afzal-Shah, M., Villar, H.E. and Livermore, D.M., (1999), Biochemical characteristics of a carbapenemase from an *Acinetobacter baumannii* isolate collected in Buenos Aires, Argentina. *J Antimicrob Chemother*, **43**, 127–131.

Ahmad, M., Urban, C., Mariano, N., Bradford, P.A., Calcagni, E., Projan, S.J., Bush, K. and Rahal, J.J., (1999), Clinical characteristics and molecular epidemiology associated with imipenem-resistant *Klebsiella pneumoniae*. *Clin Infect Dis*, **29**, 352–355.

Aires, J.R., Kohler, T., Nikaido, H. and Plesiat, P., (1999), Involvement of an active efflux system in the natural resistance of *Pseudomonas aeruginosa* to aminoglycosides. *Antimicrob Agents Chemother*, **43**, 2624–2628.

Aires, J.R. and Nikaido, H., (2005), Aminoglycosides are captured from both periplasm and cytoplasm by the AcrD multidrug efflux transporter of *Escherichia coli*. *J Bacteriol*, **187**, 1923–1929.

Akama, H., Kanemaki, M., Yoshimura, M., Tsukihara, T., Kashiwagi, T., Yoneyama, H., Narita, S., Nakagawa, A. and Nakae, T., (2004a), Crystal structure of the drug discharge outer membrane protein, OprM, of *Pseudomonas aeruginosa*: dual modes of membrane anchoring and occluded cavity end. *J Biol Chem*, **279**, 52816–52819.

Akama, H., Matsuura, T., Kashiwagi, S., Yoneyama, H., Narita, S., Tsukihara, T., Nakagawa, A. and Nakae, T., (2004b), Crystal structure of the membrane fusion protein, MexA, of the multidrug transporter in *Pseudomonas aeruginosa*. *J Biol Chem*, **279**, 25939–25942.

Alekshun, M.N. and Levy, S.B., (1997), Regulation of chromosomally mediated multiple antibiotic resistance: the *mar* regulon. *Antimicrob Agents Chemother*, **41**, 2067–2075.

Allen, N.E., Alborn, W.E., Jr, Hobbs, J.N., Jr and Kirst, H.A., (1982), 7-Hydroxytropolone: an inhibitor of aminoglycoside-2"-O-adenylyltransferase. *Antimicrob Agents Chemother*, **22**, 824–831.

Ambler, R.P., (1980), The structure of β-lactamases. *Philos Trans R Soc Lond [Biol]*, **289**, 321–331.

Azucena, E. and Mobashery, S., (2001), Aminoglycoside-modifying enzymes: mechanisms of catalytic processes and inhibition. *Drug Resist Updat*, **4**, 106–117.

Babinchak, T., Ellis-Grosse, E.J., Dartois, N., Rose, G.M. and Loh, E., (2005), The efficacy and safety of tigecycline in the treatment of complicated intra-abdominal infections: analysis of pooled clinical trial data. *Clin Infect Dis*, **41**, S354–S367.

Bandoh, K., Watanabe, K., Muto, Y., Tanaka, Y., Kato, N. and Ueno, K., (1992), Conjugal transfer of imipenem resistance in *Bacteroides fragilis*. *J Antibiot*, **45**, 542–547.

Barclay, M.L. and Begg, E.J., (2001), Aminoglycoside adaptive resistance: importance for effective dosage regimens. *Drugs*, **61**, 713–721.

Barclay, M.L., Begg, E.J. and Chambers, S.T., (1992), Adaptive resistance following single doses of gentamicin in a dynamic *in vitro* model. *Antimicrob Agents Chemother*, **36**, 1951–1957.

Barclay, M.L., Begg, E.J., Chambers, S.T., Thornley, P.E., Pattemore, P.K. and Grimwood, K., (1996), Adaptive resistance to tobramycin in *Pseudomonas aeruginosa* lung infection in cystic fibrosis. *J Antimicrob Chemother*, **37**, 1155–1164.

Barnaud, G., Arlet, G., Verdet, C., Gaillot, O., Lagrange, P.H. and Philippon, A., (1998), *Salmonella enteritidis*: AmpC plasmid-mediated inducible β-lactamase (DHA-1) with an *ampR* gene from *Morganella morganii*. *Antimicrob Agents Chemother*, **42**, 2352–2358.

Bartlett, J.G., (2003), *2003–2004 Pocket Book of Infectious Disease Therapy*. Baltimore MD, Lipinncott Williams and Wilkins.

Bastida, T., Perez-Vazquez, M., Campos, J., Cortes-Lletget, M.C., Roman, F., Tubau, F., de la Campa, A.G. and Alonso-Tarres, C., (2003), Levofloxacin treatment failure in *Haemophilus influenzae* pneumonia. *Emerg Infect Dis*, **9**, 1475–1478.

Baucheron, S., Chaslus-Dancla, E. and Cloeckaert, A., (2004), Role of TolC and parC mutation in high-level fluoroquinolone resistance in *Salmonella enterica* serotype Typhimurium DT204. *J Antimicrob Chemother*, **53**, 657–659.

Baucheron, S., Imberechts, H., Chaslus-Dancla, E. and Cloeckaert, A., (2002), The AcrB multidrug transporter plays a major role in high-level fluoroquinolone resistance in *Salmonella enterica* serovar typhimurium phage type DT204. *Microb Drug Resist*, **8**, 281–289.

Bauer, G., Berens, C., Projan, S.J. and Hillen, W., (2004), Comparison of tetracycline and tigecycline binding to ribosomes mapped by dimethylsulphate and drug-directed Fe^{2+} cleavage of 16S rRNA. *J Antimicrob Chemother*, **53**, 592–599.

Bauernfeind, A., Casellas, J.M., Goldberg, M., Holley, M., Jungwirth, R., Mangold, P., Rohnisch, T., Schweighart, S. and Wilhelm, R., (1992), A new plasmidic cefotaximase from patients infected with *Salmonella typhimurium*. *Infection*, **20**, 158–163.

Bauernfeind, A., Stemplinger, I., Jungwirth, R., Ernst, S. and Casellas, J.M., (1996a), Sequences of β-lactamase genes encoding CTX-M-1 (MEN-1) and CTX-M-2 and relationship of their amino acid sequences with those of other β-lactamases. *Antimicrob Agents Chemother*, **40**, 509–513.

Bauernfeind, A., Schneider, I., Jungwirth, R., Sahly, H. and Ullmann, U., (1999), A novel type of AmpC β-lactamase, ACC-1 produced by a *Klebsiella pneumoniae* strain causing nosocomial pneumonia. *Antimicrob Agents Chemother*,1 **43**, 1924–1931.

Bauernfeind, A., Stemplinger, I., Jungwirth, R., Wilhelm, R. and Chong, Y., (1996b), Comparative characterization of the cephamycinase bla_{CMY-1} gene and its relationship with other β-lactamase genes. *Antimicrob Agents Chemother*, **40**, 1926–1930.

Bayer, A.S., Norman, D.C. and Kim, K.S., (1987), Characterization of impermeability variants of *Pseudomonas aeruginosa* isolated during unsuccessful therapy of experimental endocarditis. *Antimicrob Agents Chemother*, **31**, 70–75.

Beauclerk, A.A. and Cundliffe, E., (1987), Sites of action of two ribosomal RNA methylases responsible for resistance to aminoglycosides. *J Mol Biol*, **193**, 661–671.

Belland, R.J., Morrison, S.G., Ison, C. and Huang, W.M., (1994), *Neisseria gonorrhoeae* acquires mutations in analogous regions of *gyrA* and *parC* in fluoroquinolone-resistant isolates. *Mol Microbiol*, **14**, 371–380.

Berens, G. and Hillen, W., (2003), Gene regulation by tetracyclines: Constraints of resistance regulation in bacteria shape TetR for applicatin in eukaryotes. *Eur J Biochem*, **270**, 3109–3121.

Bergeron, J., Ammirati, M., Danley, D., James, L., Norcia, M., Retsema, J., Strick, C.A., Su, W.G., Sutcliffe, J. and Wondrack, L., (1996), Glycylcyclines bind to the high-affinity tetracycline ribosomal binding site and evade Tet(M)- and Tet(O)-mediated ribosomal protection. *Antimicrob Agents Chemother*, **40**, 2226–2228.

Bergogne-Berezin, E. and Towner, K.J., (1996), *Acinetobacter* spp. as nosocomial pathogens: microbiological clinical, and epidemiological features. *Clin Microbiol Rev*, **9**, 148–165.

Bermudes, H., Jude, F., Chaibi, E.B., Arpin, C., Bebear, C., Labia, R. and Quentin, C., (1999), Molecular characterization of TEM-59 (IRT-17), a novel inhibitor-resistant TEM-derived β-lactamase in a clinical isolate of *Klebsiella oxytoca*. *Antimicrob Agents Chemother*, **43**, 1657–1661.

Bert, F. and Lambert-Zechovsky, N., (1996), Comparative distribution of resistance patterns and serotypes in *Pseudomonas aeruginosa* isolates from intensive care units and other wards. *J Antimicrob Chemother*, **37**, 809–813.

Bhavnani, S.M., Callen, W.A., Forrest, A., Gilliland, K.K., Collins, D.A., Paladino, J.A. and Schentag, J.J., (2003), Effect of fluoroquinolone expenditures on susceptibility of *Pseudomonas aeruginosa* to ciprofloxacin in U.S. hospitals. *Am J Health Syst Pharm*, **60**, 1962–1970.

Bonnet, R., Sampaio, J.L.M., Labia, R., Champs, C.D., Sirot, D., Chanel, C. and Sirot, J., (2000), A novel CTX-M β-lactamase (CTX-M-8) in cefotaxime-resistant Enterobacteriaceae isolated in Brazil. *Antimicrob Agents Chemother*, **44**, 1936–1942.

Bonomo, R.A., Currie-McCumber, C. and Shlaes, D.M., (1992), OHIO-1 β-lactamase resistant to mechanism-based inactivators. *FEMS Microbiol Lett*, **71**, 79–82.

Bonomo, R.A., Rudin, S.A. and Shlaes, D.M., (1997), Tazobactam is a potent inactivator of selected inhibitor-resistant class A β-lactamases. *FEMS Microbiol Lett*, **148**, 59–62.

Bou, G., Oliver, A., Ljeda, M., Monzo'n, C. and Marti'nez-Beltra'n, J., (2000), Molecular characterization of FOX-4, a new AmpC-type plasmid-mediated β-lactamase from an *Escherichia coli* strain isolated in Spain. *Antimicrob Agents Chemother*, **44**, 2549–2553.

Bradford, P.A., Bratu, S., Urban, C., Visalli, M., Mariano, N., Landman, D.L., Rahal, J.J., Brooks, S., Cebular, S. and Quale, J., (2004), Emergence of carbapenem-resistant *Klebsiella* spp. possessing the class A carbapenem-hydrolyzing KPC-2 and inhibitor-resistant TEM-30 β-lactamases in New York City. *Clin Infect Dis*. **39**, 55–60

Bradford, P.A., Urban, C., Jaiswal, A., Mariano, N., Rasmussen, B.A., Projan, S.J., Rahal, J.J. and Bush, K., (1995), SHV-7, a novel cefotaxime-hydrolyzing β-lactamase, identified in *Escherichia coli* isolates from hospitalized nursing home patients. *Antimicrob Agents Chemother*, **39**, 899–905.

Bradford, P.A., Urban, C., Mariano, N., Projan, S.J., Rahal, J.J. and Bush, K., (1997), Imipenem resistance in *Klebsiella pneumoniae* is associated with the combination of ACT-1, a plasmid-mediated AmpC β-lactamase and the loss of an outer membrane protein. *Antimicrob Agents Chemother*, **41**, 563–569.

Bradford, P.A., Weaver-Sands, D.T. and Petersen, P.J., (2005), *In vitro* activity of tigecycline against isolates from patients enrolled in phase 3 clinical trials for complicated skin and skin structure infections and complicated intra-abdominal infections. *Clin Infect Dis*, **41** (Suppl. 5), S315–332.

Bradford, P.A., Yang, Y., Sahm, D., Grope, I., Gardovska, D. and Storch, G., (1998), CTX-M-5, a novel cefotaxime-hydrolyzing β-lactamase from an outbreak of *Salmonella typhimurium* in Latvia. *Antimicrob Agents Chemother*, **42**, 1980–1984.

Bratu, S., Landman, D.L., Alam, M., Tolentino, E. and Quale, J., (2005a), Detection of KPC carbapenemase-hydrolyzing enzymes in *Enterobacter* spp. from Brooklyn, New York. *Antimicrob Agents Chemother*, **49**, 776–778.

Bratu, S., Landman, D.L., Haag, R., Recco, R., Eramo, A., Alam, M. and Quale, J., (2005b), Rapid spread of carbapenem-resistant *Klebsiella pneumoniae* in New York City. *Antimicrob Agents Chemother*, **165**, 1430–1435.

Brazas, M.D. and Hancock, R.E., (2005), Ciprofloxacin induction of a susceptibility determinant in *Pseudomonas aeruginosa. Antimicrob Agents Chemother*, **49**, 3222–3227.

Bret, L., Chanel, C., Sirot, D., Labia, R. and Sirot, J., (1996), Characterization of an inhibitor-resistant enzyme IRT-2 derived from TEM-2 β-lactamase produced by *Proteus mirabilis* strains. *J. Antimicrob Chemother*, **38**, 183–191.

Brown, S.D. and Traczewski, M.M., (2005), Comparative in vitro antimicrobial activity of a new carbapenem, doripenem: tentative disc diffusion criteria and quality control. *J Antimicrob Chemother*, **55**, 944–949.

Bryan, L.E., O'Hara, K. and Wong, S., (1984), Lipopolysaccharide changes in impermeability-type aminoglycoside resistance in *Pseudomonas aeruginosa. Antimicrob Agents Chemother*, **26**, 250–255.

Buerra, B., Soto, S., Cal, S. and Mendoza, M.C., (2000), Antimicrobial resistance and spread of class 1 integrons among *Salmonella* serotypes. *Antimicrob Agents Chemother*, **44**, 2166–2169.

Burdett, V., (1991), Tet(M)-promoted release of tetracylcine from ribosomes is GTP dependant. *J. Bacteriol*, **178**, 3246–2351.

Burk, D.L., Hon, W.C., Leung, A.K. and Berghuis, A.M., (2001), Structural analyses of nucleotide binding to an aminoglycoside phosphotransferase. *Biochemistry*, **40**, 8756–8764.

Burns J.L., Van Dalfsen J.M., Shawar R.M., Otto K.L., Garber R.L., Quan J.M., Montgomery A.B., Albers G.M., Ramsey B.W. and Smith A.L., (1999), Effect of chronic intermittent administration of inhaled tobramycin on respiratory microbial flora in patients with cystic fibrosis. *J Infect Dis.* 179(5):1190–6.

Bush, K., (1998), Metallo-β-lactamases: a class apart. *Clin Infect Dis*, **27** (Suppl. 1), S46–S53.

Bush, K., Jacoby, G.A. and Medeiros, A.A., (1995), A functional classification scheme for β-lactamases and its correlation with molecular structure. *Antimicrob Agents Chemother*, **39**, 1211–1233.

Butaye, P., Cloeckaert, A. and Schwarz, S., (2003), Mobile genes coding for efflux-mediated antimicrobial resistance in Gram-positive and Gram-negative bacteria. *Int J Antimicrob Agents*, **22**, 205–210.

Cao, L., Srikumar, R. and Poole, K., (2004), MexAB-OprM hyperexpression in NalC-type multidrug-resistant *Pseudomonas aeruginosa*: identification and characterization of the *nalC* gene encoding a repressor of PA3720–PA3719. *Mol Microbiol*, **53**, 1423–1436.

Carlet, J., Ben Ali, A. and Chalfine, A., (2004), Epidemiology and control of antibiotic resistance in the intensive care unit. *Curr Opin Infect Dis*, **17**, 309–316.

Caroff, N., Espaze, E., Be'rard, I., Richet, H. and Reynaud, A., (1999), Mutations in the ampC promotor of *Escherichia coli* isolates resistant to oxyiminocephalosporins without extended-spectrum β-lactamase production. *FEMS Microbiol Lett*, **173**, 459–465.

Castanheira, M., Toleman, M.A., Jones, R.N., Schmidt, F.J. and Walsh, T.R., (2004), Molecular characterization of a β-lactamase gene, bla_{GIM-1}, encoding a new subclass of metallo-β-lactamase. *Antimicrob Agents Chemother*, **48**, 4654–4661.

Centers for Disease Control (CDC), (2004), National nosocomial infections surveillance (NNIS) system report, data summary from January 1992 through June 2004, issued October 2004. *Am J Infect Contr*, **32**, 470–485.

Chaibi, E.B., Sirot, D., Paul, G. and Labia, R., (1999), Inhibitor-resistant TEM-β-lactamases: phenotypic, genetic and biochemical characteristics. *J Antimicrob Chemother*, **43**, 447–458.

Champoux, J.J., (2001), DNA topoisomerases: structure, function, and mechanism. *Annu Rev Biochem*, **70**, 369–413.

Chan, Y.Y. and Chua, K.L., (2005), The *Burkholderia pseudomallei* BpeAB-OprB efflux pump: expression and impact on quorum sensing and virulence. *J Bacteriol*, **187**, 4707–4719.

Chan, Y.Y., Tan, T.M., Ong, Y.M. and Chua, K.L., (2004), BpeAB-OprB, a multidrug efflux pump in *Burkholderia pseudomallei*. *Antimicrob Agents Chemother*, **48**, 1128–1135.

Chen, C.R., Malik, M., Snyder, M. and Drlica, K., (1996), DNA gyrase and topoisomerase IV on the bacterial chromosome: quinolone-induced DNA cleavage. *J Mol Biol*, **258**, 627–637.

Chen, F.J., Lauderdale, T.L., Ho, M. and Lo, H.J., (2003), The roles of mutations in *gyrA*, *parC*, and *ompK35* in fluoroquinolone resistance in Klebsiella pneumoniae. *Microb Drug Resist*, **9**, 265–271.

Cheung, T.K., Chu, Y.W., Chu, M.Y., Ma, C.H., Yung, R.W. and Kam, K.M., (2005), Plasmid-mediated resistance to ciprofloxacin and cefotaxime in clinical isolates of *Salmonella enterica* serotype Enteritidis in Hong Kong. *J Antimicrob Chemother*, **56**, 586–589.

Chevalier, J., Bredin, J., Mahamoud, A., Mallea, M., Barbe, J. and Pages, J.M., (2004), Inhibitors of antibiotic efflux in resistant *Enterobacter aerogenes* and *Klebsiella pneumoniae* strains. *Antimicrob Agents Chemother*, **48**, 1043–1046.

Chopra, I. and Roberts, M.C., (2001), Tetracycline antibiotics: mode of action, applications, molecular biology, and epidemiology of bacterial resistance. *Microbiol Mol Biol Rev*, **65**, 232–260.

Chow, J.W., Kak, V., You, I., Kao, S.J., Petrin, J., Clewell, D.B., Lerner, S.A., Miller, G.H. and Shaw, K.J., (2001), Aminoglycoside resistance genes *aph*(2")-Ib and aac(6')-Im detected together in strains of both *Escherichia coli* and *Enterococcus faecium*. *Antimicrob Agents Chemother*, **45**, 2691–2694.

Chuanchuen, R., Karkhoff-Schweizer, R.R. and Schweizer, H.P., (2003), High-level triclosan resistance in *Pseudomonas aeruginosa* is solely a result of efflux. *Am J Infect Control*, **31**, 124–127.

Cirz, R.T., Chin, J.K., Andes, D.R., de Crecy-Lagard, V., Craig, W.A. and Romesberg, F.E., (2005), Inhibition of mutation and combating the evolution of antibiotic resistance. *PLoS Biol*, **3**, e176.

Clerch, B., Barbe, J. and Llagostera, M., (1992), The role of the excision and error-prone repair systems in mutagenesis by fluorinated quinolones in *Salmonella typhimurium*. *Mutat Res*, **281**, 207–213.

CLSI., (2005), Performance standards for antimicrobial susceptibility testing: M100-S15, Fifteenth informational supplement. In *Committee for Clinical Laboratory Standards, Wayne, PA*. Vol. 25.

Connell, S.R., Tracz, D.M., Nierhaus, K.H. and Taylor, D.E., (2003a), Ribosomal protection proteins and their mechanism of tetracycline resistance. *Antimicrob Agents Chemother*, **47**, 3675–3681.

Connell, S.R., Trieber, C.A., Einfeldt, E., Dinos, G.P., Taylor, D.E. and Nierhaus, K.H., (2003b), Mechanism of Tet(O) mediated resistance. *EMBO J*, **22**, 945–953.

Connell, S.R., Trieber, C.A., Stelzl, U., Einfeldt, E., Taylor, D.E. and Nierhaus, K.H., (2002), The tetracycline resistance protein Tet(O), perturbs the conformation of the ribosomal decoding center. *Mol Microbiol*, **45**, 1463–1472.

Cormican, M.G., Marshall, S.A. and Jones, R.N., (1996), Detection of extended-spectrum β-lactamase (ESBL)-producing strains by the Etest ESBL screen. *J Clin Microbiol*, **34**, 1880–1884.

Coronado, V.G., Edwards, J.R., Culver, D.H. and Gaynes, R.P., (1995), Ciprofloxacin resistance among nosocomial *Pseudomonas aeruginosa* and *Staphylococcus aureus* in the United States. National Nosocomial Infections Surveillance (NNIS) System. *Infect Control Hosp Epidemiol*, **16**, 71–75.

Courvalin, P., (1990), Plasmid-mediated 4-quinolone resistance: a real or apparent absence? *Antimicrob Agents Chemother*, **34**, 681–684.

Cox, M.M., Goodman, M.F., Kreuzer, K.N., Sherratt, D.J., Sandler, S.J. and Marians, K.J., (2000), The importance of repairing stalled replication forks. *Nature*, **404**, 37–41.

Croom, K.F. and Goa, K.L., (2003), Levofloxacin: a review of its use in the treatment of bacterial infections in the United States. *Drugs*, **63**, 2769–2802.

Daigle, D.M., McKay, G.A. and Wright, G.D., (1997), Inhibition of aminoglycoside antibiotic resistance enzymes by protein kinase inhibitors. *J Biol Chem*, **272**, 24755–24758.

Daikos, G.L., Jackson, G.G., Lolans, V.T. and Livermore, D.M., (1990), Adaptive resistance to aminoglycoside antibiotics from first-exposure down-regulation. *J Infect Dis*, **162**, 414–420.

Daikos, G.L., Lolans, V.T. and Jackson, G.G., (1991), First-exposure adaptive resistance to aminoglycoside antibiotics *in vivo* with meaning for optimal clinical use. *Antimicrob Agents Chemother*, **35**, 117–123.

Dantley, K.A., Dannelly, H.K. and Burdett, V., (1998), Binding interraction between Tet(M) and the ribosome: requirements for binding. *J Bacteriol*, **180**, 4089–4092.

Datta, N. and Kontomichalou, P., (1965), Penicillinase synthesis controlled by infectious R Factors in *Enterobacteriaceae*. *Nature*, **208**, 239–244.

Davies, J. and Davis, B.D., (1968), Misreading of ribonucleic acid code words induced by aminoglycoside antibiotics. The effect of drug concentration. *J Biol Chem*, **243**, 3312–3316.

Dean, C.R., Narayan, S., Daigle, D.M., Dzink-Fox, J.L., Puyang, X., Bracken, K.R., Dean, K.E., Weidmann, B., Yuan, Z., Jain, R. and Ryder, N.S., (2005), Role of the AcrAB-TolC efflux pump in determining susceptibility of *Haemophilus influenzae* to the novel peptide deformylase inhibitor LBM415. *Antimicrob Agents Chemother*, **49**, 3129–3135.

Dean, C.R., Visalli, M.A., Projan, S.J., Sum, P.E. and Bradford, P.A., (2003), Efflux-mediated resistance to tigecycline (GAR-936) in *Pseudomonas aeruginosa* PAO1. *Antimicrob Agents Chemother*, **47**, 972–978.

Delihas, N. and Forst, S., (2001), MicF: an antisense RNA gene involved in response of *Escherichia coli* to global stress factors. *J Mol Biol*, **313**, 1–12.

Doi, Y., Yokoyama, K., Yamane, K., Wachino, J., Shibata, N., Yagi, T., Shibayama, K., Kato, H. and Arakawa, Y., (2004), Plasmid-mediated 16S rRNA methylase in *Serratia marcescens* conferring high-level resistance to aminoglycosides. *Antimicrob Agents Chemother*, **48**, 491–496.

Dominguez, E.A., Smith, T.L., Reed, E., Sanders, C.C. and Sanders Jr, W.E., (2000), A pilot study of antibiotic cycling in a hemotology–oncology unit. *Infect Control Hosp Epidemiol*, **21** (Suppl. 1), S4–S8.

Doyle, D., McDowall, K.J., Butler, M.J. and Hunter, I.S., (1991), Characterization of an oxytetracycline-resistance gene, *otrA*, of *Streptomyces rimosus*. *Mol Microbiol*, **5**, 2923–2933.

Drlica, K., (1984), Biology of bacterial deoxyribonucleic acid topoisomerases. *Microbiol Rev*, **48**, 273–289.

Drlica, K. and Malik, M., (2003), Fluoroquinolones: action and resistance. *Curr Top Med Chem*, **3**, 249–282.

Drlica, K. and Zhao, X., (1997), DNA gyrase, topoisomerase IV and the 4-quinolones. *Microbiol Mol Biol Rev*, **61**, 377–392.

Eda, S., Maseda, H. and Nakae, T., (2003), An elegant means of self-protection in gram-negative bacteria by recognizing and extruding xenobiotics from the periplasmic space. *J Biol Chem*, **278**, 2085–2088.

Elkins, C.A. and Nikaido, H., (2002), Substrate specificity of the RND-type multidrug efflux pumps AcrB and AcrD of *Escherichia coli* is determined predominantly by two large periplasmic loops. *J Bacteriol*, **184**, 6490–6498.

Elkins, C.A. and Nikaido, H., (2003), 3D structure of AcrB: the archetypal multidrug efflux transporter of *Escherichia coli* likely captures substrates from periplasm. *Drug Resist Updat*, **6**, 9–13.

Ellis-Grosse, E., Babinchak, T., Dartois, N., Rose, G. and Loh, E., (2005), The efficacy and safety of tigecycline in the treatment of skin and skin structure infections: results of two double-blind phase 3 comparison studies with vancomycin/aztreonam. *Clin Infect Dis*, **41**, S341–S353.

Ena, J., Amador, C., Martinez, C. and Ortiz de la Tabla, V., (1995), Risk factors for acquisition of urinary tract infections caused by ciprofloxacin resistant *Escherichia coli*. *J Urol*, **153**, 117–120.

Endtz, H.P., Ruijs, G.J., van Klingeren, B., Jansen, W.H., van der Reyden, T. and Mouton, R.P., (1991), Quinolone resistance in campylobacter isolated from man and poultry following the introduction of fluoroquinolones in veterinary medicine. *J Antimicrob Chemother*, **27**, 199–208.

Evans, K., Passador, L., Srikumar, R., Tsang, E., Nezezon, J. and Poole, K., (1998), Influence of the MexAB-OprM multidrug efflux system on quorum sensing in *Pseudomonas aeruginosa*. *J Bacteriol*, **180**, 5443–5447.

Ferrero, L., Cameron, B., Manse, B., Lagneaux, D., Crouzet, J., Famechon, A. and Blanche, F., (1994), Cloning and primary structure of *Staphylococcus aureus* DNA topoisomerase IV: a primary target of fluoroquinolones, *Mol Microbiol* **13**, 641–653.

Fey, P.D., Safranek, T.J., Rupp, M.E., F., D.E., Ribot, E., Iwen, P.C., Bradford, P.A., Angulo, F.J. and Hinrichs, S.H., (2000), Ceftriaxone-resistant *Salmonella* infection acquired by a child from cattle. *New England J Med*, **342**, 1242–1249.

Fleischmann, R.D., Adams, M.D., White, O., Clayton, R.A., Kirkness, E.F., Kerlavage, A.R., Bult, C.J., Tomb, J.F., Dougherty, B.A., Merrick, J.M., McKenney, K., Sutton, G.G., Fitzhugh, W., Fields, C.A., Gocayne, J.D., Scott, J.D., Shirley, R., Liu, L.I., Glodek, A., Kelley, J.M., Weidman, J.F., Philips, C.A., Spriggs, T., Hedblom, E., Cotton, M.D., Utterback, T., Hanna, M.C., Nguyen, D.T., Saudek, D.M., Brandon, R.C., Fine, L.D., Fritchman, J.L., Fuhrman, J.L., Geoghagen, N.S., Gnehm, C.L., McDonald, L.A., Small, K.V., Fraser, C.M., Smith, H.O. and Venter, J.C., (1995), Whole-genome random sequencing and assembly of *Haemophilus influenzae* Rd. *Science*, **269**, 496–512.

Fluit, A.C., Florijn, A., Verhoef, J. and Milatovic, D., (2005), Presence of tetracycline resistance determinants and susceptibility to tigecycline and minocycline. *Antimicrob Agents Chemother*, **49**, 1636–1638.

Fridkin, S.K., (2003), Routine cycling of antimicrobial agents as an infection-control measure. *Clin Infect Dis*, **36**, 1438–1444.

Fritsche, T.R. and Jones, R.N., (2004), Antimicrobial activity of LBM415 (NVP PDF-713) tested against *Neisseria meningitidis* and *N. gonorrhoeae*, Abstr. F-1962. In *44th Interscience Conference on Antimicrobial Agents and Chemotherapy*, Washington, DC.

Fritsche, T.R., Rhomberg, P.R. and Jones, R.N., (2004), Comparative antimicrobial characterization of LBM415 (NVP PDF-713), a new peptide deformylase inhibitor, Abstr. F-1961. In *44th Interscience Conference on Antimicrobial Agents and Chemotherapy*, Washington, DC.

Gales, A.C., Sader, H.S. and Jones, R.N., (2002), Urinary tract infection trends in Latin American hospitals: report from the SENTRY antimicrobial surveillance program (1997–2000). *Diagn Microbiol Infect Dis*, **44**, 289–299.

Galimand, M., Courvalin, P. and Lambert, T., (2003), Plasmid-mediated high-level resistance to aminoglycosides in *Enterobacteriaceae* due to 16S rRNA methylation. *Antimicrob Agents Chemother*, **47**, 2565–2571.

Galimand, M., Sabtcheva, S., Courvalin, P. and Lambert, T., (2005), Worldwide disseminated *armA* aminoglycoside resistance methylase gene is borne by composite transposon Tn1548. *Antimicrob Agents Chemother*, **49**, 2949–2953.

Garau, J., Xercavins, M., Rodriguez-Carballeira, M., Gomez-Vera, J.R., Coll, I., Vidal, D., Llovet, T. and Ruiz-Bremon, A., (1999), Emergence and dissemination of quinolone-resistant *Escherichia coli* in the community. *Antimicrob Agents Chemother*, **43**, 2736–2741.

Garcia-Rodriguez, J.A. and Jones, R.N., (2002), Antimicrobial resistance in gram-negative isolates from European intensive care units: data from the Meropenem Yearly Susceptibility Test Information Collection (MYSTIC) programme. *J Chemother*, **14**, 25–32.

Gazouli, M., Sidorenko, S.V., Tzelepi, E., Kozlova, N.S., Gladin, D.P. and Tzouvelekis, L.S., (1998a), A plasmid-mediated β-lactamase conferring resistance to cefotaxime in a *Salmonella typhimurium* clone found in St. Petersburg, Russia. *J Antimicrob Chemother*, **41**, 119–121.

Gazouli, M., Tzelepi, E., Markogiannakis, A., Legakis, N.J. and Tzouvelekis, L.S., (1998b), Two novel plasmid-mediated cefotaxime-hydrolyzing β-lactamases (CTX-M-5 and CTX-M-6) from *Salmonella typhimurium*. *FEMS Microbiol Lett*, **165**, 289–293.

Gazouli, M., Tzouvelekis, L.S., Vatopoulos, A.C. and Tzelepi, E., (1998c), Transferable class C β-lactamases in *Escherichia coli* strains isolated in Greek hospitals and characterization of two enzyme variants (LAT-3 and LAT-4) closely related to *Citrobacter freundii* AmpC β-lactamase. *J Antimicrob Chemother*, **42**, 419–425.

Ge, B., McDermott, P.F., White, D.G. and Meng, J., (2005), Role of efflux pumps and topoisomerase mutations in fluoroquinolone resistance in *Campylobacter jejuni* and *Campylobacter coli*. *Antimicrob Agents Chemother*, **49**, 3347–3354.

Ge, Y., Wikler, M.A., Sahm, D.F., Blosser-Middleton, R.S. and Karlowsky, J.A., (2004), *In vitro* antimicrobial activity of doripenem, a new carbapenem. *Antimicrob Agents Chemother*, **48**, 1384–1396.

Georgiou, M., Munoz, R., Roman, F., Canton, R., Gomez-Lus, R., Campos, J. and De La Campa, A.G., (1996), Ciprofloxacin-resistant *Haemophilus influenzae* strains possess mutations in analogous positions of GyrA and ParC. *Antimicrob Agents Chemother*, **40**, 1741–1744.

Gellert, M., Mizuuchi, K., O'Dea, M.H., Itoh, T. and Tomizawa, J.I., (1977), Nalidixic acid resistance: a second genetic character involved in DNA gyrase activity. Proc Natl Acad Sci USA, **74**, 4772–4776.

Gerrits, M.M., De Zoete, M.R., Arents, N.L., Kuipers, E.J. and Kusters, J.G., (2002), 16S rRNA mutation-mediated tetracycline resistance in *Helicobacter pylori*. *Antimicrob Agents Chemother*, **46**, 2996–3000.

Ghuysen, J.M., (1991), Serine β-lactamases and penicillin-binding proteins. In L.N., Ornston, A. Ballows, and E.P., Greenberg, editors, *Annual Reviews of Microbiology*, Vol. 45. Palo Alto, CA, Annual Reviews Inc., pp. 37–67.

Giakkoupi, P., Petrikkos, G., Tzouvelekis, L.S., Tsonas, S., The WHONET GREECE Study Group, Legakis, N.J. and Vatopoulos, A.C., (2003), Spread of integron-associated VIM-type metallo-β-lactamase genes among imipenem-nonsusceptible *Pseudomonas aeruginosa* strains in Greek hospitals. *J Clin Microbiol*, **41**, 822–825.

Giakkoupi, P., Tzouvelekis, L.S., Diakos, G.L., Miragou, V., Petrikkos, G., Legakis, N.J. and Vatopoulos, A.C., (2005), Discrepancies and interpretation problems in susceptibility testing of VIM-1 producing *Klebsiella pneumoniae* isolates. *J Clin Microbiol*, **43**, 494–496.

Gibson, R.L., Burns, J.L. and Ramsey, B.W., (2003), Pathophysiology and management of pulmonary infections in cystic fibrosis. *Am J Respir Crit Care Med*, **168**, 918–951.

Giraud, E., Cloeckaert, A., Baucheron, S., Mouline, C. and Chaslus-Dancla, E., (2003), Fitness cost of fluoroquinolone resistance in *Salmonella enterica* serovar Typhimurium. *J Med Microbiol*, **52**, 697–703.

Girlich, D., Naas, T., Bellais, S., Poirel, L., Karim, A. and Nordman, P., (2000), Biochemical-genetic characterization and regulation of expression of an ACC-1-like chromosome-borne cephalosporinase from *Hafnia alvei* . *Antimicrob Agents Chemother*, **44**, 1470–1478.

Gmuender, H., Kuratli, K., Di Padova, K., Gray, C.P., Keck, W. and Evers, S., (2001), Gene expression changes triggered by exposure of *Haemophilus influenzae* to novobiocin or ciprofloxacin: combined transcription and translation analysis. *Genome Res*, **11**, 28–42.

Goettsch, W., van Pelt, W., Nagelkerke, N., Hendrix, M.G., Buiting, A.G., Petit, P.L., Sabbe, L.J., van Griethuysen, A.J. and de Neeling, A.J., (2000), Increasing resistance to fluoroquinolones in *Escherichia coli* from urinary tract infections in the Netherlands. *J Antimicrob Chemother*, **46**, 223–228.

Gonzalez-Zorn, B., Teshager, T., Casas, M., Porrero, M.C., Moreno, M.A., Courvalin, P. and Dominguez, L., (2005), armA and aminoglycoside resistance in *Escherichia coli*. *Emerg Infect Dis*, **11**, 954–956.

Gordon, S.M., Carlyn, C.J., Doyle, L.J., Knapp, C.C., Longworth, D.L., Hall, G.S. and Washington, J.A., (1996), The emergence of *Neisseria gonorrhoeae* with decreased susceptibility to ciprofloxacin in Cleveland, Ohio: epidemiology and risk factors. *Ann Intern Med*, **125**, 465–470.

Greenberg, W.A., Priestley, E.S., Sears, P.S., Alper, P.B., Rosenbohm, C., Hendrix, M.G., Hung, H.-C. and Wong-C-H., (1999), Design and synthesis of new aminoglycoside antibiotics containing neamine as an optimal core structure: correlation of antibiotic activity with *in vitro* inhibition of translation. *J Am Chem Soc*, **121**, 6527–6541.

Grossman, R.F., Rotschafer, J.C. and Tan, J.S., (2005), Antimicrobial treatment of lower respiratory tract infections in the hospital setting. *Am J Med*, **118**, 29S–38S.

Gruson, D., Hilbert, G., Vargas, F., Valentino, R., Bebear, C., Allery, A., Bebear, C., Gbikpi-Benissan, G. and Cardinaud, J.-P., (2000), Rotation and restricted use of antibiotics in a medical intensive care unit. Impact on the incidence of ventilator-associated pneumonia caused by antibiotic-resistant gram-negative bacteria. *Am J Respir Crit Care Med*, **162**, 837–843.

Gruson, D., Hilbert, G., Vargas, F., Valentino, R., Bui, N., Pereyre, S., Bebear, C., Bebear, C.-M. and Gbikpi-Benissan, G., (2003), Strategy of antibiotic rotation: Long-term effect on incidence and susceptibilities of Gram-negative bacilli responsible for ventilator-associated pneumonia. *Crit Care Med*, **31**, 1908–1914.

Hall, L.M.C., Livermore, D.M., Gur, D., Akova, M. and Akalin, H.E., (1993), OXA-11, an extended-spectrum variant of OXA-10 (PSE-2) β-lactamase from *Pseudomonas aeruginosa*. *Antimicrob Agents Chemother*, **37**, 1637–1644.

Hasegawa, M., Kobayashi, I., Saika, T. and Nishida, M., (1997), Drug-resistance patterns of clinical isolates of *Pseudomonas aeruginosa* with reference to their lipopolysaccharide compositions. *Chemotherapy*, **43**, 323–331.

Hassett, D.J., Cuppoletti, J., Trapnell, B., Lymar, S.V., Rowe, J.J., Yoon, S.S., Hilliard, G.M., Parvatiyar, K., Kamani, M.C., Wozniak, D.J., Hwang, S.H., McDermott, T.R. and Ochsner, U.A., (2002), Anaerobic metabolism and quorum sensing by *Pseudomonas aeruginosa* biofilms in chronically infected cystic fibrosis airways: rethinking antibiotic treatment strategies and drug targets. *Adv Drug Deliv Rev*, **54**, 1425–1443.

Hata, M., Suzuki, M., Matsumoto, M., Takahashi, M., Sato, K., Ibe, S. and Sakae, K., (2005), Cloning of a novel gene for quinolone resistance from a transferable plasmid in *Shigella flexneri* 2b. *Antimicrob Agents Chemother*, **49**, 801–803.

Hebeisen, P., Heinze-Krauss, I., Angehrn, P., Hohl, P., Page, M.G.P. and Then, R.L., (2001), *In vitro* and *in vivo* properties of Ro 63-9141, a novel broad-spectrum cephalosporin with activity against methicillin-resistant *staphylococci*. *Antimicrob Agents Chemother*, **45**, 825–836.

Heisig, P., (1996), Genetic evidence for a role of *parC* mutations in development of high-level fluoroquinolone resistance in *Escherichia coli*. *Antimicrob Agents Chemother*, **40**, 879–885.

Helling, R.B., Janes, B.K., Kimball, H., Tran, T., Bundesmann, M., Check, P., Phelan, D. and Miller, C., (2002), Toxic waste disposal in *Escherichia coli*. *J Bacteriol*, **184**, 3699–3703.

Hiasa, H., Yousef, D.O. and Marians, K.J., (1996), DNA strand cleavage is required for replication fork arrest by a frozen topoisomerase-quinolone-DNA ternary complex. *J Biol Chem*, **271**, 26424–26429.

Hierowski, M., (1965), Inhibition of protein synthesis by chlorotetracylcine in the *Escherichia coli* in vitro system. *Proc Natl Acad Sci USA*, **53**, 594–599.

Hirakata, Y., Srikumar, R., Poole, K., Gotoh, N., Suematsu, T., Kohno, S., Kamihira, S., Hancock, R.E. and Speert, D.P., (2002), Multidrug efflux systems play an important role in the invasiveness of *Pseudomonas aeruginosa*. *J Exp Med*, **196**, 109–118.

Hocquet, D., Vogne, C., El Garch, F., Vejux, A., Gotoh, N., Lee, A., Lomovskaya, O. and Plesiat, P., (2003), MexXY-OprM efflux pump is necessary for a adaptive resistance of *Pseudomonas aeruginosa* to aminoglycosides. *Antimicrob Agents Chemother*, **47**, 1371–1375.

Hoffman, L.R., D'Argenio, D.A., MacCoss, M.J., Zhang, Z., Jones, R.A. and Miller, S.I., (2005), Aminoglycoside antibiotics induce bacterial biofilm formation. *Nature*, **436**, 1171–1175.

Hon, W.C., McKay, G.A., Thompson, P.R., Sweet, R.M., Yang, D.S., Wright, G.D. and Berghuis, A.M., (1997), Structure of an enzyme required for aminoglycoside antibiotic resistance reveals homology to eukaryotic protein kinases. *Cell*, **89**, 887–895.

Honore, N. and Cole, S.T., (1994), Streptomycin resistance in mycobacteria. *Antimicrob Agents Chemother*, **38**, 238–242.

Honore, N., Marchal, G. and Cole, S.T., (1995), Novel mutation in 16S rRNA associated with streptomycin dependence in *Mycobacterium tuberculosis*. *Antimicrob Agents Chemother*, **39**, 769–770.

Hooper, D.C., (1999), Mechanisms of fluoroquinolone resistance. *Drug Resist Updat*, **2**, 38–55.

Hooper, D.C., (2000), New uses for new and old quinolones and the challenge of resistance. *Clin Infect Dis*, **30**, 243–254.

Hooper, D.C., (2001), Emerging mechanisms of fluoroquinolone resistance. *Emerg Infect Dis*, **7**, 337–341.

Hooper, D.C., Wolfson, J.S., Bozza, M.A. and Ng, E.Y., (1992), Genetics and regulation of outer membrane protein expression by quinolone resistance loci *nfxB*, *nfxC*, and *cfxB*. *Antimicrob Agents Chemother*, **36**, 1151–1154.

Hooper, D.C., Wolfson, J.S., Souza, K.S., Ng, E.Y., McHugh, G.L. and Swartz, M.N., (1989), Mechanisms of quinolone resistance in *Escherichia coli:* characterization of *nfxB* and *cfxB*, two mutant resistance loci decreasing norfloxacin accumulation. *Antimicrob Agents Chemother*, **33**, 283–290.

Hossain, A., Ferraro, M.J., Pino, R.M., Dew III, R.B., Moland, E.S., Lockhart, T.J., Thomson, K.S., Goering, R.V. and Hanson, N.D., (2004), Plasmid-mediated carbapenem-hydrolyzing enzyme KPC-2 in an *Enterobacter* spp. *Antimicrob Agents Chemother*, **48**, 4438–4440.

Hughes, M.G., Evans, H.L., Chong, T.W., Smith, R.L., Raymond, D.P., Pelletier, S.J., Pruett, T.L. and Sawyer, R.G., (2004), Effect of an intensive care unit rotating empiric antibiotic schedule on the development of hospital-acquired infections on the non-intensive care unit ward. *Crit Care Med*, **32**, 53–60.

Hujer, A.M., Bethel, C.R., Hujer, K.M. and Bonomo, R.A., (2004), Antibiotic resistance in the institutionalized elderly. *Clin Lab Med*, **24**, 343–361.

Hurley, J.C., Miller, G.H. and Smith, A.L., (1995), Mechanism of amikacin resistance in *Pseudomonas aeruginosa* isolates from patients with cystic fibrosis. *Diagn Microbiol Infect Dis*, **22**, 331–336.

Issa, N.C., Rouse, M.S., Piper, K.E., Wilson, W.R., Steckelberg, J.M. and Patel, R., (2004), In vitro activity of BAL9141 against clinical isolates of gram-negative bacteria. *Diagn Microbiol Infect Dis*, **48**, 73–75.

Jacobs, C., Huang, L., Bartowsky, E., Normark, S. and Park, T., (1994), Bacterial cell wall recycling provides cytosoli muropeptides as effectors for β-lactmase induction. *EMBO J*, **13**, 4684–4694.

Jacoby, G.A., (1999), Sequence of the MIR-1 β-lactamase gene. *Antimicrob Agents Chemother*, **43**, 1759–1760.

Jacoby, G.A., (2005), The new β-lactamases. *N Engl J Med*, **352**, 380–389.

Jacoby, G.A. and Bush, K., (2005), Amino Acid Sequences for TEM, SHV and OXA Extended-Spectrum and Inhibitor Resistant β-Lactamases. Available at http://www.lahey.org/Studies/.

Jacoby, G.A. and Medeiros, A.A., (1991), More extended-spectrum β-lactamases. *Antimicrob Agents Chemother*, **35**, 1697–1704.

Jalal, S. and Wretlind, B., (1998), Mechanisms of quinolone resistance in clinical strains of *Pseudomonas aeruginosa*. *Microb Drug Resist*, **4**, 257–261.

Jana, S. and Deb, J.K., (2005), Molecular targets for design of novel inhibitors to circumvent aminoglycoside resistance. *Curr Drug Targets*, **6**, 353–361.

Jarlier, V., Nicolas, M., Fournier, G. and Philippon, A., (1988), Extended broad-spectrum β-lactamases conferring transferable resistance to newer ÿ-lactam agents in *Enterobacteriaeceae*: Hospital prevalence and susceptibility patterns. *Rev Infect Dis*, **10**, 867–878.

Jeannot, K., Sobel, M.L., El Garch, F., Poole, K. and Plesiat, P., (2005), Induction of the MexXY efflux pump in *Pseudomonas aeruginosa* is dependent on drug-ribosome interaction. *J Bacteriol*, **187**, 5341–5346.

Jellen-Ritter, A.S. and Kern, W.V., (2001), Enhanced expression of the multidrug efflux pumps AcrAB and AcrEF associated with insertion element transposition in *Escherichia coli* mutants Selected with a fluoroquinolone. *Antimicrob Agents Chemother*, **45**, 1467–1472.

Jo, J.T., Brinkman, F.S. and Hancock, R.E., (2003), Aminoglycoside efflux in *Pseudomonas aeruginosa*: involvement of novel outer membrane proteins. *Antimicrob Agents Chemother*, **47**, 1101–1111.

Jones, C.S., Osborne, D.J. and Stanley, J., (1992), Enterobacterial tetracycline resistance in relation to plasmid incompatability. *Mol Cell Probes*, **6**, 313–317.

Karlowsky, J.A., Hoban, D.J., Zelenitsky, S.A. and Zhanel, G.G., (1997), Altered *denA* and *anr* gene expression in aminoglycoside adaptive resistance in *Pseudomonas aeruginosa*. *J Antimicrob Chemother*, **40**, 371–376.

Karlowsky, J.A., Jones, M.E., Thornsberry, C., Evangelista, A.T., Yee, Y.C. and Sahm, D.F., (2005), Stable antimicrobial susceptibility rates for clinical isolates of *Pseudomonas aeruginosa* from the 2001–2003 tracking resistance in the United States today surveillance studies. *Clin Infect Dis*, **40** (Suppl. 2), S89–98.

Karlowsky, J.A., Saunders, M.H., Harding, G.A., Hoban, D.J. and Zhanel, G.G., (1996), In vitro characterization of aminoglycoside adaptive resistance in *Pseudomonas aeruginosa*. *Antimicrob Agents Chemother*, **40**, 1387–1393.

Kato, J., Nishimura, Y., Imamura, R., Niki, H., Hiraga, S. and Suzuki, H., (1990), New topoisomerase essential for chromosome segregation in *Escherichia coli*. *Cell*, **63**, 393–404.

Kato, J., Suzuki, H. and Ikeda, H., (1992), Purification and characterization of DNA topoisomerase IV in *Escherichia coli*. *J Biol Chem*, **267**, 25676–25684.

Kaye, K.S., Gold, H.S., Schwaber, M.J., Venkataraman, L., Qi, Y., De Girolami, P.C., Samore, M.H., Anderson, G., Rasheed, J.K. and Tenover, F.C., (2004), Variety of β-lactamases produced by amoxicillin-clavulanate-resistant *Escherichia coli* isolated in the northeastern United States. *Antimicrob Agents Chemother*, **48**, 1520–1525.

Khodursky, A.B., Zechiedrich, E.L. and Cozzarelli, N.R., (1995), Topoisomerase IV is a target of quinolones in *Escherichia coli*. *Proc Natl Acad Sci USA*, **92**, 11801–11805.

King, D.E., Malone, R. and Lilley, S.H., (2000), New classification and update on the quinolone antibiotics. *Am Fam Physician*, **61**, 2741–2748.

Kohler, T., Michea-Hamzehpour, M., Plesiat, P., Kahr, A.L. and Pechere, J.C., (1997), Differential selection of multidrug efflux systems by quinolones in *Pseudomonas aeruginosa*. *Antimicrob Agents Chemother*, **41**, 2540–2543.

Kohler, T., van Delden, C., Curty, L.K., Hamzehpour, M.M. and Pechere, J.C., (2001), Overexpression of the MexEF-OprN multidrug efflux system affects cell-to-cell signaling in *Pseudomonas aeruginosa*. *J Bacteriol*, **183**, 5213–5222.

Komp Lindgren, P., Karlsson, A. and Hughes, D., (2003), Mutation rate and evolution of fluoroquinolone resistance in *Escherichia coli* isolates from patients with urinary tract infections. *Antimicrob Agents Chemother*, **47**, 3222–3232.

Kono, M. and O'Hara, K., (1976), Mechanisms of streptomycin(SM)-resistance of highly SM-resistant *Pseudomonas aeruginosa* strains. *J Antibiot (Tokyo)*, **29**, 169–175.

Koronakis, V., Sharff, A., Koronakis, E., Luisi, B. and Hughes, C., (2000), Crystal structure of the bacterial membrane protein TolC central to multidrug efflux and protein export. *Nature*, **405**, 914–919.

Kotra, L.P., Haddad, J. and Mobashery, S., (2000), Aminoglycosides: perspectives on mechanisms of action and resistance and strategies to counter resistance. *Antimicrob Agents Chemother*, **44**, 3249–3256.

Kriengkauykiat, J., Porter, E., Lomovskaya, O. and Wong-Beringer, A., (2005), Use of an efflux pump inhibitor to determine the prevalence of efflux pump-mediated fluoroquinolone resistance and multidrug resistance in *Pseudomonas aeruginosa*. *Antimicrob Agents Chemother*, **49**, 565–570.

Kumar, A. and Schweizer, H.P., (2005), Bacterial resistance to antibiotics: active efflux and reduced uptake. *Adv Drug Deliv Rev*, **57**, 1486–1513.

Lauretti, L., Riccio, M.L., Mazzariol, A., Cornaglia, G., Amicosante, G., Fontana, R. and Rossolini, G.M., (1999), Cloning and characterization of bla_{VIM}, a new integron-borne metallo-β-lactamase gene from a *Pseudomonas aeruginosa* clinical isolate. *Antimicrob Agents Chemother*, **43**, 1584–1590.

Le Thomas, I., Couetdic, G., Clermont, O., Brahimi, N., Plesiat, P. and Bingen, E., (2001), *In vivo* selection of a target/efflux double mutant of *Pseudomonas aeruginosa* by ciprofloxacin therapy. *J Antimicrob Chemother*, **48**, 553–555.

Lee, A., Mao, W., Warren, M.S., Mistry, A., Hoshino, K., Okumura, R., Ishida, H. and Lomovskaya, O., (2000), Interplay between efflux pumps may provide either additive or multiplicative effects on drug resistance. *J Bacteriol*, **182**, 3142–3150.

Lee, K., Lim, J.B., Yum, J.H., Yong, D., Chong, Y., Kim, J.M. and Livermore, D.M., (2002), bla_{VIM-2} cassette-containing novel integrons in metallo-β-lactamase-producing *Pseudomonas aeruginosa* and *Pseudomonas putida* isolates disseminated in a Korean hospital. *Antimicrob Agents Chemother*, **46**, 1053–1058.

Lemozy, J., Sirot, D., Chanal, C., Huc, C., Labia, R., Dabernat, H. and Sirot, J., (1995), First characterization of inhibitor-resistant TEM (IRT) β-lactamases in *Klebsiella pneumoniae* strains. *Antimicrob Agents Chemother*, **33**, 2580–2582.

Levin, A.S., Barone, A.A., Penço, J., Santos, M.V., Marinho, I.S., Arruda, E.A.G., Manrique, E.I. and Costa, S.F., (1998), Intraveneous colistin as therapy for nosocomial infections caused by multidrug-resistant *Pseudomonas aeruginosa* and *Acinetobacter baumannii*. *Clin Infect Dis*, **28**, 1008–1011.

Levine, C., Hiasa, H. and Marians, K.J., (1998), DNA gyrase and topoisomerase IV: biochemical activities, physiological roles during chromosome replication and drug sensitivities. *Biochim Biophys Acta*, **1400**, 29–43.

Levy, S.B., (2002), Active efflux, a common mechanism for biocide and antibiotic resistance. *J Appl Micrbiol*, **92** Symposium supplement, 65S–71S.

Lewis, K., (2005), Persister cells and the riddle of biofilm survival. *Biochemistry (Mosc)*, **70**, 267–274.

Li, X.Z., Ma, D., Livermore, D.M. and Nikaido, H., (1994), Role of efflux pump(s) in intrinsic resistance of *Pseudomonas aeruginosa*: active efflux as a contributing factor to β-lactam resistance. *Antimicrob Agents Chemother*, **38**, 1742–1752.

Li, X., Mariano, N., Rahal, J.J., Urban, C.M. and Drlica, K., (2004a), Quinolone-resistant *Haemophilus influenzae*: determination of mutant selection window for ciprofloxacin, garenoxacin, levofloxacin and moxifloxacin. *Antimicrob Agents Chemother*, **48**, 4460–4462.

Li, X., Mariano, N., Rahal, J.J., Urban, C.M. and Drlica, K., (2004b), Quinolone-resistant *Haemophilus influenzae* in a long-term-care facility: nucleotide sequence characterization of alterations in the genes encoding DNA gyrase and DNA topoisomerase IV. *Antimicrob Agents Chemother*, **48**, 3570–3572.

Li, X., Zolli-Juran, M., Cechetto, J.D., Daigle, D.M., Wright, G.D. and Brown, E.D., (2004c), Multicopy suppressors for novel antibacterial compounds reveal targets and drug efflux susceptibility. *Chem Biol*, **11**, 1423–1430.

Li, X.Z. and Nikaido, H., (2004), Efflux-mediated drug resistance in bacteria. *Drugs*, **64**, 159–204.

Li, X.Z., Zhang, L. and Poole, K., (2000), Interplay between the MexA-MexB-OprM multidrug efflux system and the outer membrane barrier in the multiple antibiotic resistance of *Pseudomonas aeruginosa*. *J Antimicrob Chemother*, **45**, 433–436.

Liu, M., Haddad, J., Azucena, E., Kotra, L.P., Kirzhner, M. and Mobashery, S., (2000), Tethered bisubstrate derivatives as probes for mechanism and as inhibitors of aminoglycoside 3'-phosphotransferases. *J Org Chem*, **65**, 7422–7431.

Lu, T., Malik, M. and Drlica-Wagner, A., (2001), C-8-methoxy fluoroquinolones. *Res Adv Antimicrob Agents Chemother*, **2**, 29–42.

Livermore, D.M., (2002), Multiple mechanisms of antimicrobial resistance in *Pseudomonas aeruginosa*: our worst nightmare? *Clin Infect Dis*, **34**, 634–640.

Livermore, D.M., (2004), The need for new antibiotics. *Clin Microbiol Infect*, **10** (Suppl. 4), 1–9.

Llano-Sotelo, B., Azucena, E.F., Jr, Kotra, L.P., Mobashery, S. and Chow, C.S., (2002), Aminoglycosides modified by resistance enzymes display diminished binding to the bacterial ribosomal aminoacyl-tRNA site. *Chem Biol*, **9**, 455–463.

Lomovskaya, O. and Totrov, M., (2005), Vacuuming the periplasm. *J Bacteriol*, **187**, 1879–1883.

Lomovskaya, O. and Watkins, O., (2001a), Inhibition of efflux pumps as a novel approach to combat drug resistance in bacteria. *J Mol Microbiol Biotechnol*, **3**, 225–236.

Lomovskaya, O. and Watkins, W.J., (2001b), Efflux pumps: their role in antibacterial drug discovery. *Curr Med Chem*, **8**, 1699–1711.

Lomovskaya, O., Lee, A., Hoshino, K., Ishida, H., Mistry, A., Warren, M.S., Boyer, E., Chamberland, S. and Lee, V.J., (1999), Use of a genetic approach to evaluate the consequences of inhibition of efflux pumps in *Pseudomonas aeruginosa*. *Antimicrob Agents Chemother*, **43**, 1340–1346.

Lomovskaya, O., Warren, M.S., Lee, A., Galazzo, J., Fronko, R., Lee, M., Blais, J., Cho, D., Chamberland, S., Renau, T., Leger, R., Hecker, S., Watkins, W., Hoshino, K., Ishida, H. and Lee, V.J., (2001), Identification and characterization of inhibitors of multidrug resistance efflux pumps in *Pseudomonas aeruginosa*: novel agents for combination therapy. *Antimicrob Agents Chemother*, **45**, 105–116.

Lucet, J.-C., Decré, D., Fichelle, A., Joly-Guillou, M.-L., Pernet, M., Deblangy, C., Kosmann, M.-J. and Régnier, B., (1999), Control of a prolonged outbreak of extended-spectrum β-lactamase-producing *Enterobacteriaceae* in a university hospital. *Clin Infect Dis*, **20**, 1411–1418.

Lucet, J.-C., Chevret, S., Durand-Zleski, I., Chastang, C. and Regnier, B., (2003), Prevalence and risk factors for carriage of methicillin-resistant *Staphylococcus aureus* at admission to the intensive care unit. *Arch Int Med*, **163**, 181–188.

Ma, L., Ishii, Y., Ishiguro, M., Matsuzawa, H. and Yamaguchi, K., (1998), Cloning and sequencing of the gene encoding Toho-2, a class A β-lactamase preferentially inhibited by tazobactam. *Antimicrob Agents Chemother*, **42**, 1181–1186.

MacLeod, D.L., Nelson, L.E., Shawar, R.M., Lin, B.B., Lockwood, L.G., Dirk, J.E., Miller, G.H., Burns, J.L. and Garber, R.L., (2000), Aminoglycoside-resistance mechanisms for cystic fibrosis *Pseudomonas aeruginosa* isolates are unchanged by long-term, intermittent, inhaled tobramycin treatment. *J Infect Dis*, **181**, 1180–1184.

Macone, A., Donatelli, J., Dumont, T., Levy, S.B. and Tanaka, S.K., (2003), In vitro activity of PTK 0796 (BAY 73-6944) against gram-positive and gram-negative organisms. In *Interscience Conference on Antimicrobial Agents and Chemotherapy*. Vol. Abst. 2439 Chicago, IL.

Magnet, S., Smith, T.A., Zheng, R., Nordmann, P. and Blanchard, J.S., (2003), Aminoglycoside resistance resulting from tight drug binding to an altered aminoglycoside acetyltransferase. *Antimicrob Agents Chemother*, **47**, 1577–1583.

Mah, T.F., Pitts, B., Pellock, B., Walker, G.C., Stewart, P.S. and O'Toole, G.A., (2003), A genetic basis for *Pseudomonas aeruginosa* biofilm antibiotic resistance. *Nature*, **426**, 306–310.

Mallea, M., Chevalier, J., Eyraud, A. and Pages, J.M., (2002), Inhibitors of antibiotic efflux pump in resistant *Enterobacter aerogenes* strains. *Biochem Biophys Res Commun*, **293**, 1370–1373.

Mallea, M., Mahamoud, A., Chevalier, J., Alibert-Franco, S., Brouant, P., Barbe, J. and Pages, J.M., (2003), Alkylaminoquinolines inhibit the bacterial antibiotic efflux pump in multidrug-resistant clinical isolates. *Biochem J*, **376**, 801–805.

Mamber, S.W., Kolek, B., Brookshire, K.W., Bonner, D.P. and Fung-Tomc, J., (1993), Activity of quinolones in the Ames Salmonella TA102 mutagenicity test and other bacterial genotoxicity assays. *Antimicrob Agents Chemother*, **37**, 213–217.

Mamelli, L., Amoros, J.P., Pages, J.M. and Bolla, J.M., (2003), A phenylalanine-arginine beta-naphthylamide sensitive multidrug efflux pump involved in intrinsic and acquired resistance of *Campylobacter* to macrolides. *Int J Antimicrob Agents*, **22**, 237–241.

Mammeri, H., Van De Loo, M., Poirel, L., Martinez-Martinez, L. and Nordmann, P., (2005), Emergence of plasmid-mediated quinolone resistance in *Escherichia coli* in Europe. *Antimicrob Agents Chemother*, **49**, 71–76.

Mao, W., Warren, M.S., Black, D.S., Satou, T., Murata, T., Nishino, T., Gotoh, N. and Lomovskaya, O., (2002), On the mechanism of substrate specificity by resistance nodulation division (RND)-type multidrug resistance pumps: the large periplasmic loops of MexD from *Pseudomonas aeruginosa* are involved in substrate recognition. *Mol Microbiol*, **46**, 889–901.

Marians, K.J. and Hiasa, H., (1997), Mechanism of quinolone action. A drug-induced structural perturbation of the DNA precedes strand cleavage by topoisomerase IV. *J Biol Chem*, **272**, 9401–9409.

Martinez-Martinez, L., Pascual, A., Garcia, I., Tran, J. and Jacoby, G.A., (2003), Interaction of plasmid and host quinolone resistance. *J Antimicrob Chemother*, **51**, 1037–1039.

Martínez-Martínez, L., Pascual, A., Hernández-Allés, S., Alvarez-Díaz, D., Suárez, A.I., Tran, J., Benedí, V.J. and Jacoby, G.A., (1999), Roles of β-lactamases and porins in activities of carbapenems and cephalosporins agains *Klebsiella pneumoniae*. *Antimicrob Agents Chemother*, **43**, 1669–1673.

Martinez-Martinez, L., Pascual, A. and Jacoby, G.A., (1998), Quinolone resistance from a transferable plasmid. *Lancet*, **351**, 797–799.

McPhee, J.B., Lewenza, S. and Hancock, R.E., (2003), Cationic antimicrobial peptides activate a two-component regulatory system, PmrA-PmrB, that regulates resistance to polymyxin B and cationic antimicrobial peptides in *Pseudomonas aeruginosa*. *Mol Microbiol*, **50**, 205–217.

Meier, A., Kirschner, P., Bange, F.C., Vogel, U. and Bottger, E.C., (1994), Genetic alterations in streptomycin-resistant *Mycobacterium tuberculosis*: mapping of mutations conferring resistance. *Antimicrob Agents Chemother*, **38**, 228–233.

Menard, R., Molinas, C., Arthur, M., Duval, J., Courvalin, P. and Leclercq, R., (1993), Overproduction of 3'-aminoglycoside phosphotransferase type I confers resistance to tobramycin in *Escherichia coli*. *Antimicrob Agents Chemother*, **37**, 78–83.

Mhand, R.A., Brahimi, N., Moustaoui, N., Mdaghri, N.E., Amaruch, H., Grimont, F., Bingen, E. and Benbachir, M., (1999), Characterization of extended-spectrum β-lactamase-producing *Salmonella typhimurium* by phenotypic and genotypic typing methods. *J Clin Microbiol*, **37**, 3769–3773.

Michel, B., Ehrlich, S.D. and Uzest, M., (1997), DNA double-strand breaks caused by replication arrest. *Embo J*, **16**, 430–438.

Michel, B., Grompone, G., Flores, M.J. and Bidnenko, V., (2004), Multiple pathways process stalled replication forks. *Proc Natl Acad Sci U S A*, **101**, 12783–12788.

Middlemiss, J.K. and Poole, K., (2004), Differential impact of MexB mutations on substrate selectivity of the MexAB-OprM multidrug efflux pump of *Pseudomonas aeruginosa*. *J Bacteriol*, **186**, 1258–1269.

Miller, G.H., Sabatelli, F.J., Hare, R.S., Glupczynski, Y., Mackey, P., Shlaes, D., Shimizu, K. and Shaw, K.J., (1997), The most frequent aminoglycoside resistance mechanisms—changes with time and geographic area: a reflection of aminoglycoside usage patterns? Aminoglycoside Resistance Study Groups. *Clin Infect Dis*, **24** (Suppl. 1), S46–S62.

Miller, G.H., Sabatelli, F.J., Naples, L., Hare, R.S. and Shaw, K.J., (1995), The changing nature of aminoglycoside resistance mechanisms and the role of isepamicin—a new broad-spectrum aminoglycoside. The Aminoglycoside Resistance Study Groups. *J Chemother*, **7** (Suppl. 2), 31–44.

Mine, T., Morita, Y., Kataoka, A., Mizushima, T. and Tsuchiya, T., (1999), Expression in *Escherichia coli* of a new multidrug efflux pump, MexXY, from *Pseudomonas aeruginosa*. *Antimicrob Agents Chemother*, **43**, 415–417.

Mingeot-Leclercq, M.P., Glupczynski, Y. and Tulkens, P.M., (1999), Aminoglycosides: activity and resistance. *Antimicrob Agents Chemother*, **43**, 727–737.

Miriagou, V., Tzouvelekis, L.S., Rossiter, S., Tzelepi, E., Angulo, F.J. and Whichard, J.M., (2003), Imipenem resistance in a *Salmonella* clinical strain due to plasmid-mediated Class A carbapenemase KPC-2. *Antimicrob Agents Chemother*, **47**, 1297–1300.

Miro, E., Verges, C., Garcia, I., Mirelis, B., Navarro, F., Coll, P., Prats, G. and Martinez-Martinez, L., (2004), Resistance to quinolones and β-lactams in *Salmonella enterica* due to mutations in topoisomerase-encoding genes, altered cell permeability and expression of an active efflux system. *Enferm Infecc Microbiol Clin*, **22**, 204–211.

Miró, E., del Cuerpo, M., Navarro, F., Sabaté, M., Mireleis, B. and Prats, G., (1998), Emergence of clinical isolates with decreased susceptibility to ceftazidime and synergistic effect with co-amoxiclav due to SHV-1 hyperproduction. *J Antimicrob Chemother*, **42**, 535–538.

Moazed, D. and Noller, H.F., (1987), Interaction of antibiotics with functional sites in 16S ribosomal RNA. *Nature*, **327**, 389–394.

Montero, C., Mateu, G., Rodriguez, R. and Takiff, H., (2001), Intrinsic resistance of *Mycobacterium smegmatis* to fluoroquinolones may be influenced by new pentapeptide protein MfpA. *Antimicrob Agents Chemother*, **45**, 3387–3392.

Moore, R.A., DeShazer, D., Reckseidler, S., Weissman, A. and Woods, D.E., (1999), Efflux-mediated aminoglycoside and macrolide resistance in *Burkholderia pseudomallei*. *Antimicrob Agents Chemother*, **43**, 465–470.

Morais Cabral, J.H., Jackson, A.P., Smith, C.V., Shikotra, N., Maxwell, A. and Liddington, R.C., (1997), Crystal structure of the breakage-reunion domain of DNA gyrase. *Nature*, **388**, 903–906.

Morita, Y., Murata, T., Mima, T., Shiota, S., Kuroda, T., Mizushima, T., Gotoh, N., Nishino, T. and Tsuchiya, T., (2003), Induction of *mexCD-oprJ* operon for a multidrug efflux pump by disinfectants in wild-type *Pseudomonas aeruginosa* PAO1. *J Antimicrob Chemother*, **51**, 991–994.

Murakami, S., Nakashima, R., Yamashita, E. and Yamaguchi, A., (2002), Crystal structure of bacterial multidrug efflux transporter AcrB. *Nature*, **419**, 587–593.

Mushtaq, S., Ge, Y. and Livermore, D.M., (2004), Doripenem versus *Pseudomonas aeruginosa in vitro*: activity against characterized mutants, and transconjugants and resistance seletion potential. *Antimicrob Agents Chemother*, **48**, 3086–3092.

Mutnick, A.H., Rhomberg, P.R., Sader, H.S. and Jones, R.N., (2004), Antimicrobial usage and resistance trend relationships from the MYSTIC Programme in North America (1999–2001). *J Antimicrob Chemother*, **53**, 290–296.

Murakami, S., Nakashima, R., Yamashita, E., Matsumoto T. and Yamaguchi, A., (2006), Crystal structures of a multidrug transporter reveal a functionally rotating Mechanism. Nature **443**: 173–179.

Nachamkin, I., Ung, H. and Li, M., (2002), Increasing fluoroquinolone resistance in *Campylobacter jejuni*, Pennsylvania, USA, 1982–2001. *Emerg Infect Dis*, **8**, 1501–1503.

National Nosocomial Infections Surveillance (NNIS) System, (1999), Hospital infection program. NNIS antimicrobial resistance ICU surveillance report, 1999. Centers for disease Control and Prevention, Atlanta, GA.

Nadjar, D., Rouveau, M., Verdet, C., Donay, J.-L., Herrmann, J.-L., Lagrange, P.H., Philippon, A. and Arlet, G., (2000), Outbreak of *Klebsiella pneumoniae* producing transferable AmpC-type β-lactamase (ACC-1) originating from *Hafnia alvei*. *FEMS Microbiol Lett*, **187**, 35–40.

Nakayama, K., Takashima, K., Ishihara, H., Shinomiya, T., Kageyama, M., Kanaya, S., Ohnishi, M., Murata, T., Mori, H. and Hayashi, T., (2000), The R-type pyocin of *Pseudomonas aeruginosa* is related to P2 phage, and the F-type is related to lambda phage. *Mol Microbiol*, **38**, 213–231.

Nazir, J., Urban, C., Mariano, N., Burns, J., Tommasulo, B., Rosenberg, C., Segal-Maurer, S. and Rahal, J.J., (2004), Quinolone-resistant *Haemophilus influenzae* in a long-term care facility: clinical and molecular epidemiology. *Clin Infect Dis*, **38**, 1564–1569.

Nelson, E.C. and Elisha, B.G., (1999), Molecular basis of AmpC hyperproduction in clinical isolates of *Escherichia coli* . *Antimicrob Agents Chemother*, **43**, 957–959.

Nelson, M.L. and Levy, S.B., (1999), The reversal of tetracycline resistance mediated by different bacterial tetracycine resistance determinants by an inhibitor of the Tet(B) antiport protein. *Antimicrob Agents Chemother*, **43**, 1719–1724.

Neuhauser, M.M., Weinstein, R.A., Rydman, R., Danziger, L.H., Karam, G. and Quinn, J.P., (2003), Antibiotic resistance among gram-negative bacilli in US intensive care units: implications for fluoroquinolone use. *JAMA*, **289**, 885–888.

Nikaido, H., (1998), Multiple antibiotic resistance and efflux. *Curr Opin Microbiol*, **1**, 516–523.

Nikaido, H. and Thanassi, D.G., (1993), Penetration of lipophilic agents with multiple protonation sites into bacterial cells: tetracyclines and fluoroquinolones as examples. *Antimicrob Agents Chemother*, **37**, 1393–1399.

Nikaido, H., Basina, M., Nguyen, V. and Rosenberg, E.Y., (1998), Multidrug efflux pump AcrAB of *Salmonella typhimurium* excretes only those beta-lactam antibiotics containing lipophilic side chains. *J Bacteriol*, **180**, 4686–4692.

Nordmann, P., Mariotte, S., Nass, T., Iabia, R. and Nicolas, M.-H., (1993), Biochemical properties of a carbapenem-hydrolyzing β-lactamase from *Enterobacter cloacae* and cloning of the gene into *Escherichia coli* . *Antimicrob Agents Chemother*, **37**, 939–946.

Nordmann, P. and Poirel, L., (2002), Emerging carbapenemases in Gram-negative aerobes. *Clin Microbiol Infect*, **8**, 321–331.

Norrby, S.R. and Lietman, P.S., (1993), Safety and tolerability of fluoroquinolones. *Drugs*, **45** (Suppl. 3), 59–64.

O'Hara, K., Haruta, S., Sawai, T., Tsunoda, M. and Iyobe, S., (1998), Novel metallo-β-lactamase mediated by a *Shigella flexneri* plasmid. *FEMS Microbiol Lett*, **162**, 201–206.

Oethinger, M., Kern, W.V., Jellen-Ritter, A.S., McMurry, L.M. and Levy, S.B., (2000), Ineffectiveness of topoisomerase mutations in mediating clinically significant fluoroquinolone resistance in *Escherichia coli* in the absence of the AcrAB efflux pump. *Antimicrob Agents Chemother*, **44**, 10–13.

Ogle, J.W., Reller, L.B. and Vasil, M.L., (1988), Development of resistance in *Pseudomonas aeruginosa* to imipenem, norfloxacin, and ciprofloxacin during therapy: proof provided by typing with a DNA probe. *J Infect Dis*, **157**, 743–748.

Oh, H., Stenhoff, J., Jalal, S. and Wretlind, B., (2003), Role of efflux pumps and mutations in genes for topoisomerases II and IV in fluoroquinolone-resistant *Pseudomonas aeruginosa* strains. *Microb Drug Resist*, **9**, 323–328.

Olliver, A., Valle, M., Chaslus-Dancla, E. and Cloeckaert, A., (2005), Overexpression of the multidrug efflux operon *acrEF* by insertional activation with *IS1* or *IS10* elements in *Salmonella enterica serovar typhimurium DT204* acrB mutants selected with fluoroquinolones. *Antimicrob Agents Chemother*, **49**, 289–301.

Orth, P., Schnappinger, D., Hillen, W., Saenger, W. and Hinrichs, W., (2000), Structural basis of gene regulation by the tetracycline inducible Tet repressor-operator system. *Nature Struct Biol*, **7**, 215–219.

Osano, E., Arakawa, Y., Wacharotayankun, R., Ohta, M., Horii, T., Ito, H., Yosimura, F. and Kato, N., (1994), Molecular characterization of an enterobacterial metallo β-lactamase found in a clinical isoalte of *Serratia marcescens* that shows imipenem resistance. *Antimicrob Agents Chemother*, **38**, 71–78.

Palumbo, J.D., Kado, C.I. and Phillips, D.A., (1998), An isoflavonoid-inducible efflux pump in *Agrobacterium tumefaciens* is involved in competitive colonization of roots. *J Bacteriol*, **180**, 3107–3113.

Parr, T.R., Jr. and Bayer, A.S., (1988), Mechanisms of aminoglycoside resistance in variants of *Pseudomonas aeruginosa* isolated during treatment of experimental endocarditis in rabbits. *J Infect Dis*, **158**, 1003–1010.

Payne, D.J., Bateson, J.H., Gasson, B.C., Proctor, D., Khushi, T., Farmer, T.H., Tolson, D.A., Bell, D., Skett, P.W., Marshall, A.C., Reid, R., Ghosez, L., Combret, Y. and Marchand-Brynaert, J., (1997), Inhibition of metallo-β-lactamases by a series of mercaptoacetic acid thiol ester derivatives. *Antimicrob Agents Chemother*, **41**, 135–140.

Pearson, J.P., Van Delden, C. and Iglewski, B.H., (1999), Active efflux and diffusion are involved in transport of *Pseudomonas aeruginosa* cell-to-cell signals. *J Bacteriol*, **181**, 1203–1210.

Pena, C., Albareda, J.M., Pallares, R., Pujol, M., Tubau, F. and Ariza, J., (1995), Relationship between quinolone use and emergence of ciprofloxacin-resistant *Escherichia coli* in bloodstream infections. *Antimicrob Agents Chemother*, **39**, 520–524.

Peng, H. and Marians, K.J., (1993a), *Escherichia coli* topoisomerase IV. Purification, characterization, subunit structure, and subunit interactions. *J Biol Chem*, **268**, 24481–24490.

Peng, H. and Marians, K.J., (1993b), Decatenation activity of topoisomerase IV during oriC and pBR322 DNA replication in vitro. *Proc Natl Acad Sci U S A*, **90**, 8571–8575.

Petersen, P.J. and Bradford, P.A., (2005), Effect of medium age and supplementation with the biocatalytic oxygen-reducing reagent oxyrase on *in vitro* activities of tigecycline against recent clinical Isolates. *Antimicrob Agents Chemother*, **49**, 3910–3918.

Petersen, P.J., Jacobus, N.V., Weiss, W.J., Sum, P.E. and Testa, R.T., (1999), *In vitro* and *in vivo* antimicrobial activities of a novel glycylcycline, the 9-t-butylglycylamido derivative of minocycline (GAR-936). *Antimicrob Agents Chemother*, **43**, 738–744.

Peterson, A.A., Hancock, R.E. and McGroarty, E.J., (1985), Binding of polycationic antibiotics and polyamines to lipopolysaccharides of *Pseudomonas aeruginosa*. *J Bacteriol*, **164**, 1256–1261.

Poirel, L., Girlich, D., Naas, T. and Nordmann, P., (2001a), OXA-28, an extended-spectrum variant of OXA-10 β-lactamase from *Pseudomonas aeruginosa* and its plasmid- and integron-located gene. *Antimicrob Agents Chemother*, **45**, 447–453.

Poirel, L., Lambert, T., Turkoglu, S., Ronco, E., Gaillard, J. and Nordmann, P., (2001b), Characterization of Class 1 integrons from *Pseudomonas aeruginosa* that contain the bla(VIM-2) carbapenem-hydrolyzing β-lactamase gene and of two novel aminoglycoside resistance gene cassettes. *Antimicrob Agents Chemother*, **45**, 546–552.

Poirel, L., Weldhagen, G.F., Naas, T., Champs, C.D., Dove, M.G. and Nordmann, P., (2001c), GES-2, a class A β-lactamase from *Pseudomonas aeruginosa* with increased hydroloysis of imipenem. *Antimicrob Agents Chemother*, **45**, 2598–2603.

Poole, K., (2000), Efflux-mediated resistance to fluoroquinolones in gram-negative bacteria. *Antimicrob Agents Chemother*, **44**, 2233–2241.

Poole, K., (2001), Multidrug efflux pumps and antimicrobial resistance in *Pseudomonas aeruginosa* and related organisms. *J Mol Microbiol Biotechnol*, **3**, 255–264.

Poole, K., (2002), Outer membranes and efflux: the path to multidrug resistance in Gram-negative bacteria. *Curr Pharm Biotechnol*, **3**, 77–98.

Poole, K., (2004), Efflux-mediated multiresistance in Gram-negative bacteria. *Clin Microbiol Infect*, **10**, 12–26.

Prammananan, T., Sander, P., Brown, B.A., Frischkorn, K., Onyi, G.O., Zhang, Y., Bottger, E.C. and Wallace, R.J., Jr. (1998) A single 16S ribosomal RNA substitution is responsible for resistance to amikacin and other 2-deoxystreptamine aminoglycosides in *Mycobacterium abscessus* and *Mycobacterium chelonae*. *J Infect Dis*, **177**, 1573–1581.

Prinarakis, E.E., Miriagou, V., Tzelepi, E., Gazouli, M. and Tzouvelekis, L.S., (1997), Emergence of an inhibitor-resistant β-lactamase (SHV-10) derived from an SHV-5 variant. *Antimicrob Agents Chemother*, **41**, 838–840.

Projan, S.J., (2003), Why is big Pharma getting out of antibacterial drug discovery? *Curr Opin Microbiol*, **6**, 427–430.

Purohit, P. and Stern, S., (1994), Interactions of a small RNA with antibiotic and RNA ligands of the 30S subunit. *Nature*, **370**, 659–662.

Rafii, F., Park, M. and Novak, J.S., (2005), Alterations in DNA gyrase and topoisomerase IV in resistant mutants of *Clostridium perfringens* found after *in vitro* treatment with fluoroquinolones. *Antimicrob Agents Chemother*, **49**, 488–492.

Rahal, J.J., Urban, C. and Horn, D., (1998), Class restriction of cephalosporin use to control total cephalosporin resistance in nosocomial *Klebsiella*. *JAMA*, **280**, 1233–1237.

Ramsey, B.W., Pepe, M.S., Quan, J.M., Otto, K.L., Montgomery, A.B., Williams-Warren, J., Vasiljev, K.M., Borowitz, D., Bowman, C.M., Marshall, B.C., Marshall, S. and Smith, A.L., (1999), Intermittent administration of inhaled tobramycin in patients with cystic fibrosis. Cystic Fibrosis Inhaled Tobramycin Study Group. *N Engl J Med*, **340**, 23–30.

Randall, L.P. and Woodward, M.J., (2002), The multiple antibiotic resistance (*mar*) locus and its significance. *Res Vet Sci*, **72**, 87–93.

Randall, L.P., Cooles, S.W., Piddock, L.J. and Woodward, M.J., (2004), Effect of triclosan or a phenolic farm disinfectant on the selection of antibiotic-resistant *Salmonella enterica*. *J Antimicrob Chemother*, **54**, 621–627.

Rasheed, J.K., Jay, C., Metchock, B., Berkowitz, F., Weigel, L., Crellin, J., Steward, C., Hill, B., Medeiros, A.A. and Tenover, F.C., (1997), Evolution of extended-spectrum β-lactam resistance (SHV-8) in a strain of *Escherichia coli* during multiple episodes of bacteremia. *Antimicrob Agents Chemother*, **41**, 647–653.

Rasmussen, B. and Bush, K., (1997), Carbapenem-hydrolyzing ÿ-lactamases. *Antimicrob Agents Chemother*, **41**, 23–232.

Rasmussen, B.A., Bush, K., Keeney, D., Yang, Y., hare, R., O'Gara, C. and Medeiros, A.A., (1996), Characterization of IMI-1 β-lactamase, a novel class A carbapenem-hydrolyzing enzyme from *Enterobacter cloacae*. *Antimicrob Agents Chemother*, **40**, 2080–2086.

Renau, T.E., Leger, R., Flamme, E.M., Sangalang, J., She, M.W., Yen, R., Gannon, C.L., Griffith, D., Chamberland, S., Lomovskaya, O., Hecker, S.J., Lee, V.J., Ohta, T. and Nakayama, K., (1999), Inhibitors of efflux pumps in *Pseudomonas aeruginosa* potentiate the activity of the fluoroquinolone antibacterial levofloxacin. *J Med Chem*, **42**, 4928–4931.

Renau, T.E., Leger, R., Flamme, E.M., She, M.W., Gannon, C.L., Mathias, K.M., Lomovskaya, O., Chamberland, S., Lee, V.J., Ohta, T., Nakayama, K. and Ishida, Y., (2001), Addressing the stability of C-capped dipeptide efflux pump inhibitors that potentiate the activity of levofloxacin in *Pseudomonas aeruginosa*. *Bioorg Med Chem Lett*, **11**, 663–667.

Renau, T.E., Leger, R., Filonova, L., Flamme, E.M., Wang, M., Yen, R., Madsen, D., Griffith, D., Chamberland, S., Dudley, M.N., Lee, V.J., Lomovskaya, O., Watkins, W.J., Ohta, T., Nakayama, K. and Ishida, Y., (2003), Conformationally-restricted analogues of efflux pump inhibitors that potentiate the activity of levofloxacin in *Pseudomonas aeruginosa*. *Bioorg Med Chem Lett*, **13**, 2755–2758.

Renau, T.E., Leger, R., Yen, R., She, M.W., Flamme, E.M., Sangalang, J., Gannon, C.L., Chamberland, S., Lomovskaya, O. and Lee, V.J., (2002), Peptidomimetics of efflux pump inhibitors potentiate the activity of levofloxacin in *Pseudomonas aeruginosa*. *Bioorg Med Chem Lett*, **12**, 763–766.

Rice, L.B., Willey, S.H., Papanicolaou, G.A., Medieros, A.A., Eliopoulos, G.M., R.C Moellering, J. and Jacoby, G.A., (1990), Outbreak of ceftazidime resistance caused by extended-spectrum β-lactamases at a Massachusetts chronic-care facility. *Antimicrob Agents Chemother*, **34**, 2193–2199.

Rice, L.B., Eckstein, E.C., DeVente, J. and Shlaes, D.M., (1996), Ceftazidime-resistant *Klebsiella pneumoniae* isolates recovered at the Cleveland Department of Veterans Affairs Medical Center. *Clin Infect Dis*, **23**, 118–124.

Rivera, M., Hancock, R.E., Sawyer, J.G., Haug, A. and McGroarty, E.J., (1988), Enhanced binding of polycationic antibiotics to lipopolysaccharide from an aminoglycoside-supersusceptible, tolA mutant strain of *Pseudomonas aeruginosa*. *Antimicrob Agents Chemother*, **32**, 649–655.

Roberts, M.C., (1996), Tetracycline resistance determinants: mechanisms of action, regulation of expression, genetic mobility and distribution. *FEMS Microbiol Rev*, **19**, 1–24.

Robicsek, A., Sahm, D.F., Strahilevitz, J., Jacoby, G.A. and Hooper, D.C., (2005), Broader distribution of plasmid-mediated quinolone resistance in the United States. *Antimicrob Agents Chemother*, **49**, 3001–3003.

Robicsek, A., Strahilevits, J., Jacoby, G.A., Macielag, M., Abbanat, D., Park, C.H., Bush, K. and Hooper D.C., (2006),. Fluoroquinolone-modifying enzyme: a new adaptation of a common aminoglycoside acetyltransferase. *Nature Med*, **12**,83–88

Rodriguez Esparragon, F., Gonzalez Martin, M., Gonzalez Lama, Z., Sabatelli, F.J. and Tejedor Junco, M.T., (2000), Aminoglycoside resistance mechanisms in clinical isolates of *Pseudomonas aeruginosa* from the Canary Islands. *Zentralbl Bakteriol*, **289**, 817–826.

Rodriguez-Martinez, J.M., Pascual, A., Garcia, I. and Martinez-Martinez, L., (2003), Detection of the plasmid-mediated quinolone resistance determinant qnr among clinical isolates of *Klebsiella pneumoniae* producing AmpC-type β-lactamase. *J Antimicrob Chemother*, **52**, 703–706.

Roestamadji, J. and Mobashery, S., (1998), The use of neamine as a molecular template: inactivation of bacterial antibiotic resistance enzyme aminoglycoside 3'-phosphotransferase type IIa. *Bioorg Med Chem Lett*, **8**, 3483–3488.

Rosenberg, E.Y., Ma, D. and Nikaido, H., (2000), AcrD of *Escherichia coli* is an aminoglycoside efflux pump. *J Bacteriol*, **182**, 1754–1756.

Ross, J.I., Eady, E.A., Cove, J.H. and Cunliffe, W.J., (1998), 16S rRNA mutation associated with tetracycline resistance in a gram-positive bacterium. *Antimicrob Agents Chemother*, **42**, 1702–1705.

Rothstein, D.M., McGlynn, M., Bernan, V., McGahren, J., Zaccardi, J., Cekleniak, N. and Bertrand, K.P., (1993), Detection of tetracyclines and efflux pump inhibitors. *Antimicrob Agents Chemother*, **37**, 1624–1629.

Rubens, C.E., McNeill, W.F. and Farrar, W.E., Jr., (1979), Transposable plasmid deoxyribonucleic acid sequence in *Pseudomonas aeruginosa* which mediates resistance to gentamicin and four other antimicrobial agents. *J Bacteriol*, **139**, 877–882.

Ruzin, A., Visalli, M.A., Keeney, D. and Bradford, P.A., (2005), Influence of transcriptional activator RamA on expression of multidrug efflux pump AcrAB and tigecycline susceptibility in *Klebsiella pneumoniae*. *Antimicrob Agents Chemother*, **49**, 1017–1022.

Ryder, N.S., Dzink-Fox, J., Kubik, B., Mlineritsch, W., Alavarez, S., Bracken, K., Dean, K., Jain, R., Sundaram, A., Weidmann, B. and Yuan, Z., (2004), LBM415, a new peptide deformylase inhibitor with potent in vitro activity agains drug-resistant bacteria, Abstr. F-1959. In *44th Interscience Conference on Antimicrobial Agents and Chemotherapy*, Washington, DC.

Sabaté, M., Tarragó, R., Navarro, F., Miró, E., Vergés, C., Barbé, J. and Prats, G., (2000), Cloning and sequence of the gene encoding a novel cefotaxime-hydrolyzing β-lactamase (CTX-M-9) from *Escherichia coli* in Spain. *Antimicrob Agents Chemother*, **44**, 1970–1973.

Sader, H.S., Jones, R.N. and Silva, J.B., (2002), Skin and soft tissue infections in Latin American medical centers: four-year assessment of the pathogen frequency and antimicrobial susceptibility patterns. *Diagn Microbiol Infect Dis*, **44**, 281–288.

Saenz, Y., Zarazaga, M., Lantero, M., Gastanares, M.J., Baquero, F. and Torres, C., (2000), Antibiotic resistance in *Campylobacter* strains isolated from animals, foods, and humans in Spain in 1997–1998. *Antimicrob Agents Chemother*, **44**, 267–271.

Sahm, D.F., Critchley, I.A., Kelly, L.J., Karlowsky, J.A., Mayfield, D.C., Thornsberry, C., Mauriz, Y.R. and Kahn, J., (2001), Evaluation of current activities of fluoroquinolones against gram-negative bacilli using centralized in vitro testing and electronic surveillance. *Antimicrob Agents Chemother*, **45**, 267–274.

Saika, T., Hasegawa, M., Kobayashi, I. and Nishida, M., (1999), Ionic binding of 3H-gentamicin and short-time bactericidal activity of gentamicin against *Pseudomonas aeruginosa* isolates with different lipopolysaccharide structures. *Chemotherapy*, **45**, 296–302.

Sakon, J., Liao, H.H., Kanikula, A.M., Benning, M.M., Rayment, I. and Holden, H.M., (1993), Molecular structure of kanamycin nucleotidyltransferase determined to 3.0-A resolution. *Biochemistry*, **32**, 11977–11984.

Sanchez, L., Pan, W., Vinas, M. and Nikaido, H., (1997), The *acrAB* homolog of *Haemophilus influenzae* codes for a functional multidrug efflux pump. *J Bacteriol*, **179**, 6855–6857.

Sanchez-Pescador, R., Brown, J.T., Roberts, M.C. and Urdea, M.S., (1998), Homology of the TetM with translational elongation factors: implications for potential modes of the *tetM*-conferred tetracycline resistance. *Nucleic Acids Res*, **16**, 1218.

Sanders, C.C., Barry, A.L., Washington, J.A., Shubert, C., Moland, E.S., Traczewski, M.M., Knapp, C. and Mulder, R., (1996), Detection of extended-spectrum-β-lactamase-producing members of the family *Enterobacteriaceae* with the Vitek ESBL test. *J Clin Microbiol*, **34**, 2997–3001.

Schiappa, D.A., Hayden, M.K., Matushek, M.G., Hashemi, F.N., Sullivan, J., Smith, K.Y., Miyashiro, D., Quinn, J.P., Weinstein, R.A. and Trenholme, G.M., (1996), Ceftazidime-resistant *Klebsiella pneumoniae* and *Escherichia coli* bloodstream infection: a case-control and molecular epidemiologic infection. *J Infect Dis*, **174**, 529–536.

Schlessinger, D., (1988), Failure of aminoglycoside antibiotics to kill anaerobic, low-pH, and resistant cultures. *Clin Microbiol Rev*, **1**, 54–59.

Schnappinger, D. and Hillen, W., (1996), Tetracyclines: antibiotic action, uptake, and resistance mechanisms. *Archives of Microbiology*, **165**, 359–369.

Schweizer, H.D., (1993), Small broad-host-range gentamicin resistance gene cassettes for site-specific insertion and deletion mutagenesis. *Biotechniques*, **15**, 831–834.

Seeger, M.A., Schiefner A., Eicher, T., Verrey, F., Diederichs, K. and Pos, K.M., (2006), Structural asymmetry of AcrB trimer suggests a peristaltic pump mechanism. *Science* **313**, 1295–1298.

Shafer, W.M., Balthazar, J.T., Hagman, K.E. and Morse, S.A., (1995), Missense mutations that alter the DNA-binding domain of the MtrR protein occur frequently in rectal isolates of *Neisseria gonorrhoeae* that are resistant to faecal lipids. *Microbiology*, **141** (Pt 4), 907–911.

Shah, A.A., Hasan, F., Ahmed, S. and Hameed, A., (2004), Characteristics, epidemiology and clinical importance of emerging strains of Gram-negative bacilli producing extended-spectrum β-lactamases. *Res Microbiol*, **155**, 409–421.

Shams, W.E. and Evans, M.E., (2005), Guide to selection of fluoroquinolones in patients with lower respiratory tract infections. *Drugs*, **65**, 949–991.

Shaw, K.J., Hare, R.S., Sabatelli, F.J., Rizzo, M., Cramer, C.A., Naples, L., Kocsi, S., Munayyer, H., Mann, P., Miller, G.H., Verbist, L., Van Landuyt, H., Glupczynski, Y., Catalano, M. and Woloj, M., (1991), Correlation between aminoglycoside resistance profiles and DNA hybridization of clinical isolates. *Antimicrob Agents Chemother*, **35**, 2253–2261.

Shaw, K.J., Rather, P.N., Hare, R.S. and Miller, G.H., (1993), Molecular genetics of aminoglycoside resistance genes and familial relationships of the aminoglycoside-modifying enzymes. *Microbiol Rev*, **57**, 138–163.

Shea, M.E. and Hiasa, H., (1999), Interactions between DNA helicases and frozen topoisomerase IV-quinolone-DNA ternary complexes. *J Biol Chem*, **274**, 22747–22754.

Shearer, B.G. and Legakis, N.J., (1985), *Pseudomonas aeruginosa*: evidence for the involvement of lipopolysaccharide in determining outer membrane permeability to carbenicillin and gentamicin. *J Infect Dis*, **152**, 351–355.

Shen, L.L., (1994), Molecular mechanisms of DNA gyrase inhibition by quinolone antibacterials. *Adv Pharmacol*, **29A**, 285–304.

Shen, L.L., (2001), Quinolone interactions with DNA and DNA gyrase. *Methods Mol Biol*, **95**, 171–184.

Shen, L.L., Baranowski, J. and Pernet, A.G., (1989a), Mechanism of inhibition of DNA gyrase by quinolone antibacterials: specificity and cooperativity of drug binding to DNA. *Biochemistry*, **28**, 3879–3885.

Shen, L.L., Kohlbrenner, W.E., Weigl, D. and Baranowski, J., (1989b), Mechanism of quinolone inhibition of DNA gyrase. Appearance of unique norfloxacin binding sites in enzyme-DNA complexes. *J Biol Chem*, **264**, 2973–2978.

Shen, L.L., Mitscher, L.A., Sharma, P.N., O'Donnell, T.J., Chu, D.W., Cooper, C.S., Rosen, T. and Pernet, A.G., (1989c), Mechanism of inhibition of DNA gyrase by quinolone antibacterials: a cooperative drug-DNA binding model. *Biochemistry*, **28**, 3886–3894.

Shultz, T.R., Tapsall, J.W. and White, P.A., (2001), Correlation of *in vitro* susceptibilities to newer quinolones of naturally occurring quinolone-resistant *Neisseria gonorrhoeae* strains with changes in GyrA and ParC. *Antimicrob Agents Chemother*, **45**, 734–738.

Sirot, D., Recule, C., Chaibi, E.B., Bret, L., Croize, J., Chanal-Claris, C., Labia, R. and Sirot, J., (1997), A complex mutant of TEM-1 β-lactamase with mutations encountered in both IRT-4 and extended-spectrum TEM-15, produced by an *Escherichia coli* clinical isolate. *Antimicrob Agents Chemother*, **41**, 1322–1325.

Siu, L.K., Lo, J.Y.C., Yuen, K.Y., Chau, P.Y., Ng, M.H. and Ho, P.L., (2000), β-lactamases in *Shigella flexneri* isolates from Hong Kong and Shanghai and a novel OXA-1-like β-lactamase, OXA-30. *Antimicrob Agents Chemother*, **44**, 2034–2038.

Smith, C.A. and Baker, E.N., (2002), Aminoglycoside antibiotic resistance by enzymatic deactivation. *Curr Drug Targets Infect Disord*, **2**, 143–160.

Smith, K.E., Besser, J.M., Hedberg, C.W., Leano, F.T., Bender, J.B., Wicklund, J.H., Johnson, B.P., Moore, K.A. and Osterholm, M.T., (1999), Quinolone-resistant *Campylobacter jejuni* infections in Minnesota, 1992–1998. Investigation Team. *N Engl J Med*, **340**, 1525–1532.

Sobel, M.L., Hocquet, D., Cao, L., Plesiat, P. and Poole, K., (2005), Mutations in PA3574 (nalD) lead to increased MexAB-OprM expression and multidrug resistance in laboratory and clinical isolates of *Pseudomonas aeruginosa*. *Antimicrob Agents Chemother*, **49**, 1782–1786.

Sobel, M.L., McKay, G.A. and Poole, K., (2003), Contribution of the MexXY multidrug transporter to aminoglycoside resistance in *Pseudomonas aeruginosa* clinical isolates. *Antimicrob Agents Chemother*, **47**, 3202–3207.

Sougakoff, W., Papadopoulou, B., Nordmann, P. and Courvalin, P., (1987), Nucleotide sequence and distribution of gene tetO encoding tetracycline resistance in *Campylobacter coli*. *FEMS Microbiol Lett*, **44**, 153–159.

Spahn, C.M.T., Blaha, G., Agrawal, R.K., Penczek, P., Grassucci, R.A., Trieber, C.A., Connell, S.R., Taylor, D.E., Nierhaus, K.H. and Frank, J., (2001), Localization of the ribosomal protection protein Tet(O) on the ribosome and mechanism of tetracycline resistance. *Mol Cell*, **7**, 1037–1045.

Speer, B.S. and Salyers, A.A., (1989), Novel aerobic tetracycline resistance gene that chemically modifies tetracycline. *J Bacteriol*, **171**, 148–153.

Springer, B., Kidan, Y.G., Prammananan, T., Ellrott, K., Bottger, E.C. and Sander, P., (2001), Mechanisms of streptomycin resistance: selection of mutations in the 16S rRNA gene conferring resistance. *Antimicrob Agents Chemother*, **45**, 2877–2884.

Stover, C.K., Pham, X.Q., Erwin, A.L., Mizoguchi, S.D., Warrener, P., Hickey, M.J., Brinkman, F.S., Hufnagle, W.O., Kowalik, D.J., Lagrou, M., Garber, R.L., Goltry, L., Tolentino, E., Westbrock-Wadman, S., Yuan, Y., Brody, L.L., Coulter, S.N., Folger, K.R., Kas, A., Larbig, K., Lim, R., Smith, K., Spencer, D., Wong, G.K., Wu, Z., Paulsen, I.T., Reizer, J., Saier, M.H., Hancock, R.E., Lory, S. and Olson, M.V., (2000), Complete genome sequence of *Pseudomonas aeruginosa* PA01, an opportunistic pathogen. *Nature*, **406**, 959–964.

Strahilevitz, J. and Hooper, D.C., (2005), Dual targeting of topoisomerase IV and gyrase to reduce mutant selection: direct testing of the paradigm by using WCK-1734, a new fluoroquinolone, and ciprofloxacin. *Antimicrob Agents Chemother*, **49**, 1949–1956.

Suarez, G. and Nathans, D., (1965), Inhibition of aminoacyl tRNA binding to ribosomes by tetracylcline. *Biochem Biophys Res Commun*, **18**, 743–750.

Sugino, A., Peebles, C.L., Kreuzer, K.N. and Cozzarelli, N.R., (1977), Mechanism of action of nalidixic acid: purification of *Escherichia coli* nalA gene product and its relationship to DNA gyrase and a novel nicking-closing enzyme. *Proc Natl Acad Sci USA*, **74**, 4767–4771.

Sum, P.-E. and Petersen, P., (1999), Synthesis and structure-activity relationship of novel glycylcycline derivatives leading to the discovery of GAR-936. *Bioorg Med Chem Lett*, **9**, 1459–1462.

Takehasi, M., Altschmied, L. and Hillen, W., (1986), Kinetic and equilibrium characterization of the Tet repressor–tetracycline complex by fluorescence measurements. Evidence for divalent metal ion requirement and energy transfer. *J Mol Biol*, **187**, 641–348.

Taylor, D.E., (1986), Plasmid-mediated tetracycline resistance in *Campylobacter jejuni* and *Campylobacter coli*: expresssion in *Escherichia coli* and identification of homology with streptococcal class M determinant. *J Bacteriol*, **165**, 1037–1039.

Tenover, F.C., Raney, P.M., Williams, P.P., Rasheed, J.K., Biddle, J.W., Oliver, A., Fridkin, S.K., Jevitt, L. and McGowan, J.E., Jr., (2003), Evaluation of the NCCLS extended-spectrum β-lactamase confirmation methods for *Escherichia coli* with isolates collected during project ICARE. *J Clin Microbiol*, **41**, 3142–3146.

Testa, R.T., Petersen, P.J., Jacobus, N.V., Sum, P.E., Lee, V.J. and Tally, F.P., (1993), In vitro and in vivo antibacterial activities of the glycylcyclines, a new class of semisynthetic tetracyclines. *Antimicrob Agents Chemother*, **37**, 2270–2277.

Thanassi, D.G., Cheng, L.W. and Nikaido, H., (1997), Active efflux of bile salts by *Escherichia coli*. *J Bacteriol*, **179**, 2512–2518.

Thomson, K.S. and Sanders, C.C., (1992), Detection of extended-spectrum β-lactamases in members of the family *Enterobacteriaceae*: Comparison of the double-disk and three dimensional tests. *Antimicrob Agents Chemother*, **36**, 1877–1882.

Toleman, M.A., Simm, A.M., Murphy, T.A., Gales, A.C., Biedenbach, D.J., Jones, R.N. and Walsh, T.R., (2002), Molecular characterization of SPM-1, a novel metallo-β-lactamase isolated in Latin America: report from the SENTRY antimicrobial surveillance programme. *J Antimicrob Chemother*, **50**, 673–679.

Toltzis, P., Dul, M.J., Hoyen, C., Salvator, A., Walsh, M., Zetts, L. and Toltzis, H., (2002), The effect of antibiotic rotation on colonization with antibiotic-resistant bacilli in a neonatal intensive care unit. *Pediatrics*, **110**, 707–711.

Traczewski, M.M. and Brown, S.D., (2003), PTK 0796 (BAY 73-6944): *In vitro* potency and spectrum of activity compared to ten other antimicrobial compounds. In *43rd Interscience Conference on Antimicrobial Agents and Chemotherapy*. Vol. Abstract 2458, Chicago, IL.

Tran, J.H. and Jacoby, G.A., (2002), Mechanism of plasmid-mediated quinolone resistance. *Proc Natl Acad Sci USA*, **99**, 5638–5642.

Tran, J.H., Jacoby, G.A. and Hooper, D.C., (2005a), Interaction of the plasmid-encoded quinolone resistance protein QnrA with *Escherichia coli* topoisomerase IV. *Antimicrob Agents Chemother*, **49**, 3050–3052.

Tran, J.H., Jacoby, G.A. and Hooper, D.C., (2005b), Interaction of the plasmid-encoded quinolone resistance protein Qnr with *Escherichia coli* DNA gyrase. *Antimicrob Agents Chemother*, **49**, 118–125.

Trieber, C.A. and Taylor, D.E., (2002), Mutations in the 16S ribosomal RNA genes of *Helicobacter pylori* mediate resistance to tetracycline. *J Bacteriol*, **184**, 2131–2140.

Tygacil, (2005), Tygacil™ [package insert]. Philadelphia, PA, Wyeth Pharmaceuticals Inc. 2005. Available at http://www.fda.gov/cder/foi/label/2005/021821lbl.pdf. Accessed June 20, 2005.

Tzouvelekis, L.S., Gazouli, M., Markogiannakis, A., Paraskaki, I., Legakis, N.J. and Tzelepi, E., (1998), Emergence of resistance to third-generation cephalosporins amongst *Salmonella typhimurium* isolates in Greece: report of the first three cases. *J Antimicrob Chemother*, **42**, 273–275.

Tzouvelekis, L.S., Tzelepi, E., Tassios, P.T. and Legakis, N.J., (2000), CTX-M-type β-lactamases: an emerging group of extended-spectrum enzymes. *Intern J Antimicrob Agents*, **14**, 137–143.

Unal, S., Masterton, R. and Goossens, H., (2004), Bacteraemia in Europe—antimicrobial susceptibility data from the MYSTIC surveillance programme. *Int J Antimicrob Agents*, **23**, 155–163.

Van Bambeke, F., Michot, J.M., Van Eldere, J. and Tulkens, P.M., (2005), Quinolones in 2005: an update. *Clin Microbiol Infect*, **11**, 256–280.

Visalli, M.A., Murphy, E., Projan, S.J. and Bradford, P.A., (2003), AcrAB multidrug efflux pump is associated with reduced levels of susceptibility to tigecycline (GAR-936) in *Proteus mirabilis*. *Antimicrob Agents Chemother*, **47**, 665–669.

Vogne, C., Aires, J.R., Bailly, C., Hocquet, D. and Plesiat, P., (2004), Role of the multidrug efflux system MexXY in the emergence of moderate resistance to aminoglycosides among *Pseudomonas aeruginosa* isolates from patients with cystic fibrosis. *Antimicrob Agents Chemother*, **48**, 1676–1680.

Walsh, T.R., Bolmstrom, A., Qwarnstrom, A. and Gales, A., (2002), Evaluation of a new Etest for detecting metallo-β-lactamases in routine clinical testing. *J Clin Microbiol*, **40**, 2755–2759.

Walsh, T.R., Toleman, M.A., Poirel, L. and Nordmann, P., (2005), Metallo-β-lactamases: the quiet before the storm? *Clin Microbiol Rev*, **18**, 306–325.

Wang, F., Zhu, D. and Hu, F., (2001a), Surveillance of bacterial resistance in Shanghai. *Zhonghua Yi Xue Za Zhi*, **81**, 17–19.

Wang, H., Dzink-Fox, J.L., Chen, M. and Levy, S.B., (2001b), Genetic characterization of highly fluoroquinolone-resistant clinical *Escherichia coli* strains from China: role of *acrR* mutations. *Antimicrob Agents Chemother*, **45**, 1515–1521.

Wang, M., Sahm, D.F., Jacoby, G.A. and Hooper, D.C., (2004), Emerging plasmid-mediated quinolone resistance associated with the *qnr* gene in *Klebsiella pneumoniae* clinical isolates in the United States. *Antimicrob Agents Chemother*, **48**, 1295–1299.

Wang, M., Tran, J.H., Jacoby, G.A., Zhang, Y., Wang, F. and Hooper, D.C., (2003), Plasmid-mediated quinolone resistance in clinical isolates of *Escherichia coli* from Shanghai, China. *Antimicrob Agents Chemother*, **47**, 2242–2248.

Watanabe, M., Iyobe, S., Inoue, M. and Mitsuhashi, S., (1991), Transferable imipenem resistance in *Pseudomonas aeruginosa*. *Antimicrob Agents Chemother*, **35**, 147–151.

Weinstein, R.A., (2001), Controlling antimicrobial resistance in hospitals: infection control and use of antibiotics. *Emerg Infect Dis*, **7**, 188–192.

Westbrock-Wadman, S., Sherman, D.R., Hickey, M.J., Coulter, S.N., Zhu, Y.Q., Warrener, P., Nguyen, L.Y., Shawar, R.M., Folger, K.R. and Stover, C.K., (1999), Characterization of a *Pseudomonas aeruginosa* efflux pump contributing to aminoglycoside impermeability. *Antimicrob Agents Chemother*, **43**, 2975–2983.

Wiener, J., Quinn, J.P., Bradford, P.A., Goering, R.V., Nathan, C., Bush, K. and Weinstein, R.A., (1999), Multiple antibiotic-resistant *Klebsiella* and *Escherichia coli* in nursing homes. *JAMA*, **281**, 517–523.

Williams, R.J., Livermore, D.M., Lindridge, M.A., Said, A.A. and Williams, J.D., (1984), Mechanisms of β-lactam resistance in British isolates of *Pseudomonas aeruginosa*. *J Med Microbiol*, **17**, 283–293.

Willmott, C.J., Critchlow, S.E., Eperon, I.C. and Maxwell, A., (1994), The complex of DNA gyrase and quinolone drugs with DNA forms a barrier to transcription by RNA polymerase. *J Mol Biol*, **242**, 351–363.

Willmott, C.J. and Maxwell, A., (1993), A single point mutation in the DNA gyrase A protein greatly reduces binding of fluoroquinolones to the gyrase-DNA complex. *Antimicrob Agents Chemother*, **37**, 126–127.

Wolf, E., Vassilev, A., Makino, Y., Sali, A., Nakatani, Y. and Burley, S.K., (1998), Crystal structure of a GCN5-related N-acetyltransferase: *Serratia marcescens* aminoglycoside 3-*N*-acetyltransferase. *Cell*, **94**, 439–449.

Woodcock, J., Moazed, D., Cannon, M., Davies, J. and Noller, H.F., (1991), Interaction of antibiotics with A- and P-site-specific bases in 16S ribosomal RNA. *Embo J*, **10**, 3099–3103.

Woodford, N., Tierno, P.M., Jr, Young, K., Tysall, L., Palepou, M.-F.I., Ward, E., Painer, R.E., Suber, D.F., Shungu, D., Silver, L.L., Inglima, K., Kornblum, J. and Livermore, D.M., (2004), Outbreak of *Klebsiella pneumoniae* producing a new carbapenem-hydrolyzing class A β-lactamase, KPC-3, in a New York medical center. *Antimicrob Agents Chemother*, **48**, 4793–4799.

Wright, G.D., (1999), Aminoglycoside-modifying enzymes. *Curr Opin Microbiol*, **2**, 499–503.

Wybenga-Groot, L.E., Draker, K., Wright, G.D. and Berghuis, A.M., (1999), Crystal structure of an aminoglycoside 6'-N-acetyltransferase: defining the GCN5-related N-acetyltransferase superfamily fold. *Structure Fold Des*, **7**, 497–507.

Wyeth Pharmaceuticals, (2005), Tygacil™ [package insert]. Available at http://www.fda.gov/cder/foi/label/2005/021821lbl.pdf. Vol. Accessed June 20, 2005., Collegeville, PA, Wyeth Pharmaceuticals Inc.

Xiong, Y.Q., Caillon, J., Kergueris, M.F., Drugeon, H., Baron, D., Potel, G. and Bayer, A.S., (1997), Adaptive resistance of *Pseudomonas aeruginosa* induced by aminoglycosides and killing kinetics in a rabbit endocarditis model. *Antimicrob Agents Chemother*, **41**, 823–826.

Yamaguchi, A., Ono, N., Akasaka, T., Noumi, T. and Sawai, T., (1990), Metal-tetracycline/H+ antiporter of *Escherichia coli* encoded by a transposon, Tn*10*. *J Biol Chem*, **265**, 15525–15530.

Yamane, K., Wachino, J., Doi, Y., Kurokawa, H. and Arakawa, Y., (2005), Global spread of multiple aminoglycoside resistance genes. *Emerg Infect Dis*, **11**, 951–953.

Yan, J.J., Wu, J.J., Tsai, S.-H. and Chuang, C.-L., (2004a), Comparison of the double-disk, combined disk, and Etest methods for detecting metallo-β-lactamases in gram negative bacilli. *Diagn Microbiol Infect Dis*, **49**, 5–11.

Yan, J.J., Wu, J.J., Ko, W.C., Tsai, S.H., Chuang, C.L., Wu, H.M., Lu, Y.J. and Li, J.D., (2004b), Plasmid-mediated 16S rRNA methylases conferring high-level aminoglycoside resistance in *Escherichia coli* and *Klebsiella pneumoniae* isolates from two Taiwanese hospitals. *J Antimicrob Chemother*, **54**, 1007–1012.

Yang, Y., Rasmussen, B.A. and Shlaes, D.M., (1999), Class A β-lactamases-enzyme-inhibitor interactions and resistance. *Pharmacology & Therapeutics*, **83**, 141–151.

Yang, Y., Wu, P.C. and Livermore, D.M., (1990), Biochemical characterization of a β-lactamase that hydrolyzes penems and carbapenems from two *Serratia marcescens* isolates. *Antimicrob Agents Chemother*, **1990**, 755–758.

Yigit, H., Queenan, A.M., Anderson, G.J., Domenech-Sanchez, A., Biddle, J.W., Steward, C.D., Alberti, S., Bush, K. and Tenover, F.C., (2001), Novel carbapenem-hydrolyzing β-lactamase, KPC-1 from a carbapenem-resistant strain of *Klebsiella pneumoniae*. *Antimicrob Agents Chemother*, **45**, 1151–1161.

Yigit, H., Queenan, A.M., Rasheed, J.K., Biddle, J.W., Domenech-Sanchez, A., Alberti, S., Bush, K. and Tenover, F.C., (2003), Carbapenem-resistant strain of *Klebsiella oxytoca* harboring carbapenem-hydrolyzing β-lactamase KPC-2. *Antimicrob Agents Chemother*, **47**, 3881–3889.

Yokoyama, K., Doi, Y., Yamane, K., Kurokawa, H., Shibata, N., Shibayama, K., Yagi, T., Kato, H. and Arakawa, Y., (2003), Acquisition of 16S rRNA methylase gene in *Pseudomonas aeruginosa*. *Lancet*, **362**, 1888–1893.

Yoneyama, H., Sato, K. and Nakae, T., (1991), Aminoglycoside resistance in *Pseudomonas aeruginosa* due to outer membrane stabilization. *Chemotherapy*, **37**, 239–245.

Yoon, S.S., Hennigan, R.F., Hilliard, G.M., Ochsner, U.A., Parvatiyar, K., Kamani, M.C., Allen, H.L., DeKievit, T.R., Gardner, P.R., Schwab, U., Rowe, J.J., Iglewski, B.H., McDermott, T.R., Mason, R.P., Wozniak, D.J., Hancock, R.E., Parsek, M.R., Noah, T.L., Boucher, R.C. and Hassett, D.J., (2002), *Pseudomonas aeruginosa* anaerobic respiration in biofilms: relationships to cystic fibrosis pathogenesis. *Dev Cell*, **3**, 593–603.

Yoshida, H., Bogaki, M., Nakamura, M. and Nakamura, S., (1990), Quinolone resistance-determining region in the DNA gyrase *gyrA* gene of *Escherichia coli*. *Antimicrob Agents Chemother*, **34**, 1271–1272.

Yoshida, H., Bogaki, M., Nakamura, M., Yamanaka, L.M. and Nakamura, S., (1991), Quinolone resistance-determining region in the DNA gyrase *gyrB* gene of *Escherichia coli*. *Antimicrob Agents Chemother*, **35**, 1647–1650.

Yoshizawa, S., Fourmy, D. and Puglisi, J.D., (1999), Recognition of the codon–anticodon helix by ribosomal RNA. *Science*, **285**, 1722–1725.

Yu, E.W., Aires, J.R., McDermott, G. and Nikaido, H., (2005), A periplasmic drug-binding site of the AcrB multidrug efflux pump: a crystallographic and site-directed mutagenesis study. *J Bacteriol*, **187**, 6804–6815.

Yu, E.W., McDermott, G., Zgurskaya, H.I., Nikaido, H. and Koshland, D.E., Jr., (2003), Structural basis of multiple drug-binding capacity of the AcrB multidrug efflux pump. *Science*, **300**, 976–980.

Zbindien, R., Pünter, V. and von Graevenitz, A., (2002), In vitro activities of BAL9141, a novel broad-spectrum pyrrolidinone cephalosporin, against gram-negative nonfermenters. *Antimicrob Agents Chemother*, **46**, 871–874.

Zechiedrich, E.L. and Cozzarelli, N.R., (1995), Roles of topoisomerase IV and DNA gyrase in DNA unlinking during replication in *Escherichia coli*. *Genes Dev*, **9**, 2859–2869.

Zechiedrich, E.L., Khodursky, A.B., Bachellier, S., Schneider, R., Chen, D., Lilley, D.M. and Cozzarelli, N.R., (2000), Roles of topoisomerases in maintaining steady-state DNA supercoiling in *Escherichia coli*. *J Biol Chem*, **275**, 8103–8113.

Zembower, T.R., Noskin, G.A., Postelnick, M.J., Nguyen, C. and Peterson, L.R., (1998), The utility of aminoglycosides in an era of emerging drug resistance. *Int J Antimicrob Agents*, **10**, 95–105.

Zervos, M.J., Hershberger, E., Nicolau, D.P., Ritchie, D.J., Blackner, L.K., Coyle, E.A., Donnelly, A.J., Eckel, S.F., Eng, R.H., Hiltz, A., Kuyumjian, A.G., Krebs, W., McDaniel, A., Hogan, P. and Lubowski, T.J., (2003), Relationship between fluoroquinolone use and changes in susceptibility to fluoroquinolones of selected pathogens in 10 United States teaching hospitals, 1991–2000. *Clin Infect Dis*, **37**, 1643–1648.

Zgurskaya, H.I. and Nikaido, H., (2000), Multidrug resistance mechanisms: drug efflux across two membranes. *Mol Microbiol*, **37**, 219–225.

Zhao, X. and Drlica, K., (2001), Restricting the selection of antibiotic-resistant mutants: a general strategy derived from fluoroquinolone studies. *Clin Infect Dis*, **33** (Suppl. 3), S147–156.

Zhao, X. and Drlica, K., (2002), Restricting the selection of antibiotic-resistant mutant bacteria: measurement and potential use of the mutant selection window. *J Infect Dis*, **185**, 561–565.

Zhao, X., Malik, M., Chn, N., Drlica-Wagner, A., Wang, J-Y., Li, X. and Drlica K., (2006), Lethal action of quinolones against a temperature sensitive dnaB replication mutant of *Escherichia coli*. *Antimicrob Agents. Chemother*, **50**, 362–364.

Zhao, X., Xu, C., Domagala, J. and Drlica, K., (1997), DNA topoisomerase targets of the fluoroquinolones: a strategy for avoiding bacterial resistance. *Proc Natl Acad Sci USA*, **94**, 13991–13996.

Chapter 5
Mycobacterial Antimicrobial Resistance

Peter D. O. Davies and Richard Cooke

5.1 Definitions and Causation

Drug-resistant tuberculosis (TB) and, particularly, multidrug resistant (MDR) TB (defined as strains of *Mycobacterium tuberculosis* resistant to at least isoniazid and rifampicin, the two most powerful anti-TB drugs), have recently received heightened attention in the media. This has not only helped advancement within the field of MDR-TB, but it has also brought TB to the forefront of discussions in public health. Drug resistance in TB is primarily created by two man-made mechanisms. The first is erroneous prescribing practices combined with the inability to implement systems of patient adherence to treatment. This is often seen when private practitioners are not fully aware of TB control procedures (Mahmoudi and Iseman, 1993). For instance, one study in India revealed that 80 different treatment regimens were being recommended for difficult-to-trace patients in a single slum in Bombay (Uplekar and Shepard, 1991). The second is patient-related, as they fail to follow the prescribed regimen, or are affected by other diseases (such as co-infection with HIV) that interfere with the mechanisms of anti-TB drugs (Kochi et al., 1993). These factors place selective pressure upon the bacilli in a patient, thereby increasing the growth of resistant bacilli (Pablos-Mendez and Lessnau, 2000). However, these two factors simply reflect the principle that drug resistance tends to arise when national TB program do not adhere to international recommendations for TB control (World Health Organization, 1997).

5.2 Global Epidemiology of Tuberculosis

The study of the epidemiology of tuberculosis has been greatly facilitated by the reports produced by the WHO. The two published in 1997 and 2001 have now been updated in the 2004 report (World Health Organization, 2004).

This report includes new data from 77 settings or countries collected in the third phase of the project, between 1999 and 2002, representing 20% of the global total of new smear-positive TB cases. It includes 39 settings not previously included in the Global Project and reports trends for 46 settings.

I. W. Fong and K. Drlica (eds.), *Antimicrobial Resistance and Implications for the Twenty-First Century.* © Springer 2008

Data were included if they adhered to the following principles:

1. The sample was representative of all TB cases in the setting under evaluation.
2. New patients were clearly distinguished from those with previous treatment.
3. Optimal laboratory performance was assured and maintained through links with a supranational reference laboratory (SRL). Data were obtained through routine or continuous surveillance of all TB cases (38 settings) or from specific surveys of sampled patients, as outlined in approved protocols (39 settings). Data were reported on a standard reporting form, either annually or at the completion of the survey.

The Supranational Reference Laboratory Network (SRLN) was formed in 1994 to ensure optimal performance of the national reference laboratories participating in the Global Project. The network comprises 20 laboratories in five WHO regions and is coordinated by the Prince Léopold Institute of Tropical Medicine in Antwerp, Belgium.

The coordinating center ensures the quality of the SRLN by conducting annual proficiency testing, through the exchange of a panel of 30 pretested and coded isolates with resistance to any of the following four first-line drugs—isoniazid (INH), streptomycin (SM), rifampicin (RMP), and ethambutol (EMB).

5.2.1 Principal Conclusions of Report

1. Drug-resistant TB was found among TB patients surveyed in 74 of 77 settings between 1999 and 2002. As in the two previous surveys, drug-resistant TB, including MDR-TB, was found in all regions of the world. The prevalence of MDR-TB was exceptionally high in almost all former Soviet Union countries surveyed, including Estonia, Kazakhstan, Latvia, Lithuania, the Russian Federation, and Uzbekistan. Proportions of isolates resistant to three or four drugs were also significantly higher in this region. High prevalences of MDR-TB were also found among new cases in China (Henan and Liaoning provinces), Ecuador, and Israel. Central Europe and Africa, in contrast, reported the lowest median levels of drug resistance.
2. The proportion of retreatment of all TB cases is an indicator of program performance. As in previous phases of the Global Project, a link was found between poor program performance, or insufficient coverage of a good program, and drug resistance. Previously treated cases, worldwide, are not only more likely to be drug-resistant, but also to have resistance to more drugs than untreated patients.
3. Significant increases in prevalence of any resistance and MDR were detected in a number of settings. Increases in MDR-TB are especially worrying, since such cases are significantly more difficult to treat, and mortality is higher than for drug-susceptible cases. Increases in prevalence of any resistance may reflect an environment that favors the acquisition of additional resistance and can lead to future increases in MDR.

4. Between 1994 and 2002, the Global Project surveyed areas representing over one-third of notified TB cases worldwide. However, enormous gaps still exist in many crucial areas, especially countries with a large TB burden or where available data strongly suggest that there may be a much larger problem, particularly China, India, and countries of the former Soviet Union.
5. The ability to conduct a drug resistance survey is indicative of a reasonable level of capacity of the TB control services, most importantly the laboratory service. Thus it is likely that TB control in some unsurveyed areas is worse than in those surveyed.

In addition to the comprehensive WHO survey many papers on the incidence of drug-resistant tuberculosis have been published from individual countries and locations.

5.2.2 Asia

5.2.2.1 Russia

For larger countries such as Russia only parts can be thoroughly surveyed. The most comprehensive reports on drug resistance probably come form the Samara region (Balabanova et al., 2005; Ruddy et al., 2005).

A cross-sectional survey was undertaken of 600 patients (309 civilians, 291 prisoners) with bacteriologically confirmed pulmonary TB over a 1-year period during 2001–2002 in Samara Oblast, Russia. The prevalence of isoniazid, rifampicin, streptomycin, ethambutol, and pyrazinamide resistance in new TB cases (civilian and prison patients) was 38.0, 25.2, 34.6, 14.7, and 7.2%, respectively. The prevalence of MDR-TB was 22.7, 19.8, and 37.3% in all new cases, new civilian cases, and new prison cases, respectively, with an overall prevalence of 45.5 and 55.3% in previously treated cases. Factors associated with resistance included previous TB treatment for more than 4 weeks, smoking (for isoniazid resistance), the presence of cavitations on the chest radiograph, and imprisonment. HIV was not associated with resistance in all patients. The rates of resistance were significantly higher in prisoners, with rate ratios (RR) of 1.9 (95% CI 1.1–3.2) for MDR-TB, 1.9 (95% CI 1.1–3.2) for rifampicin, and 1.6 (95% CI 1.0–2.6) for isoniazid.

5.2.2.2 Countries of the Former Russian Republic

To determine levels of drug resistance within a directly observed treatment strategy (DOTS) program supported by Medicines Sans frontiers in two regions in Uzbekistan and Turkmenistan, Central Asia, the authors conducted a cross-sectional survey of smear-positive TB patients in selected districts of Karakalpakstan (Uzbekistan) and Dashoguz (Turkmenistan). High levels of MDR-TB were found in both regions. In Karakalpakstan, 14 (13%) of 106 new patients were infected with

MDR-TB; 43 (40%) of 107 previously treated patients were similarly infected. The proportions for Dashoguz were 4% (four of 105 patients) and 18% (18 of 98 patients), respectively. Overall, 27% of patients with positive smear results whose infections were treated through the DOTS program in Karakalpakstan and 11% of similar patients in Dashoguz were infected with multidrug resistant strains of TB on admission (Cox et al., 2004).

5.2.2.3 Indian Subcontinent Bangladesh

A TB surveillance system was set up in a population of 106,000 in rural Matlab and in a TB clinic in urban Dhaka. Of 657 isolates, resistance to one or more drugs was observed in 48.4% of isolates. Resistance to streptomycin, isoniazid, ethambutol, and rifampicin was observed in 45.2, 14.2, 7.9 and 6.4% of isolates, respectively. Multidrug resistance (MDR) was observed in 5.5% of isolates. MDR-TB was significantly higher among persons who had received tuberculosis treatment previously. The authors conclude that the incidence of drug resistance in Bangladesh is high (Zaman et al., 2005).

5.2.2.4 Indian Subcontinent India

Reports from areas as far apart as Gujarat and Chennai show high incidences of drug resistance.

Of the 1472 patients evaluated, 804 (54.6%) were treatment failure cases and 668 (45.4%) were relapse cases; 822 patients (373 failure and 449 relapse) were culture-positive. Of these 822 patients, 482 (58.64%, 261 failure and 221 relapse) were resistant to one or more drugs. Resistance to one drug was observed in 86 patients (10.46%), to two drugs in 149 (18.13%), to three drugs in 122 (14.84%), and to four drugs in 125 (15.21%). Single-drug resistance was most commonly seen with isoniazid (62 patients, 7.5%), followed by streptomycin (12 patients, 1.4%), rifampicin (eight patients, 0.97%), and ethambutol (four patients, 0.4%). Resistance to isoniazid plus rifampicin alone was seen in 76 patients (9.2%) (Shah et al., 2002).

Sputum specimens of 618 patients (61.8%) of a total of 1000 examined had shown culturable *Mycobacterium tuberculosis*. Four hundred and ninety-five patients (49.5%) were found to expectorate tubercle bacilli resistant to one or more anti-TB drugs. MDR-TB was detected in 339 patients (33.9%). HIV seropositivity among MDR-TB was 4.42%. Significantly, 245 patients (24.5%) had tubercle bacilli resistant to one or more reserve drugs too (ethionamide, kanamycin, and/or ofloxacin) (Deivanayagam et al., 2002).

5.2.2.5 Indian Subcontinent Pakistan

A report from Northern Pakistan shows an alarming increase in resistance.

Mycobacterium tuberculosis was isolated from 325 clinical specimens. Antimicrobial susceptibility of the isolates was tested against the four first-line anti-tuberculosis drugs (rifampicin, isoniazid, streptomycin, and ethambutol). Fifteen per cent of the isolates were resistant to a single drug, 28% were multidrug resistant including 7% which were resistant to all the four drugs. The overall resistance against individual drugs was rifampicin 32%, isoniazid 37%, streptomycin 19%, and ethambutol 17% (Butt et al., 2004).

A report from Sri Lanka also suggests increasing resistance (Senaratne, 2004).

5.2.2.6 South East Asia Thailand

Between January 1995 and December 2000, 899 isolates of *M. tuberculosis* were recovered. Rifampicin (RIF) resistance was the most common finding (8.2%). Twenty-two patients (2.4%) were infected with MDR-TB. Other susceptibility results showed resistance to isoniazid (INH) (4.2%), ethambutol (EMB) (4.3%), streptomycin (SM) (3.7%), kanamycin (Kana) (3.0%), and ofloxacin (Oflox) (2.3%) (Reechaipichitkul, 2002).

5.2.2.7 South East Asia Philippines

Despite high resistance rates cure rates in one report were reasonably good at 73%. One hundred and forty-nine patients with MDR-TB were enrolled from April 1999 to 30 May 2002 at the McCarthy Medical Centre DOTS Clinic. Referrals were from private institutions and practicing physicians in 73.2% of cases. Approximately 30% of isolates tested were resistant to all five first-line drugs, 39.4% to four, 16.8% to three, 12.1% to two. Fluoroquinolone resistance was noted in 40.9% of all the isolates, including 54.5% of those resistant to five drugs and 34.6% of those resistant to four drugs (Tupasi et al., 2003).

5.2.2.8 South East Asia Taiwan

A high prevalence of drug resistance has been shown from Taiwan.

Among the 183 patients, the prevalence of resistance to any drug was 42.6%; 14.2% were resistant to one drug, 13.7% to two drugs, 7.1% to three drugs, 7.7% to four drugs or more, and 24.6% had MDR-TB. The prevalence of any drug resistance among patients with relapse, treatment after default, and treatment after failure was 33.3, 42.1, and 69.7%, respectively, while the prevalence of MDR-TB in these groups was 12.9, 19.3 and 66.7%, respectively (Chiang et al., 2004).

In the 90 culture proven cases in which anti-tuberculosis drug susceptibility was tested, 39 patients showed resistance to at least one drug, nine patients were resistant to only one drug, nine patients were resistant to two drugs and 21 patients were resistant to more than three drugs. The common reasons for failure of treatments were: (1) poor patient compliance to medication: 50 cases, (2) multiple drug resistance: 30 cases, and (3) delayed treatment: 19 cases (Tsai et al., 2004).

5.2.2.9 South East Asia Hong Kong

In Hong Kong resistance rates are low and control better. Of 1921 patients with a previous history of treatment and positive cultures, 1425 (74.2%) had isolates susceptible to all four first-line drugs, while 176 (9.2%) were multidrug resistant (MDR-TB). For the MDR-TB group, 101 (57.4%) isolates were sensitive to all second-line drugs, while 30 (17.0%) were resistant to three or more second-line drugs (Kam and Yip, 2004).

5.2.2.10 South East Asia Vietnam

A study from Vietnam suggests that though MDR-TB is emerging it is not yet a great problem. A cohort of 2901 patients with new smear-positive tuberculosis was enrolled in Vietnam. Of 40 failure cases, 17 had MDR at enrolment. At failure, 15 of the 23 (65%) patients without primary MDR had acquired MDR. Of 39 relapse cases and 143 controls, none had primary MDR (Quy et al., 2003).

5.2.2.11 South East Asia Korea

Korea reports further resistance. In one study drug resistance (DR) to at least one drug was identified in 75 cases (24.4%); the rate of primary DR being 18.7% and acquired DR, 39.3%. MDR was identified in 31 (10.1%); primary MDR in 7.0% and acquired MDR in 21.4%. The risk factors of DR were previous TB treatment, pulmonary involvement, and associated medical illness. The DR group showed lesser adherence to treatment than the drug-sensitive group. The DR group showed more frequent self-interruption of medication, lower completion rate of treatment, and a higher failure rate of follow-up than the drug-sensitive group. In previously treated tuberculosis patients, a higher rate of overall DR and MDR, larger number of resistant drugs, and more-frequent self-interruption of medication were observed than in the newly diagnosed patients. Among the DR group, the acquired DR (ADR) group was older, less educated, was treated for longer duration, and had more advanced disease than the primary DR group (Lee and Chang, 2001).

5.2.2.12 South East Asia Cambodia

Resistance may be low in Cambodia but this study of HIV-positive patients only detected 41 cases with active tuberculosis (Kimerling et al., 2002). Primary isoniazid resistance was detected in six new cases (15%) and no MDR-TB was identified. Evidence from East Timor is that resistance is low (Kelly et al., 2002). Of the *M. tuberculosis* isolates, 82.2% were fully sensitive, 17.2% were resistant to isoniazid, and 8.6% were resistant to isoniazid and streptomycin.

5.2.2.13 Mongolia

A similar situation is present in Mongolia. Resistance to any of the four major drugs (streptomycin, isoniazid, rifampicin, and ethambutol) was as high as 28.9% (95% CI 24.7–33.5), primarily due to high streptomycin resistance of 24.2% (95% CI 20.3–28.6). Isoniazid resistance was also high, at 15.3% (95% CI 12.1–19.1). Resistance levels to ethambutol and rifampicin were relatively low, at 1.7% (95% CI 0.8–3.5) and 1.2% (95% CI 0.5–2.9), presumably because these drugs were only recently introduced into Mongolia. MDR was also rare, at 1.0% (95% CI 0.1–1.8) (Tsogt et al., 2002).

5.2.2.14 Japan

Resistance levels to a single drug in Japan were also quite high whereas MDR levels were low.

Among patients with no prior treatment, resistance to any of the four drugs was found in 10.3%, and the prevalence of primary MDR was 0.8%. The prevalence of acquired resistance was 42.4% for any of the four drugs and 19.7% for MDR, indicating a high prevalence rate compared with those reported in the WHO/IUATLD global project. About 73% of resistant isolates from new cases were resistant to one drug, while 64.3% of resistant isolates from the re-treatment cases were resistant to two or more drugs ($P < 0.0001$). No significant differences in resistance rates by sex, age group, nationality, district, and/or accompanying diseases were observed in any of the new or re-treatment cases (Hirano et al., 2001).

5.2.2.15 Saudi Arabia

A review of drug resistance from Saudi gives a variable picture. All published material on the prevalence of drug-resistant tuberculosis within Saudi Arabia over the period 1979–1998 was reviewed. The prevalence of single-drug-resistant tuberculosis ranged from 3.4 to 41% for isoniazid, 0 to 23.4% for rifampicin, 0.7 to 22.7% for streptomycin, and 0 to 6.9% for ethambutol. The prevalence of MDR-TB (defined by WHO as resistance to two or more first-line anti-tuberculosis drugs) ranged from 1.5 to 44% in different regions (Abe and Abu, 2002).

5.2.2.16 Qatar

Data from Qatar is similar.

Mycobacterium tuberculosis culture reported to the Division of Public Health TB Unit from January 1996 to December 1998. There were 406 isolates with positive *M. tuberculosis* culture. Sixty-one (15%) were resistant to one or more of the four anti-tuberculosis drugs, of which 58 (95%) were from newly diagnosed cases (primary) and three (5%) were from previously treated cases (acquired). Primary resistance was as follows: any resistance 15%, INH 12.4%, RMP 2%, SM 5.2%, EMB 0.8%, and MDR (resistance to INH and RMP at least) was found in 0.8%. Acquired resistance was as follows: any resistance 15%, INH 15%, RMP 5%, SM 5%, and MDR 5% (Al-Marri, 2001).

5.2.3 Africa

Drug resistance in Africa presents a mixed picture. Poorer countries generally have less drug resistance than richer ones perhaps reflecting the poorer regimens and supervision provided by private health practitioners which cannot be afforded in poor countries. Also because the poorest countries have not been able to afford rifampicin (RMP) containing regimens until relatively recently, MDR-TB has not yet had time to develop.

5.2.3.1 Central African Republic

The Central African Republic for example had little resistance (Kassa et al., 2004). Overall drug resistance and MDR were 15.2 and 0.6%, respectively. Isoniazid (INH) and streptomycin (SM) were the only drugs associated with TB monoresistance. No significant difference was found in the epidemiological or clinical data of children infected with a resistant strain and those infected with a susceptible strain.

5.2.3.2 South Africa

In contrast, South Africa, a richer country had very high rates. A high prevalence of drug-resistant and multiresistant TB was observed in this region. At least 55% of previously treated and 19% of new cases from all areas were resistant to at least one of the drugs tested. New patients from Ngwelezane and Manguzi area had a high prevalence of any RMP resistance (11.0%) and ethambutol (ETH) resistance (3.9%), respectively (Lin, 2004).

5.2.3.3 Malawi

In Malawi resistance among patients with recurrent disease was moderately high. In 164 patients with cultures of *Mycobacterium tuberculosis*, 122 (81%) were fully sensitive, 25 (15%) had resistance to INH and/or SM, and 6 (4%) had resistance to INH and RMP (MDR-TB) (Salaniponi et al., 2003).

5.2.3.4 The Gambia

In the Gambia drug resistance is still uncommon. Only nine (4%) of the patients had strains that were resistant to one or more drugs. None of the patients with drug-resistant *M. tuberculosis* had previously been treated for tuberculosis (Adegbola et al., 2003).

5.2.3.5 Botswana

Though higher in Botswana drug resistance was not nearly as high as in South Africa.

Drug resistance occurred in 6.3% of new patients (95% CI 4.6–8.6) and 22.8% of retreatment patients (95% CI 16.5–30.1). Resistance to at least INH and RMP was found in 0.5% of new (95% CI 0.1–1.3) and 9.0% of retreatment patients (95% CI 5.1–14.5) (Talbot et al., 2003).

5.2.3.6 Ethiopia

Drug resistance across Ethiopia shows wide variation. In studies from 1984 to 2001, the initial resistance to isoniazid ranges from 2 to 21% and initial resistance to streptomycin ranges from 2 to 20%. MDR-TB defined as resistance to at least INH and RMP was also reported in about 1.2% of new cases and 12% of retreatment cases. In all studies which included ethambutol susceptibility test, EMB resistance is either nil or very low (below 0.5%). All MDR isolates were susceptible to ethambutol (Abate, 2002).

5.2.4 The Americas

5.2.4.1 Canada

Canada reports overall resistance patterns of 12.5% for any resistance and 1.5% for MDR (http://www.phac-aspc.gc.ca/publicat/tbdrc03/pdf/tbdrc03e.pdf).

5.2.4.2 United States

Resistance patterns in the United States are similar being 13.0% for any resistance and 1.4% for MDR-TB (Mahmoudi and Iseman, 1993) (http://www.hpa.org.uk/infections/topics_az/tb/epidemiology/figures/figure5.htm).

5.2.4.3 Peru

Some of the highest resistance rates in the world have been described in Peru. Of 1680 isolates tested, 1144 (68%) were resistant to at least one anti-tuberculosis drug and 926 (55%) were MDR-TB strains. Of 926 MDR isolates, 50 (5%) were resistant to INH and RMP alone, while 367 (40%) were resistant to at least five first-line drugs; 146 unique drug resistance profiles were identified, the most common of which accounted for 11% of drug-resistant isolates. The annual prevalence of isolates with resistance to at least five first-line drugs rose significantly during the study period, from 29 to 37% ($P = 0.00086$) (Timperi et al., 2005).

5.2.4.4 Mexico

Evidence from Mexico has come from the USA/Mexico border surveillance.

In this study it was found that the Mexican-born patients were 3.6 times as likely as the US-born patients have resistance to at least INH and RMP (i.e., to have MDR-TB) and twice as likely to have isoniazid resistance (Schneider et al., 2004).

5.2.4.5 Argentina

Data from Argentina has been influenced by a large nosocomial outbreak.

Of 291 HIV-negative MDR-TB patients, 79 were initially MDR.

An ascending trend of initial MDR-TB during this decade was observed ($P = 0.0033$). The M strain, which was responsible for an institutional AIDS-associated outbreak that peaked in 1995–1997, caused 24 of the 49 initial MDR-TB cases available for restriction fragment length polymorphism. Of those, 21 were diagnosed in 1997 or later. Hospital exposure increased the risk of acquiring M strain-associated MDR-TB by approximately two and a half times. The emergence of initial MDR-TB among HIV-negative patients after 1997 was apparently a sequel of the AIDS-related outbreak (Palmero et al., 2003).

5.2.5 Europe

5.2.5.1 The Baltic States

These countries have been shown to have the highest rates of resistance in the world. A study in Lithuania showed that of 1163 isolates, 475 (41%) were resistant to at least one first-line drug and 263 (23%) were resistant to at least INH and RMP (MDR); this included 76 of 818 (9.3%) from new patients and 187 of 345 (54%) from previously treated patients. Of 52 MDR isolates randomly selected for extended testing at an international reference laboratory, 27 (51%, 95% CI 38–66) had resistance to pyrazinamide (PZI), 21 (40%, 95% CI 27–55) to kanamycin (KAN), and 9 (17%, 95% CI 8–30) to ofloxacin (OFL) (Dewan et al., 2005).

Of the 204 patients assessed, in a study from Estonia, 55 (27%) had been newly diagnosed with MDR-TB and 149 (73%) had earlier been treated with first-line or second-line drugs for this disease. Assessment of treatment outcomes showed that 135 (66%) patients were cured or completed therapy, 14 (7%) died, 26 (13%) defaulted, and treatment failed in 29 (14%). Of the 178 adherent patients, 135 (76%) achieved cure or treatment completion (Leimane et al., 2005; Lockman et al., 2003).

5.2.5.2 Italy

Data from HIV patients in Italy shows low levels of resistance. Prevalence of resistance to INH or RIF was 12.8 and 4.3% among new cases and 17.4 and 26.1% among previously treated cases, respectively. Prevalence rates of DR and MDR were 14.5 and 2.6% among new cases and 30.4 and 12.5% among previously treated cases, respectively. No statistically significant risk factors associated with DR or MDR-TB emerged in the analysis (Vanacore et al., 2004).

5.2.5.3 France

Drug resistance levels in France are also low (Robert et al., 2003).

A total of 264 distinct patients were reported to the National Reference Centre for Resistance of Mycobacteria to Anti-tuberculosis Drugs during the 8-year surveillance period resulting in a mean annual prevalence of MDR-TB of 0.6%. A mean of 16% of the MDR-TB patients were reported over several subsequent years. The majority of patients were male (69.7%), foreign-born (55.7%), with a previous history of treatment (65.9%), and pulmonary involvement (92.8%) with smear-positive results (59.1%). HIV co-infection was present in 20.8% of the patients. Strains were resistant only to isoniazid and rifampin in 37.9% of the cases, and additional resistance to both streptomycin and ethambutol was present in 25.8%. HIV co-infection and female status were statistically associated with primary resistance, whereas smear-positive results were associated with secondary resistance. Foreign birth and smear-positive results were associated with a chronic status.

5.2.5.4 Greece

In Greece resistance is low.

Of one hundred and one initial isolates of *M. tuberculosis* 48.79% were susceptible to all first-line anti-tuberculosis drugs, INH, SM, EMB and RMP. The prevalence of primary mono- and poly-drug resistance was lower (INH 5.79%, SM 4.34%, EMB 1.93%, INH + SM 3.38%, INH + EMB 0.9%, INH + SM + EMB 1.44%) when compared with the prevalence of secondary (acquired) mono- and poly-drug resistance (INH 8.69%, SM 10.14%, EMB 3.86%, INH + SM 3.86%, INH + RMP 1.44%, INH + SM + EMB 4.83%, INH + SM + RMP 1.44%). No primary mono-resistance to RMP or primary MDR (defined here as *M. tuberculosis* resistant to at least INH and RMP) were observed (Trakada et al., 2004).

5.2.5.5 Bulgaria

In Bulgaria levels are higher.

Resistance to at least one anti-tuberculosis drug was established in 24.7% of the patients. Mono-resistance was found in 13.5% of the cases. The median prevalence of combined resistance to RMP and INH was 6.7%. The prevalence of resistance to RMP or INH was 21.4% (Torosyan et al., 2000).

5.2.5.6 Poland

Poland has shown an increase in anti-tuberculosis drug resistance though rates are low. Compared with a previous survey in 1997, the current survey showed a twofold increase in tuberculosis resistance in new cases: any resistance was 3.6% in 1997 vs. 6.1% in 2000 ($P < 0.001$), MDR was 0.6% vs. 1.2% ($P < 0.01$), and no cases of four-drug resistance in 1997 vs. 15 cases in 2000. No statistical differences were observed in the rate of acquired resistance in both surveys (Augustynowicz et al., 2003).

5.2.5.7 Turkey

Turkey has high rates of resistance.

Primary resistance to one or more drugs was detected in 87 (23.8%) patients; resistance to INH was most common (54 patients) followed by resistance to EMB ($n = 39$), RMP ($n = 11$), and SM ($n = 9$). One-drug resistance was detected in 69 patients; two-drug resistance in 11, three-drug resistance in 6, and four-drug resistance in 1. MDR (resistance to at least INH and RMP) was detected in 10 patients. In logistic-regression analysis, primary drug resistance was associated with radiological advanced tuberculosis ($P < 0.001$) (Kartaloglu et al., 2002).

5.2.5.8 United Kingdom

In the United Kingdom resistance is low with 9% resistance to any drug and 1.2% MDR-TB (http://www.hpa.org.uk/infections/topics_az/tb/menu.htm).

5.3 The Bacterium

Mycobacteria, together with other mycolic acid-containing organisms, form part of the order *Actinomycetales*. They are therefore related to the genera *Rhodococcus, Nocardia, Corynebacterium, Gordona,* and *Tsukamurella*.

The genus *Mycobacterium* currently comprises about 100 different well-described species. However, about 95% of mycobacterial infections in humans can be attributed to about 20 species. Mycobacteria are non-motile, obligatory aerobic, acid-fast rod-shaped bacilli that are Gram-positive and contain mycolic acids. The genome of mycobacteria is large ($3–5.5 \times 10^9$ Da) and like other prokaryotes is arranged as a closed circle. The nucleotide base composition of mycobacterial DNA is extraordinarily rich in guanine (G) and cytosine (C) with a G–C ratio of 61–71% (Levy-Frebault and Portaels, 1992).

Mycobacteria are significantly smaller in size than other types of bacteria and cell shape varies from cocco-bacilli to elongated rods but pleomorphic morphology is common. The most prominent feature of mycobacteria, that is uniformly present and distinctive to the genus, is the complex lipid-rich cell envelope. This confers upon these bacteria the property of "acid-fastness" (i.e., resistance to decolorization when stained with carbol fuchsin and decolorized with dilute hypochloric acid). Some other closely related genera (e.g., *Nocardia* and *Rhodococcus*) may exhibit weak acid-fastness but usually decolorize when acid-alcohol (e.g., Kinyoun method) is applied.

Though mycobacteria are classified as Gram-positive bacteria, most do not stain well with Gram's stain. Mycobacteria possess a cell wall polysaccharide that resembles that of Gram-positive bacteria. However, the mycobacterial peptidoglycan contains lipids in place of proteins and polysaccharides (Brennan and Draper, 1994). Furthermore, the mycobacterial envelope contains a plasma membrane that is quite similar in structure and function to the plasma membrane of other bacteria except for the presence of lipoarabinomannan, lipomannan, and phosphatidylinositol mannosides. The peptidoglycan confers cell shape while the next layer of the envelope, arabinogalactan esterified to the mycolic acids, is largely believed to restrict permeability. At the exterior surface of the cell envelope are the mycosides, a mixture of glycolipids. Cord factor, considered responsible for the distinct growth of tubercle bacilli in the shape of serpentine cords, appears to be a mixture of mycolate-containing molecules.

The *Mycobacterium* genus can be divided into three groups according to their pathogenicity: the mycobacterium tuberculosis complex (MTBC), the nontuberculous mycobacteria and a third group consisting of only one named species, *Mycobacterium leprae*. Although genetically closely related, these strains differ considerably in their epidemiology, pathogenicity, and host spectrum.

5.3.1 Mycobacterium tuberculosis Complex

The MTBC consists of the closely related species *M. tuberculosis* (MTB), *M. bovis*, *M. africanium*, *M. microtti* and "*M. cannetti*." MTB is the most significant pathogen for humans in Europe and North America while *M. africanum* is widely distributed among African patients (David et al., 1978). *M. africanum* includes two sub-types, I and II, which are found in different geographic regions.

Mycobacterium bovis comprises *M. bovis* subsp. *bovis*, *M. bovis* subsp. *caprae* and the *M. bovis*-derived BCG vaccine strain. *M. bovis* is often referred to as the bovine tubercle bacillus. Both sub-species of *M. bovis* are reported to infect humans yet they have a broad host range including wildlife and domestic livestock. Cattle in particular are susceptible to infection and subsequent tuberculous lung disease. Recent studies have demonstrated *M. bovis* infection to be endemic in some wild species in the United Kingdom. Infection in the badger population particularly is self-maintaining and represents a significant obstacle to the eradication of *M. bovis* from domestic cattle herds.

An accurate assessment of the proportion of TB caused by *M. bovis* is difficult. Studies in England and Wales in 1931, 1937, and 1941 estimate that around 6% of deaths, for all human forms of human TB, were due to *M. bovis* infection. During the 1930s, 40% of slaughtered cattle in England and Wales had obvious tuberculosis. Rates of disease have fallen significantly since then to about 0.4% of UK cattle herds but appear to be on the increase again. The overall reduction in human *M. bovis* infection reflects the widespread implementation of infection control measures: particularly milk pasteurization, meat inspection, tuberculin skin testing of cattle, and the slaughter of infected animals. The prevalence of TB caused by *M. bovis* in developing countries is largely unknown. However, the organism is known to be widely distributed and its zoonotic importance should not be underestimated.

It is not possible to clinically differentiate between TB caused by *M. bovis* and that caused by MTB. Though standard anti-tuberculosis chemotherapy is effective against TB caused by *M. bovis*, pyrazinamide (PYR) resistance is the main criterion from differentiating the subspecies *M. bovis* ssp. *bovis* (PYR resistant) from *M. bovis* ssp. *caprae* (PYR sensitive) (Konno et al., 1967). The vaccination strain *M. bovis* BCG is more frequently used for bladder cancer immunotherapy and can then be detected in human urine specimens from bladder cancer patients (Durek et al., 2001).

Mycobacterium microtti is the vole strain of TB. "*M. canetti*" has only been rarely isolated in humans and may be merely a subspecies of MTB (van Soolingen et al., 1997, 1998).

5.3.2 Environmental Mycobacteria

The environmental mycobacteria, also referred to as "mycobacteria other than the tubercle bacilli" (MOTT), are a heterogeneous group of pigmented and non-pigmented acid-fast, rod-shaped bacilli which do not belong to the MTBC. These mycobacteria occur ubiquitously and, as classical opportunists, may infect patients with pre-existing lung diseases or impaired immune functions. However, an increase in mycobacteriosis has also been observed in immunocompetent patients.

5.3.3 Mycobacterium leprae

Mycobacterium leprae differs from other mycobacteria in that its G and C content is 54–57% and has been shown to be only remotely related to other mycobacteria by DNA hybridization (Grosskinsky et al., 1989). It is on obligatory pathogen and causes the clinical picture of leprosy. The organism cannot be cultivated in vitro. Although eradicated in industrialized countries, it remains an endemic problem in many developing countries.

5.4 Mechanisms of Development of Drug Resistance

5.4.1 Introduction

Resistance to any given chemotherapeutic agent occurs by mutation at a low but constant rate so that treatment with a single drug, however powerful, will inevitably lead to selection of mutants (Mitchell et al., 1995). Two forms of drug resistance are recognized in anti-tuberculous chemotherapy: acquired and primary or initial. Acquired resistance is the result of suboptimal theory that encourages the selective growth of mutants to one or more anti-tuberculous drugs. Primary resistance is due to infection from a source case who has drug-resistant disease. This division of resistance into acquired and primary is important epidemiologically. The former indicates that the drug regimens or the supervision of therapy is suboptimal whilst the latter implies that transmission of disease in the community is not being adequately controlled.

The problem of treatment failure caused by the emergence of drug resistance became apparent soon after the introduction of anti-tuberculosis therapy and led to the advocacy of multiple drug regimens.

Although resistance to isoniazid (INH) or streptomycin is not uncommon, patients whose disease is caused by such resistant strains usually respond to short-course combination chemotherapy. Resistance to rifampicin (RIF) is much more serious particularly as such strains are often resistant to INH and these drugs are the key components of modern chemotherapeutic regimens. Thus the term multidrug resistance (MDR) has been adopted by the World Health Organization (WHO) to refer to MTB strains that are resistant to RIF and INH with or without resistance to additional drugs (Kochi et al., 1993).

As a consequence, much scientific attention has been paid to determining the molecular mechanism of rifampicin resistance and thereby establishing laboratory tests to rapidly identify such strains. However, the mechanism of action of other anti-tuberculosis drugs has not been so well defined at the genetic level. Indeed, it has been estimated that if the best available techniques were used to predict INH, aminoglycoside, or fluoroquinolone resistance roughly 10, 40, and 25% of resistant isolates would be missed (Drobniewski, 1997).

5.4.2 Rifampicin Resistance

RMP is a semi-synthetic derivative of rifamycin. Its mechanism of activity and resistance was originally extensively studied in *Escherichia coli* (Gale et al., 1981; McClure and Cech, 1978). RMP binds to the β-subunit of ribonucleic acid (RNA) polymerase resulting in inhibition of transcription initiation. A similar mechanism of action has been demonstrated in *Mycobacterium smegmatis* (Levin and Hatfull, 1993).

Missense mutations and short deletions in the central region of the RNA polymerase β-subunit gene (*rpoB*) result in rifampicin resistance (RR) in *E. coli*. This insight has led to the characterization of the *MTB rpoB* gene and to the identification of a wide variety of mutations conferring RR. An analysis of approximately 500 MTB RR strains from global sources found that about 96% of RR clinical isolates have mutations in an 81 bp core region corresponding to codons 507–533 of the *rpoB* gene (Ramaswamy and Musser, 1998; Telenti et al., 1993). As a result, the primary structure of the *rpoB* gene is altered thereby diminishing rifampicin-binding affinity. Such mutations are absent in susceptible strains.

The codons 507–533 encode 27 amino acids. Although minor discrepancies have been reported, there is generally a strong correlation between specific amino acid substitutions and rifampicin minimum inhibitory concentrations (MICs). Missense mutations in codons 513, 526 or 531 result in low level RR (Ramaswamy and Musser, 1998). By exploiting the codons predominately associated with RR MTB isolates, commercial deoxyribonucleic acid (DNA) probes have been developed and incorporated into DNA strip assays for the rapid detection of RR (Hillemann et al., 2000, 2005).

The molecular mechanism of resistance in approximately 4% of RR tuberculosis isolates is not known. Such strains have no detected mutations in either the 81 bp core region of *rpoB* or elsewhere in the *rpoB* gene.

5.4.3 Isoniazid Resistance

In contrast to RR, the genetic mutations associated with isoniazid resistance (INR) are located over a much longer gene sequence (Slayden and Barry, 2000). Laboratory detection is also technically more difficult.

Following an observation between INR and loss of catalase-peroxidase activity, the structural gene *katG* was cloned and sequenced. Genetic studies confirmed that *katG* participated in mediating susceptibility to isoniazid. Although a diverse array of distinct *katG* changes are uniquely represented among INR MTB strains, amino acid substitutions at the codon position 315 are the most frequent: approximately 50–95% of INR isolates contain mutations in codon 315 (Mokrousov et al., 2002; Telenti et al., 1997). The missense change usually involves a change from serine to threonine. The amino acid substitution appears to balance the need to maintain active catalase-peroxidase activity to detoxify host antibacterial radicals whilst reducing the conversion of the prodrug INH to activated INH (isonictonic acid), a process that would otherwise normally kill the bacterium (Ramaswamy and Musser, 1998).

Ethambutol (ETH) is a structural analog of INH and is thought to inhibit mycolic acid biosynthesis in MTB. Molecular studies of INH–ETH resistant mutants of *M. smegmatis* and *M. bovis* identified a two-gene operon with contiguous open reading frames designated *mabA* and *inhA*, coding for products that participated in resistance to both INH and ETH (Lefford, 1966). It is now thought that amino acid substitutions in the nicotinamide adenine dinucleotide hydrogenase binding site of *inhA* results in reduced affinity for enoyl reductase, which confers INR by preventing the inhibition of mycolic acid biosynthesis. Approximately 20–35% of INR strains contain mutations in the *inhA* regulatory region (Piatek et al., 2000; Telenti et al., 1997).

In *E. coli* and *Salmonella typhimurium*, *katG* is part of an oxidative stress regulon induced by the gene *oxyR* in response to the detrimental effects of hydrogen peroxide. The *oxyR* gene controls the expression of *katG* and several other genes including *ahpC*, which encodes the small subunit of alkylhydroperoxide reductase. Mutations in either *oxyR* or *ahpC* in *E. coli* confer isoniazid susceptibility to this otherwise INR bacterium (Rosner, 1993). Genetic studies into INR MTB suggest that mutations in the *oxyR–ahpC* intergenic region are selected due to reduction in catalase or peroxidase activity attributable to *katG* changes arising during INH therapy. It is estimated that 10–15% of INR strains with *inhA* mutations have additional mutations in the *ahpC–oxyR* region, often in conjunction with *katG* mutations outside of codon 315 (Sreevatson et al., 1997).

5.4.4 Other Drug-Resistant Mechanisms

Streptomycin resistance is due mainly to mutations in the 16S-RNA gene or the *rpsL* gene encoding ribosomal protein S12. Resistance to pyrazinamide, in the great majority of organisms, is caused by mutations in the gene *pncA* encoding pyrazinamidases that result in enhanced enzyme activity. Ethambutol resistance in approximately 60% of organisms is due to amino acid replacements at position 306 of an arabinosyl transferase encoded by the *embB* gene. Amino acid changes in the A subunit of DNA gyrase cause fluoroquinolone resistance in host organisms. Kanamycin resistant is due to nucleotide substitutions in the *rrs* gene encoding 16Sr RNA (Ramaswamy and Musser, 1998).

5.5 Laboratory Diagnosis

Drug sensitivity testing (DST) is only of value when it is quality controlled for accuracy (Drobniewski et al., 2003). Hence great emphasis is placed by the WHO on submitting mycobacterial cultures to laboratories with extensive experience in this aspect of clinical mycobacteriology. In general, DST analysis should be therefore performed at specialist national and regional centers only.

Drug resistance detection methods for MTB can be broadly classified into the following groups: culture on solid media, automated liquid-based culture, phenotype detection, and molecular amplification methods.

5.5.1 Culture on Solid Media

The most common methods for DST for MTB are based on the "critical concentration" of anti-tuberculous drugs and the percentage of resistant bacilli within a test population. Critical concentrations for anti-tuberculous agents have been established on an empirical clinical basis. Therapeutic success is considered unlikely if the proportion of drug-resistant strains within a population of MTB exceeds a threshold of 1%, at a concentration of the anti-tuberculous agent known to be therapeutically effective against a fully susceptible strain. The susceptibility test method used to measure the percentage of resistance is, therefore, referred to as the "proportion method."

The "agar proportion method" can be applied as either a direct or indirect test. In the direct test, a smear-positive specimen is used as the source of the inoculum and the specimen is inoculated directly on to the test media agar with or without anti-tuberculous drugs. In an indirect test, a mycobacterial isolate is used as the inoculum source for the susceptibility test. Inoculating several dilutions of a standardized suspension of mycobacteria onto agar (e.g., Middlebrook 7H10 plates) is the basis of the "agar proportion method." The number of colony forming units (CFUs) that grow on the drug-resistant plates are compared with the number of CFUs on the drug-free plate. If the number of CFUs on the drug-containing medium exceeds 1% of the total number of CFUs on the drug-free medium, then the isolate is considered "resistant" to that drug at that concentration. It should be noted, however, that pyrazinamide (PZA) is only active at an acidic pH and therefore DSTs for this drug must be performed in media with a pH of 5.5–6.0.

The "absolute concentration method" consists of inocoluating media with and without anti-mycobacterial agents with a carefully controlled inoculum containing 2×10^3–1×10^4 CFUs of mycobacteria. Resistance is defined as growth that is greater than a certain number of CFU (usually 20) of a particular drug concentration and this must be precisely confirmed for each batch of media.

The "resistant ratio method" is similar to the absolute concentration method except that a second identical series of tubes are inoculated with the standard MTB H37 Rv strain. The DST results are then expressed in terms of the ratio of the MIC of drug necessary to inhibit the growth of the test isolate of MTB to that of the standard H37 Rv strain (Friedland, 2004).

5.5.2 Automated Liquid-Based Culture

Using solid culture, DST reports should be available within 30 days of receipt of the mycobacterial isolate (Centers for Disease Control and Prevention, 2005). To improve turnaround times, there has been considerable interest in the role of automated mycobacterial liquid culture systems for DST. These systems measure changes in gas pressure, carbon dioxide production, or oxygen consumption either fluorometrically, colorimetrically, or radiometrically. They need no further operator input after loading until the systems signals positive. For routine mycobacterial detection they have proved more rapid than solid media with most mycobacteria being isolated within 2–3 weeks (Drobniewski, et al., 2003).

The "agar proportion method" was first adapted to the Bactec 460 TB automated liquid-based system used for the radiometric detection of growth inhibition. In the "radiometric proportion method," the test isolate is inoculated into Bactec 460TB 12B media with or without the addition of test drug. The concentration of mycobacteria inoculated into media without drug is 100-fold less than the concentration incolulated into Bactec 460 TB media with drug. If a drug inhibits the growth of a test strain, such that the growth of the control (no drug with 100-fold fewer organisms) reaches a growth threshold before growth in the drug-containing medium reaches the same threshold, then the test isolate is considered susceptible to that drug at the concentration tested (Friedland, 2004).

Until recently, only two broth systems were accepted by the United States Food and Drug Administration (FDA) for drug susceptibility testing: the radiometric BACTEC 460 (Becton Dickinson, Sparks, MD, USA) and the non-radiometric ESPII system (Trek diagnostic system, Westlake, OH, USA). Both are suitable for rapid DST of streptomycin, INH, RMP, and EMB though only the BACTEC 460 system can detect PZA resistance: PZA susceptibility testing is technically difficult because the bacterial activity of the drug is optimal only in an acid environment that inhibits the growth of most MTB isolates. The BACTEC Mycobacteria Growth Indicator Tube (MGIT) 960 (Becton Dickinson) is a non-radiometric fully automated system that has been recently FDA-cleared for DST and by-passes some of the radiometric disadvantages related to the BACTEC 460 system. Two recent evaluations suggest that the MGIT system is a rapid and reliable alternative for MTB DST to first-line drugs including PZA (Johansen et al., 2004; Scarparo et al., 2004). Mean times for DST by MGIT 960 were approximately 5 and 7 days in the two studies.

Rapid colorimetric culture systems have been applied both for the detection and DST of MTB. A rapid solid culture system has shown promising results in comparison with BACTEC 460 though MTB isolate numbers were small (Baylan et al., 2004). Similarly, a colorimetric liquid system, for the majority of MTB strains, produced reliable DST results available within 5 days (Syre et al., 2003). For resource-poor countries, rapid colorimetric DST may therefore offer a satisfactory alternative to the more expensive automated liquid-based DST systems.

5.5.3 Phenotypic Methods

Though genotypic methods have been widely used for MTB DST, they are expensive, labor intensive and time consuming. In developing countries, alternative phenotypic assays based on mycobacteriophages have been developed and applied to the determination of drug resistance. The incorporation of the "*lux*" gene, which encodes for luciferase in the phage, allows further detection of mycobacteriophage through emission of light and this indirectly allows detection of viable

MTB. This principle can be applied to the detection of drug-resistant strains (Jacobs et al., 1993).

FAST Plaque TB-RMP test (Biotech Laboratories, UK) is a commercial diagnostic kit that is based on the ability of a bacteriophage to specifically infect viable, RR MTB. After the infection period, bacteriophage then "reports" the presence of infected cells by the release of progeny phage zones of clearing on a lawn of rapidly growing host cells, *Mycobacterium smegatis* (Kisa et al., 2003). The method requires neither highly skilled personnel nor expense. Results have shown good correlation with the BACTEC 460 TB system and the conventional proportion method, and allows the detection of RR 48 h from culture (Krishnamurthy et al., 2002).

5.5.4 Molecular Methods

Rapid detection of patients with MDR-TB may result in a more prompt change to effective therapy in the individual patient, shorten the period of infectiousness, reduce the number of additional new MDR-TB patients, and allow for the institution of early appropriate hospital infection control measures (e.g., single room isolation with negative-pressure ventilation). Cost savings using rapid molecular methods to diagnose MDR-TB have been estimated at between £50,000 and £150,000 per annum (Drobniewski et al., 2000). However, nucleic acid amplification tests (NAATS) are not necessary or appropriate for all cases and should only be used when the result is likely to change management.

The molecular methods developed for the detection of antimicrobial resistance in mycobacteria are based on the knowledge of the genetic basis of resistance, with particular attention to RR as this is predictive of MDR.

Several in-house NAATS have been evaluated for the rapid detection of drug resistance and include single-strand conformation polymorphism, DNA sequencing analysis, and ligase detection reaction on oligonucleotide microchips. High cost and demand for technical expertise are limitations that all these methods share. The commercially available INNO-LIPA Rif. TB test (Innogenetics, Belgium) is a PCR-based hybridization assay which is able to simultaneously identify isolates for MTB complex and susceptibility to RIF. Comparisons with BACTEC 460 (Juréen et al., 2004) and direct DNA sequencing (Mortilla et al., 1998; Rossau et al., 1997) suggest high accuracy and technical simplicity. An alternative commercial DNA assay strip, Genotype MTBDR (Hain Lifescience, Germany), has been recently evaluated for its ability to detect mutations conferring resistance to RIF and INH in clinical MTB complex isolates. The assay identified 102 of 103 MDR-TB strains mutations in the *rpoB* gene (99%) and 91 (88.4%) with INH mutations in codon 315 of *kat*G (Hillemann et al., 2000, 2005).

The direct detection of RIF MTB in respiratory samples has been recently studied by PCR DNA sequencing (Yam et al., 2004), real-time PCR (Marín et al., 2004), and the INNO-LIPA Rif. TB assay (Johansen et al., 2003). Though there are currently only a small number of studies, the reported sensitivities in comparison with conventional methods are high warranting further evaluations (Wada et al., 2004).

5.6 The Development of Drug Resistant Tuberculosis

Unfortunately the very success of the drug treatment of tuberculosis has been the catalyst for the emergence of a new wave of drug resistance. Patients have been allowed to take their medication at home completely unsupervised. The experience of the early single use of SM taught us that taking one drug on its own for tuberculosis would result in resistance to that drug. There is a danger that if the patient is sent home with three separate drugs, he or she might take a single drug at a time. In the patient with extensive lung disease, taking a single drug for just a few days may allow drug resistance to emerge. If a patient happens to be resistant to one drug and takes a combination of two drugs including the one to which he is resistant, drug resistance to the second drug will emerge. Similarly if the patient is resistant to two drugs, and takes these two drugs and a third one, then resistance to the third will emerge and so on. In this way a combination of poor compliance and poor medical supervision may result in MDR.

The author (PDOD) was recently asked to see a patient, who had developed MDR-TB as a result of an unfortunate series of events, principally due to poor medical supervision. He was a 65-year-old Pakistani man who came to the United Kingdom about 30 years ago. A year before he was referred to the author he was admitted to his local district general hospital with a pneumonia. A chest X-ray showed a cavitating shadow in the upper lobe of his left lung and sputum samples were positive on smear and (later) culture for *M. tuberculosis*. A diagnosis of tuberculosis was made and he was started on INH, RMP, and PZI. He improved satisfactorily initially as an inpatient and then at home. After two months he was seen in the clinic and the PZI was stopped *before sensitivity results were available.* Three weeks later the sensitivity results were available showing him to have bacteria resistant to INH. He was recalled to clinic and ethambutol alone was added. Over the next few months, although not particularly unwell he did not improve much either. Sputum was again taken for smear and culture. These remained positive and when the sensitivity results came back for the second time the patient's bacteria were found to be resistant to INH, RMP, PZI, and SM. He was again seen in the clinic when PZI alone was reintroduced. The initial three-drug starting regimen was correct for the current guidelines at the time. In this situation it did not matter that four drugs were not given initially. The medical service's first mistake was to change the regimen before sensitivity results were available. Had the doctor waited for the results, which showed resistance to INH, the patient could have been continued on RMP and PZI and cure would probably have resulted after 6–9 months of therapy.

The second mistake was to add a single drug to a regimen of INH and RMP when it was known that the patient had been resistant to INH from the start and had effectively been on a single agent, RMP, for 3 weeks. At this stage at least three drugs should have been added, not one. The result of adding only one was that the bacteria became resistant to that one too. It was at this point that the doctor made the third mistake, again adding in a single drug, PZI, to a regimen, which was failing. The result at the end of a year is that instead of a cured patient with former resistance to the relatively treatable combination of SM and INH, we have a patient with continuing disease, now resistant to SM, INH, RMP, EMB, and probably PZI. All first-line drugs are likely to be useless.

5.7 Clinical Presentation of Drug-Resistant Tuberculosis

In principle there should be no difference in the way drug resistant or MDR-TB presents over fully susceptible tuberculosis. The bacteria cause disease in the human in exactly the same way. There have, however, been a number of studies which have sought to find a difference. In essence it is likely to be due to the inability of drugs to effect a cure, rather than any difference in host/organism interaction.

A study from Korea matched 47 MDR-TB patients with 47 fully susceptible patients matched by age and gender. All were HIV negative. On univariate analysis the MDR patients were more likely to have bronchiectasis, lung destruction, calcified granuloma, and cavitation. On multivariate analysis cavitation was the only difference. The authors concluded that MDR patients were more likely to present with cavitation. Multiple cavities suggested drug resistance (Kim et al., 2004).

5.8 AIDS in the Presentation of MDR-TB

A study of eight patients with HIV-positive MDR-TB showed all developed meningitis as a terminal complication. Intracerebral mass lesions were found in three patients, hydrocephalus in three, and infarcts in two. Seven died 1–16 weeks after diagnosis and the eighth was lost to follow-up (Daikos et al., 2003).

A study from Thailand showed a higher than expected prevalence of MDR-TB amongst HIV-positive cases. This suggested that HIV increases transmission of MDR-TB in the outbreak situation (Punnotok et al., 2000).

5.9 HIV and TB

The biggest single risk factor for developing tuberculosis is HIV infection. In some countries in the world especially sub-Saharan Africa the co-infection rate of HIV and TB is estimated to be over 1000 per 100,000 of population. In spite of some excellent TB control programs, many of these countries are still experiencing increases in tuberculosis case rates because of HIV co-infection. Few countries have universal HIV testing or comprehensive TB culture and drug sensitivity reporting and so the global epidemiology of MDR-TB in HIV-positive patients is as yet mostly unmeasured. In the United States in 1998 it was estimated that 20% of all patients with TB were HIV co-infected but the proportion with MDR-TB was not known.

The reason that HIV is strongly associated with MDR-TB is through outbreaks. These occur because HIV-positive patients have an increased risk of developing active tuberculosis disease once infected with *Mycobacterium tuberculosis*. In HIV non-infected people for every 10 people who are exposed and infected with TB (whether drug sensitive or resistant), only one will develop the disease during their lifetime. For those with HIV infection this risk of around 10% in a whole lifetime is telescoped down to only 1–2 years. Highly active antiretroviral therapy does have an impact on rates of developing tuberculosis, however, and may reduce the rates of tuberculosis in countries where patients are offered anti-HIV treatment.

5.10 MDR-TB Outbreaks

MDR-TB outbreaks have occurred in European hospitals (Breathnach et al., 1998; Coronado et al., 1993; Hannan et al., 2002; Moro et al., 2000) and USA clinics (Pitchenik et al., 1990) for HIV-positive patients and for substance abuse users (Conover et al., 2001) in prisons (CDC, 1992) and homeless shelters. From 1990 to 1992 nine large outbreaks of MDR-TB were reported from the USA, all organisms isolated were resistant to INH and RMP (as this is a definition of MDR-TB) and most had SM and EMB resistance also. The HIV infection rate amongst these patients was from 20 to 100% and the mortality was from 60 to 89%. The interval from TB diagnosis to death was between 4 and 16 weeks. Various strains of MDR-TB were circulating around during this time, especially the notorious "W" strain, which infected 199 patients in New York from 1991 to 1994 and involved 30 hospitals (Shafer et al., 1995). Centres for Disease Control and Prevention (1993). Other types named "N2," "W1," and "AB" infected a large number of patients in 10–16 hospitals. The majority of these outbreaks were brought under control because of public health administrative measures including infection control policies for patients with HIV who have a cough, segregation of potential infectious patients, the use of negative pressure rooms, and submicron masks (Maloney et al., 1995; Moro et al., 1998; Stroud et al., 1995). Procedures such as nebulization of pentamidine for PCP prophylaxis, saline for induced sputa, or even salbutamol for those with obstructive airway disease were no longer performed in open areas but confined to negative pressure rooms.

5.11 The Treatment of Multidrug Resistant Tuberculosis

5.11.1 Introduction

With no new drugs for the treatment of tuberculosis for over a decade there has essentially been no change in the treatment of drug-resistant or drug-susceptible disease for that period of time.

The specific management of drug-resistant patients is only possible where facilities exist for both mycobacterial culture and for drug susceptibility testing, which excludes most of the developing world. Treatment guidelines in the United Kingdom, and elsewhere, are predicated on the drug-resistance data prevailing in the circumstances of their use. In developed countries the inclusion of the fourth drug (EMB but occasionally SM) depends on the level of isoniazid resistance expected or known in a given patient group.

In the United Kingdom, the British Thoracic Society guideline (Joint Tuberculosis Committee of the British Thoracic Society, 1998), recommend the omission of EMB in the initial phase only if the patient meets all the following criteria, based on the drug resistance data held (Hayward et al., 1996; PHLS, 1999):

- White ethnic origin
- Previously untreated for tuberculosis
- Known, or thought likely to be on risk assessment, HIV-negative
- Not a known contact of drug-resistant disease.

The European Respiratory Society (Migliori et al., 1999) recommends the inclusion of EMB in the initial phase for those in WHO Treatment Group I, that is, new sputum smear-positive tuberculosis, new smear-negative tuberculosis with extensive parenchymal involvement, and new cases of severe extra-pulmonary tuberculosis (not defined). A three-drug initial phase is recommended for those in WHO Category III: new smear-negative pulmonary tuberculosis (except in Category I) and new less severe forms of extra-pulmonary tuberculosis. The American Thoracic Society (American Thoracic Society, 1994) advises the inclusion of EMB or SM in the initial phase of a daily regimen at an INH resistance prevalence of 4%. Most parts of the world do not have the capabilities of mycobacterial culture and drug susceptibility testing. Here therefore the "standard" advised regimen has to cover the possibility of the commoner drug resistances. Studies in Hong Kong using regimens of RMP, INH, and PZI, with SM or EMB and both SM and EMB (Hong Kong Chest Service/British Medical Research Council, 1981, 1982, 1987), and using RMP and INH throughout with PZI for 2, 4, or 6 months (Hong Kong Chest Service/British Medical Research Council, 1991) showed that they were all effective in patients with initial INH and/or SM resistance. Treatment failures and relapse rates were low. If routine drug susceptibility tests are not available, these regimens can be assumed to be highly effective if such resistances are present. The results of these studies provide the rationale for using four drugs in the initial phase, with RMP and INH in the continuation phase, in areas or in population subgroups with significant incidences of INH and/or SM resistance.

An analysis of the influence of initial drug resistance on response to short-course regimens in the Medical Research Council (MRC) collaborative trials in Hong Kong, Singapore, and Africa was reported in 1986 (Mitchison and Nunn, 1986). In those trials, patients with initial INH and/or SM resistance had a failure rate of 12% when given a 6-month RMP and INH regimen and a failure rate of 17% in those given RMP in the 2-month initial phase. As the number of drugs given in the regimen and the duration of RMP treatment increased, the failure rate fell, reaching only 2% of those receiving four to five drugs including RMP throughout a 6-month regimen. Relapse rates after chemotherapy were only slightly increased with initially resistant organisms. The key exception was that of RMP resistance, where the outcome was much poorer.

Although these data apply to countries without drug susceptibility testing, the view taken in the United Kingdom is that where drug susceptibility tests are available, they should be followed and treatment modified according to the Joint Tuberculosis Committee of the British Thoracic Society (BTS) (Joint Tuberculosis Committee of the British Thoracic Society, 1998). The strength of the scientific evidence to support various recommendations (Petrie et al., 1995) is also given in the 1998 BTS treatment guidelines.

5.11.2 Management of Non-MDR Resistance

5.11.2.1 Isolated Resistances

1. *SM resistance.* Some of the drug resistance reported, particularly in ethnic minority groups, is to streptomycin alone. This is not clinically important since streptomycin is not often used as a first-line drug in developed countries and the efficacy of the regimen recommended for both respiratory and non-respiratory tuberculosis is not affected.
2. *SM and IHN resistance.* Combined SM and INH resistance is the commonest dual resistance in the United Kingdom. Management should be as for INH resistance found after treatment is commenced but with treatment fully supervised throughout (Joint Tuberculosis Committee of the British Thoracic Society, 1998.)
3. *Other combinations.* Other combinations of non-MDR-TB resistance are uncommon. Treatment needs to be individualized depending on the combination involved and is best determined after discussion with a highly experienced clinician and mycobacterial services. Patients with drug resistance (excluding isolated SM resistance) should be followed up for 12 months after cessation of therapy.

5.11.3 Multi-drug-Resistant Tuberculosis

The risk factors for MDR-TB are those for ordinary drug resistance but are even more exaggerated. The United Kingdom data shows that the odds ratios for risk factors were previous treatment 11.7 (95% CI 5.5–20), HIV-positivity 8.9 (2.1–30.7), birth in India 4.6 (1.2–6.1), residence in London 4.0 (1.7–9.3), and male sex 2.2 (1.1–4.3). In countries without drug susceptibility testing a patient having treatment failure should be considered to be at risk of MDR-TB, and anyone failing a supervised retreatment with a WHO Category II Retreatment regimen should be assumed to have MDR-TB.

5.11.3.1 Infection Control

Patients with suspected MDR-TB should have molecular testing of samples with RMP resistance probes where possible. Those with clinical or microbiological suspicion/proof of MDR-TB should be isolated in a negative pressure room and have infection control and HIV assessments made (Joint Tuberculosis Committee of the British Thoracic Society, 2000; The Inter Departmental Working Group on Tuberculosis, 1998). Criteria for the removal from strict respiratory isolation are also given.

5.11.3.2 Clinical Management

The drug treatment of MDR-TB is time consuming and demanding on both patient and physician. In the United Kingdom the advice is that treatment should only be carried out by:

> Physicians with substantial experience in managing complex resistant cases.
> Only in hospitals with appropriate isolation facilities.
> In very close liaison with mycobacteriology reference centers.

This may require the transfer of patients to an appropriate unit where the above criteria are met. Treatment of such patients has to be planned on an individual basis (Goble et al., 1993; Iseman, 1993) and needs to include second-line drugs (see Table 5.1). Such treatments must be closely monitored because of the increased toxicity but, more importantly, full compliance is essential to prevent the emergence of further drug resistance, so that all such treatment must be directly observed throughout, both as an inpatient and as an outpatient.

Treatment should start with five or more drugs to which the organism is, or is likely to be, susceptible and continued until sputum cultures become negative. Drug treatment then has to be continued with at least three drugs to which the organism is susceptible on in vitro testing for a minimum of nine further months, and perhaps up to or beyond 24 months, depending on the in vitro drug resistance profile, the available drugs (Drobniewski, 1997), and the patient's HIV status. Consideration may also have to be given to resection of pulmonary lesions under drug cover (Goble et al., 1993).

For example when initiating treatment for a patient who may be drug resistant a clear history of any previous treatment, specifically which drugs the patient may have taken previously, should be obtained. If the RMP resistant gene results can be obtained rapidly (within a day or two) treatment may usually be delayed until results are available. If not the patient should be started on INH and RMP, and at least four drugs they have not had previously until resistance patterns are known. An injectable drug should be given, such as capreomycin or amikacin. SM should not be used unless sensitivity to this is known, as SM resistance is relatively common so that if a patient has had EMB and PZI previously these should not be given. A possible oral combination in addition to the injectable could then be prothionamide

Table 5.1 Second-line drugs: dosages and side-effects

Drug	Children	Adults	Main side-effects
Streptomycin	15 mg/kg	15 mg/kg (max dose 1 g)	Tinnitus, ataxia, vertigo, renal impairment
Amikacin	15 mg/kg	15 mg/kg	As for streptomycin
Kanamycin		15 mg/kg	As for streptomycin
Capreomycin		15mg/kg	As for streptomycin
Ethionamide or Prothionamide	15–20 mg/kg	<50 kg 375 mg bd >50 kg 500 mg bd	Gastrointestinal, hepatitis avoid in pregnancy
Cycloserine		250–500 mg bd	Depression: fits
Ofloxacin		400 mg bd	Abdominal distress, headache, tremulousness
Ciprofloxacin		750 mg bd	As ofloxacin plus drug interactions
Moxifloxacin		400 mg od	As ofloxacin plus drug interactions
Azithromycin		500 mg od	Gastrointestinal upset
Clarithromycin		500 mg bd	As azithromycin
Rifabutin		300–450 mg	As for rifampicin: uveitis can occur with drug interactions, e.g., macrolides. Often cross-resistance with rifampicin
Thiacetazone	4 mg/kg	150 mg od	Gastrointestinal, vertigo, rash, conjunctivitis. AVOID if HIV-positive (Stevens–Johnson syndrome)
Clofazimine		300 mg od	Headache, diarrhea, red skin discoloration
PAS sodium	300 mg/kg	10 g od or 5 g bd	Gastrointestinal, hepatitis, rash, fever

or ethionamide, cycloserine, and moxifloxacin. Thus where resistance to INH and RMP is not known initially a total of six drugs should be given at the start. These can be tailored when sensitivity results are known but a single drug should not be added unless sensitivities are known and the patient is improving. Drugs, if added, should be added at least two at a time.

The outcome in MDR-TB depends on how rapidly the diagnosis is made, what treatment and facilities are available, and the patients' HIV status. Results in patients who are HIV-positive have been poor with high mortalities often because of late diagnosis (Small et al., 1993; Telzak et al., 1995), but the outcome in those who are HIV-negative can be much better where appropriate facilities exist and where the drug resistance profile is less extensive. After treatment all MDR-TB patients require long-term follow-up.

The cardinal rule in the treatment of tuberculosis is that a single drug should never be added to a failing regimen. It can therefore be argued that to give a WHO retreatment regimen (World Health Organisation, 1991) (Category II), which adds only a single drug to the combination that has failed (Category I) regimen, breaks this rule, and may actually be adding to the incidence of MDR-TB.

A recent review of the management of MDR-TB concludes that in high-income countries individual regimens should be adapted for each patient. In other settings, restricted resources could justify the implementation of standardized therapeutic guidelines with second-line drugs to make management easier and reduce costs (Caminero, 2005).

5.11.3.3 The Place of Surgery in the Management of Drug Resistant Tuberculosis

There have been many recent studies putting forward the success of surgery in the treatment of MDR-TB. There may be a selection bias in that the less-fit patients may not have been suitable for surgery.

Chan et al. (2004) compared their results with an earlier cohort. The current cohort had greater long-term success rates in terms of sputum conversion, 75 vs. 56%, and lower tuberculosis death rates, 12 vs. 22%, than the earlier one. Surgical resection and fluoroquinolone therapy were associated with improved microbiological and clinical outcomes in the 205 patients studied after adjusting for other variables. The improvement was statistically significant for surgery and among older patients for fluoroquinolone therapy.

A 20-year study of 80 patients with MDR-TB from Tokyo concluded that pulmonary resection combined with chemotherapy achieves high cure rates with acceptable morbidity and remains the treatment of choice for MDR-TB (Shiraishi et al., 2004).

A study of 49 patients from Korea came to the same conclusion (Park et al., 2002).

A series of 172 patients over 17 years from Denver concluded that surgery remains an important adjunct to medical therapy for the treatment of multidrug-resistant *Mycobacterium tuberculosis* (Pomerantz et al., 2001).

There are three basic selection criteria for adjunctive surgical treatment of patients with MDR-TB. First, profound drug resistance in vitro is present leading to a high probability of failure or relapse with medical therapy alone. Secondly, the disease is sufficiently localized so that its great preponderance could be resected with expectation of adequate postoperative cardiopulmonary capacity. Thirdly, there is sufficient drug activity to suppress the mycobacterial burden to facilitate healing of the bronchial stump. Patients must receive chemotherapy prior to surgery for a minimum duration of 3 months. If possible, patients should achieve sputum culture conversion to negativity before surgery. Unfortunately, this may not always occur.

5.11.4 Outcomes in the Treatment of MDR-TB: The Place of Directly Observed Therapy (DOT)

DOT is recommended by WHO for all patients receiving treatment for tuberculosis whether drug susceptible or not.

There have been many studies which have shown that DOT is successful not only at preventing the emergence of drug resistance but in reducing the prevalence of established drug resistance.

A study of DOTS introduced to a setting in Mexico after 5 years concluded that even in settings with moderate rates of MDR tuberculosis, DOTS can rapidly reduce the transmission and incidence of both drug-susceptible and drug-resistant tuberculosis (De Riemer et al., 2005).

A study of over 200 patients from Latvia showed good outcomes from DOTS. Assessment of treatment outcomes showed that 135 (66%) patients were cured or completed therapy, 14 (7%) died, 26 (13%) defaulted, and treatment failed in 29 (14%). Of the 178 adherent patients, 135 (76%) achieved cure or treatment completion. Those given more drugs did better but the study did not give the resistance patterns of the patients (Leimane et al., 2005).

DOT has also been found to be of benefit in Peru and Greece (Shin et al., 2004; Trakada et al., 2004).

Using sensitivity testing to determine appropriate therapy has now been given the title DOTS plus. If resources permit for sensitivity testing the drug combination for the treatment of resistant tuberculosis can be tailored individually. Several studies have shown good results and lowered levels of resistance using this DOTS plus (Kam and Yip, 2004).

Results of treating MDR cases seem to have been less successful in Russia. One report details 66.7% failure rates (Toungoussova et al., 2004).

5.11.5 Outcomes in Children

MDR-TB in children is particularly problematic as bacteriological specimens are usually not obtainable for sensitivity testing. An early index of suspicion and good contact history taking is therefore important (Mukherjee et al., 2003; Schaaf et al., 2003).

5.12 New Drugs in the Management of MDR-TB

New drug classes which might potentially merit further evaluation include nitroim-idazopyrans, oxazolidinones (Cynamon et al., 1999), polaxamers, ubiquinone analogs (Schraufnagel, 1999), and 2-pyridones (Stover et al., 2000). Very scanty data concerning their antimycobacterial activities are currently available. Methods to design novel drugs based on structure of a cellular component or genetic material have been attempted. One possible target of intervention is the mycobacterial glyoxylate pathway which involves key enzymes like isocitrate lyase.

Aside from new drugs, new methods of delivery of old drugs may facilitate better treatment of tuberculosis. Such possible innovative approaches include liposomal delivery of aminoglycosides and fluoroquinolones, implant delivery (Oleksijew et al., 1998) of isoniazid, and aerosolization of cytokines (Stover et al., 2000), aminoglycosides (Stover et al., 2000), and rifampicin microparticles (Suarez et al., 2001) for treatment of tuberculosis.

With the advent of gene therapy, a new strategy of antimycobacterial treatment might be developed in the future. Mycobacteriophages can be used to deliver anti-sense nucleic acids or other materials to combat mycobacteria including the drug-resistant strains. Although antimycobacterial agents can be directed at the microorganisms, conventional gene therapy is targeted more at the humans. The latest developments in gene transfer technology have revealed the possibility of developing preventive or therapeutic vaccines for tuberculosis (Romano et al., 2000).

5.13 Immunotherapy of Multidrug Resistant Pulmonary Tuberculosis

In humans, cell-mediated immunological protective response is based on macrophage activation and granuloma formation that require the cytokines especially interferon-gamma (IFN-γ) and tumor necrosis factor. IFN-γ was shown to have efficacy in lowering bacillary load in patients with MDR-TB and non-tuberculous mycobacteriosis in anecdotal reports (Condos et al., 1997; Holland et al., 1994). Adjunctive immunotherapy with low-dose recombinant human interleukin-2 was also found to stimulate immune activation and may enhance the antimicrobial response in MDR-TB (Johnson et al., 1997). Aerosolized interferon-alpha treatment was shown in a study to reduce the sputum culture colony counts of patients with MDR-TB (Giosue et al., 2000). Preliminary data concerning the use of heat-killed *Mycobacterium vaccae* (NCTC 11659) in patients with MDR-TB in several centers in different continents have suggested possible efficacy (Stanford et al., 2001). A randomized clinical trial of this form of immunotherapy in MDR-TB patients appears warranted. Other potential candidates of immunotherapy may include vitamin D and interleukin-12. It is clear that the current data on the role of immunotherapy in MDR-TB are limited, and further evaluation of this modality of therapy is required.

Strategies in the management of multidrug-resistant tuberculosis

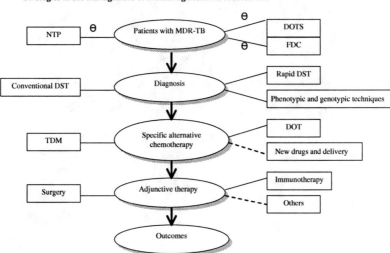

DOTS = Directly observed therapy, short-course
DOT = Directly observed therapy
DST = Drug susceptibility testing

FDC = Fixed-dose combination drugs
NTP = National tuberculosis programme
TDM = Therapeutic drug monitoring

Fig. 5.1 Strategies in the management of multidrug-resistant tuberculosis.

5.14 Summary of Management Plan of Multidrug Resistant Pulmonary Tuberculosis

A proposed plan is depicted in Figure 5.1. The crucial elements for its prevention are shown at the highest level.

5.15 Therapeutic Drug Monitoring in Anti-tuberculosis Chemotherapy

Therapeutic drug monitoring (TDM) can be viewed as an arena of multidisciplinary service of clinicians, clinical pharmacologists, and chemical pathologists in their provision of care to patients. There are a number of potentially useful areas for the application of TDM in anti-tuberculosis chemotherapy. The three major ones are (Yew, 2001): (1) optimization of drug therapy to ensure/improve success in specific clinical settings, (2) management of pharmacokinetic drug interactions, and (3) management of drug–disease interactions. Some important examples of the first indication may include study of drug malabsorption in tuberculosis patients who respond poorly to chemotherapy (particularly HIV-infected subjects), guidance of drug dosing, and scheduling in the therapy of patients with MDR-TB, as well as facilitation of refined dosing in some patients with tuberculous empyema and tuberculous meningitis. The second indication focuses largely

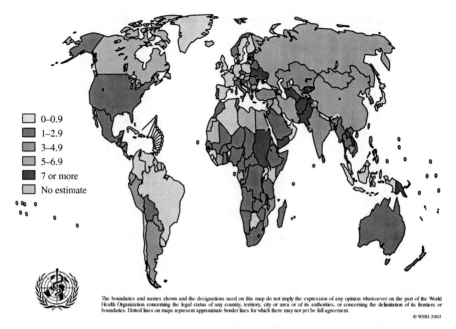

0–0.9
1–2.9
3–4.9
5–6.9
7 or more
No estimate

The boundaries and names shown and the designations used on this map do not imply the expression of any opinion whatsoever on the part of the World Health Organization concerning the legal status of any country, territory, city or area or of its authorities, or concerning the delimitation of its frontiers or boundaries. Dotted lines on maps represent approximate border lines for which there may not yet be full agreement.

© WHO 2002

Fig. 5.2 Estimated percent of new TB cases with MDR, 2000.
Source: Dye et al.(2002) *J. Infect. Dis.*, **185** (8), 1197–1202.

on the pharmacokinetic drug–drug interactions of rifampicin or isoniazid. The third indication is especially relevant for patients with renal dysfunction or failure (Figure 5.2).

The TDM of anti-tuberculosis agents apparently provides a new opportunity in the clinical management of selected patients with tuberculosis bearing a goal to improve their outcome. It might represent a novel paradigm of care for these patients in the future, given the possibility of tremendous advancement in pertinent knowledge, technology, and expertise. A recent report of TDM seems to imply that results were mixed but the study was not carried out in a randomized fashion (Li et al., 2004).

5.16 Environmental (Atypical) Mycobacteria

5.16.1 Introduction

5.16.1.1 Nomenclature

The first problem is what to call these bacteria collectively. The term "atypical mycobacteria" is inappropriate as of all the mycobacteria, *M. tuberculosis*, with its Asian and African variants, and M. *leprae* are the only ones which are obligative parasites. All the others can live freely in the environment, so that

Table 5.2 Major environmental mycobacteria causing human disease (data on other mycobacteria which may cause disease rarely is too inconclusive to provide helpful information on likely site)

Species	Sites of infection
M. avium-intracellulare	Lungs
	Lymph Glands
	Disseminated (in HIV)
M. kansasii	Lungs
M. xenopi	Lungs
M. malmoense	Lungs
M. fortuitum	Soft tissues
M. chelonei	Soft tissues
M. ulcerans	Skin
M. marinum	Soft tissues

"environmental mycobacteria" may be a better term (Table 5.2). The Americans usually use the term "mycobacteria other than tuberculosis" (MOTT), which is cumbersome, or "non-tuberculous mycobacteria," which is inaccurate, as these bacteria form tubercles in the infected host identical to those produced by *M. tuberculosis* (Davies, 1994).

5.16.1.2 Distribution

Mycobacterium avium-intracellulare (MAC) used to be the least common of the four major species causing lung disease. The advent of HIV, which for reasons unknown, seems to attract MAC infection specifically, has now made MAC the commonest in developed countries with an appreciable HIV-infected population. Among HIV-negative individuals, *M. kansasii* is the commonest species found in Western Europe and parts of the United States. It is also the commonest species in the United Kingdom, although *M. xenopi* may be more common in the South East and *M. malmoense* in the North. These bacteria are free living occurring in soil, house dust, and tap water. They may infect wild or domestic animals as well as humans.

5.16.2 Bacteriology and Pathology

Environmental mycobacteria do not cause the disease we call tuberculosis. Infection with these organisms is not transmissible, therefore the diseases they cause do not have public health implications for contact tracing, unlike those caused by *M. tuberculosis*.

Histological tissues affected by these mycobacteria appear identical to those in which disease arises due to *M. tuberculosis* with characteristic granulomata formation leading to caseous necrosis.

On direct smear all mycobacteria stain positive to acid and alcohol fast Zehl–Neelsen and are therefore indistinguishable from one another. Only on culture do the species-specific characteristics permit identification. The result is that patients with sputum smear-positive pulmonary disease caused by environmental mycobacteria are usually first diagnosed as having tuberculosis with the public health implications which that brings. Diagnosis is only revised when cultures are available some 6 weeks later, which may result in confusion of patients and doctors alike. Molecular techniques such as the polymerase chain reaction (PCR) can provide more rapid species identification than culture and may be used more frequently in future by those countries able to afford it (Banks and Campbell, 2003).

5.16.3 Clinical Presentation

Disease commonly occurs in the latter half of life among those with pre-existing lung disease such as previous tuberculosis, chronic bronchitis, and emphysema, and less commonly bronchiectasis. In the author's experience infection with mycobacteria is strongly associated with smoking. A survey of over 40 patients from Liverpool showed only one to be a life-long non-smoker (Davies and Davies, 1995).

Patients may present with acute or chronic illness that is radiographically and clinically indistinguishable from tuberculosis. Symptoms are usually chronic and include cough with sputum and sometimes hemoptysis, weight loss, malaise, and night sweats. Symptoms may be insidious and occur over many months before the patient presents. A study looking at the symptomatic presentation of *M. kansasii* infection compared with tuberculosis found that with the exception of diabetes and alcohol intake, which favored a diagnosis of tuberculosis, there were no features between the two, likely to be diagnostically helpful (Evans et al., 1996).

The American Thoracic Society bases diagnosis on a combination of symptoms, radiographic changes, and at least two isolates of the bacteria (American Thoracic Society, 1997).

Some patients have disease presenting insidiously over more than a year. Weight loss and cachexia are often accompanied by long-standing respiratory symptoms. These are often attributed to the pre-existing pulmonary disease. Mycobacterial infection may be missed if it is not considered and the appropriate investigations requested.

5.16.4 Chest Radiography and Diagnosis

The chest radiograph may be difficult to interpret, often showing chronic abnormalities which may be attributed to the pre-existing lung disease. Genuine infection is likely if several specimens show the same strain of mycobacteria from symptomatic patients whose chest X-ray shows abnormalities consistent with mycobacterial disease. Chest radiography usually shows upper lobe shadowing, which is more often unilateral than bilateral. Cavitation tends to be more frequent than in tuberculosis and fibrosis more marked, but these features cannot be used to distinguish infection by one organism from the other (Evans et al., 1996).

Sometimes the cavities can be very thick walled making penetration by drugs difficult. A single isolate from sputum which is not repeated is unlikely to be of significance as these mycobacteria from the environment may easily contaminate specimens or cultures.

5.16.5 Disseminated Disease

Before the HIV epidemic, disseminated infection with environmental mycobacteria was rare. MAC and *M. kansasii* have been described as causing disseminated disease in some patients. In the presence of HIV infection, co-infection with MAC usually occurs at very low CD4 counts and dissemination is usual.

Patients may have pronounced immunodeficiency and complain of fever with weight loss and general malaise. Gastrointestinal symptoms usually predominate with pulmonary disease less marked. Enlarged lymph glands, hepatosplenomegaly, and skin lesions are common. The chest X-ray may show hilar gland lymphadenopathy, pulmonary nodules, or patchy alveolar infiltrates. Cavitation in disseminated pulmonary disease is uncommon. The diagnosis may be made from tissue biopsy particularly bone marrow, and even by obtaining bacteria from blood cultures. Infected tissue typically shows poorly formed granulomata teeming with acid-fast bacilli. Serially performed blood cultures can be used as a way of monitoring the efficacy of treatment.

More common than blood-born spread of disease to distant sites, environmental mycobacteria may infect soft tissue by puncture wounds of the skin, including sites of surgical incision. Soft tissue infections from swimming pools or fish tanks as result of *M. balnei* (*M. marinum*) infections may cause soft tissue inflammation and possible ulceration.

Treatment of pulmonary disease caused by environmental mycobacteria is much more problematic than the treatment of *M. tuberculosis*. In vitro sensitivity testing is thought to be a poor guide to success. Environmental mycobacteria are usually resistant to all first-line anti-tuberculosis drugs with the exception of ethambutol. Uncontrolled trials have shown that treatment with RMP and EMB is effective in most patients infected with *M. kansasii* and *M. xenopi* (Subcommittee of the Joint Tuberculosis Committee of the British Thoracic Society, 2000). The length of treatment has been a subject of debate but a recent study showed a 9% relapse rate after as little as 9 months of RMP and EMB in the treatment of *M. kansasii* (American Thoracic Society, 1997).

The same agents have been said to be effective against the other common environmental pulmonary infectors, *M. malmoense* and MAC but a treatment length of 2 years is recommended. Recently the newer drugs, particularly clarithromycin and ciprofloxacin have been said to be effective but trial data is scarce.

Treatment is not always successful and the patient may remain sputum smear or culture positive despite months or years of treatment. It may be necessary to treat concomitant lung disease such as chronic bronchitis with broad-spectrum antibiotics and bronchodilators. A number of different factors must be taken into account when assessing the efficacy of treatment. These will include general well being and symptomatology of the patient, weight gain, sputum conversion to a smear negative and culture negative state, and radiological improvement. The last is probably the poorest guide to overall improvement as disease is often long standing and fibrosis permanent. In the author's experience cigarette smoking is strongly associated with environmental pulmonary infections and it is very difficult to eliminate bacteria from patients who continue to smoke. If the patient is relatively fit and the area of infection well contained, in an abscess cavity for example, removal of the infected area by surgery is an option. In one case of a male patient aged 73 with a cavitating lesion in the right lung apex, who grew *M. malmoense* from the sputum, 2 years of therapy with rifampicin and ethambutol eliminated the bacteria for nearly a year but he relapsed with a return of symptoms and smear and culture presence of *M. malmoense*. Surgical excision of the right apical cavity affected a cure.

Sometimes the patient may be relatively asymptomatic, despite being heavily sputum smear positive for one of these organisms. In these cases a period of observation, while withholding treatment may be advisable bearing in mind the possible adverse effects of the treatment. In this respect *M. malmoense* and MAC in HIV serum-negative individuals may have a very low virulence and may not be causing any additional pathology. *M. xenopi*, *M. kansasii*, and the much rarer *M. szulgei* tend to be more virulent and will probably result in spread of disease if not actively treated. Other organisms such as *M. gordonai* or *M. fortuitum*, which have been isolated in pulmonary disease, are unlikely to cause active pathology and may be due to specimen contamination.

Sensitivity testing often shows bacterial susceptibility to the older first-line drugs such as ethionamide and cycloserine. The efficacy of these in vivo remains very difficult to quantify and the common adverse events they cause probably precludes their use.

Treatment of soft tissue infection is even more problematical. Drug treatment is rarely effective and surgical excision is often the only treatment. In children with lymphatic glands infected with these organisms total excision is often used. Alternatively they may be left untreated as they usually cause no systemic disease, and regress in time.

5.16.6 Clarithromycin and MAC in HIV Infection

Disseminated *Mycobacterium avium* complex (DMAC) infection is a common complication of advanced HIV disease and is an independent predictor of mortality. The clinical features of DMAC infection are fever, weight loss, abdominal pain, anemia, elevated serum alkaline phosphatase, and elevated serum lactate dehydrogenase. The diagnosis is made by blood cultures; clinical diagnosis is unreliable. Chemoprophylaxis of DMAC infection with azithromycin is recommended when the CD4 lymphocyte count is below 50 cells/mm^3. Established DMAC infection is treated with clarithromycin plus ethambutol, unless the isolate is macrolide-resistant, in which case the optimal therapy is uncertain. Highly active antiretroviral therapy is important in both prevention and treatment of DMAC infection (Shafran, 1998).

In a randomized, placebo-controlled, double-blind study of clarithromycin in patients with AIDS in the United States and Europe, *M. avium* complex infection developed in 19 of the 333 patients (6%) assigned to clarithromycin and in 53 of the 334 (16%) assigned to placebo (adjusted hazard ratio, 0.31; 95% CI, 0.18–0.53; $P < 0.001$). During the follow-up period of about 10 months, 32% of the patients in the clarithromycin group died and 41% of those in the placebo group died (hazard ratio, 0.75; $P = 0.026$). In the clarithromycin group, isolates from 11 of the 19 patients with *M. avium* complex infection were resistant to clarithromycin (Pierce et al., 1996). Thus in patients with advanced AIDS, the prophylactic administration of clarithromycin is well tolerated, prevents *M. avium* complex infection, and reduces mortality. A randomized, placebo-controlled trial was conducted to evaluate the efficacy of clarithromycin in the prevention of DMAC infection in patients with AIDS. Special attention was given to the development of clarithromycin resistance. The median time to documented MAC bacteremia was 199 days for placebo-treated patients, 217 days for clarithromycin-treated patients infected with clarithromycin-susceptible MAC, and 385 days for clarithromycin-treated patients infected with clarithromycin-resistant MAC. Most of the patients with clarithromycin-resistant isolates (91%) had a baseline CD4 T-cell count of < 20/µl, while these low counts occurred in only 25% of patients having clarithromycin-susceptible breakthrough isolates. The emergence of clarithromycin resistance did not affect the total period of survival. Resistance to clarithromycin in breakthrough MAC isolates emerges most likely when the patient is extremely immunodeficient at the time of initiation of the preventative therapy (Craft et al., 1998).

Susceptibility testing for baseline MAC isolates from AIDS patients not previously treated with clarithromycin or azithromycin does not appear to be useful in guiding therapy (Shafran et al., 1998).

Because results of recent randomized trials indicate minimal efficacy of continuing MAC prophylaxis in patients who respond to potent combination antiretroviral therapy, the observed high incidence of macrolide-resistant bacterial colonization of the respiratory tract in this trial supports the discontinuation of macrolide prophylaxis in all AIDS patients whose CD4 counts have risen above 100 cells/µL (Aberg et al., 2001).

References

Abate, G., (2002), Drug-resistant tuberculosis in Ethiopia: problem scenarios and recommendation. *Ethiopian Med J*, **40**, 79–86.

Abe, C. and Abu, A.K.K., (2002), Status of antituberculosis drug resistance in Saudi Arabia 1979–98. *East Mediterranean Hlth J*, **8**, 664–670.

Aberg, J.A., Wong, M.K., Flamm, R., Notario, G.F. and Jacobson, M.A., (2001), Presence of macrolide resistance in respiratory flora of HIV-Infected patients receiving either clarithromycin or azithromycin for *Mycobacterium avium* complex prophylaxis. *HIV Clin Trials*, **2**, 453–459.

Adegbola, R.A., Hill, P., Baldeh, I., Out, J., Sarr, R., Sillah, J., Lienhardt, C., Corrah, T., Manneh, K., Drobniewski, F. and McAdam, K.P.J.W., (2003), Surveillance of drug-resistant *Mycobacterium tuberculosis* in The Gambia. *Int J Tuberc Lung Dis*, 7, 390–393.

Al-Marri, M.R., (2001), Pattern of mycobacterial resistance to four anti-tuberculosis drugs in pulmonary tuberculosis patients in the state of Qatar after the implementation of DOTS and a limited expatriate screening programme. *Int J Tuberc Lung Dis*, 5, 1116–1121.

American Thoracic Society, (1994), Treatment of tuberculosis and tuberculosis infection in adults and children. *Am J Respir Crit Care Med*, 149, 1359–1374.

American Thoracic Society, (1997), Diagnosis and treatment of diseases caused by nontuberculous mycobacteria. *Am J Respir Crit Care Med*, 156, S1–S25.

Augustynowicz, K.E., Zwolska, Z., Jaworski, A., Kostrzewa, E. and Klatt, M., (2003), Drug-resistant tuberculosis in Poland in 2000: second national survey and comparison with the 1997 survey. *Int J Tuberc Lung Dis*, 7, 645–651.

Balabanova, Y., Ruddy, M., Hubb, J., Yates, M., Malomanova, N., Fedorin, I. and Drobniewski, F., (2005), Multidrug-resistant tuberculosis in Russia: clinical characteristics, analysis of second-line drug resistance and development of standardized therapy. *Eur J Clin Microbiol Infect Dis* 24, 136–139.

Banks, J. and Campbell, I., (2003), Environmental mycobacteria. In P.D.O. Davies, editor, *Clinical Tuberculosis*. London, Arnold, pp. 439–448.

Baylan, O., Kisa, O., Albay, A. and Doganci, L., (2004), Evaluation of a new automated, rapid colormetric culture system using solid medium for laboratory diagnosis of tuberculosis and determination of anti-tuberculosis drug susceptibility. *Int J Tuberc Lung Dis*, 8, 772–777.

Breathnach, A.S., de Ruiter, A., Holdsworth, G.M., Bateman, N.T., O'Sullivan, D.G., Rees, P.J., Snashall, D., Milburn, H.J., Peters, B.S., Watson, J., Drobniewski, F.A. and French, G.L., (1998), An outbreak of multi-drug-resistant tuberculosis in a London teaching hospital. *J Hosp Infect*, 92, 111–117.

Brennan, P.J. and Draper, P., (1994), Ultrastructure of *Mycobacterium tuberculosis*: In Bloom, E.R., editor, *Tuberculosis: Pathogenesis, Protection and Control*. Washington, DC, American Society for Microbiology, pp. 271–284.

Butt, T., Ahmad, R.N., Kazmi, S.Y. and Rafi, N., (2004), Multi-drug resistant tuberculosis in Northern Pakistan. *J Pak Med Assoc*, 54, 469–472.

Caminero, J.A., (2005), Management of multidrug resistant tuberculosis and patients in retreatment. *Eur Respir J*, 25, 928–936.

Centres for Disease Control and Prevention, (1992), Transmission of multidrug-resistant tuberculosis among immunocompromised persons in a correctional system: New York, 1991. *MMWR Morb Mortal Wkly Rep*, 41 (28), 507–509.

Centres for Disease Control and Prevention, (1993), Outbreak of multidrug-resistant tuberculosis at a hospital: New York City, 1991. *MMWR Morb Mortal Wkly Rep*, 42 (22), 427, 433–434.

Centres for Disease Control and Prevention, (2005), National plan for reliable tuberculosis laboratory services using a systems approach: recommendations from CDC and the Association of Public Health Laboratories. Task Force on Tuberculosis Laboratory Services. *MMWR Morb Mortal Wkly Rep*, 54 [RR-6], 1–12.

Chan, E.D., Laurel, V., Strand, M.J., Chan, J.F., Huynh, M.L.N., Goble, M. and Iseman, M.D., (2004), Treatment and outcome analysis of 205 patients with multidrug-resistant tuberculosis. *Am J Respir Crit Care Med*, 169, 1103–1109.

Chiang, Y.C., Hsu, C.J., Huang, M.R., Lin, P.T. and Luh, K.T., (2004), Antituberculosis drug resistance among retreatment tuberculosis patients in a referral centre in Taipei. *J Formosa Med Assoc*, 103, 411–541.

Condos, R., Rom, W.N. and Schluger, N.W., (1997), Treatment of multidrug-resistant pulmonary tuberculosis with interferon-g via aerosol. *Lancet*, 349, 1513–1515.

Conover, C., Ridzon, R., Valway, S., Schoenstadt, L., McAuley, J., Onorato, I. and Paul, W., (2001), Outbreak of multidrug-resistant tuberculosis at a methadone treatment program. *Int J Tuberc Lung Dis*, 1, 59–64.

Coronado, V.G., Beck-Sague, C.M., Hutton, M.D., Davis, B.J., Nicholas, P., Villareal, C., Woodley, C.L., Kilburn, J.O., Crawford, J.T. and Frieden, T.R., (1993), Transmission of multidrug-resistant *Mycobacterium tuberculosis* among persons with human immunodeficiency virus infection in an urban hospital: epidemiologic and restriction fragment length polymorphism analysis. *J Infect Dis*, **1684**, 1052–1055.

Cox, H.S., Orozco, J.D., Male, R., Ruesch, G.S., Falzon, D., Small, I., Doshetov, D., Kebede, Y. and Aziz, M., (2004), Multidrug-resistant tuberculosis in central Asia. *Emerg Infect Dis*, **10**, 865–872.

Craft, J.C., Notario, G.F., Grosset, J.H. and Heifets, L.B., (1998), Clarithromycin resistance and susceptibility patterns of *Mycobacterium avium* strains isolated during prophylaxis for disseminated infection in patients with AIDS. *Clin Infect Dis*, **27**, 807–812.

Cynamon, M.H., Klemens, S.P., Sharpe, C.A. and Chase, S., (1999), Activities of several novel oxazolidinones against *Mycobacterium tuberculosis* in a murine model. *Antimicrob. Agents Chemother*, **43**, 1189–1191.

Daikos, G.L., Cleary, T., Rodriguez, A. and Fischl, M.A., (2003), Multidrug-resistant tuberculous meningitis in patients with AIDS. *Int J Tuberc Lung Dis*, **7**, 394–398.

David, H.L., Jahan, M.T., Jumin, J., Grandry, J. and Lehman, E.H., (1978), Numerical taxonomy analysis of *Mycobacterium africanum*. *Int J Syst Bacteriol*, **28**, 464–472.

Davies, L. and Davies, P.D.O., (1995), Non-tuberculosis mycobacterial infections: no fire without a smoke. *Am J Respir Crit Care Med*, **151**, A715.

Davies, P.D.O., (1994), Infection with *Mycobacterium kansasii*. (Editorial). *Thorax*, **49**, 435–436.

Deivanayagam, C.N., Rajasekaran, S., Venkatesan, S., Mahilmaran, A., Ahmed, P.R.K., Annadurai, S., Kumar, S., Chandrasekar, C., Ravichandran, N. and Pencillaiah, R., (2002), *Ind J Chest Dis Allied Sci*, **44**, 237–242.

De Riemer, K., Garcia, G.L., Bobadilla-del-Valle, M., Palacios, M.M., Martinez, G.A., Small, P.M., Sifuentes, O.J. and Ponce-de-Leon, A., (2005), Does DOTS work in populations with drug-resistant tuberculosis? *Lancet*, **365**, 1239–1245.

Dewan, P., Sosnovskaja, A, Thomsen, V., Cicenaite, J., Laserson, K., Johansen, I., Davidaviciene, D. and Wells, C., (2005), High prevalence of drug-resistant tuberculosis, Republic of Lithuania, 2002. *Int J Tuberc Lung Dis*, **9**, 170–174.

Drobniewski, F.A., (1997), Is death inevitable with multiresistant TB plus HIV infection? *Lancet*, **349**, 71–72.

Drobniewski, F.A., Caws, M., Gibson, A. and Young, D., (2003), Modern laboratory diagnosis of tuberculosis. *Lancet Infect Dis*, **3**, 141–147.

Drobniewski, F.A., Watterson, S.A., Wilson, S.M. and Harris, G.S., (2000), A clinical, microbiological and economic analysis of national service for the rapid diagnosis of tuberculosis and rifampicin resistance in *Mycobacterium tuberculosis*. *J Med Microbiol*, **49**, 271–278.

Durek, C., Richter, E., Basteck, A., Rüsch-Gerdes, R., Gerdes, J., Jocham, D. and Bohle, A., (2001), The fate of bacillus Calmette-Guerin after intravesical instillation. *J Urol*, **165**, 1765–1768.

Dye, (2002), *J. Infect Dis*, **185**, 1197–1202.

Evans, S.A., Colville, A., Evans, A.L., Crisp A.L. and Johnston, I.D.A., (1996), Pulmonary *Mycobacterium kansasii* infection: comparison of the clinical features, treatment and outcome with pulmonary tuberculosis. *Thorax*, **51**, 1248–1252.

Friedland, J.S., (2004), Tuberculosis. In J. Cohen, W.G. Powderly, editors, *Infectious Diseases*, 2nd Edition, Volume 1. New York, Mosby, pp. 401–418.

Gale, E.F., Cundliff, E., Reynolds, P.E., Richmond, M.H., Waring, M.J., (1981), *The Molecular Basis of Antibiotic Action*, 2nd Edition. New York, Wiley.

Giosue, S., Casarini, M., Ameglio, F., Zangrilli, P., Palla, M., Attieri, A.M. and Bisette, A., (2000), Aerosolized interferon-alpha treatment in patients with multidrug-resistant pulmonary tuberculosis. *Eur Cytokine Netw*, **11**, 99–104.

Goble, M., Iseman, M.D., Madsen, L.A., Waite, D., Ackerson, L. and Horsburgh, C.R, (1993), Treatment of 171 patients with pulmonary tuberculosis resistant to isoniazid and rifampin. *N Engl J Med*, **328**, 527–532.

Grosskinsky, C.M., Jacobs, W.R., Jr, Clark-Curtiss, J.E. and Bloom, B.R., (1989), Genetic relationships among *Mycobacterium tuberculosis* and candidate leprosy vaccine strains determined by

DNA hybridization: identification of an *M. leprae*-specific repetitive sequence. *Infect Immun*, **57**, 1535–1541.

Hannan, M.M., Peres, H., Maltez, F., Hayward, A.C., Machado, J., Morgado, A., Proenca, R., Nelson, M.R., Bico, J., Young, D.B. and Gazzard, B.S., (2002), Investigation and control of a large outbreak of multi-drug resistant tuberculosis at a central Lisbon hospital. *J Hosp Infect*, **72**, 91–97.

Hayward, A.C., Bennett, D.E., Herbert. J. and Watson, J.M., (1996), Risk factors for drug resistance in patients with tuberculosis in England and Wales 1993–4. *Thorax*, **51**(Suppl 3), S32.

Hillemann, D., Weizenegger, M., Kubica, T., Richter, E. and Niemann, S., (2000), Use of genotype DNA MTBDR assay for the rapid detection of rifampicin and isoniazid resistance in *Mycobacterium tuberculosis*. *Microbes Infect*, **2**, 659–669.

Hillemann, D., Weizenegger, M., Kubica, T., Richter, E. and Niemann, S., (2005), Use of genotype MTB DR assay for rapid detection of rifampicin and isoniazid resistance in *Mycobacterium tuberculosis* complex isolates. *J Clin Microbiol*, **43**, 3699–3703.

Hirano, K., Wada, M. and Aoyagi, T., (2001), Resistance of *Mycobacterium tuberculosis* to four first-line anti-tuberculosis drugs in Japan, 1997. *Int J Tuberc Lung Dis*, **5**, 46–52.

Holland, S.M., Eisenstein, E.M., Kuhns, D.B., Turner, M.L., Fleisher, T.A., Strober, W. and Gallin, J.I., (1994), Treatment of refractory disseminated nontuberculous mycobacterial infection with interferon-gamma: a preliminary report. *N Engl J Med*, **330**, 1348–1355.

Hong Kong Chest Service/British Medical Research Council, (1981), First report: controlled trial of four thrice-weekly regimens and a daily regimen all given for 6 months for pulmonary tuberculosis. *Lancet*, **1**, 171–174.

Hong Kong Chest Service/British Medical Research Council, (1982), Second report: controlled trial of four thrice-weekly regimens and a daily regimen all given for 6 months. The results up to 24 months. *Tubercle*, **63**, 89–98.

Hong Kong Chest Service/British Medical Research Council, (1987), Five year follow-up of a controlled trial of five 6-month regimens of chemotherapy for pulmonary tuberculosis. *Am Rev Respir Dis*, **136**, 1339–1342.

Hong Kong Chest Service/British Medical Research Council, (1991), Controlled trial of 2, 4 and 6 months of pyrazinamide in 6 month, three-times weekly regimens for smear-positive tuberculosis, including an assessment of a combined preparation of isoniazid, rifampicin and pyrazinamide. Results at 30 months. *Am Rev Respir Dis*, **143**, 700–706.

Iseman, M., (1993), Treatment of multidrug resistant tuberculosis. *N Engl J Med*, **329**, 784–790.

Jacobs, W.R., Jr, Barletta, R.G., Udani, R., Chan, J., Kalkut, G., Sosne, G., Kieser, T., Sarkis, G.J., Hatfull, G.F. and Bloom, B.R., (1993), Rapid assessment of drug susceptibilities of *Mycobacterium tuberculosis* by means of luciferase reported phages, *Science*, **260**, 819–822.

Johansen, I.S., Lundgern, B., Sosnovskaja, A. and Thomsen, V.O., (2003), Direct detection of multidrug-resistant *Mycobacterium tuberculosis* in clinical specimens in low-and high-incidence countries by line probe assay. *J Clin Microbiol*, **41**, 4454–4456.

Johansen, I.S., Thomsen, V.O., Marjamäkj, M., Sosnovskaja, A. and Lungern, B., (2004), Rapid, automated, non-radiometric susceptibility testing of *Mycobacterium tuberculosis* complex to four first-line anti-tuberculous drugs used in standard short-course chemotherapy. *Diagn Microbiol Infect Dis*, **50**, 103–107.

Johnson, B.J., Bekker, L.G., Rickman, Brown, S., Lesser, M., Ress, S., Willcox, P., Steyn, L. and Kaplan, G., (1997), rhuIL-2 adjunctive therapy in multidrug resistant tuberculosis: a comparison of two treatment regimens and placebo. *Tuber Lung Dis*, **78**, 195–203.

Joint Tuberculosis Committee of the British Thoracic Society, (1998), Chemotherapy and management of tuberculosis in the United Kingdom: recommendations 1998. *Thorax*, **53**, 536–548.

Joint Tuberculosis Committee of the British Thoracic Society, (2000), Control and prevention of tuberculosis in the United Kingdom: Code of Practice 2000. *Thorax*, **55**, 887–901.

Juréen, P., Warngren, J. and Hoffner, S.E., (2004), Evaluation of the line probe assay (Li PA) for rapid detection of rifampicin resistance in *Mycobacterium tuberculosis*. *Tuberculosis*, **84**, 311–316.

Kam, K.M. and Yip, C.W., (2004), Surveillance of *Mycobacterium tuberculosis* susceptibility to second-line drugs in Hong Kong, 1995–2002, after the implementation of DOTS-plus. *Int J Tuberc Lung Dis*, **8**, 760–766.

Kartaloglu, Z., Bozkanat, E., Ozturkeri, H., Okutan, O. and Ilvan, A., (2002), Primary antituber-culosis drug resistance at Turkish military chest diseases hospital in Istanbul. *Med Princ Pract*, **11**, 202–205.

Kassa, K.E., Bobossi, S.G., Takeng, E.C., Nambea, K.T.B., Yapou, F. and Talarmin, A., (2004), Surveillance of drug-resistant childhood tuberculosis in Bangui, Central African Republic. *Int J Tuberc Lung Dis*, **8**, 574–578.

Kelly, P.M., Scott, L. and Krause, V.L., (2002), Tuberculosis in East Timorese refugees: implications for health care needs in East Timor. *Int J Tuberc Lung Dis*, **6**, 980–987.

Kim, H.C., Goo, M.J., Lee, H.J., Park, H.S., Park, M.C., Kim, T.J. and Im, J.G., (2004), Multidrug-resistant tuberculosis versus drug-sensitive tuberculosis in human immunodeficiency virus-negative patients: computed tomography features. *J Comput Assist Tomogr*, **28**, 366–371.

Kimerling, M.E., Schuchter, J, Chanthol, E., Kunthy, T., Stuer, F., Glaziou, P. and Ee, O., (2002), Prevalence of pulmonary tuberculosis among HIV-infected persons in a home care program in Phnom Penh, Cambodia. *Int J Tuberc Lung Dis*, **6**, 988–994.

Kisa, O., Abley, A., Bedir, O., Baylon, O. and Doganci, L., (2003), Evaluation of FAST Plaque to TB-RIF for determination of rifampicin resistance in *Mycobacterium tuberculosis* complex isolates. *Int J Tuberc Lung Dis*, **7**, 284–288.

Kochi, A., Vareldzis, B. and Styblo, K., (1993), Multidrug-resistant tuberculosis and its control. *Res Microb*, **144**, 104–110.

Konno, K., Feldmann, F.M. and McDermott, W., (1967), Pyrazinamide susceptibility and amidase activity of tubercle bacilli. *Am Rev Respir Dis*, **95**, 461–469.

Krishnamurthy, A., Rodriques, C. and Mehta, A.P., (2002), Rapid detection of rifampicin resistance in *M. tuberculosis* by phage assay. *Ind J Med Microbiol*, **20**, 211–214.

Lee, J. and Chang, J.H., (2001), Drug-resistant tuberculosis in a tertiary referral teaching hospital of Korea. *Kor J Int Med*, **16**, 173–179.

Lefford, M.J., (1966), The ethionamide sensitivity of British pre-treatment strains of *Mycobacterium tuberculosis*. *Tubercle Lond*, **47**, 198–206.

Leimane, V., Riekstina, V., Holtz, T.H., Zarovska, E., Skripconoka, V., Thorpe, L.E., Laserson, K.F. and Wells, C.D., (2005), Clinical outcome of individualised treatment of multidrug-resistant tuberculosis in Latvia: a retrospective cohort study. *Lancet*, **365**, 318–326.

Levin, M.E. and Hatfull, G.F., (1993), *Mycobacterium smegmatis* RNA polymerase: DNA super-coiling, action of rifampicin and mechanism of rifampicin resistance. *Mol Microbiol*, **8**, 277–285.

Levy-Frebault, V.V. and Portaels, F., (1992), Proposed minimal standards for the genus *Mycobacteria* and the description of new slowly growing *Mycobacterium* species. *Int J Syst Bacteriol*, **42**, 315–323.

Li, J., Burzynski, J.N., Lee, Y.A., Berg, D., Driver, R.C., Ridzon, R. and Munsiff, S.S., (2004), Use of therapeutic drug monitoring for multidrug-resistant tuberculosis patients. *Chest*, **126**, 1770–1776.

Lin, J., Sattar, A.N. and Puckree, T., (2004), An alarming rate of drug-resistant tuberculosis at Ngwelezane Hospital in northern KwaZulu Natal, South Africa. *Int J Tuberc Lung Dis*, **8**, 568–573.

Lockman, S., Kruuner, A., Binkin, N., Levina, K., Wang, Y., Danilovitsh, M., Hoffner, S. and Tappero, J., (2003), Clinical outcomes of Estonian patients with primary multidrug-resistant versus drug-susceptible tuberculosis. *Clin Infect Dis*, **32**, 373–380.

Mahmoudi, A. and Iseman, M.D., (1993), Pitfalls in the care of patients with tuberculosis. *JAMA*, **270**, 65–68.

Maloney, S.A., Pearson, M.L., Gordon, M.T., Del Castillo, R., Boyle, J.F. and Jarvis, W.R., (1995), Efficacy of control measures in preventing nosocomial transmission of multidrug-resistant tuberculosis to patients and health care workers. *Ann Intern Med*, **122**, 90–95.

Marín, M., de Viedma, D.G., Ruiz-Serrano, M.J. and Bouza, E., (2004), Rapid detection of multiple rifampicin and isoniazid resistance mutations in *Mycobacterium tuberculosis* in respiratory samples by real time PCR. *Antimicrob Agents Chemother*, **48**, 4293–4300.

Mortilla, H. J., Soini, H., Vyshnevskiy, B. I., Otten, T.F., Vasilyef, A.V., Huovinen, P. and Vil-janen, M.K., (1998), Rapid detection of rifampicin-resistant *Mycobacterium tuberculosis* by sequencing and line probe assay. *Scand J Infect Dis*, **30**, 129–132.

McClure, W.R. and Cech, C.L., (1978), On the mechanism of rifampicin inhibition if RNA synthesis. *J Biol Chem*, **253**, 8949–8956.

Mitchell, I., Wendon, J., Fi, H.S. and Williams, R., (1995), Anti-tuberculous therapy and acute liver failure. *Lancet*, **345**, 555–556.

Mitchison, D.A. and Nunn, A.J., (1986), Influence of initial drug resistance on the response to short-course chemotherapy of pulmonary tuberculosis. *Am Rev Respir Dis*, **133**, 423–430.

Migliori, G.B., Raviglione, M.C., Schaberg, T., Davies, P.D.O., Zellweger, J.P., Grzemska, M., Mihaescu, T., Clancy, L. and Casali, L., (1999), Tuberculosis management in Europe. *Eur Respir J*, **14**, 978–992.

Mokrousov, L.O., Narvskaya, T., Otten, T., Limeshenke, E., Steklova, L. and Vyshnevskiy, B., (2002), High prevalence of *katG* Ser 315Thr substitution among isoniazid-resistant *Mycobacterium tuberculosis* clinical isolates from Northwestern Russia 1996 to 2001. *Antimicrob Agents Chemother*, **46**, 1417–1424.

Moro, M.L., Errante, I., Infuso, A., Sodano, L., Gori, A., Orcese, C.A., Salamina, G., D'Amico, C., Besozzi, G. and Caggese, L., (2000), Effectiveness of infection control measures in controlling a nosocomial outbreak of multidrug-resistant tuberculosis among HIV patients in Italy. *Int J Tuberc Lung Dis*, **41**, 61–68.

Moro, M.L., Gori, A., Errante, I., Infuso, A., Franzetti, F., Sodano, L. and Iemoli, E., (1998), An outbreak of multidrug-resistant tuberculosis involving HIV-infected patients of two hospitals in Milan, Italy. Italian Multidrug-Resistant Tuberculosis Outbreak Study Group. *AIDS*, **129**, 1095–1102.

Mukherjee, J.S., Joseph, J.K., Rich, M.L., Shin, S.S., Furin, J.J., Seung, K.J., Sloutsky, A., Socci, A.R., Vanderwarker, C., Vasquez, L., Palacios, E., Guerra, D., Viru, F.A., Farmer, P. and Del-Castillo, H.E., (2003), Clinical and programmatic considerations in the treatment of MDR-TB in children: a series of 16 patients from Lima, Peru. *Int J Tuberc Lung Dis*, **7**, 637–644.

Oleksijew, A., Meulbroek, J., Ewing, P., Jarvis, K., Mitten, M., Paige, L., Tovcimak, A., Nukkula, M., Chu, D. and Alder, J.D., (1998), *In vivo* efficacy of ABT-255 against drug-sensitive and -resistant *Mycobacterium tuberculosis* strains. *Antimicrob Agents Chemother*, **42**, 2674–2677.

Pablos-Mendez, A. and Lessnau, K., (2000), Clinical mismanagement and other factors producing antituberculosis drug resistance. In I. Bastian and F. Portaels, editors, *Multidrug-resistant Tuberculosis*. Dordrecht, Kluwer, pp. 59–76.

Palmero, D., Ritacco, V., Ambroggi, M., Marcela, N., Barrera, L., Capone, L., Dambrosi, A., di-Lonardo, M., Isola, N., Poggi, S., Vescovo, M. and Abbate, E., (2003), Multidrug-resistant tuberculosis in HIV-negative patients, Buenos Aires, Argentina. *Emerg Inf Dis*, **9**, 965–969.

Park, S.K., Lee, C.M., Heu, J.P. and Song, S.D., (2002), A retrospective study for the outcome of pulmonary resection in 49 patients with multidrug-resistant tuberculosis. *Int J Tuberc Lung Dis*, **6**, 143–149.

Petrie, J.G., Barnwell, E. and Grimshaw, J., (1995), on behalf of the Scottish InterCollegiate Guidelines Network. *Clinical Guidelines: Criteria for Appraisal for National Use.* Edinburgh: Royal College of Physicians.

PHLS (1999) *Tuberculosis Update*. Public Health Laboratory Service Communicable Disease Surveillance Centre. December 1999.

Piatek, A.S., Telenti, A., Murray, M.R., El-Hajj, H. and Jacobs, W.R., Jr., (2000), Genotypic analysis of *Mycobacterium tuberculosis* in two distinct populations using molecular beacons: implications for rapid susceptibility testing. *Antimicrob Agents Chemother*, **44**, 103–110.

Pierce, M., Crampton, S., Henry, D., Heifets, L., LaMarca, A., Montecalvo, M., Wormser, G.P., Jablonowski, H., Jemsek, J., Cynamon, M., Yangco, B.G., Notario, G. and Craft, J.C., (1996), A randomized trial of clarithromycin as prophylaxis against disseminated *Mycobacterium avium* complex infection in patients with advanced acquired immunodeficiency syndrome. *N Engl J Med*, **335**, 384–391.

Pitchenik, A.E., Burr, J., Laufer, M., Miller, G., Cacciatore, R., Bigler, W.J., Witte, J.J. and Cleary, T., (1990), Outbreaks of drug-resistant tuberculosis at AIDS centre. *Lancet*, **336**, 440–441.

Pomerantz, B.J., Cleveland, J.C., Olson, H.K. and Pomerantz, M., (2001), Pulmonary resection for multi-drug resistant tuberculosis. *J Thorac Cardiovasc Surg,* **121**, 448–453.

Punnotok, J., Shaffer, N., Naiwatanakul, T., Pumprueg, U., Subhannachart, P., Ittiravivongs, A., Chuchotthaworn, C., Ponglertnapagorn, P., Chantharojwong, N., Young, N.L., Limpakarnjanarat, K. and Mastro, T.D., (2000), Human immunodeficiency virus-related tuberculosis and primary drug resistance in Bangkok, Thailand. *Int J Tuberc Lung Dis,* **4**, 537–543.

Quy, H.T.W., Lan, N.T.N., Borgdorff, M.W., Grosset, J., Linh, P.D., Tung, L.B., van-Soolingen, D., Raviglione, M., Co, N.V. and Broekmans, J., (2003), Drug resistance among failure and relapse cases of tuberculosis: is the standard re-treatment regimen adequate? *Int J Tuberc Lung Dis,* **7**, 631–636.

Ramaswamy, S. and Musser, J.M., (1998), Molecular genetic-basis of antimicrobial agent resistance in *Mycobacterium tuberculosis,* 1998 update. *Tuber Lung Dis,* **79**, 3–29.

Romano, G., Michell, P., Pacilio, C. and Giordano, A., (2000), Latest development in gene transfer technology, achievements, perspectives and controversies over therapeutic applications. *Stem Cells,* **18**, 19–39.

Rosner, J.L., (1993), Susceptibilities of *oxyR* regulon mutants of *Escherichia coli* and *Salmonella typhimurium* to isoniazid. *Antimicrob Agents Chemother,* **37**, 2251–2253

Rossau, R., Traore, H., De Beenhouwer, H., Mijs, W., Jannes, G., de Rijk, P. and Portaels, F., (1997), Evaluation of the INNO-LIPA Rif. TB assay, a reverse hybridization assay for the simultaneous detection of *Mycobacterium tuberculosis* complex and its resistance to rifampicin. *Antimicrob Agents Chemother,* **41**, 2093–2098.

Ruddy, M., Balabanova, Y., Graham, C., Fedorin, I., Malomanova, N., Elisarova, E., Kuznet-znov, S., Gusarova, G., Zakharova, S., Melentyev, A., Krukova, E., Golishevskaya, V., Erokhin, V., Dorozhkova, I. and Drobniewski, F., (2005), Rates of drug resistance and risk factor analysis in civilian and prison patients with tuberculosis in Samara Region, Russia. *Thorax,* **60**, 130–135.

Reechaipichitkul, W., (2002), Multidrug-resistant tuberculosis at Srinagarind Hospital, Khon Kaen, Thailand. *Southeast Asian J Trop Med Pub Hlth,* **33**, 570–574.

Robert, J., Trystram, D., Truffot, P.C. and Jarlier, V., (2003), Multidrug-resistant tuberculosis, eight years of surveillance in France. *Eur Respir J,* **22**, 833–837.

Salaniponi, F.M., Nyirenda, T.E., Kemp, J.R., Squire, S.B., Godfrey-Faussett, P. and Harries, A.D., (2003), Characteristics, management and outcome of patients with recurrent tuberculosis under routine programme conditions in Malawi. *Int J Tuberc Lung Dis,* **7**, 948–952.

Scarparo, C., Ricordi, P., Ruggiero, G. and Piccoli, P., (2004), Evaluation of the fully Automated BACTEC MGIT 960 system for testing susceptibility to *Mycobacterium tuberculosis* to pyrazinamide, streptomycin, isoniazid, rifampicin and ethembutol and comparison with the Radiometric BACTEC 460 TB method. *J Clin Microbiol,* **42**, 1109–1114.

Schaaf, H.S., Shean, K. and Donald, P.R., (2003), Culture confirmed multidrug resistant tuberculosis: diagnostic delay, clinical features, and outcome. *Arch Dis Child,* **88**, 1106–1111.

Schneider, E., Laserson, K.F., Wells, C.D. and Moore, M., (2004), Tuberculosis along the United States–Mexico border, 1993–2001. *Rev Panam Salud Pub,* **16**, 23–34.

Schraufnagel, D.E., (1999), Tuberculosis treatment for the beginning of the next century. *Int J Tuberc Lung Dis,* **3**, 651–662.

Senaratne, W.V., (2004), Outcome of treatment of multidrug resistant tuberculosis. *Ceylon Med J,* **49**, 86–87.

Shafer, R.W., Small, P.M., Larkin, C., Singh, S.P., Kelly, P., Sierra, M.F., Schoolnik, G. and Chirgwin, K.D., (1995), Temporal trends and transmission patterns during the emergence of multidrug-resistant tuberculosis in New York City: a molecular epidemiologic assessment. *J Infect Dis,* **171**, 170–176.

Shafran, S.D., (1998), Prevention and treatment of disseminated *Mycobacterium avium* complex infection in human immunodeficiency virus-infected individuals. *Int J Infect Dis,* **3**, 39–47.

Shafran, S.D., Talbot, J.A., Chomyc, S., Davison, E., Singer, J., Phillips, P., Salit, I., Walmsley, S.L., Fong, I.W., Gill, M.J., Rachlis, A.R. and Lalonde, R.G., (1998), Does in vitro susceptibility to rifabutin and ethambutol predict the response to treatment of

Mycobacterium avium complex bacteremia with rifabutin, ethambutol, and clarithromycin? Canadian HIV Trials Network Protocol 010 Study Group. *Clin Infect Dis*, **27**, 1401–1405.

Shah, A.R., Agarwal, S.K. and Shah, K.V., (2002), Study of drug resistance in previously treated tuberculosis patients in Gujarat, India. *Int J Tuberc Lung Dis*, **6**, 1098–1010.

Shin, S., Furin, J., Bayona, J., Mate, K., Kim, J.Y. and Farmer, P., (2004), Community-based treatment of multidrug-resistant tuberculosis in Lima, Peru: 7 years of experience. *Soc Sci Med*, **59**, 1529–1539.

Shiraishi, Y., Nakajima, Y., Katsuragi, N., Kurai, M. and Takahashi, N., (2004), Resectional surgery combined with chemotherapy remains the treatment of choice for multidrug-resistant tuberculosis. *J Thorac Cardovasc Surg*, **128**, 523–528.

Slayden, R.A., Barry, C.E., 3rd, (2000), The genetics and biochemistry of isoniazid resistance in *Mycobacterium tuberculosis*. *Microbes Infect*, **2**, 659–669.

Small, P.M., Shafer, R.W., Hopewell, P.C., Singh, S.P., Murphy, M.J., Desmond, E., Sierra, M.F. and Schoolnik, G.K., (1993), Exogenous re-infection with multi-drug resistant *Mycobacterium tuberculosis* in patients with advanced HIV infection. *N Engl J Med*, **328**, 1137–1144.

Sreevatson, S., Pan, X., Zhang, Y., Deretic, V. and Musser, J.M., (1997), Analysis of the *oxyR–ahpC* region in isoniazid-resistant and susceptible *Mycobacterium tuberculosis* complex organisms recovered from diseased humans and animals in diverse localities. *Antimicrob Agents Chemother*, **41**, 600–606.

Stanford, J.L., Stanford, C.A., Grange, J.M., Lan, N.N. and Etemadi, A., (2001), Does immunotherapy with heat-killed *Mycobacterium vaccae* offer hope for the treatment of multidrug-resistant pulmonary tuberculosis? *Respir Med*, **95**, 444–447.

Stover, C.K., Warrener, P., Van Devanter, D.R., Sherman, D.R., Arain, T.M., Langhorne, M.H., Anderson, S.W., Towell, J.A., Yuan, Y., McMurray, D.N., Kreiswirth, B.N., Barry, C.E. and Baker, W.R., (2000), A small-molecule nitroimidazopyran drug candidate for the treatment of tuberculosis. *Nature*, **405**, 962–966.

Stroud, L.A., Tokars, J.I., Grieco, M.H., Crawford, J.T., Culver, D.H., Edlin, B.R., Sordillo, E.M., Woodley, C.L., Gilligan, M.E. and Schneider, N., (1995), Evaluation of infection control measures in preventing the nosocomial transmission of multidrug-resistant *Mycobacterium tuberculosis* in a New York City hospital. *Infect Control Hosp Epidemiol*, **3**, 141–147.

Suarez, S., O'Hara, P., Kazantseva, M., Newcomer, C.E., Hopfer, R., McMurray, D.N., Hickey, A.J., (2001), Airways delivery of rifampicin microparticles for the treatment of tuberculosis. *J Antimicrob Chemother*, **48**, 431–434.

Subcommittee of the Joint Tuberculosis Committee of the British Thoracic Society, (2000), Management of opportunist mycobacterial infections: Joint Tuberculosis Committee guidelines 1999. *Thorax*, **55**, 210–218.

Syre, H., Phyu, S., Sandven, P., Bjorvatn, B. and Grewd, H.M.S., (2003), Rapid colorimetric method for testing susceptibility of *Mycobacterium tuberculosis* to isoniazid and rifampicin in liquid cultures. *J Clin Microbiol*, **41**, 5173–5177.

Talbot, E.A., Kenyon, T.A., Mwasekaga, M.J., Moeti, T.L., Mallon, V. and Binkin, N.J., (2003), Control of anti-tuberculosis drug resistance in Botswana. *Int J Tuberc Lung Dis*, **7**, 72–77.

Telenti, A., Honoré, N., Bernasconi, J., March, J., Ortega, A., Heym, B., Takiff, H.E. and Cole, S.T., (1997), Genotyping assessment of isoniazid and rifampicin resistance in *Mycobacterium tuberculosis*: a blind study at reference laboratory level. *J Clin Microbiol*, **35**, 719–723.

Telenti, A., Imboden, P., Marchesi, F., Lowrie, D., Cole, S., Colston, M.J., Matter, L., Schopfer, K. and Bodmer, T., (1993), Detection of rifampicin-resistance mutations in *Mycobacterium tuberculosis*. *Lancet*, **341**, 647–650.

Telzak, E.E., Sepowitz, K., Alpert, P., Mannheimer, S., Medard, F., el-Sadr, W., Blum, S., Gagliardi, A., Salomon, N. and Turett, G., (1995), Multidrug-resistant TB in patients without HIV infection. *N Engl J Med*, **333**, 907–911.

The Inter Departmental Working Group on Tuberculosis, (1998), The Prevention and Control of Tuberculosis in the United Kingdom: UK Guidance on the Prevention and Control of 1) HIV-Related Tuberculosis and 2) Drug Resistant, Including Multiple Drug Resistant, Tuberculosis. Department of Health, The Scottish Office, The Welsh Office, September 1998.

Timperi, R, Han, L.L., Sloutsky, A, Becerra, M.C., Nardell, E.A., Salazar, J.J. and Smith, F.M.C., (2005), Drug resistance profiles of *Mycobacterium tuberculosis* isolates: five years' experience and insight into treatment strategies for MDR-TB in Lima, Peru. *Int J Tuberc Lung Dis*, **9**, 175–180.

Torosyan, A.A., Gurova, A.A. and Mandulova, P.V., (2000), A fifteen-month survey of directly observed therapy—short course and antibiotic drug resistance in the region of Plovdiv. *Folia Med*, **42**, 10–14.

Toungoussova, O.S., Nizovtseva, N.I., Mariandyshev, A.O., Caugant, D.A., Sandven, P. and Bjune, G., (2004), Impact of drug-resistant *Mycobacterium tuberculosis* on treatment outcome of culture-positive cases of tuberculosis in the Archangel oblast, Russia, in 1999. *Eur J Clin Microbiol Infect Dis*, **23**, 174–179.

Trakada, G., Tsiamita, M. and Spiropoulos, K., (2004), Drug-resistance of *Mycobacterium tuberculosis* in Patras, Greece. *Monaldi Arch Chest Dis*, **61**, 65–70.

Tsai, M.S., Chong, I.W., Hwang, J.J., Wang, T.H. and Huang, S.M., (2004), A subset of clinical status of pulmonary tuberculosis in Southern Taiwan. *Southeast Asian J Trop Med Pub Hlth*, **35**, 136–139.

Tsogt, G., Naranbat, N., Buyankhisig, B., Batkhuyag, B., Fujiki, A. and Mori, T., (2002), The nationwide tuberculosis drug resistance survey in Mongolia, 1999. *Int J Tuberc Lung Dis*, **6**, 289–294.

Tupasi, T.E., Quelapio, M.I.D., Orillaza, R.B., Alcantar, C., Mira, N.R.C., Abeleda, M.R., Belen, V.T., Arnisto, N.M., Rivera, A.B., Grimaldo, E.R., Derilo, J.O., Dimarucut, W., Arabit, M. and Urboda, D., (2003), DOTS-Plus for multidrug-resistant tuberculosis in the Philippines: global assistance urgently needed. *Tuberculosis*, **83**, 52–58.

Uplekar, M.W. and Shepard, D.S., (1991), Treatment of tuberculosis by private practitioners in India. *Tubercle*, **72**, 284–290.

Vanacore, P., Koehler, B., Carbonara, S., Zacchini, F., Bassetti, D., Antonucci, G., Ippolito, G. and Girardi, E., (2004), Drug-resistant tuberculosis in HIV-infected persons: Italy 1999–2000. *Infection*, **32**, 328–332.

van Soolingen, D., Hoogenboezem, T., de Haas P.E.W., Hermans, P.W., Koedam, M.A., Teppema, K.S., Brennan, P.J., Besra, G.S., Portaels, F., Top, J., Schouls, L.M. and Embden, J.D., (1997), A novel pathogen taxon of the *Mycobacterium tuberculosis* complex, Canetti: characterization of an exceptional isolate from Africa. *Int J Syst Bacteriol*, **47**, 1236–1245.

van Soolingen, D., von der Zanden, A.G.M., de Haas, P.E.W., Noordhoek, G.T., Kiers, A., Foudraine, N.A., Portaels, F., Kolk, A.H.J., Kremer, K. and van Embden, J.D.A., (1998), Diagnosis of *Mycobacterium microti* infections among humans by using novel genetic markers. *J Clin Microbiol*, **36**, 1840–1845.

Wada, T., Maeda, S., Tamaru, A., Shigeyoshi, L., Hase, A. and Kobayashi, K., (2004), Dual-probe assay for rapid detection of drug-resistant *Mycobacterium tuberculosis* by real-time PCR. *J Clin Microbiol*, **42**, 5277–5285.

World Health Organization, (1997), Treatment of Tuberculosis: Guidelines for National Programmes. Geneva, WHO. WHO/TB/97.220.

World Health Organisation Tuberculosis Unit, (1991), Division of Communicable Diseases. Guidelines for Tuberculosis Treatment in Adults and Children in National Tuberculosis Programmes. Geneva, WHO. WHO/TB/91, pp 1–61.

World Health Organization, (2004), Anti-tuberculosis Drug Resistance in the World. Report No.3. Geneva, WHO.

Yam, W.E., Tem, C.M., Leung, C.C., Tong, L., Chan, K. H., Leung, E.T.Y., Wong, K.C., Yew, W.W., Seto, W.H., Yuen, K. and Ho, P.L., (2004), Direct detection of rifampicin-resistant *Mycobacterium tuberculosis* in respiratory specimens by PCR-DNA sequencing. *J Clin Microbiol*, **42**, 4438–4443.

Yew, W.W., (2001), Therapeutic drug monitoring in antituberculosis chemotherapy: clinical perspectives. *Clin Chim Acta*, **313**, 31–36.

Zaman, K., Rahim, Z., Yunus, M., Arifeen, S., Baqui, A., Sack, D., Hossain, S., Banu, S., Islam, A.M., Ahmed, J., Breiman, R. and Black, R., (2005), Drug resistance of *Mycobacterium tuberculosis* in selected urban and rural areas in Bangladesh. *Scand J Infect Dis*, **37**, 21–26.

Chapter 6
Antimicrobial Resistance of Anaerobic Bacteria

Ellie J. C. Goldstein, Diane M. Citron, and David W. Hecht

6.1 Introduction

While Leeuwenhoek may have been the first to see anaerobes under the microscope and Pasteur coined the term "anaerobies" (Goldstein, 1995), it was not until 1891 that the first clinical case of anaerobic infection was described and not until 1893 when Veillon finally isolated and propagated the first anaerobes from human clinical infections (Finegold, 1994). Since then, anaerobes have remained somewhat hidden pathogens from medical scrutiny, and recognition of their important role in human infections has often been underappreciated. While there have been periods of explosive advances in the recovery of anaerobes from human infections and a revolution in their taxonomy, it is surprising that their patterns of resistance remain inaccessible to everyday clinicians and are studied only by a small cadre of researchers scattered worldwide. While automated methods of antimicrobial susceptibility testing (AST) are available in every clinical laboratory for aerobes, many labs, even in some large medical centers, are either unable or unwilling to culture anaerobes except from very specific sites and even fewer perform AST.

In 1990, Goldstein et al. (1992) surveyed 120 US hospitals with bed capacities of 200–1,000 and found that 30% did not perform anaerobic ASTs. Of those that did perform ASTs 16% used a reference laboratory, and 54% performed them in house. Blood culture isolates were tested by 97% of the laboratories, sterile body sites by 73%, and pure culture isolates by 47%. If a physician requested an anaerobic AST, it was performed by only 39% of laboratories. For laboratories doing testing, the broth disk method, no longer sanctioned by the Clinical and Laboratory Standards Institute (CLSI), was used most often (56%), followed by microdilution (33%), β-lactamase testing (25%), macro-tube dilution (2%), and agar dilution (2%). The antimicrobial agents tested were as follows: penicillin–ampicillin, 94%; clindamycin, 94%; metronidazole, 90%; chloramphenicol, 80%; cefoxitin, 76%; tetracyclines, 51%; and erythromycin, 45%. A follow-up study (Goldstein et al., 1995a) done three years later, showed a dramatic shift in attitudes regarding anaerobic ASTs with only 23% performing such tests and most laboratories (68%) did not include this information in annual susceptibility reports to the hospital antibiogram. Thirty percent of laboratories noted that they relied on the published literature to supply clinicians data on anaerobic isolate susceptibilities. Recent unpublished data

I. W. Fong and K. Drlica (eds.), *Antimicrobial Resistance and Implications for the Twenty-First Century.* © Springer 2008

from the R.M. Alden Research Laboratory suggests that currently (2006), fewer than 30% of laboratories perform anaerobic ASTs. It is therefore no surprise that recognition of anaerobic resistance has lagged behind that of aerobic pathogens. In addition, one must be aware that there may be differences in breakpoints promulgated by CLSI, the FDA, EUCAST and other national bodies resulting in differences in the determination of susceptibility and resistance of isolates.

6.2 Anaerobic Antimicrobial Susceptibility Testing (AST)

Currently both CLSI (2007) and the ASM Manual of Microbiology (Citron and Hecht, 2003) suggest that hospitals should test individual patient isolates to assist in their care and periodically establish patterns of resistance for certain anaerobes and to include these data in the hospital antibiogram. Strains from members of the *Bacteroides fragilis* group, *Prevotella* spp., *Fusobacterium* spp., and *Clostridium* spp. should be tested with at least 10 representative isolates from each genus. Since *B. fragilis* is a more common pathogen, more isolates, up to 30 for each species, should be included. The antimicrobial agents tested should be those available on the hospital's formulary to help clinicians in their selection of empirical therapy. CLSI provides recommendations for testing of various antibiotics based upon the identification of the anaerobic isolate. Individual isolates AST from specific patients should be performed to assist patient care when (1) selecting an active agent is critical for disease management, (2) the likelihood of long-term therapy, (3) isolates recovered from the blood, brain, bone, or joint, (4) in cases of therapeutic failure, or (5) organisms with unpredictable susceptibility patterns (Table 6.1).

6.2.1 Methods

The CLSI has designated the agar dilution procedure using supplemented Brucella agar as the reference method (NCCLS, 1993, 1997, 2001; CLSI/NCCLS, 2004). This method is neither simple, practical, nor economical to perform, but allows

Table 6.1 Antimicrobial susceptibility testing methods for anaerobic Bacteria

Method	Medium	Inoculum	Advantages	Disadvantages
Agar dilution	Supplemented Brucella blood agar [vitamin K_1 and hemin]	10^5 cfu/spot	Reference method, suitable for annual study and surveillance	Research setting, Labor intensive
Broth microdilution	Supplemented Brucella broth [blood]	10^6 cfu/mL	Multiple antibiotics/isolate, commercially available	Limited to *B. fragilis* group,
E-test	Brucella blood agar	Swab plate with 0.5 McFarland standard	Convenient for single isolate and a few drugs	

different laboratories to compare MIC results (Roe et al., 2002a,b). Acceptable alternative testing methods include broth microdilution (reserved only for the *B. fragilis* group), and the E-test (A–B Biodisk, Solna, Sweden), (Citron et al., 1991). β-lactamase testing provides very limited data, but can be useful if penicillin therapy is being considered. Broth disk elution and disk diffusion tests should not be performed for anaerobic bacteria as their susceptibility results do not correlate with the agar dilution reference method (Citron et al., 1991).

6.2.1.1 β-lactamase Testing

BLA testing is a simple test that allows one to determine production of β-lactamase by an isolate that is therefore resistant to penicillin and ampicillin. This in not a true AST but is used when limited information is required. It is not recommended for *B. fragilis* group species, as most are likely β-lactamase producers or resistant to penicillin by other mechanisms. Because alternative mechanisms of resistance to β-lactams exist, a negative test does not assure susceptibility to penicillin.

Testing is performed using either a chromogenic assay as a nitrocefin disk assay (Cefinase; BBL, Cockeysville, MD) or an S1 chromogenic disk (International Bioclinical, Inc., Portland, OR) according to manufacturer's directions. A positive test exhibits a change of color from yellow to red usually within 5–10 min with the exception of some *Bacteroides* strains that may require up to 30 min to react (CLSI/NCCLS, 2004).

6.2.1.2 E-Test

The E-test uses a plastic strip with an interpretative MIC scale on one side and predetermined antimicrobial concentration gradient on the other side. The individual isolate is suspended in broth or saline to the turbidity of the 0.5 McFarland Standard and applied with a cotton swab to a 150 mm diameter petri dish of supplemented Brucella blood agar. Up to six antimicrobials may be tested by applying the strips in a radial fashion. Smaller plates can be used if fewer strips are needed. The plates should be incubated for 48 h at 35°C in an anaerobic environment. MICs for *B. fragilis* and other rapidly growing strains may be read at 24 h. The MIC is read at the point where the elliptical zone of inhibition intersects the strip. It has become an excellent and convenient choice for testing individual anaerobic organisms.

Citron et al. (1991) first reported the E-Test with anaerobes in 1991 in a study that compared it to the agar dilution method using both supplemented Brucella agar and Wilkins–Chalgren agar, and found a correlation of 85% within one dilution and 98% within two-dilutions. Discrepancies occurred with some strains against cefoxitin and some when agar depth varied. Rosenblatt and Gustafson have noted that some *Prevotella* and *Bacteroides* strains show false susceptibility when testing penicillin and ceftriaxone that is minimized if β-lactamase-producing strains are eliminated (Rosenblatt and Gustafson, 1995). A more significant warning is potential false resistance to metronidazole as a result of test conditions and medium quality (Cormicon et al., 1996) which can be avoided by reducing the test plates in an anaerobic chamber overnight before testing.

6.2.1.3 Agar Dilution

The agar dilution procedure is considered the reference method. It is neither simple nor inexpensive and is mostly performed by reference laboratories and in large medical center laboratories. The method is described in CLSI/NCCLS reference documents (2004).

The most common method for inoculum preparation is direct colony suspension from 24–48-h growth on an anaerobic blood agar plate. Well-isolated colonies are taken up by rolling a cotton swab over them and then suspended in reduced Brucella broth to achieve turbidity equivalent to the turbidity of the 0.5 McFarland standard. Alternatively five or more isolated colonies of an organism grown on anaerobic media are inoculated into enriched thioglycolate medium (without indicator) and incubated for 6–24 h at 37°C. The turbidity of the mixture is adjusting to 0.5 McFarland standard by addition of reduced Brucella broth. The suspensions are pipetted into the individual wells of the replicator device and applied to the antimicrobial-containing plates with the multipronged inoculator. The final concentration of organisms is approximately 10^5 cfu/spot. For both recommended *Bacteroides* quality control strains, an inoculum of 10^5 cfu/spot is achieved while only 10^4 cfu/spot is obtained for the *C. difficile* quality control strain, due to the large size of cells (CLSI M11-A7, 2007).

Antimicrobial susceptibility powders are reconstituted according to the manufacturers' instructions. Serial twofold dilutions of the various antimicrobial agents can be prepared according to an algorithm (CLSI) or twofold dilutions using specified solvents and diluents (CLSI/NCCLS, 2007) to achieve the desired final test concentrations. The high concentration stock solutions often may be stored at –70°C, except for clavulanic acid, until the day of use or may be prepared on the day of use. Diluted solutions cannot be stored at –70°C.

Brucella blood agar supplemented with 5µg/ml hemin, 1 µg/ml vitamin K_1, and 5% laked sheep blood is the recommended testing medium. The use of different animal blood products may affect results. The 17 ml supplemented Brucella agar blanks, with hemin and vitamin K_1, are prepared in advance and stored until the day of the test. On the day of testing these blanks may be flash autoclaved, boiled or microwaved to melt the agar and then placed in a ~50°C water bath. One milliliter of laked sheep blood is then added to each melted blank.

Two milliliters of each test antimicrobial concentration is added to the molten agar and mixed by gently inverting the tubes three times. Each mixture is then poured into separate petri dishes and allowed to dry. Once the agar has solidified, the plates are inverted and placed into a 37°C incubator to dry. This step is necessary because moist plates will not absorb the inoculum, and condensate on lids increases the risk of contamination. Once dry, the plates are ready for use. The antimicrobial containing plates should be used the same day, but may be stored at 4°C for periods of up to 72 h if necessary. Plates containing any antimicrobial that may be unstable, such as clavulanic acid or imipenem, must be used on the day of preparation.

Once the plates are ready for use, and the inoculum is prepared, a clean and adequate sized space, away from breezes and air currents should be used for the test procedure. The inoculum-replicating apparatus, such as a Steers–Foltz replicator, is used and will have either 32 or 36 prongs, which deliver 1 to 2 μl on to the agar surface, corresponding to 1×10^5 cfu/spot. Previously sterilized inoculum wells are filled using a pipette with the test isolate[s] inoculum. The stamping procedure consists of first immersing the prongs into the inoculum and immediately stamping them onto the agar plate. A Brucella blood agar without antibiotics should be stamped prior to and after each set of antibiotics containing plates for removal of residual antibiotics and for growth control. Plates with the lowest antimicrobial concentration should be stamped first and subsequently continuing with the next highest dilution of each antibiotic set. One set of antimicrobial free plates should be stamped and incubated aerobically to act as a control for contamination by aerobic bacteria. After inoculation, the stamped plates should sit until inoculum liquid is absorbed into the medium ($\sim$10 min) and then incubated in an anaerobic environment at 35°C for 42–48 h.

After 42–48 h of incubation, the results are read by comparing each plate to the growth control plate. If there is any aerobic growth suggestive of contamination on the control set of plates, the MIC may not be interpreted for that test organism. Some anaerobes may grow slightly on the aerobic plate. The MIC is defined, as the antimicrobial concentration at which there is a marked reduction in the appearance of growth compared with the control plate. This includes the change from good growth to a haze, multiple tiny colonies, or one to several normal-sized colonies. Color photo examples of endpoint readings are available (CLSI/NCCLS, 2004).

6.2.1.4 Broth Microdilution

The broth microdilution method as described in CLSI M11-A7 (2007) has been validated only for testing of *B. fragilis* group members and is not applicable to other anaerobes. To perform this test, Brucella broth supplemented with hemin (5 μg/ml), vitamin K_1, and lysed cleared horse blood is the recommended medium. Commercial trays are available. Trays should be stored at –70°C. For in house tray preparation, antibiotics are diluted in volumes of 15–100 ml (Aldridge and Schiro, 1994), and delivered using a device that can simultaneously dispense aliquots of 0.05–0.1 ml per well; if the volume is 0.05 ml per well, then the antibiotic concentrations should be prepared at twice the final desired concentration, since addition of 0.05 μl of the inoculum will add another twofold dilution. Final test volumes of less than 0.1 ml are not recommended. The inoculum is standardized as equivalent to a turbidity of 0.5 McFarland standard and diluted to achieve a final concentration of 1×10^6 cfu/ml in the test wells.

If frozen trays are used then they should be thawed at room temperature, and inoculated within 15 min of inoculum preparation. A colony count and purity check of the inoculum may be performed by removing 10 μl aliquot from the growth control well, diluting it into 10 ml of saline, and streaking 0.1 ml onto the surface of an anaerobic blood agar plate, and incubating anaerobically for 48 h. Inoculated trays should be incubated for 44–48 h at 35°C in an anaerobic atmosphere in an environment that ensures sufficient humidity to prevent drying. The MIC values are read by viewing the plates from the bottom using a stand and a mirror and the MIC is considered the concentration where there is no growth, or a significant reduction of growth. Appearance in the growth control is required to interpret results. A trailing effect may be observed for some drug/organism combinations.

6.2.1.5 Quality Control

Quality control strains that should be used for anaerobic bacteria are *B. fragilis* ATCC 25285 and *B. thetaiotaomicron* ATCC 29741 with *C. difficile* ATCC 700057 limited to use with agar dilution or *E. lentum* ATCC 43055 limited to use with either agar dilution or E-test methods. Many laboratories have experienced repeated difficulty reading endpoints for *E. lentum* ATCC 43055 prompting the recent testing and approval of *C. difficile* ATCC 700057 (CLSI/NCCLS, 2007).

Most information about resistance in anaerobic bacteria emanates from a plethora of in vitro studies that compare MICs of a variety of agents and sometimes correlate them to the accepted clinical breakpoints. Heseltine et al. (1983) analyzed the effect of several antimicrobial agents on the clinical outcomes of patients with appendicitis. They found that patients who received an agent active against anaerobes had a better outcome than those who did not.

In the late 1980s, two studies evaluated the efficacy of cefoxitin in patients with intra-abdominal infections and correlated the clinical outcome with various in vitro testing methods (Goldstein et al., 1990; Snydman et al., 1992). Studying 13 patients and their anaerobic isolates, they noted that the broth elution method showed poor correlation with clinical outcome. No differences were found with the other seven methods studied. They reported that patients with organisms having an MIC ≤ 32 to cefoxitin were usually cured and failure occurred when the MIC was ≥ 128 µg/ml. Snydman et al. (1992) reviewed data on 19 patients, 11 cured, and 8 failures. They noted that cefoxitin showed time-dependent killing and that the time over the MIC was an important predictor of clinical success. Regarding the different media and methods employed, they found that Wilkins–Chalgren agar yielded higher MICs for the *B. fragilis* group species and that of the 11 methods tested that brain heart infusion agar supplemented with blood used in the agar dilution method and the Micromedia commercial broth microdilution system were the best at predicting outcomes and that the broth elution test did not predict success or failure. These studies supported the association of antibiotic resistance among *Bacteroides* sp. and clinical failure

Nguyen et al. (2000) showed a clinical correlation between in vitro susceptibility testing and clinical outcome for 128 patients with *Bacteroides* bacteremias. In a prospective observational multicenter study performed between January 1991 and May 1995, they found that 30-day mortality was 45% for patients with inactive antimicrobial therapy compared to 16% ($P = 0.04$) with active therapy (Nguyen et al., 2000). Of the two other end points studied, inactive therapy was associated with an 82% failure rate compared to 22% ($P = 0.002$) for active therapy and micro-biological persistence was 42% compared to 12% ($P = 0.06$). The specificity of active therapy directed against *Bacteroides* species was reliably predicted with a specificity of 97% and a positive predictive value of 82%. This confirmed the impor-tance and reliability of in vitro susceptibility studies in predicting clinical outcome for anaerobic infections. These recommendations are also echoed in other recent publications (Citron and Hecht, 2003; Jacobus et al., 2004).

6.3 Antimicrobial Resistance among Anaerobes

While the role of antimicrobial resistance in aerobic bacteria has been headline news around the world, the steadily rising rate of antibiotic resistance among some anaerobes has increased significantly over the last few decades but has garnered little attention. Most significantly members of the *B. fragilis* group have developed resistance to all classes of antimicrobials (Hecht, 2002, 2004, 2006; Hecht and Vedantam, 1999) and this has occurred worldwide (Katsandri et al., 2006; Schapiro et al., 2004). Unfortunately, due to issues in the isolation, identification, and suscep-tibility testing of anaerobes, data is often not available on local levels but we must rely on a few large-scale reports and anecdotal cases of resistance and extrapolate this information.

The Tufts-New England Medical Center has coordinated ongoing in vitro surveil-lance studies that have reported significant increases in resistance among the *B. frag-ilis* group strains since the 1980s (Snydman et al., 1999, 2002a,b, 2007). These and other surveillance studies note that there are center-to-center variations in suscepti-bility patterns, which are not predictable. For example, a study based at Loyola Med-ical Center, noted variations in clindamycin activity amongst different Chicago area hospitals (Hecht et al., 1993). There may also be differences in various reports that are inapparent due to differences in methodology as seen in a national survey coor-dinated at Louisiana State University Health Science Center which utilized the broth dilution method testing for *B. fragilis* group and other anaerobes (Aldridge et al., 2001). However, all three surveys confirm that resistance is increasing with hospital-to-hospital variation, even within the same geographic area. In addition, sources of isolates or quality control (QC) reading variations may also cause discrepancies even when the same methodology is utilized. Goldstein et al. (2006a) reported on the comparative activity of moxifloxacin against intra-abdominal isolates and in their review of the literature found wide variations with their results and those reported by Snydman et al. (2002a) and Edmiston et al. (2004). One must always be cautious in extrapolating survey report susceptibility data to an individual patient situation.

Snydman et al. (2007) reported on a survey of 5225 *B. fragilis* group isolates from 10 geographically distributed US medical centers and analyzed trends from 1997 to 2004 using the reference agar dilution method. The distribution of species isolated was *B. fragilis*, 52.1%; *B. thetaiotaomicron*, 18.7%; *B. ovatus*, 10.4%; *B. vulgatus*, 5.9%; *B. distasonis*, 5.2%; *B. uniformis*, 3.2%; and other species (*B. eggerthii, B. merdae*, and *B. stercoris*) 4.5%. Surprisingly, they noted that there was an increased susceptibility of the isolates tested over the study period with decreases in geometric mean MICs for imipenem, meropenem, piperacillin–tazobactam, and cefoxitin. *B. fragilis* was more susceptible to antimicrobials than the other species, while *B. distasonis* was the most resistant to β-lactams. In contrast to β-lactams, there was in increase in geometric mean MICs to clindamycin and moxifloxacin for some of the species. Still this study does not report specific QC results and their distribution, nor does it analyze susceptibility by the clinical sources of the isolates.

6.3.1 Future Directions of Research

Since the current level of funding and hence research on anaerobes has been less than that for aerobes, future trends will be based, in part, on the application of mechanisms of resistance found in aerobes to anaerobes of particular interest because of their emergence and recognition in either severe or fatal infections. The national and international surveys on anaerobic susceptibility have at least drawn attention to the increasing problem of resistance amongst pathogenic anaerobes. Currently, the means of toxin production in the "super-toxin" producers of *C. difficile* have spurred interest in that organism and its quinolone resistance pattern. Should further *B. fragilis* resistance to metronidazole develop that too will become a fertile field of research. Several groups have been exploring porin channels and the outer membrane proteins of selected anaerobes and that too is expected to continue the advancement of our knowledge. The future awaits changes in the clinical present.

6.3.2 Mechanisms and Prevalence of Resistance

Table 6.2 summarizes the current known mechanisms of resistance and resistance genes for anaerobic bacteria. Antimicrobial resistance mechanisms vary for each class of agent and their prevalence and expression varies by species. One must also recognize that there may be marked geographical variation in the prevalence of resistance.

6.3.3 Antibiotic Resistance by Genus and Species

6.3.3.1 Gram-Negative Bacilli and Cocci

Bacteroides fragilis Group

The *B. fragilis* group consists of 13 species with *B. fragilis* generally being the most susceptible to antimicrobials; greater than 95% are resistant to penicillin principally due to β-lactamase production. Piperacillin is the most active of the ureidopenicillins against the *B. fragilis* group, although susceptibility has fallen from approximately 90–70% over the prior 10 years (Snydman et al., 1999, 2002).

Table 6.2 Anaerobic bacterial gene transfer factors contributing to antibiotic resistance

Bacterial group	Antibiotic	Gene designation	Transferable	Transfer factor
Bacteroides fragilis *group*	Clindamycin	*ermF, ermS, ermG*	+	Plasmid
	Tetracycline	*tetQ, tetX*[a]	+	Plasmid
	Cephalosporin	*cepA, cblA*	N.D.[b]	
	Cefoxitin	*cfxA*	+	Transposon
	Carbapenems	*ccrA, cfiA*	+	Plasmid [c]
	Metronidazole	*NimA-F*	+ (*nimA,C,D*)	Plasmid
	Quinolones	*gyrA, gyrB, parC, parE*	??	
	Streptomycin	*AadS*[a]	+	Transposon
Clostridium perfringens	Chloramphenicol	*catQ*[d], *catP*	+	Plasmid
	Clindamycin	*ermQ, ermP*	N.D.	
	Tetracycline	*tetA(P), tetB(P)*	+	Plasmid
Clostridium difficile	Tetracycline	*Tcr, tetM*	+	Transposon?[e]
	Chloramphenicol	*CatD*	+	Transposon
	Clindamycin	*ErmB, ermZ, ermBZ*	+	Transposon
Clostridium butyricum	Chloramphenicol	*catA, catB*	N.D.	
Prevotella spp	Tetracycline	*tetQ, tetO, tetM*	+	Transposon (*tetQ*)
	Clindamycin	*ErmF*	+	Transposon
Fusobacterium spp	Tetracycline	*TetM, tetW*	+	Transposon
Peptostre-ptococcus spp	Clindamycin	*ErmTR*	??	
	Tetracycline	*tetM,, tetK, tetO*	??	
Porphyromonas	Tetracycline	*tetQ*	??	
Bifidobacterium	Tetracycline	*tetW*		

[a] cryptic.
[b] Not determined.
[c] A plasmid-borne imipenem resistance determinant has been isolated, but the gene has not been characterized.
[d] *catQ* was characterized from a non-conjugative strain.
[e] The exact nature of the transfer factor is unknown.

The penicillinase-resistant penicillins such as oxacillin and nafcillin, and first generation cephalosporins are not active against these organisms. β-lactam–β-lactamase inhibitor combinations, such as ampicillin/sulbactam, amoxicillin/clavulanate, ticarcillin/clavulanate, and piperacillin/tazobactam are active against nearly all strains of *B. fragilis* group, with <2% resistance in most reports (Behra-Miellet et al., 2003b; Goldstein, 2000; Hecht and Osmolski, 2003; Hedberg and Nord, 2003; Koeth et al., 2004; Snydman et al., 1999, 2002, Teng et al., 2002).

Among the cephalosporins, cefoxitin and cefotetan are generally active against *B. fragilis* but the latter is much less active against the other members of the *B. fragilis* group (Aldridge et al., 2001). Ceftizoxime has some activity against most members of the *B. fragilis* group, inhibiting <50% of isolates (Snydman et al., 1999) but susceptibility data varies widely among published studies (60–90%), in part due to different testing methods (Aldridge and Schiro, 1994). Cefotetan and ceftizoxime are no longer available in the United States.

Clindamycin resistance is widely reported against *Bacteroides* spp. worldwide (Behra-Miellet et al., 2003b; Hecht and Osmolski, 2003; Hedberg and Nord, 2003; Koeth et al., 2004; Snydman et al., 1999, 2002; Teng et al., 2002) and may be linked to transferable tetracycline resistance. Chloramphenicol, metronidazole, tinidazole, and the carbapenems (imipenem, ertapenem, and meropenem) are generally uniformly active against all members of the *B. fragilis* group (Aldridge et al., 2001; Snydman et al., 2002) although imipenem resistant strains have been reported (Polglajen et al., 1994; Snydman et al., 1999). Metronidazole resistance has been reported in the United States, United Kingdom, and continental Europe. Trovafloxacin, moxifloxacin, and gatifloxacin have good activity against some members of the *B. fragilis* group (Snydman et al., 2002a) and a broad range of other anaerobes (Ednie, Jacobs and Appelbaum, 1998; Hoellman et al., 2001; Snydman et al., 2002b). Resistance to fluoroquinolones in *Bacteroides* and other anaerobes is increasing with the widespread use of this class of antimicrobial agents. Tigecycline, although without CLSI defined breakpoints for anaerobes but with FDA approved breakpoints has in vitro activity against a variety of anaerobic organisms (Betriu et al., 2005; Goldstein et al., 2006b; Jacobus et al., 2004) (Table 6.3).

Table 6.3 Susceptibility MIC breakpoints [CLSI and/or FDA] for anaerobic bacteria.[a]

Antimicrobial Agent	MIC(µg/mL)		
	Susceptible	Intermediate	Resistant
Beta-lactams			
Penicillin	≤ 0.5	1	≥ 2
Ampicillin	≤ 0.5	1	≥ 2
Ampicillin/sulbactam	≤ 8/4	16/8	≥ 32/16
Amoxicillin/clavulanic acid	≤ 4/2	8/4	≥ 16/8
Piperacillin	≤ 32	64	≥ 128
Piperacillin/tazobactam	≤ 32/4	64/4	≥ 128/4
Cefoxitin	≤ 16	32	≥ 64
Carbapenems			
Ertapenem	≤ 4	8	≥ 16
Imipenem	≤ 4	8	≥ 16
Meropenem	≤ 4	8	≥ 16
Fluoroquinolones			
Moxifloxacin	<2	4	≥ 8
Miscellaneous Agents			
Chloramphenicol	≤ 8	16	≥ 32
Clindamycin	≤ 2	4	≥ 8
Metronidazole	≤ 8	16	≥ 32
Tetracycline	≤ 4	8	≥ 16
Tigecycline[a]	≤ 4	8	≥ 16

[a] Compiled from recommendations of FDA, Citron and Hecht (2003) and Clinical Laboratory Standards Institute/National Committee for Clinical laboratory Standards (2004).

Prevotella and *Porphyromonas*

Data on *Prevotella* and *Porphyromonas* species susceptibility is limited but comparatively both genera are more susceptible than the *B. fragilis* group. Currently, about 50% of *Prevotella* spp. strains are resistant to penicillin and ampicillin due to β-lactamase production, while susceptibility to piperacillin, cefoxitin, and cefotetan ranges from 70 to 90% in most published studies (Hecht and Osmolski, 2003; Koeth et al., 2004; Matto et al., 1990). Approximately 8% of *Porphyromonas* spp. strains produce β-lactamase (Tanaka et al., 1999) which can rise to 17% of strains recovered from serious pelvic infections (Pelak et al., 2002). Susceptibility of *Porphyromonas* isolates are usually reported in general surveys but currently β-lactamase production is considered rare (Goldstein et al., 1995a). Both genera are nearly uniformly susceptible to carbapenems, metronidazole, and chloramphenicol, although clindamycin resistance has been observed in a minority of strains (Goldstein et al., 1995a).

Other Gram-Negative Bacilli

Penicillin resistance in *Fusobacterium* nucleatum has been observed; increasing resistance rates in children is due to β-lactamase production and related to age, day care attendance and exposure to antimicrobial agents (Nyfors et al., 2003). Fatal sepsis due to a β-lactamase producing strain occurred in a compromised patient (Goldstein et al. 1995b). In general, >90% of *Fusobacterium* spp. are susceptible to cephalosporins and cephamycins (Goldstein et al., 1995a; Koeth et al., 2004). *Bilophila wadsworthia* frequently produces β-lactamase, and therefore is resistant to penicillin and ampicillin but generally susceptible to clindamycin, cefoxitin, β-lactam/β-lactamase inhibitor combinations, carbapenems, and metronidazole (Baron et al., 1993). *Sutterella wadsworthensis*, may demonstrate resistance to clindamycin, piperacillin, and/or metronidazole (Molitoris et al., 1997).

6.3.3.2 Gram Positives

Non-sporeforming Gram-Positive Bacilli

The *Eubacterium* group, *Actinomyces* , *Propionibacterium* , and *Bifidobacterium* are usually susceptible to β-lactam agents, including the penicillins, cephalosporins, carbapenems, and β-lactam/β-lactamase inhibitor combinations. *Lactobacillus* spp. show species variations resulting in widely variably susceptible patterns to cephalosporins and other agents; penicillin and ampicillin are often active (Goldstein et al., 2003a,b). There are no breakpoints for vancomycin and anaerobes, but this agent has very good in vitro activity against all *Propionibacterium* spp., *Actinomyces* spp., *Eubacterium* group species, peptostreptococci, and some *Lactobacillus* spp., while much less active against *L. casei* and several other species (Goldstein et al., 2003a). Newer Gram-positive spectrum agents such as linezolid, daptomycin, dalbavancin, telavancin, and ramoplanin also exhibit excellent in vitro activity against most anaerobic gram-positive species (Behra-Miellet et al., 2003a; Citron et al., 2003; Goldstein et al., 2003a,b, 2004). Most non-spore forming gram-positive anaerobes are resistant to metronidazole. Moxifloxacin has exhibited good activity against *Actinomyces* species, including *A. ondontolyticus* and *A. viscosus*, *Eubacterium alactolyiticum* , and *E. limosum* and a variety of lactobacilli with MIC$_{90}$s $\leq$2 µg/ml (Goldstein et al., 2006a,b). While most were susceptible to moxifloxacin, there was strains variability and resistance in some isolates of *E. lentum* and *Lactobacillus plantarum* (Goldstein et al., 2006a,b).

Spore Forming Gram-Positive Bacilli

Clostridium perfringens is generally very susceptible to most anti-anaerobic antimicrobials, as well as fluoroquinolones (Behra-Miellet et al., 2003b; Hecht and Osmolski, 2003). However, non-perfringens *Clostridium* spp. as *C. clostridioforme* and *C. innocuum* and *C. difficile* have variable susceptibility (Credito and Appelbaum, 2004; Goldstein et al., 2003a; Teng et al., 2002) with potential resistance to clindamycin, fluoroquinolones, and β-lactams; while chloramphenicol and metronidazole remain active. *C. difficile* may be resistant to many β-lactams, including cephalosporins, fluoroquinolones, and clindamycin, but retain susceptibility to metronidazole and vancomycin; *C. innocuum* has vancomycin MICs of 8–32 µg/ml (Goldstein et al., 2003a).

Gram-Positive Cocci

These gram-positive cocci are highly susceptible to all β-lactams, β-lactam/β-lactamase inhibitors, cephalosporins, carbapenems, chloramphenicol, and metronidazole (Goldstein et al., 2003a,b; Koeth et al., 2004; Teng et al., 2002). Fluoroquinolone resistance is increasing among the species isolated from skin and soft tissue infections (Citron, Abstract E-1440, Interscience Conference on Antimicrobial Agents and Chemotherapy, 2005) and some strains may be resistant to clindamycin. The rarity of metronidazole resistance in *Peptostreptococcus* spp. should prompt further attention to correct identification, because anaerobic strains of streptococci in the milleri group are always resistant to metronidazole.

6.3.4 Resistance to β-Lactam Antibiotics

Resistance to β-lactam antibiotics may occur by one of three major mechanisms: inactivating enzymes particularly β-lactamases (BLAs), which include penicillinases and more commonly, cephalosporinases, low affinity penicillin binding proteins (PBPs), or decreased permeability (porin channel alterations). BLAs are the most common mechanism causing resistance to β-lactam antibiotics in anaerobes and are common amongst all the *B. fragilis* group species and the *Prevotella* species. The cephalosporinases are usually of the type 2e class but are inhibited by the three β-lactamase inhibitors, sulbactam, clavulanic acid, and tazobactam. Each cephalosporin agent may have either a class or individual inhibitor enzyme that will inactive it. Cefoxitin inactivating enzymes occur in many *B. fragilis* group species (Rogers et al., 1993). The zinc metalloenzymes, also known as carbapenemases, are encoded by either *ccrA* or *cfiA* genes of *B. fragilis* group (97) and are active against all β-lactam antibiotics including the carbapenems. In addition, current BLA inhibitors do not inactivate them. Although carbapenem resistance occurs is <1% of isolates in the US, up to 3% of *Bacteroides* strains carry one of the genes. Selective pressure caused by a promoter (contained in an insertion sequence) that has inserted upstream of the *ccrA* or *cfiA* genes (Edwards and Reed, 2000; Polglajen et al., 1994) has been experimentally noted to induced to higher level resistance. Soki et al. (2004) reported on the molecular characterization of 15 strains of imipenem-resistant, *cifA*-positive *B. fragilis* from the United States, Hungary, and Kuwait, and noted that some isolates possessed "novel activation mechanisms" which suggests more than one mechanism of activation exists.

Production of BLAs by other anaerobic bacteria has been generally less well studied, but strains of *Clostridium*, *Porphyromonas*, and *Fusobacterium* express resistance by one or more of these enzymes. Penicillinase producing *Fusobacterium* species and *Clostridium* species express penicillinases that are typically inhibited by clavulanic acid, although exceptions amongst some *Clostridium* sp. occur (Appelbaum et al., 1994; Rasmussen et al., 1997).

Uncommonly, other mechanisms of resistance to β-lactam antibiotics occur but are less well studied and include changes in the OMP/porin channels and decreased PBP affinity. Decreased binding to PBP2 or PBP1 complex has been reported in rare clinical isolates in cefoxitin resistance among *B. fragilis* (Fang et al., 2002). *B. fragilis* has been reported to have 16 efflux pumps (BmeB1-16) from the resistance nodulation division (RND) several of which are prevalent in wild type strains (Pumbwe et al., 2006). Mutant strains have been capable of over-expression of one or more of these efflux pumps resulting in antimicrobial resistance. Such alterations in pore-forming proteins has been associated with high MICs to ampicillin/sulbactam in some strains of *B. fragilis* (Wexler, 2002; Wexler and Halebian, 1990). Specifically, the bmeB efflux pump has been noted to contribute to "high-level clinically relevant resistance to beta lactams" (Pumbwe et al., 2006).

6.3.4.1 β-lactam Antibiotics

Amongst the *B. fragilis* group species, resistance to β-lactam agents such as the penicillins and cephamycins such as cefoxitin and cefotetan, is fairly widespread. Third generation cephalosporins exhibit consistently poor activity against the *B. fragilis* group species (Snydman et al. 1999). More than 95% of *B. fragilis* group species are resistant to penicillin G by virtue of BLA production. In contrast, cefoxitin retains activity against most *B. fragilis* group members, although resistance has ranged between 8–22% over the period of 1987–2000. Cefotetan activity is very similar to that of cefoxitin against *B. fragilis*, but is much less potent against other members of the *B. fragilis* group (30–87% resistant) (Snydman et al., 2002). Resistance to piperacillin, the most active ureidopenicillin against anaerobes, is also now widespread among members of the *B. fragilis* group (average 25% resistant). Resistance is also found among non-*Bacteroides* anaerobes (Aldridge et al., 2001; Hecht and Vedantam, 1999).

The β-lactam/ BLA inhibitor combinations and carbapenems have maintained their excellent activity. The three combination agents of ampicillin/sulbactam, ticarcillin/clavulanate, and piperacillin/tazobactam are generally active against members of the *B. fragilis* group (Citron and Hecht, 2003; Snydman et al., 2002a). However, species to species variation in susceptibility has been identified, with the many non-BLA producing *B. distasonis* demonstrating elevated MICs at or approaching the susceptible breakpoint (Snydman et al., 2002a). Resistance rates for piperacillin–tazobactam are generally <1% for all *B. fragilis* group species (Snydman et al., 2007). In contrast, the rate of ampicillin–sulbactam resistance has risen to 20% for *B. distasonis* isolates in 2002–2004 but has remained low for the other *B. fragilis* group species. The carbapenems, imipenem, meropenem, and ertapenem, remain very potent against all members of the *B. fragilis* group, with only rare (<0.1%) resistance identified (Aldridge et al., 2001; Hecht et al., 1993; Snydman et al., 2002a, 2007). Geometric mean MICs for ertapenem for *B. distasonis*, *B. thetaiotaomicron* and *B. ovatus* have been reported to be onefold dilution higher (2 μg/ml) than those for imipenem and meropenem (<1μg/ml) (Snydman et al., 2007) for 2004. In a study of anaerobes isolated from pediatric intra-abdominal infections, Goldstein et al. (2006c) reported that all *Bacteroides* isolates produced β-lactamases and were susceptible to carbapenems and β-lactam/β-lactamase inhibitor combinations but cefoxitin was poorly active against *B. thetaiotaomicron* isolates.

Resistance to β-lactam agents among non-*B. fragilis* anaerobes is generally low. *Prevotella* sp. may also produce BLAs, resulting in >50% resistant to penicillin. Aldridge et al. (2001) has reported penicillin resistance for *Fusobacterium* sp., *Porphyromonas* sp., and *Peptostreptococcus* sp. at 9, 21, and 6%, respectively. Resistance to cefoxitin, cefotetan, β-lactam/ BLA inhibitor combinations, and carbapenems was 0% in the same survey, with the exception of *Peptostreptococcus* sp. and *Porphyromonas* sp. (4 and 5% resistance to ampicillin/sulbactam, respectively). Goldstein et al. (2006c) reported that several *Prevotella and Porphyromonas* species isolated from pediatric intra-abdominal infections produced β-lactamases.

6.3.4.2 Clindamycin Resistance

Clindamycin resistance is mediated by a macrolide–lincosamide–streptogram (MLS) type 23S methylase (Jimenez-Daz et al., 1992), typically encoded by one of several *erm* genes that are regulated and expressed at high levels (Roberts et al., 1999; Whittle et al., 2002). Some isolates may contain *erm* genes that result in only slightly elevated MICs that do not get classified as resistant. Some of these strains can be detected by testing for erythromycin resistance; however, there is currently no specific recommendation to screen for clindamycin-resistance using erythromycin. Laboratory transfer of *erm* genes by conjugation in *B. fragilis* group organisms is reported, and may explains the rapid emergence of this resistance phenotype (Privitera et al., 1979; Tally et al., 1979; Welch et al., 1979). Because of increasing worldwide, and geographically variable resistance, clindamycin is not recommended as empiric therapy for intra-abdominal infections (Dalmau et al., 1997; Solomkin et al., 2003).

Clindamycin resistance among *B. fragilis* group species has increased in the last two decades (Snydman et al., 2007). In an 8-year study (1997–2004) 19.3% of 2721 *B. fragilis* isolates were clindamycin resistant compared to 29.6% for *B. distasonis*, 33.4% for *B. ovatus*, 33/3% for *B. thetaiotaomicron* and 35.6% for *B. vulgatus* strains. This is compared to only 3% clindamycin resistance in 1987 (Hecht and Vedantam, 1999; Snydman et al., 1999). In a small study of pediatric intra-abdominal isolates, there was only 6% clindamycin-resistance in *B. fragilis* isolates compared to 80% for *B. thetaiotaomicron* and 45% for other *B. fragilis* group isolates. For many non-*Bacteroides* anaerobes, resistance has also increased, albeit not as significantly as the *B. fragilis* group (Hecht and Vedantam, 1999). A recent report found up to 10% resistance for *Prevotella* sp., *Fusobacterium* sp., *Porphyromonas* sp., and *Peptostreptococcus* sp., with higher rates for some *Clostridium* sp. (especially *C .difficile*) (Ackermann et al., 2003; Aldridge et al., 2001). Clindamycin resistance of *P. acnes* isolates has become more frequent and has been associated with prior therapy for acne (Nord and Oprica, 2006). Of note is that EUCAST and CLSI breakpoints are different for *P. acnes.*

6.3.4.3 Metronidazole Resistance

Metronidazole resistance among *B. fragilis* group isolates has now been reported worldwide (Breuil et al., 1989; Urban et al., 2002). Metronidazole resistance among Gram-positive anaerobes is far more common, especially for most isolates of *P. acnes* and *Actinomyces* sp. Metronidazole-resistant *B. fragilis* group isolates, although rare, may carry one of six known *nim* genes (*nim* A–F) either on the chromosome or on a mobilizable plasmid that appear to encode a nitroimidazole reductase that converts 4- or 5-nitroimidazole to 4- or 5-aminoimidazole preventing the formation of toxic nitroso residues necessary for the agents' activity (Carlier et al., 1997; Haggoud et al., 1994). Gal and Brazier (2004) found 50/206 (24%) human isolates submitted to their anaerobic reference laboratory possessed *nim* genes and resulted in MICs of 1.5 to >256 µg/ml for metronidazole, including 16 isolates with MICs $\geq$ 32 µg/ml. This suggested incomplete mobilization of *nim* gene associated resistance. High level expression of the *nim* genes requires an insertion sequence with a promoter, similar to that of carbapenem resistance (Trinh et al., 1995). Gal and Brazier (2004) also speculated that other mechanisms of resistance are likely to occur and that prolonged exposure to metronidazole "can select for therapeutic resistance."

Table 6.4 Summary of clinical reports of metronidazole resistance in *Bacteroides fragilis* group species

References	Country Isolate	Source	Demographics		Prior flagyl therapy
Katsandri et al. (2006)	Greece	B. vulgatus nim −Abdominal drainage	75 years old male	gastric Cancer	yes
Schapiro et al. (2004)	US	B. fragilis nim +ankle wound	60 years old woman	fracture	no
Snydman et al. (in press)/Hecht, 2000 [plus personal communication]	US	B fragilis Abdominal		Crohns disease	yes
Turner et al. (1995)	UK	B fragilis blood/ empyema			
Wareham et al. (2005)	UK	B. fragilis blood/ abdominal	48 years old female	pancreatitis	yes

The mechanism of metronidazole-resistance for non-*Bacteroides* anaerobes is unknown. Mutations in the *rdxA* gene (van der Wouden et al., 2001), or due to flavin oxidoreductase (frxA), ferridoxin-like proteins (*fdxA*, *fdxB*) and pyruvate oxidoreductase (*porA*, *porB*) have been postulated (Breuil et al., 1989; Carlier et al., 1997).

Clinical reports of cases of metronidazole resistance are summarized in Table 6.4. Snydman (2007) has noted a single isolate of *B. fragilis* in 2002 with an MIC of 64 μg/ml. Schapiro et al. (2004) reported a case of 60-year-old female with a metronidazole resistant *B. fragilis* (*nim* A) isolated from an ankle abscess. The patient had traveled through Ghana and sustained a compound fracture of her ankle and had previously received multiple antibiotics, but not metronidazole. Katsandri et al. (2006) reported a case of a 75 year old male with gastric cancer who had been previously treated with metronidazole as well as other agents that grew a metronidazole, multidrug-resistant *B. vulgatus* (MIC > 256 μg/ml) from abdominal drainage fluid. The isolate was negative for *nim* class genes suggesting another mechanism of metronidazole resistance was present but the isolate did carry the *cif*A gene. Several other unpublished cases are also known to have occurred.

6.3.4.4 Fluoroquinolone Resistance

Fluoroquinolone resistance among anaerobes has been a topic of interest and controversy. Trovafloxacin was FDA approved for treatment of anaerobic infections but has limited clinical utility due to hepatic toxicity concerns. Trovafloxacin resistance among *Bacteroides* species increased from 3 to 8% during the period 1994–1996, prior to the antibiotics release in 1997 (Snydman et al., 1999, 2002a,b) and resistance increased to 25% in 2001 (Golan et al., 2003). Moxifloxacin has been approved by the FDA for the therapy of complicated skin and skin structure infections including those due to *B. fragilis* and for mixed intra-abdominal infections due to *B. fragilis, B. thetaiotaomicron*, Peptostreptococcus species and *Clostridium perfringens*.

Fluoroquinolone resistance among *Bacteroides* sp. has been attributed to either a mutation in the quinolone resistance determining region (QRDR) of the gyrase A gene (*gyrA*) from single or multiple mutations, or an alteration in efflux of the antibiotic (Oh et al., 2002; Onodera and Sato, 1999; Ricci and Piddock, 2003). High-level resistance may be secondary to both mechanisms in the same cell, although only a few strains have been tested to date. Both of these mechanisms appear to be responsible for the cross class resistance to newer quinolones.

In 2003, Golan et al. (2003) tested 4434 *B. fragilis* group isolates collected from 12 US medical centers between 1994 and 2001 and found that resistance to fluoro-quinolones varied by species and source of isolation with *B. vulgatus* isolates from decubitus ulcers to be the most resistant (71%). Moxifloxacin resistance rates varied from 17% for *B. fragilis* isolated from the female GU tract to 52% for all blood culture isolates (moxifloxacin MIC breakpoint, 4 µg/ml). Their most recent survey (Snydman et al., 2007) reported 27% of *B. fragilis*, 26% of *B. thetaiotaomicron*, 38% of *B. ovatus* and only 54.7% of *B. vulgatus* to be moxifloxacin resistant. In contrast, Goldstein et al. (2006a) noted 87% of *B. fragilis* and 87% of *B. thetaiotaomicron* strains isolated from intra-abdominal infections (2001–2004) to be susceptible to moxifloxacin. Overall, 86% 303/363) of *B. fragilis* group isolates and 417/450 of all other anaerobic genera and species, including *Fusobacterium*, *Prevotella*, *Porphyromonas*, *C. perfringens*, *Eubacterium*, and *Peptostreptococcus* species, were susceptible to <2 µg/ml of moxifloxacin. Wexler et al. (2000) tested 179 respiratory anaerobes and found only one strain of *C. clostridioforme* to be resistant. Edmiston et al. (2004) tested 550 anaerobes isolated from intra-abdominal and diabetic foot infections and reported 97% were susceptible to moxifloxacin. Factors potentially causing this disparity include variations in QC readings and readings of results, geographic variability with clonal populations or differences in susceptibility related to the sources of isolation and antimicrobial usage patterns. Consistent with this latter theory, Goldstein et al. (2006c) reported 41/42 *B. fragilis* group strains isolated from pediatric intra-abdominal infections were susceptible to moxifloxacin, an agent that is relatively contraindicated in children so that there is scant pediatric usage. One of fifteen *B. thetaiotaomicron* isolates was found to be resistant. *Fusobacterium canifelinum*, isolated from cat and dog bite wounds in humans, is intrinsically resistant to fluoroquinolones due to Serine79 replacement with leucine and Gly83 replacement with argenine on *gyrA* (Conrads et al., 2005).

6.3.4.5 Tetracycline Resistance

Tetracycline resistance is almost universal among *Bacteroides* species, *Prevotella* species and many other anaerobes, limiting its use in therapy (Nikolich et al., 1992). Several genes encoding resistance have been identified among various anaerobes, which encode protective proteins resulting in protection of the ribosomes. Tetracycline resistance and the inducible transfer of this resistance determinant occur upon exposure to low levels of tetracycline. The emergence of tetracycline resistance in *Propionibacterium acnes* has been correlated with prior therapy (Nord and Oprica, 2006). Other antimicrobials, such as doxycycline and minocycline are more active than tetracycline. Tigecycline, a new glycylcycline derivative of minocycline, has been noted to be active against anaerobic bacteria (Goldstein et al., 2006b; Jacobus et al., 2004). Jacobus et al. (2004) reported that 89.7% of 831 *B. fragilis* group isolates were susceptible to <8 µg/ml of tigecycline and that *B. distasonis* isolates were the most resistant. Snydman et al. (2007) noted 5% of *B. fragilis*, 3.6% of *B. thetaiotaomicron*, 3.3% of *B. ovatus* and 7.2% of the unusual *B. fragilis* group species were resistant to tigecycline. Goldstein et al. (2006b) tested 396 unusual anaerobes and found all gram-positive anaerobes and 228/232 gram-negative anaerobes to be susceptible to <1µg/ml of tigecycline; only one isolate of *Prevotella oralis* had an MIC of 8 µg/ml. Tigecycline has been approved by the FDA for therapy of complicated skin and skin structure infections including those due to *B. fragilis* and intra-abdominal infections including those due to *B. fragilis*, *B. thetaiotaomicron*, *B. uniformis*, *B. vulgatus*, *C. perfringens* and *Peptostreptococcus micros*.

6.3.4.6 Resistance to Other Agents

Resistance to aminoglycosides is universal among anaerobes, due to the lack of uptake by bacteria under anaerobic conditions, and subsequent failure to attach to their ribosome targets (Bryan et al., 1979). Chloramphenicol resistance is very rare although there is clustering of MICs for some strains around the breakpoint. When resistance is found, it is due to inactivation of the drug by nitroreduction or acetyltransferase (Britz and Wilkinson, 1978); this agent is also rarely used in the clinical setting due to potential hematopoetic toxicity (Rasmussen et al., 1997). Sulfamethoxazole–trimethoprim (TMP–SMX) is a folate synthesis inhibitor and has no activity against anaerobes.

6.3.5 Conclusion

Overall, resistance in anaerobic bacteria is generally rising but goes unrecognized. Increased education of microbiologists in proper isolation and identification of anaerobes and increased performance of AST on a local level for individual patient care could improve the situation. Hopefully, as molecular probes and techniques evolve, they will simplify and enhance reporting of anaerobic bacteria, and will provide clinicians with reports that will ultimately enhance patient care.

References

Ackermann, G., Degner, A., Cohen. S.H., Silva, J., Jr and Rodloff, A.C., (2003), Prevalence and association of macrolide–lincosamide–streptogramin B (MLS(B)) resistance with resistance to moxifloxacin in Clostridium difficile. *J Antimicrob Chemother* **51**, 599–603.
Aldridge, K.E., Ashcraft, D., Cambre, K., Pierson, C.L., Jenkins, S.G. and Rosenblatt, J.E., (2001), Multicenter survey of the changing in vitro antimicrobial susceptibilities of clinical isolates of

Bacteroides fragilis group, Prevotella, Fusobacterium, Porphyromonas, and Peptostreptococcus species. *Antimicrob Agents Chemother*, **45**, 1238–1243.

Aldridge, K.E. and Schiro, D.D., (1994), Major methodology-dependent discordant susceptibility results for *Bacteroides fragilis* group isolates but not other anaerobes. *Diagn Microbil Infect Dis*, **20**, 135–142.

Appelbaum, P.C., Spangler, S.K., Pankuch, G.A., Philippon, A., Jacobs, M.R., Goldstein, E.J.C. and Citron, D.M., (1994), Characterization of a beta-lactamase from *Clostridium clostridioforme. J Antimicrob Chemother*, **33**, 33–40.

Baron, E.J., Ropers, G., Summanen. P. and Courcol, R.J., (1993), Bactericidal activity of selected antimicrobial agents against *Bilophila wadsworthia* and *Bacteroides gracilis. Clin Infect Dis*, **16**, S339–S343.

Behra-Miellet, J., Calvet, L. and Dubreuil, L., (2003a), Activity of linezolid against anaerobic bacteria. *Int J Antimicrob Agents*, **22**, 28–34.

Behra-Miellet, J., Dubreuil, L. and Jumas-Bilak, E., (2003b), Antianaerobic activity of moxifloxacin compared with that of ofloxacin, ciprofloxacin, clindamycin, metronidazole and beta-lactams. *Int J Antimicrob Agents*, **20**, 366–374.

Betriu, C., Culebras, E., Gomez, M., Rodriguez-Avial, I. and Picazo, J.J., (2005), In vitro activity of tigecycline against *Bacteroides* species. *J Antimicrob Chemother*, **56**, 349–352.

Breuil, J., Dublanchet, A., Truffaut, N. and Sebald, M., (1989), Transferable 5-nitroimidazole resistance in the *Bacteroides fragilis* group. *Plasmid*, **21**, 151–154.

Britz, M.L. and Wilkinson, R.G., (1978), Chloramphenicol acetyltransferase of *Bacteroides fragilis. Antimicrob Agents Chemother*, **14**, 105–111.

Bryan, L.E., Kowand, S.K. and Van Den Elzen, H.M., (1979), Mechanism of aminoglycoside antibiotic resistance in anaerobic bacteria: Clostridium perfringens and *Bacteroides fragilis. Antimicrob Agents Chemother*, **15**, 7–13.

Carlier, J.P., Sellier, N., Rager, M.N. and Reysset, G., (1997), Metabolism of a 5-nitroimidazole in susceptible and resistant isogenic strains of *Bacteroides fragilis. Antimicrob Agents Chemother*, **41**, 1495–1499.

Citron, D.M. and Hecht, D.W., (2003), Susceptibility test methods: anaerobic bacteria. In P.R. Murray, E.J. Baron, J.H. Jorgensen, M.A. Pfaller and R.H. Yolken, editors, 8th edition. Washington, D.C., ASM Press, pp. 1141–1148.

Citron, D.M., Merriam, C.V., Tyrrell, K.L., Warren, Y.A., Fernandez, H. and Goldstein, E.J.C., (2003), In vitro activities of ramoplanin, teicoplanin, vancomycin, linezolid, bacitracin, and four other antimicrobials against intestinal anaerobic bacteria. *Antimicrob Agents Chemother*, **47**, 2334–2338.

Citron, D.M., Ostovari, M.I., Karlsson, A. and Goldstein, E.J.C., (1991), Evaluation of the E test for susceptibility testing of anaerobic bacteria. *J Clin Microbiol*, **29**, 2197–2203.

Clinical and Laboratory Standards Institute (CLSI)/National Committee for Clinical Laboratory Standards (NCCLS), (2004), Methods for antimicrobial susceptibility testing of anaerobic bacteria. Approved standard M11-A6. Wayne, PA, Clinical and Laboratory Standards Institute.

Clinical and Laboratory Standards Institute (CLSI)/National Committee for Clinical Laboratory Standards (NCCLS), (2007), Methods for antimicrobial susceptibility testing of anaerobic bacteria. Approved standard M11-A7. Wayne, PA, Clinical and Laboratory Standards Institute.

Conrads, G., Citron, D.M. and Goldstein, E.J.C., (2005), Genetic determinant of intrinsic quinolone resistance in *Fusobacterium canifelinum. Antimicrob Agents Chemother*, **49**, 434–437.

Cormican, M.G., Erwin, M.E. and Jones, R.N., (1996), False resistance to metronidazole by E-test among anaerobic bacteria investigations of contributing test conditions and medium quality. *Diagn Microbiol Infect Dis*, **24**, 117–119.

Credito, K.L. and Appelbaum, P.C., (2004), Activity of OPT-80, a novel macrocycle, compared with those of eight other agents against selected anaerobic species. *Antimicrob Agents Chemother*, **48**, 4430–4434.

Dalmau, D., Cayouette, M., Lamothe, F., Vincelette, J., Lachance, N., Bourgault, A.M., Gaudreau, C. and Turgeon, P.L., (1997), Clindamycin resistance in the *Bacteroides fragilis* group: association with hospital-acquired infections. *Clin Infect Dis*, **24**, 874–877.

Edmiston, C.E., Krepel, C.J., Seabrook, G., Somberg, L.R., Nakeeb, A., Cambria, R.A. and Town, J.B., (2004), *In vitro* activities of moxifloxacin against 900 aerobic and anaerobic surgical isolates from patients with intra-abdominal and diabetic foot infections. *Antimicrob Agents Chemother*, **48**, 1012–1016.

Ednie, L.M., Jacobs, M.R. and Appelbaum, P.C., (1998), Activities of gatifloxacin compared to those of seven other agents against anaerobic organisms. *Antimicrob Agents Chemother*, **42**, 2459–2462.

Edwards, R. and Read, P.N., (2000), Expression of the carbapenemase gene (cfiA) in *Bacteroides fragilis. J Antimicrob Chemother*, **46**, 1009–1012.

Fang, H., Edlund, C., Nord, C.E. and Hedberg, M., (2002), Selection of cefoxitin-resistant *Bacteroides thetaiotaomicron* mutants and mechanisms involved in beta-lactam resistance. *Clin Infect Dis*, **35**, S47–S53.

Finegold, S.M., (1994), Review of early research in anaerobes. *Clin Infect Dis*, **18**, (Suppl. 4), S248–S249.

Gal, M. and Brazier, J.S., (2004), Metronidazole resistance in *Bacteroides* spp. carrying *nim* genes and the selection of slow-growing metronidazole-resistant mutants. *J Antimicrob Chemother*, **54**, 109–116.

Golan, Y., McDermott, L.A., Jacobus, N.V., Goldstein, E.J.C., Finegold, S.M., Harrell, L.J., Hecht, D.W., Jenkins, S.G., Pierson, C., Venezia, R., Rihs, J., Iannini, P., Gorbach, S.L. and Snydman, D.R., (2003), Emergence of fluoroquinolone resistance among *Bacteroides* species. *J Antimicrob Chemother*, **52**, 208–213.

Goldstein, E.J.C., (1995), Anaerobes under assault: from a cottage industry to the industrial revolution of medicine & microbiology. *Clin Infect Dis* **20** (Suppl. 2),112–116.

Goldstein, E.J.C, Citron, D.M., Cole, R.E., Rangel, D.M., Seid, A.S. and Ostovari, M.I., (1990), Cefoxitin in the treatment of aerobic/anaerobic infections: prospective correlation of in vitro susceptibility methods with clinical outcome. *Hospital Practice*, Symposium Supplement, **25** (Suppl. 4), 38–45.

Goldstein, E.J.C., Citron, D.M. and Goldman, R.J., (1992), National hospital survey of anaerobic culture and susceptibility testing methods: results and recommendations for improvement. *J Clin Microbiol*, **30**, 1529–1534.

Goldstein, E.J.C., Citron, D.M., Goldman, R.J., Claros, M.C. and Hunt-Gerardo, S., (1995a), United States national hospital survey of anaerobic culture and susceptibility methods. *Anaerobe*, **1**, 309–314.

Goldstein, E.J.C., Summanen, P.H., Citron, D.M., Rosove, M.H. and Finegold. S.M., (1995b), Fatal sepsis due to a beta-lactamase-producing strain of *Fusobacterium nucleatum* subspecies *polymorphum. Clin Infect Dis*, **20**, 797–800.

Goldstein, E.J.C., Citron, D.M., Merriam, C.V., Warren, Y. and Tyrrell, K.L., (2000), Comparative In vitro activities of ertapenem (MK-0826) against 1,001 anaerobes isolated from human intra-abdominal infections. *Antimicrob Agents Chemother*, **44**, 2389–2394.

Goldstein, E.J.C., Citron, D.M., Merriam, C.V., Warren, Y., Tyrrell, K. and Fernandez, H.T., (2003a), In vitro activities of dalbavancin and nine comparator agents against anaerobic gram-positive species and corynebacteria. *Antimicrob Agents Chemother*, **47**, 1968–1971.

Goldstein, E.J.C., Citron, D.M., Merriam, C.V., Warren, Y.A., Tyrrell, K.L. and Fernandez, H.T., (2003b), In vitro activities of daptomycin, vancomycin, quinupristin–dalfopristin, linezolid, and five other antimicrobials against 307 gram-positive anaerobic and 31 Corynebacterium clinical isolates. *Antimicrob Agents Chemother*, **47**, 337–341.

Goldstein, E.J.C., Citron, D.M., Merriam, C.V., Warren, Y.A,, Tyrrell, K.L. and Fernandez, H.T., (2004), In vitro activities of the new semisynthetic glycopeptide telavancin (TD-6424), vancomycin, daptomycin, linezolid, and four comparator agents against anaerobic gram-positive species and *Corynebacterium* spp. *Antimicrob Agents Chemother*, **48**, 2149–2152.

Goldstein, E.J.C., Citron, D.M., Warren, Y.A., Merriam, C.V. and Fernandez, H., (2006a), In vitro activity of moxifloxacin against 923 anaerobes isolated from human intra-abdominal infections. *Antimicrob Agents Chemother*, **50**, 148–155.

Goldstein, E.J.C., Citron, D.M., Warren, Y.A., Tyrrell, K.L., Merriam, C.V. and Fernandez, H.T., (2006b), The comparative in vitro susceptibilities of 396 unusual anaerobic strains to tigecycline and eight other antimicrobial agents. *Antimicrob Agents Chemother*, **50**, 3507–3513.

Goldstein, E.J.C., Citron, D.M., Vaidya, S.A., Warren, Y.A., Tyrrell, K.L., Merriam, C.V. and Fernandez, H., (2006c), *In vitro* activity of 11 antibiotics against 74 anaerobes isolated from pediatric intra-abdominal infections. *Anaerobe*, **12**, 63–66.

Haggoud, A., Reysset, G., Azeddoug, H. and Sebald, M., (1994), Nucleotide sequence analysis of two 5-nitroimidazole resistance determinants from Bacteroides strains and of a new insertion sequence upstream of the two genes. *Antimicrob Agents Chemother*, **38**, 1047–1051.

Hecht, D.W., (2002), Evolution of anaerobe susceptibility testing in the United States. *Clin Infect Dis*, **35**, S28–S35.

Hecht, D.W., (2004), Prevalence of antibiotic resistance in anaerobic bacteria: worrisome developments. *Clin Infect Dis*, **39**, 92–97.

Hecht, D.W., (2006), Anaerobes: antibiotic resistance, clinical significance, and the role of susceptibility testing. *Anaerobe*, **12**, 115–121.

Hecht, D.W. and Osmolski, J.R., (2003), Activities of garenoxacin (BMS-284756) and other agents against anaerobic clinical isolates. *Antimicrob Agents Chemother* **47**, 910–916.

Hecht, D.W., Osmolski, J.R. and O'Keefe, J.P., (1993), Variation in the susceptibility of Bacteroides fragilis group isolates from six Chicago hospitals. *Clin Infect Dis*, **16** (Suppl 4), S357–S360.

Hecht, D.W. and Vedantam, G., (1999), Anaerobe resistance among anaerobes: what now? *Anaerobe*, **5**, 421–429.

Hedberg, .M. and Nord, C.E., (2003), Antimicrobial susceptibility of *Bacteroides fragilis* group isolates in Europe. *Clin Microbiol Infect*, **9**, 475–488.

Heseltine, P.N., Yellin, A.E., Applaman, M.D., Gill, M.A., Chenella, F.C., Kern, J.W. and Berne, T.V., (1983), Perforated and gangenrous appendicitis: an analysis of antibiotic failures. *J Infect Dis*, **148**, 322–329.

Hoellman, D.B., Kelly, L.M., Jacobs, M.R. and Appelbaum, P.C., (2001), Comparative antianaerobic activity of BMS 284756. *Antimicrob Agents Chemother*, **45**, 589–592.

Jacobus, N.V., McDermott, L.A., Ruthazer, R. and Snydman, D.R., (2004), In vitro activities of tigecycline against the *Bacteroides fragilis* group. *Antimicrob Agents Chemother*, **48**, 1034–1036.

Jimenez-Diaz, A., Reig, M., Baquero, F. and Ballesta, J.P., (1992), Antibiotic sensitivity of ribosomes from wild-type and clindamycin resistant *Bacteroides vulgatus* strains. *J Antimicrob Chemother*, **30**, 295–301.

Katsandri, A., Paparaskevas, J., Pantazatou, A., Petrikkos, G.L., Thomopoulos, G., Houhoula, D.P. and Avlamis, A., (2006), Two cases of infections due to multidrug-resistant *Bacteroides fragilis* group strains. *J Clin Microbiol*, **44**, 3465–3467.

Koeth, L.M., Good, C.E., Appelbaum, P.C., Goldstein, E.J.C, Rodloff, A.C., Claros, M. and Dubreuil, L.J., (2004), Surveillance of susceptibility patterns in 1297 European and US anaerobic and capnophilic isolates to co-amoxiclav and five other antimicrobial agents. *J Antimicrob Chemother*, **53**, 1039–1044.

Matto, J., Asikainen, S., Vaisanen, M.L., Von Troil-Linden, B., Kononen, E., Saarela, M., Salminen, K., Finegold, S.M and Jousimies-Somer, H., (1999), Beta-lactamase production in *Prevotella intermedia*, *Prevotella nigrescens*, and *Prevotella pallens* genotypes and in vitro susceptibilities to selected antimicrobial agents. *Antimicrob Agents Chemother*, **43**, 2383–2388.

Molitoris, E., Wexler, H.M. and Finegold, S.M., (1997), Sources and antimicrobial susceptibilities of *Campylobacter gracilis* and *Sutterella wadsworthensis*. *Clin Infect Dis*, **25**, S264–S265.

National Committee for Clinical Laboratory Standards (NCCLS), (1993), *Methods for antimicrobial susceptibility testing of anaerobic bacteria*, 3rd edition. Approved standard M11-A3. Villanova, PA, National Committee for Clinical Laboratory Standards.

National Committee for Clinical Laboratory Standards (NCCLS), (1997), *Methods for antimicrobial susceptibility testing of anaerobic bacteria*, 4th edition. M11-A4. Villanova, PA, NCCLS.

National Committee for Clinical Laboratory Standards (NCCLS), (2001), *Methods for antimicrobial susceptibility testing of anaerobic bacteria*, 5th edition. Approved standard M11-A5. Wayne, PA, National Committee for Clinical Laboratory Standards.

Nguyen, M.H., Yu, V.L., Morris, A.J. McDermott, L., Wagener, M.W., Harrell, L. and Snydman, D.R., (2000), Antimicrobial resistance and clinical outcome of Bacteroides bacteremia: findings of a multicenter prospective observational trial. *Clin Infect Dis*, **30**, 870–876.

Nikolich, M.P., Shoemaker, N.B. and Salyers, A.A., (1992), A *Bacteroides* tetracycline resistance gene represents a new class of ribosome protection tetracycline resistance. *Antimicrob Agents Chemother* **36**, 1005–1012.

Nord, C.E. and Oprica, C., (2006), Antibiotic resistance in *Propionibacterium acnes*, microbiological and clinical aspects. *Anaerobe*, **12**, 207–210.

Nyfors, S., Kononen, E., Syrjanen, R., Komulainen, E. and Jousimies-Somer. H., (2003), Emergence of penicillin resistance among *Fusobacterium nucleatum* populations of commensal oral flora during early childhood. *J Antimicrob Chemother*, **51**, 107–112.

Oh, H., Hedberg, M. and Edlund, C., (2002), Efflux-mediated fluoroquinolone resistance in the *Bacteroides fragilis* group. *Anaerobe*, **8**, 277–282.

Onodera, Y. and Sato, K., (1999), Molecular cloning of the gyrA and gyrB genes of *Bacteroides fragilis* encoding DNA gyrase. *Antimicrob Agents Chemother*, **43**, 2423–2429.

Pelak, B.A., Citron, D.M., Motyl, M., Goldstein, E.J.C., Woods, G.L. and Teppler, H., (2002), Comparative in vitro activities of ertapenem against bacterial pathogens from patients with acute pelvic infection. *J Antimicrob Chemother*, **50**, 735–741.

Polglajen, I., Breuil, J. and Collatz, E., (1994), Insertion of a novel DNA sequence, 1S1186, upstream of the silent carbapenemase gene cfiA, promotes expression of carbapenem resistance in clinical isolates of *Bacteroides fragilis*. *Mol Microbiol*, **12**, 105–114.

Privitera, G., Dublanchet, A. and Sebald, M., (1979), Transfer of multiple antibiotic resistance between subspecies of *Bacteroides fragilis*. *J Infect Dis*, **139**, 97–101.

Pumbwe, L., Chang, A., Smith, R.L. and Wexler, H.M., (2006), Clinical significance of overexpression of multiple RND-family efflux pumps in *Bacteroides fragilis* isolates. *J Antimicrob Chemother*, **58**, 543–548.

Rasmussen, B.A., Bush, K. and Tally, F.P., (1997), Antimicrobial resistance in anaerobes. *Clin Infect Dis*, **24** (Suppl 1), S110–S120.

Ricci, V. and Piddock, L., (2003), Accumulation of garenoxacin by *Bacteroides fragilis* compared with that of five fluoroquinolones. *J Antimicrob Chemother*, **52**, 605–609.

Roberts, M.C., Sutcliffe, J., Courvalin, P., Jensen, L.B., Rood, J. and Seppala, H., (1999), Nomenclature for macrolide and macrolide-lincosamide-streptogramin B resistance determinants. *Antimicrob Agents Chemother*, **43**, 2823–2830.

Roe, D.E., Finegold, S.M., Citron, D.M., Goldstein, E.J.C., Wexler, H.M., Rosenblatt, J.E., Cox, M.E., Jenkins, S.G. and Hecht D.W., (2002a), Multilaboratory comparison of anaerobe susceptibility results using 3 different agar media. *Clin Infect Dis*, **35**, S40–S46.

Roe, D.E., Finegold, S.M., Citron, D.M. Goldstein, E.J.C., Wexler, H.M., Rosenblatt, J.E., Cox, M.E., Jenkins, S.G. and Hecht D.W., (2002b), Multilaboratory comparison of growth characteristics for anaerobes, using 5 different agar media. *Clin Infect Dis*, **35**, S36–S39.

Rogers, M.B., Parker, A.C. and Smith, C.J., (1993), Cloning and characterization of the endogenous cephalosporinase gene, cepA, from *Bacteroides fragilis* reveals a new subgroup of Ambler class A beta-lactamases. *Antimicrob Agents Chemother*, **37**, 2391–2400.

Rosenblatt, J.E. and Gustafson, D.R., (1995), Evaluation of the Etest for susceptibility testing of anaerobic bacteria. *Diagn Microbiol Infect Dis*, **22**, 279–284.

Schapiro, J.M., Gupta, R. Stefanson, E., Fang, F.C. and Limaye, A.P., (2004), Isolation of metronidazole-resistant *Bacteroides fragilis* carrying the nimA nitroreductase gene from a patient in Washington State. *J Clin Microbiol*, **42**, 4127–4129.

Snydman, D.R., Cuchural, G.J., Jr, McDermott, L. and Gill, M., (1992), Correlation of various in vitro testing methods with clinical outcomes in patients with *Bacteroides fragilis* group infections treated with cefoxitin: a retrospective analysis. *Antimicrob Agents Chemother*, **36**, 540–544.

Snydman, D.R., Jacobus, N.V., McDermott, L.A., Ruthazer, R., Goldstein, E.J.C., Finegold, S.M., Harrell, L.J., Hecht, D.W. Jenkins, S.G., Pierson, C., Venezia, R., Rihs, J. and Gorbach, S.L., (2002a), National survey on the susceptibility of *Bacteroides fragilis* Group: report and analysis of trends for 1997–2000. *Clin Infect Dis*, **35**, S126–S134.

Snydman, D.R., Jacobus, N.V., McDermott, L.A., Ruthazer, R., Goldstein, E.J.C., Finegold, S.M., Harrell, L., Hecht, D.W., Jenkins, S., Pierson, C., Venezia, R., Rihs, J. and Gorbach, S.L., (2002b), In vitro activities of newer quinolones against *Bacteroides* group organisms. *Antimicrob Agents Chemother*, **46**, 3276–3279.

Snydman, D.R., Jacobus, N.V., McDermott, L.A., Ruthazer, R., Goldstein, E.J.C., Finegold, S.M., Harrell, L., Hecht, D.W., Jenkins, S., Pierson, C., Venezia, R., Rihs, J. and Gorbach, S.L., (2007), National survey on the susceptibility of *Bacteroides fragilis* Group: report and analysis of trends for 1997–2004, a US Survey. *Antimicrob Agents Chemother*, **51**, 1649–1655.

Snydman, D.R., Jacobus, N.V., McDermott, L.A., Supran, S., Cucheral, G.J., Jr, Finegold, S.M., Harrell, L.J., Hecht, D.W., Iannini, P., Jenkins, S.G., Pierson, C., Rihs, J. and Gorbach, S.L., (1999), Multicenter study of in vitro susceptibility of the *Bacteroides fragilis* group, 1995 to 1996, with comparison of resistance trends from 1990 to 1996. *Antimicrob Agents Chemother*, **43**, 2417–2422.

Soki, J., Fodor, E., Hecht, D.W., Edwards, R., Rotimi, V.O., Kerkes, I., Urban, E. and Nagy, E., (2004), Molecular characterization of imipenem-resistant, cifA-positive *Bacteroides* isolates from the USA, Hungary and Kuwait. *J Med Microbiol*, **53**, 413–419.

Solomkin, J.S., Mazuski, J.E., Baron, E.J., Sawyer, R.G., Nathens, A.B., DiPiro, J.T., Buchman, T., Dellinger, E.P., Jernigan, J., Gorbach, S. Chow, A.W., Bartlett, J. and the Infectious Diseases Society of America. (2003) Guidelines for the selection of anti-infective agents for complicated intra-abdominal infections. *Clin Infect Dis*, **37**, 997–1005.

Tally, F.P., Snydman, D.R., Gorbach, S.L. and Malamy, M.H., (1979), Plasmid-mediated, transferable resistance to clindamycin and erythromycin in *Bacteroides fragilis*. *J Infect Dis*, **139**, 83–88.

Tanaka, K., Kawamura, C., Fukui, K., Kato, H., Kato, N., Nakamura, T., Watanabe, K. and Ueno, K., (1999), Antimicrobial Susceptibility and [beta]-lactamase production of *Prevotella* spp. and *Porphyromonas* spp. *Anaerobe*, **5**, 461–463

Teng, L.J., Hsueh, P.R., Tsai, J.C., Liaw, S.J., Ho, S.W. and Luh, K.T.(2002) High incidence of cefoxitin and clindamycin resistance among anaerobes in Taiwan. *Antimicrob Agents Chemother*, **46**, 2908–2913.

Trinh, S., Haggoud, A., Reysset, G. and Sebald, M. (1995) Plasmids pIP419 and pIP421 from *Bacteroides*: 5-nitroimidazole resistance genes and their upstream insertion sequence elements. *Microbiology*, **141** (Pt 4), 927–935.

Turner, P., Edwards, R., Weston, V., Gazis, A., Ispahani, P. and Greenwood, D. (1995) Simultaneous resistance to metronidazole, co-amoxiclav, and imipenem in clinical isolate of *Bacteroides fragilis*. *Lancet*, **345**, 1275–1277.

Urban, E., Soki, J., Brazier, J. S., Nagy, E. and Duerden, B. I., (2002), Prevalence and characterization of *nim* gnes of *Bacteroides* sp. isolated in Hungary. *Anaerobe*, **8**, 175–179.

van der Wouden, E.J., Thijs, J.C., Kusters, J.G., van Zwet, A.A. and Kleibeuker, J.H. (2001) Mechanism and clinical significance of metronidazole resistance in *Helicobacter pylori*. *Scand J Gastroenterol* (Suppl.), **234**, 10–14.

Wareham, D.W., Wilks, M., Ahmed, D., Brazier, J.S. and Miller, M. (2005) Anaerobic sepsis due to multidrug-resistant *Bacteroides fragilis*: Microbiological cure and clinical response with linezolid therapy. *Clin Infect Dis*, 40e67–40e68.

Welch, R.A., Jones, K.R. and Macrina, F.L., (1979), Transferable lincosamide-macrolide resistance in *Bacteroides*. *Plasmid*, **2**, 261–268.

Wexler, H.M., (2002), Outer-membrane pore-forming proteins in gram-negative anaerobic bacteria. *Clin Infect Dis*, **35**, S65–S71.

Wexler, H.M. and Halebian, S., (1990), Alterations to the penicillin-binding proteins in the *Bacteroides fragilis* group: a mechanism for non-β-lactamase mediated cefoxitin resistance. *J Antimicrob Chemother*, **26**, 7–20.

Wexler, H.M., Molitoris, E., Molitoris, D. and Finegold, S.M., (2000), In vitro activity of moxifloxacin against 179 strains of anaerobic bacteria found in pulmonary infections. *Anaerobe*, **6**, 227–231.

Whittle, G., Shoemaker, N.B. and Salyers, A.A., (2002), The role of *Bacteroides* conjugative transposons in the dissemination of antibiotic resistance genes. *Cell Mol Life Sci*, **59**, 2044–2054.

Yang, Y., Rasmussen, B.A. and Bush, K., (1992), Biochemical characterization of the metallo-beta-lactamase CcrA from *Bacteroides fragilis* TAL3636. *Antimicrob Agents Chemother*, **36**, 1155–1157.

Chapter 7
Clinical Significance and Biological Basis of HIV Drug Resistance

Jorge L. Martinez-Cajas, Marco Petrella, and Mark A. Wainberg

7.1 Introduction

HIV-1 drug resistance has emerged as a major factor that limits the effectiveness of antiviral drugs in treatment regimens. Many studies have shown that the development and transmission of drug-resistant HIV-1 is largely a consequence of incomplete suppressive antiretroviral regimens; HIV-1 drug resistance can significantly diminish the effectiveness and duration of benefit associated with combination therapy for the treatment of HIV/AIDS (D'Aquila et al., 2003; Lorenzi et al., 1999; Quiros-Roldan et al., 2001; Rousseau et al., 2001; Winters et al., 2000; Yeni et al., 2002). Resistance-conferring mutations in both the HIV-1 reverse transcriptase (RT) and protease (PR) genes may precede the initiation of therapy due to both spontaneous mutagenesis and the spread of resistant viruses by sexual and other means. However, it is also generally believed that multiple drug mutations to any single or combination of antiretroviral agents (ARVs) are required in order to produce clinical resistance to most ARVs and that these are in fact selected following residual viral replication in the presence of incompletely suppressive drug regimens (Balotta et al., 2000; de Jong et al., 1996; Mayers, 1999).

In the case of the protease inhibitors (PIs) (Condra, 1998; Deeks, 1999; Molla et al., 1996), and most nucleoside analogue reverse transcriptase inhibitors (NRTIs), the development of progressive high-level phenotypic drug resistance follows the accumulation of primary resistance-conferring mutations in each of the HIV-1 PR and RT genes (Frost et al., 2000; Gotte and Wainberg, 2000; Loveday, 2001). Non-nucleoside reverse transcriptase inhibitors (NNRTIs) have low genetic barriers for the development of drug resistance and, frequently, a single primary drug resistance mutation to any one NNRTI may be sufficient to confer high-level phenotypic drug resistance to this entire class of ARVs (Bacheler et al., 2001; Deeks, 2001).

Furthermore, differences have also been reported in regard to the development and evolution of antiretroviral drug resistance between subtype B HIV-1 and several group M non-B subtypes. Non-B subtypes, for example, subtype C HIV-1 variants, are known to possess naturally occurring polymorphisms at several RT and PR codons that are implicated in drug resistance (Holguin and Soriano, 2002; Kantor et al., 2002). In some studies, the presence of these polymorphisms did not significantly reduce susceptibility to ARVs in phenotypic resistance assays or limit the

I. W. Fong and K. Drlica (eds.), *Antimicrobial Resistance and Implications for the Twenty-First Century.* © Springer 2008

Common NRTI mutations associated with HIV drug resistance

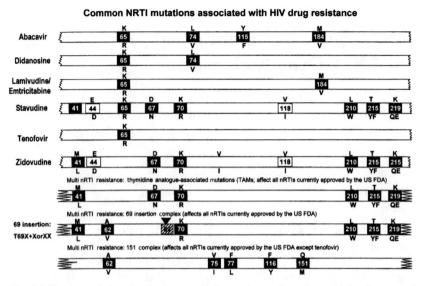

Fig. 7.1 The letter designation at the top of a box refers to the amino acid that is present in the wild-type sequence of RT. The letters at the bottom refer to substitutions associated with drug resistance. Sometimes, several different amino acid changes at the same codon can confer drug resistance, as for example both Y and F in the case of position 215. Based on figures available through the website of IAS-USA. Johnson VA et al. Update of the Drug Resistance Mutations in HIV-1: Fall, 2005. Top HIV Med, **13**, 125–131.

Common NNRTI mutations associated with HIV drug resistance

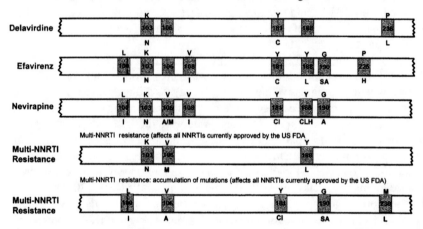

Fig. 7.2 The letter designation at the top of a box refers to the amino acid that is present in the wild-type sequence of RT. The letters at the bottom refer to substitutions associated with drug resistance. Sometimes, several different amino acid changes at the same codon can confer drug resistance, as for example both C and I in the case of position 181. Based on figures available through the website of IAS-USA. Johnson VA et al. Update of the Drug Resistance Mutations in HIV-1: Fall, 2005. Top HIV Med, **13**, 125–131.

effectiveness of an initial antiretroviral therapy (ART) regimen for a period of up to 18 months (Alexander et al., 2002; Holguin and Soriano, 2002). However, it has also been suggested that polymorphisms at resistance positions may sometimes facilitate selection of novel pathways leading to drug resistance, especially with incompletely suppressive antiretroviral regimens (Holguin and Soriano, 2002). This, in turn, may have important clinical implications with respect to choice of effective ART. This warrants increased genotypic surveillance on a worldwide basis, as the prevalence of non-B HIV-1 infection is increasing rapidly.

As illustrated in Figures 7.1– 7.3, it has been possible to select numerous drug resistance mutations for all licensed ARVs and investigational agents including the HIV-1 entry inhibitors (D'Aquila et al., 2003; Wei et al., 2002; Yeni et al., 2002). In view of the hypervariability of HIV-1 and limitation of existing antiretroviral combinations to completely suppress viral replication, it is essential that new anti-HIV drug discovery initiatives focus on the identification of new therapeutic targets and the development of antiretroviral agents with more robust genetic barriers and a broader spectrum of activity against drug-resistant HIV-1 variants.

Common PR mutations associated with HIV drug resistance

	L 10	13	G 16	K 20	L 24	30	V 32	L 33	35	M 36	43	M 46	47	G 48	I 50	53	I 54	58	D 60	I 62	63	69	A 71	G 73	74	77	V 82	83	I 84	I 85	N 88	L 90	I 93
Atazanavir	IFV		E	RMI	I		I	IFV		ILV		IL		V	L		LVMT		E	V			VITL	CSTA			AT		V	V	S	M	L
Amprenavir/ fosamprenavir	FIRV						I					IL	V		V		LVM							S			AFST		V			M	
Inidinavir	IRV			MR	I		I			I		IL					V						VT	SA		I	AFT		V			M	
Lopinavir/ ritonavir	FIRV			MR	I		I	F				IL	VA		V	L	VLAMTSE				P		VT	S			AFTS		V			M	
Nelfinavir	FI					N				I		IL											VT			I	AFTS		V		DS	M	
Ritonavir	FIRV			MR			I	F		I		IL			V		VL						VT			I	AFTS		V			M	
Saquinavir	IRV													V			VL						VT	S		I	A		V			M	
Tipranavir/ ritonavir	V	V		MR				F	G	I	T	L	V				AMV	E				K			P		LTD	V	V			M	

■ MAJOR MUTATION
☐ MINOR MUTATION

Fig. 7.3 The letter designation at the top of a box refers to the amino acid that is present in the wild-type sequence of RT. The letters at the bottom refer to substitutions associated with drug resistance. Sometimes, several different amino acid changes at the same codon can confer drug resistance, as for example both I and L in the case of position 46. Based on figures available through the website of IAS-USA. Johnson VA et al. Update of the Drug Resistance Mutations in HIV-1: Fall, 2005. Top HIV Med, **13**, 125–31.

7.2 Generation of HIV-1 Drug Resistance

Resistance mutations to ARVs may arise spontaneously as a result of the error-prone replication of HIV-1 and, in addition, are selected both in vitro and in vivo by pharmacological pressure (Menendez-Arias, 2002; Preston and Dougherty, 1996; Roberts et al., 1988). The high rate of spontaneous mutation in HIV-1 has been largely attributed to the absence of a $3' \rightarrow 5'$exonuclease proof-reading mechanism. Sequence analyses of HIV-1 DNA have detected several types of mutations including base substitutions, additions and deletions (Roberts, 1988). The frequency of spontaneous mutation for HIV-1 varies considerably as a result of differences among viral strains studied in vitro (Rezende, 1998). Overall mutation rates for wild-type laboratory strains of HIV-1 have been reported to range from 97×10^{-4} to 200×10^{-4} per nucleotide for the HXB2 clonal variant of HIV-1 to as high as 800×10^{-4} per nucleotide for the HIV-1 NY5 strain (Rezende, 1998; Roberts et al., 1988).

In addition to the low fidelity of DNA synthesis by HIV-1 RT, other interdependent factors that affect rates of HIV mutagenesis include RT processivity, viral replication capacity, viral pool size, and availability of target cells for infection (Coffin, 1995; Colgrove and Japour, 1999; Drosopoulos, 1998; Overbaugh and Bangham, 2001). It follows that an alteration in any or combination of these factors might influence the development of HIV drug resistance. There is also data showing that thymidine analog mutations (TAMs) in RT can significantly increase the likelihood of further mutant HIV-1 distributions and evolution of drug resistance; furthermore, this can happen in the presence or absence of concomitant nucleoside RT inhibitors (NRTIs) (Mansky, 2002; Mansky et al., 2003).

7.3 Inhibitors of Reverse Transcriptase

The RT enzyme is encoded by the HIV *pol* gene and is responsible for the transcription of double-stranded proviral DNA from viral genomic RNA. Two categories of drugs have been developed to block RT; these are nucleoside analog RT inhibitors (NRTIs) that act to arrest DNA chain elongation by acting as competitive inhibitors of RT and non-nucleoside RT inhibitors (NNRTIs) that act as non-competitive antagonists of enzyme activity by binding to the catalytic site of RT. NRTIs are administered to patients as precursor compounds that are phosphorylated to their active triphosphate form by cellular enzymes. These compounds lack a 3' hydroxyl group (OH) necessary for elongation of viral DNA. These analogs can compete effectively with normal deoxynucleotide triphosphate (dNTP) substrates for binding to RT and incorporation into viral DNA (Furman et al., 1986; Hart et al., 1992).

NNRTI antiviral activity is incompletely understood but is known to involve the binding of these non-competitive inhibitors to a hydrophobic pocket close to the catalytic site of RT (Ding et al., 1995; Wu et al., 1991). NNRTI inhibition reduces the catalytic rate of polymerization without affecting nucleotide binding or nucleotide-induced conformational change (Spence et al., 1995). NNRTIs are particularly active at template positions at which the RT enzyme naturally pauses. NNRTIs do not seem to influence the competition between dideoxynucleotide triphosphates (ddNTPs) and the naturally occurring dNTPs for insertion into the growing proviral DNA chain (Gu et al., 1991).

Both types of RT inhibitors have been shown to successfully diminish plasma viral burden in HIV-1 infected subjects. However, monotherapy with all drugs has led to drug resistance. Patients who receive combinations of three or more drugs are less likely to develop resistance, since these "cocktails" can suppress viral replication with much greater efficiency than single drugs or two drugs in combination. Although mutagenesis is less likely to happen in this circumstance, it can still occur, and the emergence of drug-resistant breakthrough viruses has been demonstrated in patients receiving highly active antiretroviral therapy (HAART) (Gunthard et al., 1998; Palmer et al., 1999). Furthermore, the persistence of reservoirs of latently infected cells represents another major impediment to currently applied anti-HIV chemotherapy (Finzi et al., 1999). Replication of HIV might resume once therapy is stopped or interrupted and, therefore, eradication of a latent reservoir of 10^5 cells might take as long as 60 years, a goal that is not practical with currently available drugs and technology (Finzi et al., 1999; Wong et al., 1997).

Resistance to 3TC ((-)-2', 3'-dideoxy-3'-thiacytidine, lamivudine) develops quickly whereas resistance to other NRTIs commonly appears only after about 6 months of therapy. Phenotypic resistance is detected by comparing the IC_{50} (or drug concentration capable of blocking viral replication by 50%) of pretreatment viral isolates with those obtained after therapy. Thus, higher IC_{50} values obtained after several months of treatment reflect a loss in viral susceptibility to antiretroviral agents (ARVs). Selective polymerase chain reaction (PCR) analysis of the RT genome confirms that the number of mutations associated with drug resistance increases concomitantly with increases in IC_{50} values.

Mutations associated with drug resistance have been reported in response to the use of any single NRTI or NNRTI (see Figures 7.1 and 7.2) (Schinazi et al., 2000). However, not all drugs elicit the same mutagenic response; sensitivity and resistance patterns must be considered on an individual drug basis. For example, patients on 3TC monotherapy may develop high-level, i.e. 1000-fold resistance within weeks, whereas 6 months or more are often required in order for sensitivity to zidovudine (ZDV) to drop by 50–100-fold. In contrast, HIV may appear to remain reasonably sensitive, even after prolonged monotherapy, to four of the other commonly used nucleoside analogs: ddI (didanosine), ddC (zalcitabine), d4T (stavudine), and ABC (abacavir). In the case of ZDV, increases in IC_{50} below threefold are regarded as non-significant, while 10–50-fold increases usually represent partial resistance, and increases above 50-fold denote high-level resistance.

Viral resistance to nucleoside analogs can often develop independently of the dose of drug that is administered. Tissue culture data have shown that HIV-1 resistance can be easily demonstrated against each of NRTIs, NNRTIs, and PIs, by gradually increasing the concentration of compound in the tissue culture medium (Japour et al., 1993; Jennings et al., 2004). Cell lines are especially useful in this regard, since HIV replication occurs very efficiently in such hosts. Tissue culture selection provides an effective preclinical means of studying HIV mutagenesis, especially since the same resistance-conferring mutations that arise in cell culture also appear clinically. Owing to the high turnover and mutation rate of HIV-1, the retroviral quasispecies will also include defective virions and singly mutated drug-resistant variants that are present prior to commencement of therapy. Multiply mutated variants appear later, because it requires time to accumulate multiple mutations within a single viral genome, and are not commonly found in the retroviral pool of untreated patients. An exception to this involves cases of new infection with drug-resistant viruses transmitted from extensively treated individuals. Patients with advanced infection have a higher viral load and a broader range of quasispecies than newly infected individuals (Markham et al., 1998). Such patients are often immunosuppressed and may also have diminished ability to immunologically control viral replication, possibly leading to more rapid development of drug resistance.

Site-directed mutagenesis has shown that a variety of RT mutations encode HIV resistance to both NRTIs and NNRTIs. Crystallographic and biochemical data have demonstrated that mutations conferring resistance to NNRTIs are found in the peptide residues that make contact with these compounds within their binding pocket (Ding et al., 1995; Wu et al., 1991).

Resistance-encoding mutations to NRTIs are found in different regions of the RT enzyme encoding genes, probably due to the complexity of nucleoside incorporation, which involves several distinct steps. These mutations can decrease RT susceptibility to nucleoside analogs. A summary of primary RT mutations has been published elsewhere (Schinazi et al., 2000).

It has also been shown that a family of insertion and deletion mutations between codons 67 and 70 can cause resistance to a variety of NRTIs including ZDV, 3TC, ddI, ddC, and d4T. Usually, these insertions confer multidrug resistance when present in a ZDV-resistant background. Another less frequently observed resistance mutation, K65R, has been shown to be associated with prior treatment with ABC-containing regimens and results in reduced antiviral susceptibility to both ABC and the acyclic RT nucleotide analog tenofovir (TDF). Hence, resistance to these antiretroviral agents can develop via genetic pathways involving either the TAMs or K65R as hallmark drug resistance mutations (Winston et al., 2002). In recent years, the proportion of genotyped clinical samples containing K65R has increased from less than 1% to almost 4%, reflecting the increased use of TDF in treatment regimens.

Diminished sensitivity to NNRTIs appears quickly both in culture selection protocols and in patients (Ding et al., 1995; Gu et al., 1991). NNRTIs share a common binding site, and mutations that encode NNRTI resistance are located within the binding pocket that makes drug contact (Chong et al., 1994; Ding et al., 1995; Esnouf et al., 1995; Finzi et al., 1999; Gu et al., 1991; Gunthard et al., 1998; Japour et al., 1993; Palmer et al., 1999; Richman et al., 1991; Schinazi et al., 2000; Vandamme et al., 1994; Wong et al., 1997; Wu et al., 1991). This explains the finding that extensive cross-resistance is observed among all currently approved NNRTIs (Esnouf et al., 1995; Fletcher et al., 1995). A substitution at codon 181 (tyrosine to cysteine) (Y181C) is a common mutation that encodes cross-resistance among many NNRTIs (Balzarini et al., 1993; Byrnes et al., 1993; Richman et al.,

1991). Replacement of Y181 by a serine or histidine also conferred HIV resistance to NNRTIs (Sardana et al., 1992). A mutation at amino acid 236 (proline to leucine) (P236L), conferring resistance to a particular class of NNRTIs that include delavirdine, can also diminish resistance to nevirapine and other NNRTIs, particularly if a Y181C mutation is also present in the same virus (Dueweke et al., 1993). Other important substitutions are Y188C and Y188H that can also confer NNRTI resistance.

Another drug resistance mutation, namely K103N (lysine to asparagine), is commonly observed and is responsible for reduced susceptibility to all approved NNRTIs (Balzarini et al., 1993; Byrnes et al., 1993; Richman et al., 1991). Substitution of K103N results in alteration of interactions between NNRTIs and RT. The K103N mutation shows synergy with Y181C in regard to resistance to NNRTIs, unlike antagonistic interactions involving Y181C and P236L (Nunberg et al., 1991).

Resistance to NNRTIs is also observed in cell-free enzyme assays (Boyer et al., 1993; Byrnes et al., 1993; Jonckheere et al., 1994; Loya et al., 1994; Sardana et al., 1992). Both Y181I and Y188L mediate decreased sensitivity to NNRTIs without affecting either substrate recognition or catalytic efficiency, supporting the idea that resistance to NNRTIs is attributable to diminished ability of these drugs to be bound by RT.

7.4 Protease Inhibitors

The HIV protease (PR) is an aspartic proteinase made of two identical monomers that form a central, symmetric, active site. This cavity binds the natural substrates of PR, for example, gag and gag–pol polyproteins and, is also the region of the molecule targeted by the protease inhibitor drugs (PIs).

Drug-resistant viruses have been observed for all protease inhibitors (PIs) developed to date, while some strains of HIV have displayed cross-resistance to a variety of PIs after either clinical use or in vitro drug exposure (Condra, 1998; Deeks, 1999; Murphy, 1999; Palmer et al., 1999). As many as 40 drug resistance mutations have been observed at as many as 20 different positions within the PR (Boden and Markowitz, 1998; Condra, 1998; Deeks, 1999; Hertogs et al., 2000; Johnson et al., 2005; Murphy, 1999) (see Figure 7.3). An analysis of the sequences submitted to the Stanford HIV-1 resistance database (Shafer et al., 2000; Velazquez-Campoy et al., 2003) found that the most common amino acid substitutions in the PR occurred at the following positions: Leu 10 (33.4%), Met 36 (23.1%), Met 46 (19.4%), Ile 54 (15.4%), Leu 63 (80.8%), Ala 71 (28.2%), Gly 73 (9.6%), Val 77 (28.9%), Val 82 (21.1%), Ile 84 (7.8%), Leu 90 (21.4%), and Ile 93 (33.4%). These mutations occur in the various regions of the PR molecule: the substrate binding cavity, the flaps, the dimerization domain and the β-core region. Most major mutations appear within the active site cavity, but some, for example, L90M occur at the dimerization interface. Certain mutations within the PR gene are more important than others and can confer resistance, virtually on their own, to certain PIs (Condra, 1998; Deeks, 1999; Murphy, 1999). Mutations occurring inside the active side cavity (positions D30, V32, G48, I50, V82, I84) emerge frequently and some affect most PIs as they directly alter the interaction of the inhibitors with PR (Ohtaka et al., 2002). Binding studies, for example, indicated that the double mutation V82F/I84V lowered the binding affinity of PR for saquinavir, nelfinavir, indinavir, ritonavir, amprenavir, and lopinavir, to different extents for each, with some being more affected than others (Velazquez-Campoy et al., 2003). On the other hand, there appear to exist hallmark resistance mutations for certain PIs. The mutation I50V, specifically selected by amprenavir, lowers its binding affinity by 150-fold, compared to nelfinavir, saquinavir, and ritonavir whose binding affinity was reduced by only 20–70-fold (Ohtaka et al., 2002). Atazanavir, for example, selects the mutation I50L that specifically confers resistance to this drug (IC50 increase in 2.1–5.4), and its effect can be amplified by the mutations A71V (5.7–10-fold increase in IC50) (Colonno et al., 2004). Similarly, the mutation D30N is uniquely selected by nelfinavir and results in a sixfold increase in EC90, but does not significantly affect other PIs (Patick et al., 1998).

Wide arrays of secondary mutations have been observed, that, when combined with primary mutations, can cause increased levels of resistance to occur (see Figure 7.3). Furthermore, the presence of certain secondary mutations on their own may not lead to drug resistance, and, in this context, some of these amino acid changes should be considered to represent naturally occurring polymorphisms. Nevertheless, accumulation of mutations in the protease can result in resistance to all currently used protease inhibitors. Binding studies have shown that the effect of mutations in the active site and the flaps are only additive, while the effect of the mutations at the dimerization interface amplify those of the flaps and active site resulting in a final effect greater than the addition of the individual effects (Velazquez-Campoy et al., 2003). More recent, high resolution crystallographic data from a multidrug-resistant PR carrying nine resistance mutations showed that the a partial expansion of the active site due to an amino acid of shorter side chains at positions 82 and 84 (V→A,S,T,F) (Logsdon et al., 2004) is enhanced by the mutation 90M by means of a domino effect that results in a wide-open separation of the flaps (Martin et al., 2005; Martinez et al., 2004).

In general, protease resistance mutations lead to reduced viral fitness, which in part is related to the greater number of substitutions involved and a direct alteration of the properties of the active site. However, the mutation D30N which confer resistance to nelfinavir, causes itself a great reduction of the replicative fitness of HIV (Martinez-Picado et al., 1999). Therefore it seems that PR enzyme appears to adapt more easily than RT to pressures exerted by antiviral drugs. It should be noted that a resistant mutated HIV protease can recover its catalytic activity when "compensatory" mutations within the protease itself or in its substrates—the gag and gag–pol precursor proteins—appear. For instance, the nelfinavir-induced mutations N88S causes severe loss of viral fitness and hypersusceptibility to amprenavir. The mutations L63P and V77I partially restore fitness and decrease the level of amprenavir hypersusceptibility (Resch et al., 2002). In gag, compensatory mutations have been found at or near the cleavage site mutations and they include substitutions, deletions and insertions (Doyon et al., 1996, 1998; Gatanaga et al., 2002; Liang et al., 1999; Myint et al., 2004; Tamiya et al., 2004). These mutations may have clinical relevance since they can potentially restore pathogenicity of resistant viruses. It is known that persistent viral replication in patients experiencing treatment failure is associated to disease progression and death (Zaccarelli et al., 2005).

7.5 Entry Inhibitors

Enfuvirtide (ENF) was the first entry inhibitor to be approved for clinical use and is currently available in Europe and North America for subcutaneous administration. This agent is a 36 amino acid synthetic peptide that structurally mimics the heptad repeat 2 (HR2) region of gp41 and inhibits the association of HR2 with the heptad repeat (HR1) of the same protein. This action impedes the formation of a six helix bundle that mediates the fusion of the HIV envelope and cell membranes (He et al., 2003). ENF has demonstrated superiority in clinical trials by improving the virologic and immunologic responses of patients receiving enfuvirtide plus an optimized background therapy (OB), with a median reduction of the viral load of 1.48 $\log_{10}$ at 48 weeks compared to a reduction of only 0.63 log10 in those given OB alone (Lalezari et al., 2003c; Lazzarin et al., 2003). ENF is currently indicated for treatment-experienced patients who have failed other modalities of therapy.

Unfortunately, resistance is expected to emerge. Initial in vitro studies selected ENF-resistant viruses carrying mutations G36S and V38M which affect the highly conserved region of the N terminus of HR1 (Rimsky et al., 1998). Subsequently, sequencing of virus recovered from patients enrolled Phase II and III clinical trials as well as site-directed mutagenesis experiments clarified the roles of V38A, Q40H, N42T, and N43D as resistance conferring mutations (Greenberg et al., 2002, 2003b; Lalezari et al., 2003a,b; Lazzarin et al., 2003; Mink et al., 2002; Sista et al., 2002; Wei et al., 2002; Wheat et al., 2002). It became apparent that genotypic changes from position 36–45 (GIVQQQNNLL) of the HR1 region are of principal importance in development of resistance. Only one amino acid substitution suffices to cause significant reduction of drug susceptibility which makes ENF a drug with a low genetic barrier for resistance (Mink et al., 2002; Poveda et al., 2004, 2005; Sista et al., 2004). Interestingly, there is a wide variation in the susceptibility of virus isolated from naïve and enfuvirtide-treated patients, which may be due to preexisting natural polymorphisms in the HR2 region or changes induced by drug selection in the same region (Sista et al., 2003). The rate of virological failure in the two main clinical trials at weeks 8 and 24 were 16 and 41%, respectively for TORO-I, and 19 and 49%, respectively for TORO-II (Greenberg et al., 2003b). Emergence of mutations in the amino acids 36–45 was detected in 205/218 (94%) of viruses from patients experiencing virological failure in these two clinical trials. The most frequent mutations occurred at position 38 (104/218, 47.7% most often a V38A), position 43 (58/218, 30.7%, most often a N43D) and position 36 (58/218, 26.6%, most often a G36D) (Greenberg et al., 2003a). More recent studies have described N42Q/H and N43Q as resistance mutations (Xu et al., 2005).

Enfuvirtide resistant viruses have been found to have reduced fitness in competition assays (Lu et al., 2004; Menzo et al., 2004). These experiments have allowed us to grade the fitness of some recombinant viruses carrying resistance mutations as follows: wild type > N42T > V38A > N42T/N43K approximately N42T/N43S > V38A/N42D approximately V38A/N42T (Lu et al., 2004). Interestingly, an 80-week follow-up study of four HAART-experienced patients who participated in the TORO-II trial and who had an initial 1.0 log decrease described viral rebound in all, with two patients reaching pretreatment baseline levels (Poveda et al., 2004). This suggests development of further compensatory adaptations in domains other than HR1 that would restore viral fitness. Two studies have found a possible role for the mutation S138A in HR2 as a compensatory mutation (Perez-Alvarez et al., 2006; Xu et al., 2005].

Co-receptor usage also appears to influence susceptibility to ENF (Derdeyn et al., 2000; Reeves et al., 2002). However, results from the studies have thus far been controversial. Some in vitro studies have found that CCR5 strains were more resistant to ENF (Derdeyn et al., 2000; Reeves et al., 2002), yet in vivo studies have not shown important differences in the response to ENF when comparing patients infected with R5-tropic strains and those infected with X4-tropic strains (Su et al., 2003; Whitcomb et al., 2003).

Host factors can also affect efficacy of ENF. High density of CCR5 coreceptors on the cellular surface might facilitate HIV fusion, thus reducing the time during which HIV gp41 could be targeted by ENF. This idea is supported by the fact that individuals carrying Δ32-CCR5, who express low levels of CCR5, show better responses to ENF (Dean et al., 1996; Reeves et al., 2002).

It can be of concern that enfuvirtide resistance related changes could cause cross-resistance to new ARVs in earlier phases of development. However, new data showed that HR1 mutations had minimal effect on virus sensitivity to other classes of entry inhibitors, including those targeting CD4 binding (BMS-806 and a CD4-specific monoclonal antibody [MAb]), coreceptor binding (CXCR4 inhibitor AMD3100 and CCR5 inhibitor TAK-779), or fusion (T-1249) (Reeves et al., 2005).

7.6 Antiretroviral Drug Resistance in Non-B Subtypes of HIV-1 Group M

Genotypic divergence of *pol* gene sequences between different HIV-1 subtypes is only beginning to be investigated, although the RT and PR enzymes are the main targets of ART (Becker-Pergola et al., 2000; Cornelissen et al., 1997; Gao et al., 1994; Shafer et al., 1998; Vanden Haesevelde et al., 1994). Group O and HIV-2 viruses carry natural polymorphisms Y181C and Y181I that confer intrinsic resistance to NNRTIs (Descamps et al., 1995, 2007; Tantillo et al., 1994). Subtype F isolates, showing 11% nucleotide sequence variation from subtype B and group M viruses, have also been reported to have reduced sensitivity to some NNRTIs while retaining susceptibility to others such as nevirapine and delavirdine, NRTIs and PIs (Apetrei et al., 1998; Shafer et al., 1997). In contrast, the drug sensitivity of subtype C isolates from treatment-naïve patients in Zimbabwe was reported to be similar to that of subtype B isolates (Birk and Sonnerborg, 1998; Shafer et al.,

1997). Recent studies conducted with five Ethiopian subtype C clinical isolates showed natural resistance to NNRTIs in one case and resistance to ZDV in another, due to natural polymorphisms at positions G190A and K70R, respectively (Loemba et al., 2002). Another study reported no differences in drug susceptibility among subtypes A, B, C, and E; subtype D viruses showed reduced susceptibility due to rapid growth kinetics (Palmer et al., 1998). High prevalence (i.e. 94%) of a valine polymorphism (GTG) at position 106 in RT from subtype C HIV-1 clinical isolates has also been reported (Brenner et al., 2003). In tissue culture experiments, selection of subtype C with efavirenz (EFV) was associated with development of high-level (i.e. 100–1000-fold) phenotypic resistance to all NNRTIs. This was a consequence of a V106M mutation that arose in place of the V106A substitution that is more commonly seen with subtype B viruses (Brenner et al., 2003). This V106M mutation conferred broad cross-resistance to all currently approved NNRTIs and was selected on the basis of differential codon usage at position 106 in RT, due to redundancy in the genetic code.

Genotypic diversity and drug resistance may be particularly relevant in establishing treatment strategies against African and Asian strains. First, since many antiviral drugs have been designed based on sequences of subtype B RT and PR enzymes, and drug resistance profiles, if not responses, may be different for non-B viral strains. Second, drug resistance may develop more rapidly in resource-poor countries if only sub-optimal therapeutic regimens are available. Global phenotypic and genotypic screening of non-B subtypes is warranted so as not to jeopardize the outcome of recently introduced antiretroviral treatments (Petrella et al., 2001).

7.7 Transmission of HIV Drug-Resistance

As stated, HAART, including drugs that inhibit the RT and PR enzymes of HIV-1, has resulted in declining morbidity and mortality (Palella et al., 1998). The failure to completely suppress viral replication allows for the development of genotypic changes in HIV-1 that confer resistance to each of the three major classes of antiretroviral drugs (D'Aquila et al., 2003; Hirsch et al., 2003; Wainberg and Friedland, 1998). Cumulative data indicate that single drug-resistant variants can be transmitted to approximately 10–15% of newly infected persons in western countries in which ARVs have been available for many years, with transmission of dual and triple-class multidrug resistance (MDR) observed in 3–5% of cases (Boden et al., 1999; Little et al., 2002; Salomon et al., 2000; Yerly et al., 1999).

There is concern that the transmission of MDR viruses in primary HIV-1 infection (PHI) may limit future therapeutic options. Treatment failure has been observed in several individuals harboring MDR infections (Gandhi et al., 2003; Hecht et al., 1998; Little et al., 2002). Some reports have shown an impaired fitness of transmitted MDR variants compared with wild-type (WT) infections acquired in PHI, and the mutations that were transmitted in such patients persisted in the absence of treatment (Brenner et al., 2002a). This persistence differs from the rapid outgrowth of WT viruses in established infections upon treatment interruption, due to the selective growth advantage and fitness of WT variants (Brenner et al., 2002; Devereux et al., 1999; Verhofstede et al., 1999). Taken together, these findings suggest that archival WT viruses may not exist in MDR infections transmitted during PHI.

Several reports have also documented cases of intersubtype superinfection (A/E and B) in recently infected intravenous drug users (IDU) (Jost et al., 2002; Ramos et al., 2002). Other studies have failed to confirm superinfection following IDU exposure, suggesting that superinfection is a relatively rare event (Gonzales et al., 2003; Tsui et al., 2004). Several subsequent reports demonstrated superinfection in subtype B infections. In one case, a WT superinfection arose following a primary MDR infection (Altfeld et al., 2002; Koelsch et al., 2003).

It is important to assess the virological consequences of transmission of drug-resistant variants in primary infection, as well as the time to disappearance of resistant virus in those patients not initially treated. Genotypic analysis indicates that a single dominant HIV-1 species can persist for more than 2 years in circulating plasma and peripheral blood mononuclear cells (PBMC), regardless of route of transmission. Resistant and MDR infections can persist for 2–7 years following PHI. Superinfection with a second MDR strain in a patient originally infected with an MDR strain from an identified source partner has also been described (Brenner et al., 2002). In spite of a rapid decline in plasma viremia suggestive of an effective immune response, this patient was susceptible to a second infection which occurred concomitant with a dramatic rise in viral load. Five other subtype B superinfections have been described, as well as three intersubtype A/E and B superinfections (Allen and Altfeld, 2003; Altfeld et al., 2002; Jost et al., 2002; Koelsch et al., 2003; Ramos et al., 2002; Smith et al., 2004). Six of the seven superinfections described have occurred in the first year following initial infection.

Many have attributed superinfection to coinfection during primary infection. Two longitudinal studies ($n = 37$ in both studies) involving IDU indicated that superinfection is a rare phenomenon that was not observed during 1–12 years of follow-up spanning 215 and 1072 total years of exposure (Altfeld et al., 2002; Koelsch et al., 2003). However, it is not known whether any patients were recruited within the first year of HIV-1 exposure in these studies. In the case of the MDR infections cited above, identification of the source partner of infection argues against coinfection (Brenner et al., 2002).

Findings of HIV-1 superinfection are a matter of concern insofar as such results challenge the assumption that immune responses can protect against reinfection. Of course, the impaired viral fitness of the initial MDR infection described above may be a factor in permitting superinfection. The initial MDR strain showed a 13-fold impaired replicative capacity from a WT variant strain from the isolated source partner following a treatment interruption. Fitness considerations may also have been important in a WT superinfection of an initial MDR infection and cases of subtype B superinfection following A/E infections that elicited low-level viremia (Jost et al., 2002; Ramos et al., 2002).

In newly infected individuals, multimutated viruses conferring MDR may represent a new determinant of virological outcome. Persistence of MDR in the absence of treatment raises serious issues regarding HIV-1 management. For recently infected MDR patients, drug resistance analysis and viral fitness may provide useful information in regard to ultimate therapeutic strategies.

It is interesting to note that the presence of the M184V mutation in RT, associated with high-level resistance to 3TC, seems to have been associated with the persistence of low viral load. In two PHI cases, rebounds to a high level of plasma viremia occurred only at times when the M184V mutation in RT could no longer be detected. A third PHI patient maintained low plasma viremia over 5 years, and his virus also contained the M184V mutation throughout this time. In an additional individual, high viral loads were present at times after primary infection in spite of the M184V mutation, but virus could only be isolated from this individual in coculture experiments after loss of the M184V mutation (Brenner et al., 2004). These data are consistent with previous findings on loss of fitness conferred by the M184V mutation in RT, alongside multiple other pleiotropic effects, including diminished processivity, diminished rates of nucleotide excision, and diminished rates of initiation of reverse transcription (Petrella et al., 2002; Turner et al., 2004; Wainberg et al., 1996).

Other studies suggest that despite reduced ARV susceptibility, MDR infections may be of some immunological and virological benefit due to the impaired replicative capacity of MDR variants (Baxter et al., 2000; Colgrove et al., 1998; Dickover et al., 2001; Verhofstede et al., 1999). Moreover, in all cases, RT assays and competitive fitness assays showed MDR viruses to have compromised replicative capacity. The absence of genotypic changes in these viruses over time further supports the concept of expansion of predominant MDR quasispecies during primary infection. Recombination events can also occur in this period. It is also important to point out that the replication fitness of a given virus versus its transmission fitness may represent two very different concepts.

Antiretroviral therapy, by reducing HIV-1 replication, has been shown not only to impact significantly on morbidity and mortality but also to reduce the spread of HIV-1(Quinn et al., 2000; Yerly et al., 2001). Treatment effectiveness is hampered by the development of drug-resistant (DR) strains, leading inexorably to virological failure (Wainberg and Friedland, 1998). The transmissibility of DR strains is not fully understood and may differ from that of wild-type (WT) strains for at least two reasons: first, the relative fitness of DR strains compared with WT in the absence of therapy and second, the degree to which partially active therapies can reduce viral load in persons harbouring resistant viruses (Phillips, 2001; Yerly et al., 1999). As a consequence of widespread use of ART in North America, the transmission of DR strains in recently infected (RI) individuals has increased from 3.8% in 1996 to 14% in 2000. Such an increase of primary DR is of public health concern since a clear association between DR and early treatment failure has been reported (Little et al., 2002). However, several groups in Europe and Australia have reported a recent stable or decreasing trend in DR transmission for RT and/or protease inhibitors (PI) (Ammaranond et al., 2003; Chaix et al., 2003), and have attributed this decline to the widespread use of suppressive triple ARV regimens since 1996. This presupposes that transmission of drug-resistant variants may have earlier been more common due to the widespread use of suboptimal dual therapy or even monotherapy regimens prior to 1996 and the likelihood that these suboptimal regimens may have selected for drug resistance mutations with very high frequency (Ammaranond et al., 2003).

7.8 Clinical Value of HIV Drug Resistance Testing

Genotype, real phenotype, and virtual phenotype-based antiretroviral resistance tests (GART, PART, and vPART, respectively) are available for clinicians treating HIV-infected patients as tools to diagnose HIV drug resistance.

In GART, resistance-conferring mutations are detected by sequencing of plasma viral RNA or hybridization probes. For clinicians, interpretation carried out with the help of resistance mutation lists or algorithms which define mutations as to whether or not they confer resistance and as to whether they are of major or minor importance. These algorithms integrate information collected from numerous studies in order to infer levels of resistance. Several algorithms are now available online as interactive resources that accept a nucleotide sequence or a set of mutations, and return a simple interpretation (e.g., Detroit Medical Center resistance algorithm, Stanford HIV Database). Clinicians hence use clinical and genotype data to select the subsequent drug regimen. Currently, GART is the most widely used resistant test due to its low cost and technical simplicity.

In PART, resistance assays directly measure viral replication under drug pressure. In these assays, a vector is constructed which is suitable to receive a genetic segment of interest from a clinical isolate, either through cloning or recombination. For instance, a segment containing the HIV-1 RT and protease genes is inserted into a vector that contains the genetic code of a reference laboratory wild type virus, but lacks both segments of interest, and appropriate cells (generally PBMCs) are transfected. The level of viral particles produced is measured by detecting the expression of a gene, for example, RT or p24. The virus grows under a range of drug concentrations, and the concentrations of drug that reduce viral replication by 50% (IC50) and 90% (IC90) are determined. The ratio IC50 of the mutated virus/IC50 of the wild-type virus is calculated and the results are generally reported as number of folds increase. A series of cut-off points have been established for each drug above which resistance is present. This strategy facilitates interpretation (resistance present or not), but may unnecessarily prohibit the use of drugs for viruses considered resistant by the test but that could still be suppressed by a drug in vivo. Another manner in which to optimize the interpretation of phenotypic data is the concept of inhibitory quotient (IQ). The IQ is a ratio that relates the drug level at trough (Cmin) with a measure of the viral susceptibility to a given drug (fold change). The IQ and a normalized IQ (NIQ) have been found to predict virologic response to several protease inhibitors in patients that have experienced drug failure due to resistance (Casado et al., 2003; Castagna et al., 2004; Gonzalez de Requena et al., 2004; Marcelin et al., 2003; Shulman et al., 2002; Winston et al., 2005). Higher IQs predict better responses. PART offers some advantages: it directly measures viral replication under drug pressure, accounts for the effect of polymorphisms of the inserted segment and, presents quantitative results. It is, unfortunately, more expensive, labor demanding and time consuming than GART. Nevertheless, it can be helpful for assessing resistance of viruses harboring complex mutations for which resistance correlates do not yet exist or whose effect on resistance are still equivocal. In these latter cases, GART and PART can clearly offer complementary information.

245

Large databases of paired genotype–phenotype assays have supported the construction of "virtual phenotype" (vPART) calculators that render a quantitative HIV-1 estimation of resistance to ARV drugs based on a statistical prediction of the most likely phenotype for a given genetic sequence. The accuracy of this estimation depends on the number of genotypes in the database that match the problem genotype and the variability in drug susceptibility of the phenotypes used to create the predicting pool. Rare genetic sequences and those with poorer matches will be less likely to be accurate than the more common and better matched. Although a good correlation of virtual phenotypes with "real phenotypes" has been reported (Graham et al., 2001; Larder and Hertogs, 2000) it should be kept in mind that vPART is a probability estimation.

Most studies have found a therapeutic advantage in terms of likelihood of viral suppression and mean virologic RNA reduction for resistance-test-guided treatments when compared to treatments based only on clinical judgment (DeGruttola et al., 2000; Durant et al., 1999; Gianotti et al., 2006; Ormaasen et al., 2004; Tural et al., 2002). Consequently, most international guidelines recommend resistance to assist the selection of therapeutic regimens following pharmacologic viral failure (DDHS, 2006, Hirsch et al., 2003; Vandamme et al., 2004). A meta-analysis of pertinent clinical trials estimated that the proportion of patients experiencing viral suppression at 3–6 months can be increased by just 10% with the use of GART and found no benefit in terms of CD4 cell counts (Panidou et al., 2004). However, the authors cautiously pointed out that the short-term follow-up may have impeded the detection of immunologic benefit. Additionally, due to the increased prevalence of primary drug resistance, which ranges from 6 to 16% (Bennett et al., 2005; Cane et al., 2005; Weinstock et al., 2004; Wensing et al., 2005), consideration for obtaining a resistance test prior to initiation of therapy in drug naïve patients has been encouraged (DDHS, 2006; Vandamme et al., 2004). The cost-effectiveness of obtaining GART in all treatment naïve chronically infected individuals given an HIV primary resistance prevalence over 1% has been supported by one study (Sax et al., 2005). The rational for this recommendation includes detecting primary drug resistance mutations that could threaten future therapeutic efficacy. As reversion of primary resistant mutations for NRTIs and NNRTIs may take as long as 1 year, and those for PIs may take even longer, (Barbour et al., 2004) drug-resistant HIV will be more likely detected in patients infected within the previous 2 years.

Although resistance testing represents a significant advance in HIV therapy, it adds further complexity to the management of HIV infection, since interpretation of results may not be straightforward and clinical correlates do not yet exist for all resistance mutations. Common limitations for all resistance tests are inability to detect virus archived in viral reservoirs, insensitivity to viral quasispecies that represent less than 20% of the total viral mixture, and the requirement of a minimum viral load (500–1000 plasma HIV RNA copies/milliliters) for the test to be performed. Current methods appear to have similar performance at detecting antiretroviral resistance (Ferrer et al., 2003; Gianotti et al., 2006; Mazzotta et al., 2003; Saracino et al., 2004), but GART is more sensitive for minority populations than PART. Consequently, GART can detect sentinel mutations (e.g., M184V) before changes in phenotypic resistance become evident. Finally, several studies have clearly demonstrated that expert advice adds benefit to results from resistance testing (Badri et al., 2003; Bossi et al., 2004; Clevenbergh et al., 2003; Saracino et al., 2004; Torre and Tambini, 2002; Tural et al., 2002).

7.9 Future Direction

Numerous compounds are currently under clinical or preclinical investigation. They have been designed to target different viral or host cell proteins and can be grouped as follows:

1. Novel NRTIs: PSI-5004, SPD-754, DPC-817, elvucitabine, alovudine, MIV-210, amdoxovir, DOT;
2. NNRTIs: UC-781, dapivirine (TMC-120), etravirine (TMC125), rilpivirine;
3. PIs: Darunavir (TMC-114);
4. Other viral proteins:

   a. gp120: cyanovirin N, PRO-542, BMS-806;
   b. attachment inhibitors: BMS-488043;
   c. integrase: L-870,812, PDPV-165;
   d. capsid proteins: PA-457, alpha-HCG);

5. Cellular proteins:

   a. CD4 expression downregulator: CADAs;
   b. CD4 directed antibodies: TNX-355;
   c. CXCR4 antagonists: AMD-070, KRH-1636, KRH-2731;
   d. CCR5 antagonists: TAK-220, vicriviroc (SCH-D), AK-602, maraviroc (UK-427857).

Among the molecules listed above, the protease inhibitor TMC-114, the CCR5 antagonists TAK-220 and maraviroc, the CXCR4 antagonists AMD-070, KRH-1636 and KRH-2731, the NRTIs SPD-754, DPC-817, and the NNRTIs, etravirine and seem promising for optimization of chronic infection since they can be administered orally and are active against multidrug-resistant HIV strains (Dorr et al., 2005).

Resistance to coreceptor antagonists emerges predominantly by mutations of gp120 that restores the interaction with the coreceptors despite the antagonism of the drug. CCR5-tropic viruses tend to preserve coreceptor usage and become resistant through mutations of the gp120 V3 region (Kuhmann et al., 2004; Trkola et al., 2002).

Drug resistance acquired by this mechanism appear to be drug specific (Westby et al., 2005), with the exception of vicriviroc-induced resistant mutations that may affect other compounds from the same group (Marozsan et al., 2005). Switch to a CXCR4 usage is an alternative way used by CCR5 tropic HIV to overcome this group of drugs. Animal studies with maraviroc (UK-427-857), as well as clinical studies with aplaviroc demonstrated, however, that this switch can occur and result in emergence of CXCR4 tropic virus (Kitrinos et al., 2005; Westby et al., 2006;

Westby et al., 2004b). This fact could imply severe consequences in humans, as CXCR4 tropic viruses are in general more pathogenic. However, CCR5 antagonists did not rapidly select for CXCR4 tropic viruses in vitro (Westby et al., 2004a). These issues will need to be further considered during clinical studies.

Regarding CXCR4 antagonists, mutations in the HIV gp120 V3 domain seem to mainly mediate resistance to these compounds, although changes in regions V1, V2, and V4 can also be associated with resistance (de Vreese et al., 1996; Schols et al., 1998). Switch to CCR5-tropism in resistant viruses have also been observed (Este et al., 1999).

There is increased interest in the development of topical agents to halt sexual transmission of HIV. The NRTI and NNRTIs tenofovir and TMC-120 and UC-781 are being tested in early clinical trials as vaginal microbicides (Lederman et al., 2006). Also, cyanovirin, BMS-806 and PRO 542 are agents under investigation for use as topical microbicides (Lederman et al., 2006). The potential for development of resistance to drugs used as microbicides and the ability of these drugs to inhibit transmission of resistant viruses (e.g., K65R carrying virus) are still unknown.

Combination therapy will likely be maintained as the standard approach for the treatment of AIDS as it has proved to maximize potency and minimize toxicity and lower the risk for emergence of resistance. Future research will focus on the development of oral therapies with minimal pill burden, and less toxic effects in order to maximize long-term adherence. It is of critical importance to extend the study of resistance and its effects on viral fitness to the non-B subtype of HIV-1 and HIV-2. Finally, drugs that enhance natural host defense mechanisms (APOBEC3G) (Mangeat et al., 2003; Mariani et al., 2003) and those that promote viral DNA mobilization (Lehrman et al., 2005) may offer a hope for curative therapies.

7.10 Conclusion

The accumulation of specific resistance-conferring mutations is associated with the development of phenotypic resistance to anti-HIV drugs which can significantly diminish the effectiveness and longevity of ART. Cross-resistance among drugs of the same class also occurs frequently and is most problematic with NNRTIs due to their lower genetic barrier for rapid selection of drug resistance compared to other classes of ARVs. There is now also data indicating that cross-resistance amongst the NRTIs may in fact be more widespread than was initially thought (Nijhuis et al., 2001). Furthermore, the emergence of new drug resistance mutations is helping to establish new mutant distributions with additional pathways for developing cross-resistance to ARVs (Brenner et al., 2002b; Nijhuis et al., 2001). These new patterns of cross-resistance together with increasing transmission of multidrug-resistant HIV-1 variants are problematic and seriously limit the number of effective treatment options that are now available for long-term management of HIV-infection.

Additional strategies, in addition to new drug discovery programs, are urgently required to help curb the development of drug-resistant HIV-1. One possible approach that merits further consideration is based on the maintenance of specific fitness-attenuating drug mutations in therapeutic regimens for HIV-1 infection (Ait-Khaled et al., 2002, 2003). The M184V substitution in RT has been extensively studied in this regard because of its ability to impair viral replication capacity while limiting the development of subsequent drug resistance mutations in HIV-1 RT, for example, TAMs and the Q151M multidrug complex resistance mutation associated with use of AZT and d4T (Petrella et al., 2002). Of course, restricted evolution of drug resistance in these circumstances may also result from other alterations of RT function by M184V (Wainberg et al., 1996). One recent study has shown that viruses containing the M184V mutation may be transmitted less frequently than viruses containing other mutations associated with drug resistance (Daar and Richman, 2005), perhaps because M184V compromises viral replicative capacity. Further work on these and other topics is needed to improve our understanding of HIV drug resistance in the context of clinical relevance, successful antiviral chemotherapy, and likelihood of transmission of resistant strains.

References

DDHS Panel on Antiretoviral Guidelines for Adult & Adolescents, (2006), Guidelines for the Use of Antiretroviral Agents in HIV-1-Infected Adults and Adolescents. Office of AIDS Research Advisory Council. AIDS Info/US Department of Health and Human Services.

Ait-Khaled, M., Rakik, A., Griffin, P., Stone, C., Richards, N., Thomas, D., Falloon, J. and Tisdale, M., (2003), HIV-1 reverse transcriptase and protease resistance mutations selected during 16–72 weeks of therapy in isolates from antiretroviral therapy-experienced patients receiving abacavir/efavirenz/amprenavir in the CNA2007 study. *Antivir Ther*, **8**, 111–120.

Ait-Khaled, M., Stone, C., Amphlett, G., Clotet, B., Staszewski, S., Katlama, C. and Tisdale, M., (2002), M184V is associated with a low incidence of thymidine analogue mutations and low phenotypic resistance to zidovudine and stavudine. *AIDS*, **16**, 1686–1689.

Alexander, C.S., Montessori, V., Wynhoven, B., Dong, W., Chan, K., O'Shaughnessy, M.V., Mo, T., Piasecczny, M., Montaner, J.S. and Harrigan, P.R., (2002), Prevalence and response to antiretroviral therapy of non-B subtypes of HIV in antiretroviral-naive individuals in British Columbia. *Antivir Ther*, **7**, 31–35.

Allen, T.M. and Altfeld, M., (2003), HIV-1 superinfection. *J Allergy Clin Immunol*, **112**, 829.

Altfeld, M., Allen, T.M., Yu, X.G., Johnston, M.N., Agrawal, D., Korber, B.T., Montefiori, D.C., O'Connor, D.H., Davis, B.T., Lee, P.K., Maier, E.L., Harlow, J., Goulder, P.J., Brander, C., Rosenberg, E.S. and Walker, B.D., (2002), HIV-1 superinfection despite broad CD8+ T-cell responses containing replication of the primary virus. *Nature*, **420**, 434–439.

Ammaranond, P., Cunningham, P., Oelrichs, R., Suzuki, K., Harris, C., Leas, L., Grulich, A., Cooper, D.A. and Kelleher, A.D., (2003), No increase in protease resistance and a decrease in reverse transcriptase resistance mutations in primary HIV-1 infection: 1992–2001. *AIDS*, **17**, 264–267.

Apetrei, C., Descamps, D., Collin, G., Loussert-Ajaka, I., Damond, F., Duca, M., Simon, F. and Brun-Vezinet, F., (1998), Human immunodeficiency virus type 1 subtype F reverse transcriptase sequence and drug susceptibility. *J Virol*, **72**, 3534–3538.

Bacheler, L., Jeffrey, S., Hanna, G., D'Aquila, R., Wallace, L., Logue, K., Cordova, B., Hertogs, K., Larder, B., Buckery, R., Baker, D., Gallagher, K., Scarnati, H., Tritch, R. and Rizzo, C., (2001), Genotypic correlates of phenotypic resistance to efavirenz in virus isolates from patients failing nonnucleoside reverse transcriptase inhibitor therapy. *J Virol*, **75**, 4999–5008.

Badri, S.M., Adeyemi, O.M., Max, B.E., Zagorski, B.M. and Barker, D.E., (2003), How does expert advice impact genotypic resistance testing in clinical practice? *Clin Infect Dis*, **37**, 708–713.

Balotta, C., Berlusconi, A., Pan, A., Violin, M., Riva, C., Colombo, M.C., Gori, A., Papagno, L., Corvasce, S., Mazzucchelli, R., Facchi, G., Velleca, R., Saporetti, G., Galli, M., Rusconi, S. and Moroni, M., (2000), Prevalence of transmitted nucleoside analogue-resistant HIV-1 strains and pre-existing mutations in pol reverse transcriptase and protease region: outcome after treatment in recently infected individuals. *Antivir Ther*, **5**, 7–14.

Balzarini, J., Karlsson, A., Perez-Perez, M.J., Camarasa, M.J., Tarpley, W.G. and De Clercq, E., (1993), Treatment of human immunodeficiency virus type 1 (HIV-1)-infected cells with combinations of HIV-1-specific inhibitors results in a different resistance pattern than does treatment with single-drug therapy. *J Virol*, **67**, 5353–5359.

Barbour, J.D., Hecht, F.M., Wrin, T., Liegler, T.J., Ramstead, C.A., Busch, M.P., Segal, M.R., Petropoulos, C.J. and Grant, R.M., (2004), Persistence of primary drug resistance among recently HIV-1 infected adults. *AIDS*, **18**, 1683–1689.

Baxter, J.D., Mayers, D.L., Wentworth, D.N., Neaton, J.D., Hoover, M.L., Winters, M.A., Mannheimer, S.B., Thompson, M.A., Abrams, D.I., Brizz, B.J., Ioannidis, J.P. and Merigan, T.C., (2000), A randomized study of antiretroviral management based on plasma genotypic antiretroviral resistance testing in patients failing therapy. CPCRA 046 Study Team for the Terry Beirn Community Programs for Clinical Research on AIDS. *AIDS*, **14**, F83–F93.

Becker-Pergola, G., Kataaha, P., Johnston-Dow, L., Fung, S., Jackson, J.B. and Eshleman, S.H., (2000), Analysis of HIV type 1 protease and reverse transcriptase in antiretroviral drug-naive Ugandan adults. *AIDS Res Hum Retroviruses*, **16**, 807–813.

Bennett, D., McCormick, L., Kline, R., Wheeler, W., Hemmen, M., Smith, A., Zaidi, I., Donolero, T. and the HIV Drug Resistant AVRDRP/VARHS Surveillance Group, (2005), US surveillance of HIV drug resistance at diagnosis using HIV diagnostic sera. 12th Conference on Retroviruses and Opportunistic Infections. Feb 22–25. Boston, MA. Abstract 674.

Birk, M. and Sonnerborg, A., (1998), Variations in HIV-1 pol gene associated with reduced sensitivity to antiretroviral drugs in treatment-naive patients. *AIDS*, **12**, 2369–2375.

Boden, D., Hurley, A., Zhang, L., Cao, Y., Guo, Y., Jones, E., Tsay, J., Ip, J., Farthing, C., Limoli, K., Parkin, N. and Markowitz, M., (1999), HIV-1 drug resistance in newly infected individuals. *JAMA*, **282**, 1135–1141.

Boden, D. and Markowitz, M., (1998), Resistance to human immunodeficiency virus type 1 protease inhibitors. *Antimicrob Agents Chemother*, **42**, 2775–2783.

Bossi, P., Peytavin, G., Ait-Mohand, H., Delaugerre, C., Ktorza, N., Paris, L., Bonmarchand, M., Cacace, R., David, D.J., Simon, A., Lamotte, C., Marcelin, A.G., Calvez, V., Bricaire, F., Costagliola, D. and Katlama, C., (2004), GENOPHAR: a randomized study of plasma drug measurements in association with genotypic resistance testing and expert advice to optimize therapy in patients failing antiretroviral therapy. *HIV Med*, **5**, 352–359.

Boyer, P.L., Currens, M.J., McMahon, J.B., Boyd, M.R. and Hughes, S.H., (1993), Analysis of nonnucleoside drug-resistant variants of human immunodeficiency virus type 1 reverse transcriptase. *J Virol*, **67**, 2412–2420.

Brenner, B., Routy, J.P., Quan, Y., Moisi, D., Oliveira, M., Turner, D. and Wainberg, M.A., (2004), Persistence of multidrug-resistant HIV-1 in primary infection leading to superinfection. *AIDS*, **18**, 1653–1660.

Brenner, B., Turner, D., Oliveira, M., Moisi, D., Detorio, M., Carobene, M., Marlink, R.G., Schapiro, J., Roger, M. and Wainberg, M.A., (2003), A V106M mutation in HIV-1 clade C viruses exposed to efavirenz confers cross-resistance to non-nucleoside reverse transcriptase inhibitors. *AIDS*, **17**, F1–F5.

Brenner, B.G., Routy, J.P., Petrella, M., Moisi, D., Oliveira, M., Detorio, M., Spira, B., Essabag, V., Conway, B., Lalonde, R., Sekaly, R.P. and Wainberg, M.A., (2002a), Persistence and fitness of multidrug-resistant human immunodeficiency virus type 1 acquired in primary infection. *J Virol*, **76**, 1753–1761.

Brenner, B.G., Turner, D. and Wainberg, M.A., (2002b), HIV-1 drug resistance: can we overcome? *Expert Opin Biol Ther*, **2**, 751–761.

Byrnes, V.W., Sardana, V.V., Schleif, W.A., Condra, J.H., Waterbury, J.A., Wolfgang, J.A., Long, W.J., Schneider, C.L., Schlabach, A.J., Wolanski, B.S., Graham, D.J., Gotlib, L., Rhodes, A., Titus, D.C., Roth, E., Blahy, OA., Quintero, J.C., Staszewski S. and Emini, E.A.,

(1993), Comprehensive mutant enzyme and viral variant assessment of human immunodeficiency virus type 1 reverse transcriptase resistance to nonnucleoside inhibitors. *Antimicrob Agents Chemother*, **37**, 1576–1579.

Cane, P., Chrystie, I., Dunn, D., Evans, B., Geretti, A.M., Green, H., Phillips, A., Pillay, D., Porter, K., Pozniak, A., Sabin, C., Smit, E., Weber, J. and Zuckerman, M., (2005), Time trends in primary resistance to HIV drugs in the United Kingdom: multicentre observational study. *BMJ*, **331**, 1368.

Casado, J.L., Moreno, A., Sabido, R., Marti-Belda, P., Antela, A., Dronda, F., Perez-Elias, M.J. and Moreno, S., (2003), Individualizing salvage regimens: the inhibitory quotient (Ctrough/IC50) as predictor of virological response. *AIDS*, **17**, 262–264.

Castagna, A., Gianotti, N., Galli, L., Danise, A., Hasson, H., Boeri, E., Hoetelmans, R., Nauwelaers, D. and Lazzarin, A., (2004), The NIQ of lopinavir is predictive of a 48-week virological response in highly treatment-experienced HIV-1-infected subjects treated with a lopinavir/ritonavir-containing regimen. *Antivir Ther*, **9**, 537–543.

Chaix, M.L., Descamps, D., Harzic, M., Schneider, V., Deveau, C., Tamalet, C., Pellegrin, I., Izopet, J., Ruffault, A., Masquelier, B., Meyer, L., Rouzioux, C., Brun-Vezinet, F. and Costagliola, D., (2003), Stable prevalence of genotypic drug resistance mutations but increase in non-B virus among patients with primary HIV-1 infection in France. *AIDS*, **17**, 2635–2643.

Chong, K.T., Pagano, P.J. and Hinshaw, R.R., (1994), Bisheteroarylpiperazine reverse transcriptase inhibitor in combination with 3'-azido-3'-deoxythymidine or 2',3'-dideoxycytidine synergistically inhibits human immunodeficiency virus type 1 replication in vitro. *Antimicrob Agents Chemother*, **38**, 288–293.

Clevenbergh, P., Bozonnat, M.C., Kirstetter, M., Durant, J., Cua, E., del Giudice, P., Montagne, N. and Simonet, P., (2003), Variable virological outcomes according to the center providing antiretroviral therapy within the PharmAdapt clinical trial. *HIV Clin Trials*, **4**, 84–91.

Coffin, J.M., (1995), HIV population dynamics in vivo: implications for genetic variation, pathogenesis, and therapy. *Science*, **267**, 483–489.

Colgrove, R. and Japour, A., (1999), A combinatorial ledge: reverse transcriptase fidelity, total body viral burden, and the implications of multiple-drug HIV therapy for the evolution of antiviral resistance. *Antiviral Res*, **41**, 45–56.

Colgrove, R.C., Pitt, J., Chung, P.H., Welles, S.L. and Japour, A.J., (1998), Selective vertical transmission of HIV-1 antiretroviral resistance mutations. *AIDS*, **12**, 2281–2288.

Colonno, R., Rose, R., McLaren, C., Thiry, A., Parkin, N. and Friborg, J., (2004), Identification of I50L as the signature atazanavir (ATV)-resistance mutation in treatment-naive HIV-1-infected patients receiving ATV-containing regimens. *J Infect Dis*, **189**, 1802–1810.

Condra, J., (1998), Virologic and clinical implications of resistance to HIV-1 protease inhibitors. *Drug Resist Updat*, **1**, 292–299.

Cornelissen, M., van den Burg, R., Zorgdrager, F., Lukashov, V. and Goudsmit, J., (1997), pol gene diversity of five human immunodeficiency virus type 1 subtypes: evidence for naturally occurring mutations that contribute to drug resistance, limited recombination patterns, and common ancestry for subtypes B and D. *J Virol*, **71**, 6348–6358.

D'Aquila, R.T., Schapiro, J.M., Brun-Vezinet, F., Clotet, B., Conway, B., Demeter, L.M., Grant, R.M., Johnson, V.A., Kuritzkes, D.R., Loveday, C., Shafer, R.W. and Richman, D.D., (2003), Drug resistance mutations in HIV-1. *Top HIV Med*, **11**, 92–96.

Daar, E.S. and Richman, D.D., (2005), Confronting the emergence of drug-resistant HIV type 1: impact of antiretroviral therapy on individual and population resistance. *AIDS Res Hum Retroviruses*, **21**, 343–357.

de Jong, M.D., Schuurman, R., Lange, J.M. and Boucher, C.A., (1996), Replication of a pre-existing resistant HIV-1 subpopulation in vivo after introduction of a strong selective drug pressure. *Antivir Ther*, **1**, 33–41.

de Vreese, K., Kofler-Mongold, V., Leutgeb, C., Weber, V., Vermeire, K., Schacht, S., Anne, J., de Clercq, E., Datema, R. and Werner, G., (1996), The molecular target of bicyclams, potent inhibitors of human immunodeficiency virus replication. *J Virol*, **70**, 689–696.

Dean, M., Carrington, M., Winkler, C., Huttley, G.A., Smith, M.W., Allikmets, R., Goedert, J.J., Buchbinder, S.P., Vittinghoff, E., Gomperts, E., Donfield, S., Vlahov, D., Kaslow, R., Saah, A.,

Rinaldo, C., Detels, R., O'Brien, S.J., (1996), Genetic restriction of HIV-1 infection and progression to AIDS by a deletion allele of the CKR5 structural gene. Hemophilia Growth and Development Study, Multicenter AIDS Cohort Study, Multicenter Hemophilia Cohort Study, San Francisco City Cohort, ALIVE Study. *Science*, **273**, 1856–1862.

Deeks, S.G., (1999), Failure of HIV-1 protease inhibitors to fully suppress viral replication. Implications for salvage therapy. *Adv Exp Med Biol*, **458**, 175–182.

Deeks, S.G., (2001), International perspectives on antiretroviral resistance. Nonnucleoside reverse transcriptase inhibitor resistance. *J Acquir Immune Defic Syndr*, **26** (Suppl 1), S25–S33.

DeGruttola, V., Dix, L., D'Aquila, R., Holder, D., Phillips, A., Ait-Khaled, M., Baxter, J., Clevenbergh, P., Hammer, S., Harrigan, R., Katzenstein, D., Lanier, R., Miller, M., Para, M., Yerly, S., Zolopa, A., Murray, J., Patick, A., Miller, V., Castillo, S., Pedneault, L. and Mellors, J., (2000), The relation between baseline HIV drug resistance and response to antiretroviral therapy: re-analysis of retrospective and prospective studies using a standardized data analysis plan. *Antivir Ther*, **5**, 41–48.

Derdeyn, C.A., Decker, J.M., Sfakianos, J.N., Wu, X., O'Brien, W.A., Ratner, L., Kappes, J.C., Shaw, G.M. and Hunter, E., (2000), Sensitivity of human immunodeficiency virus Type 1 to the fusion inhibitor T-20 is modulated by coreceptor specificity defined by the V3 loop of gp120. *J Virol*, **74**, 8358–8367.

Descamps, D., Collin, G., Letourneur, F., Apetrei, C., Damond, F., Loussert-Ajaka, I., Simon, F., Saragosti, S., Brun-Vezinet, F., (1997), Susceptibility of human immunodeficiency virus type 1 group O isolates to antiretroviral agents: in vitro phenotypic and genotypic analyses. *J Virol*, **71**, 8893–8898.

Descamps, D., Collin, G., Loussert-Ajaka, I., Saragosti, S., Simon, F. and Brun-Vezinet, F., (1995), HIV-1 group O sensitivity to antiretroviral drugs. *AIDS*, **9**, 977–978.

Devereux, H.L., Youle, M., Johnson, M.A. and Loveday, C., (1999), Rapid decline in detectability of HIV-1 drug resistance mutations after stopping therapy. *AIDS*, **13**, F123–F127.

Dickover, R.E., Garratty, E.M., Plaeger, S. and Bryson, Y.J., (2001), Perinatal transmission of major, minor, and multiple maternal human immunodeficiency virus type 1 variants in utero and intrapartum. *J Virol*, **75**, 2194–2203.

Ding, J., Das, K., Moereels, H., Koymans, L., Andries, K., Janssen, P.A., Hughes, S.H. and Arnold, E., (1995), Structure of HIV-1 RT/TIBO R 86183 complex reveals similarity in the binding of diverse nonnucleoside inhibitors. *Nat Struct Biol*, **2**, 407–415.

Dorr, P., Westby, M., Dobbs, S., Griffin, P., Irvine, B., Macartney, M., Mori, J., Rickett, G., Smith-Burchnell, C., Napier, C., Webster, R., Armour, D., Price, D., Stammen, B., Wood, A. and Perros, M., (2005), Maraviroc (UK-427,857), a potent, orally bioavailable, and selective small-molecule inhibitor of chemokine receptor CCR5 with broad-spectrum anti-human immunodeficiency virus Type 1 activity. *Antimicrob Agents Chemother*, **49**, 4721–4732.

Doyon, L., Croteau, G., Thibeault, D., Poulin, F., Pilote, L. and Lamarre, D., (1996), Second locus involved in human immunodeficiency virus type 1 resistance to protease inhibitors. *J Virol*, **70**, 3763–3769.

Doyon, L., Payant, C., Brakier-Gingras, L. and Lamarre, D., (1998), Novel Gag–Pol frameshift site in human immunodeficiency virus type 1 variants resistant to protease inhibitors. *J Virol*, **72**, 6146–6150.

Drosopoulos, W.C., Rezende, L.F., Wainberg, M.A. and Prasad, V.R., (1998), Virtues of being faithful: can we limit the genetic variation in human immunodeficiency virus? *J Mol Med*, **76**, 604–612.

Dueweke, T.J., Pushkarskaya, T., Poppe, S.M., Swaney, S.M., Zhao, J.Q., Chen, I.S., Stevenson, M. and Tarpley, W.G., (1993), A mutation in reverse transcriptase of bis(heteroaryl)piperazine-resistant human immunodeficiency virus type 1 that confers increased sensitivity to other non-nucleoside inhibitors. *Proc Natl Acad Sci USA*, **90**, 4713–4717.

Durant, J., Clevenbergh, P., Halfon, P., Delgiudice, P., Porsin, S., Simonet, P., Montagne, N., Boucher, C.A., Schapiro, J.M. and Dellamonica, P., (1999), Drug-resistance genotyping in HIV-1 therapy: the VIRADAPT randomised controlled trial. *Lancet*, **353**, 2195–2199.

Esnouf, R., Ren, J., Ross, C., Jones, Y., Stammers, D. and Stuart, D., (1995), Mechanism of inhibition of HIV-1 reverse transcriptase by non-nucleoside inhibitors. *Nat Struct Biol*, **2**, 303–308.

Este, J.A., Cabrera, C., Blanco, J., Gutierrez, A., Bridger, G., Henson, G., Clotet, B., Schols, D. and De Clercq, E., (1999), Shift of clinical human immunodeficiency virus Type 1 isolates from X4 to R5 and prevention of emergence of the Syncytium-inducing phenotype by blockade of CXCR4. *J Virol*, **73**, 5577–5585.

Ferrer, E., Podzamczer, D., Arnedo, M., Fumero, E., McKenna, P., Rinehart, A., Perez, J.L., Barbera, M.J., Pumarola, T., Gatell, J.M. and Gudiol, F., (2003), Genotype and phenotype at baseline and at failure in human immunodeficiency virus-infected antiretroviral-naive patients in a randomized trial comparing zidovudine and lamivudine plus nelfinavir or nevirapine. *J Infect Dis*, **187**, 687–690.

Finzi, D., Blankson, J., Siliciano, J.D., Margolick, J.B., Chadwick, K., Pierson, T., Smith, K., Lisziewicz, J., Lori, F., Flexner, C., Quinn, T.C., Chaisson, R.E., Rosenberg, E., Walker, B., Gange, S., Gallant, J. and Siliciano, R.F., (1999), Latent infection of CD4+ T cells provides a mechanism for lifelong persistence of HIV-1, even in patients on effective combination therapy. *Nat Med*, **5**, 512–517.

Fletcher, R.S., Arion, D., Borkow, G., Wainberg, M.A., Dmitrienko, G.I. and Parniak, M.A., (1995), Synergistic inhibition of HIV-1 reverse transcriptase DNA polymerase activity and virus replication in vitro by combinations of carboxanilide nonnucleoside compounds. *Biochemistry*, **34**, 10106–10112.

Frost, S.D., Nijhuis, M., Schuurman, R., Boucher, C.A. and Brown, A.J., (2000), Evolution of lamivudine resistance in human immunodeficiency virus type 1-infected individuals: the relative roles of drift and selection. *J Virol*, **74**, 6262–6268.

Furman, P.A., Fyfe, J.A., St Clair, M.H., Weinhold, K., Rideout, J.L., Freeman, G.A., Lehrman, S.N., Bolognesi, D.P., Broder, S., Mitsuya, H. and Barry, D.W., (1986), Phosphorylation of 3'-azido-3'-deoxythymidine and selective interaction of the 5'-triphosphate with human immunodeficiency virus reverse transcriptase. *Proc Natl Acad Sci USA*, **83**, 8333–8337.

Gandhi, R.T., Wurcel, A., Rosenberg, E.S., Johnston, M.N., Hellmann, N., Bates, M., Hirsch, M.S. and Walker, B.D., (2003), Progressive reversion of human immunodeficiency virus type 1 resistance mutations in vivo after transmission of a multiply drug-resistant virus. *Clin Infect Dis*, **37**, 1693–1698.

Gao, Q., Gu, Z., Salomon, H., Nagai, K., Parniak, M.A. and Wainberg, M.A., (1994), Generation of multiple drug resistance by sequential in vitro passage of the human immunodeficiency virus type 1. *Arch Virol*, **136**, 111–122.

Gatanaga, H., Suzuki, Y., Tsang, H., Yoshimura, K., Kavlick, M.F., Nagashima, K., Gorelick, R.J., Mardy, S., Tang, C., Summers, M.F. and Mitsuya, H., (2002), Amino acid substitutions in Gag protein at non-cleavage sites are indispensable for the development of a high multitude of HIV-1 resistance against protease inhibitors. *J Biol Chem*, **277**, 5952–5961.

Gianotti, N., Mondino, V., Rossi, M.C., Chiesa, E., Mezzaroma, I., Ladisa, N., Guaraldi, G., Torti, C., Tarquini, P., Castelli, P., Di Carlo, A., Boeri, E., Keulen, W., Kenna, P.M. and Lazzarin, A., (2006), Comparison of a rule-based algorithm with a phenotype-based algorithm for the interpretation of HIV genotypes in guiding salvage regimens in HIV-infected patients by a randomized clinical trial: the mutations and salvage study. *Clin Infect Dis*, **42**, 1470–1480.

Gonzales, M.J., Delwart, E., Rhee, S.Y., Tsui, R., Zolopa, A.R., Taylor, J. and Shafer, R.W., (2003), Lack of detectable human immunodeficiency virus type 1 superinfection during 1072 person-years of observation. *J Infect Dis*, **188**, 397–405.

Gonzalez de Requena, D., Gallego, O., Valer, L., Jimenez-Nacher, I. and Soriano, V., (2004), Prediction of virological response to lopinavir/ritonavir using the genotypic inhibitory quotient. *AIDS Res Hum Retroviruses*, **20**, 275–278.

Gotte, M. and Wainberg, M.A., (2000), Biochemical mechanisms involved in overcoming HIV resistance to nucleoside inhibitors of reverse transcriptase. *Drug Resist Updat*, **3**, 30–38.

Graham, N., Peeters, M., Verbiest, W., Harrigan, R. and Larder B., (2001), The Virtual Phenotype is an independent predictor of clinical response. Program and abstracts of the 8th Conference on Retroviruses and Opportunistic Infections. 4–8 February 2001, Chicago, USA. Abstract 524.

Greenberg, M., Sista, P., Miralles, G., Melby, T., Davison, D., Jin, L., Mosier, S., Mink, M., Nelson, E., Demasi, R., Fang, L., Cammack, N., Salgo, M., Duff, F., and Matthews, T. for the T1249-101 Study Group, (2002), Enfuvirtide (T-20) and T-1249 resistance: observations from Phase

II clinical trials of enfuvirtide in combination with oral antiretrovirals (ARVs) and a Phase I/II dose-ranging monotherapy trial of T-1249. In XI International HIV Drug Resistance Workshop, Seville, Spain, 2–5 July 2002. Abstract 128. *Antivir Ther*, **7**, S106.

Greenberg, M.L., Melby, T., Sista, P., DeMasi, R., Cammack, N., Salgo, M., Whitcomb, J., Petropoulos, C. and Matthews, T., (2003a), Baseline and on-treatment susceptibility to enfuvirtide seen in TORO 1 and TORO 2 through 24 weeks. 10th Conference on Retroviruses and Opportunistic Infections, Boston, USA, 10–14 February 2003. Abstract 141.

Greenberg, M.L., Melby, T., Sista, P., Demasi, R., Cammack, N., Salgo, M., Whitcomb, J., Petropoulos, C. and Matthews, T.J., (2003b), Baseline and on-treatment susceptibility to enfuvirtide seen in TORO 1 and TORO 2 through 24 weeks. 10th Conference on Retroviruses and Opportunistic Infections, Boston, USA, 10–14 February 2003. Abstract 141.

Gu, Z., Quan, Y., Li, Z., Arts, E.J. and Wainberg, M.A., (1995), Effects of non-nucleoside inhibitors of human immunodeficiency virus type 1 in cell-free recombinant reverse transcriptase assays. *J Biol Chem*, **270**, 31046–31051.

Gunthard, H.F., Wong, J.K., Ignacio, C.C., Guatelli, J.C., Riggs, N.L., Havlir, D.V. and Richman, D.D., (1998), Human immunodeficiency virus replication and genotypic resistance in blood and lymph nodes after a year of potent antiretroviral therapy. *J Virol*, **72**, 2422–2428.

Hart, G.J., Orr, D.C., Penn, C.R., Figueiredo, H.T., Gray, N.M., Boehme, R.E. and Cameron, J.M., (1992), Effects of (–)-2'-deoxy-3'-thiacytidine (3TC) 5'-triphosphate on human immunodeficiency virus reverse transcriptase and mammalian DNA polymerases alpha, beta, and gamma. *Antimicrob Agents Chemother*, **36**, 1688–1694.

He, Y., Vassell, R., Zaitseva, M., Nguyen, N., Yang, Z., Weng, Y. and Weiss, C.D., (2003), Peptides trap the human immunodeficiency virus Type 1 envelope glycoprotein fusion intermediate at two sites. *J Virol*, **77**, 1666–1671.

Hecht, F.M., Grant, R.M., Petropoulos, C.J., Dillon, B., Chesney, M.A., Tian, H., Hellmann, N.S., Bandrapalli, N.I., Digilio, L., Branson, B. and Kahn, J.O., (1998), Sexual transmission of an HIV-1 variant resistant to multiple reverse-transcriptase and protease inhibitors. *N Engl J Med*, **339**, 307–311.

Hertogs, K., Bloor, S., Kemp, S.D., Van den Eynde, C., Alcorn, T.M., Pauwels, R., Van Houtte, M., Staszewski, S., Miller, V. and Larder, B.A., (2000), Phenotypic and genotypic analysis of clinical HIV-1 isolates reveals extensive protease inhibitor cross-resistance: a survey of over 6000 samples. *AIDS*, **14**, 1203–1210.

Hirsch, M.S., Brun-Vezinet, F., Clotet, B., Conway, B., Kuritzkes, D.R., D'Aquila, R.T., Demeter, L.M., Hammer, S.M., Johnson, V.A., Loveday, C., Mellors, J.W., Jacobsen, D.M. and Richman, D.D., (2003), Antiretroviral drug resistance testing in adults infected with human immunodeficiency virus type 1: 2003 recommendations of an International AIDS Society-USA Panel. *Clin Infect Dis*, **37**, 113–128.

Holguin, A. and Soriano, V., (2002), Resistance to antiretroviral agents in individuals with HIV-1 non-B subtypes. *HIV Clin Trials*, **3**, 403–411.

Japour, A.J., Mayers, D.L., Johnson, V.A., Kuritzkes, D.R., Beckett, L.A., Arduino, J.M., Lane, J., Black, R.J., Reichelderfer, P.S., D'Aquila, R.T. and Crumpacker, C.S., (1993), Standardized peripheral blood mononuclear cell culture assay for determination of drug susceptibilities of clinical human immunodeficiency virus type 1 isolates. The RV-43 Study Group, the AIDS Clinical Trials Group Virology Committee Resistance Working Group. *Antimicrob Agents Chemother*, **37**, 1095–1101.

Jennings, C.L., Bosch, R.J., Dragavon, J., Kabat, W., Shugarts, D.L., Boone, L., Decker, W.D., Givens, M.R., Lambrecht, L., Bick, C.T.S., DiFrancesco, R., Hartman, K.E., Cooper, M., Kondo, P.K., Medvik, K.A., Sepelak, S., Tustin, N., Kaczmarek, M., Yanavich, C., Mong-Kryspin, L., Siminski, S., Vincent, C.A., Kerkau, M.G., Livnat, D., Piwowar-Manning, E.M. and Tooley, K., (2004), The virology manual for HIV laboratories. National Institute for Allergy and Infectious Disease. Available at http://aactg.s-3.com/labmanual.htm.

Johnson, V.A., Brun-Vezinet, F., Clotet, B., Conway, B., Kuritzkes, D.R., Pillay, D., Schapiro, J.M., Telenti, A. and Richman, D.D., (2005), Update of the drug resistance mutations in HIV-1: Fall 2005. *Top HIV Med*, **13**, 125–131.

Jonckheere, H., Taymans, J.M., Balzarini, J., Velazquez, S., Camarasa, M.J., Desmyter, J., De Clercq, E. and Anne, J., (1994), Resistance of HIV-1 reverse transcriptase against [2',5'-bis-O-(tert-butyldimethylsilyl)-3'-spiro-5"-(4"-amino-1",2"- oxathiole-2",2"-dioxide)] (TSAO) derivatives is determined by the mutation Glu138→Lys on the p51 subunit. *J Biol Chem*, **269**, 25255–25258.

Jost, S., Bernard, M.-C., Kaiser, L., Yerly, S., Hirschel, B., Samri, A., Autran, B., Goh, L.-E. and Perrin, L., (2002), A Patient with HIV-1 Superinfection. *N Engl J Med*, **347**, 731–736.

Kantor, R., Zijenah, L.S., Shafer, R.W., Mutetwa, S., Johnston, E., Lloyd, R., von Lieven, A., Israelski, D. and Katzenstein, D.A., (2002), HIV-1 subtype C reverse transcriptase and protease genotypes in Zimbabwean patients failing antiretroviral therapy. *AIDS Res Hum Retroviruses*, **18**, 1407–1413.

Kitrinos, K., LaBranche, C., Stanhope, M., Madsen, H. and Demarest, J., (2005), Clonal analysis detects pre-existing R5X4-tropic virus in patient demonstrating populations-level tropism shift on 873140 monotherapy. *Antivir Ther*, **10**, S68. Abstract 61.

Koelsch, K.K., Smith, D.M., Little, S.J., Ignacio, C.C., Macaranas, T.R., Brown, A.J., Petropoulos, C.J., Richman, D.D. and Wong, J.K.,(2003), Clade B HIV-1 superinfection with wild-type virus after primary infection with drug-resistant clade B virus. *AIDS*, **17**, F11–F16.

Kuhmann, S.E., Pugach, P., Kunstman, K.J., Taylor, J., Stanfield, R.L., Snyder, A., Strizki, J.M., Riley, J., Baroudy, B.M., Wilson, I.A., Korber, B.T., Wolinsky, S.M. and Moore, J.P., (2004), Genetic and phenotypic analyses of human immunodeficiency virus Type 1 escape from a small-molecule CCR5 inhibitor. *J Virol*, **78**, 2790–2807.

Lalezari, J.P., DeJesus, E., Northfelt, D.W., Richmond, G., Wolfe, P., Haubrich, R., Henry, D., Powderly, W., Becker, S., Thompson, M., Valentine, F., Wright, D., Carlson, M., Riddler, S., Haas, F.F., DeMasi, R., Sista, P.R., Salgo, M. and Delehanty, J., (2003a), A controlled Phase II trial assessing three doses of enfuvirtide (T-20) in combination with abacavir, amprenavir, ritonavir and efavirenz in non-nucleoside reverse transcriptase inhibitor-naive HIV-infected adults. *Antivir Ther*, **8**, 279–287.

Lalezari, J.P., Eron, J.J., Carlson, M., Cohen, C., DeJesus, E., Arduino, R.C., Gallant, J.E., Volberding, P., Murphy, R.L., Valentine, F., Nelson, E.L., Sista, P.R., Dusek, A. and Kilby, J.M., (2003b), A phase II clinical study of the long-term safety and antiviral activity of enfuvirtide-based antiretroviral therapy. *AIDS*, **17**, 691–698.

Lalezari, J.P., Henry, K., O'Hearn, M., Montaner, J.S.G., Piliero, P.J., Trottier, B., Walmsley, S., Cohen, C., Kuritzkes, D.R., Eron, J.J., Jr, Chung, J., DeMasi, R., Donatacci, L., Drobnes, C., Delehanty, J., Salgo, M. and the TSG, (2003c), Enfuvirtide, an HIV-1 fusion inhibitor, for drug-resistant HIV infection in North and South America. *N Engl J Med*, **348**, 2175–2185.

Larder, B.A., Kemp, S.D., and Hertogs, K., (2000), Quantitative prediction of HIV-1 phenotypic drug resistance from genotypes: the virtual phenotype (VirtualPhenotype). *Antiviral Ther*, **5** (Suppl 3), 49. Abstract 63.

Lazzarin, A., Clotet, B., Cooper, D., Reynes, J., Arasteh, K., Nelson, M., Katlama, C., Stellbrink, H.-J., Delfraissy, J.-F., Lange, J., Huson, L., DeMasi, R., Wat, C., Delehanty, J., Drobnes, C., Salgo, M. and the TSG, (2003), Efficacy of enfuvirtide in patients infected with drug-resistant HIV-1 in Europe and Australia. *N Engl J Med*, **348**, 2186–2195.

Lederman, M.M., Offord, R.E. and Hartley, O., (2006), Microbicides and other topical strategies to prevent vaginal transmission of HIV. *Nat Rev Immunol*, **6**, 371–382.

Lehrman, G., Hogue, I.B., Palmer, S., Jennings, C., Spina, C.A., Wiegand, A., Landay, A.L., Coombs, R.W., Richman, D.D., Mellors, J.W., Coffin, J.M., Bosch, R.J. and Margolis, D.M., (2005), Depletion of latent HIV-1 infection in vivo: a proof-of-concept study. *Lancet*, **366**, 549–555.

Liang, C., Rong, L., Quan, Y., Laughrea, M., Kleiman, L. and Wainberg, M.A., (1999), Mutations within four distinct gag proteins are required to restore replication of human immunodeficiency virus type 1 after deletion mutagenesis within the dimerization initiation site. *J Virol*, **73**, 7014–7020.

Little, S.J., Holte, S., Routy, J.P., Daar, E.S., Markowitz, M., Collier, A.C., Koup, R.A., Mellors, J.W., Connick, E., Conway, B., Kilby, M., Wang, L., Whitcomb, J.M., Hellmann, N.S. and Richman, D.D., (2002), Antiretroviral-drug resistance among patients recently infected with HIV. *N Engl J Med*, **347**, 385–394.

Loemba, H., Brenner, B., Parniak, M.A., Ma'ayan S, Spira, B., Moisi, D., Oliveira, M., Detorio, M. and Wainberg, M.A., (2002), Genetic divergence of human immunodeficiency virus type 1 Ethiopian clade C reverse transcriptase (RT) and rapid development of resistance against non-nucleoside inhibitors of RT. *Antimicrob Agents Chemother*, **46**, 2087–2094.

Logsdon, B.C., Vickrey, J.F., Martin, P., Proteasa, G., Koepke, J.I., Terlecky, S.R., Wawrzak, Z., Winters, M.A., Merigan, T.C. and Kovari, L.C., (2004), Crystal structures of a multidrug-resistant human immunodeficiency virus Type 1 protease reveal an expanded active-site cavity. *J Virol*, **78**, 3123–3132.

Lorenzi, P., Opravil, M., Hirschel, B., Chave, J.P., Furrer, H.J., Sax, H., Perneger, T.V., Perrin, L., Kaiser, L. and Yerly, S., (1999), Impact of drug resistance mutations on virologic response to salvage therapy. Swiss HIV Cohort Study. *AIDS*, **13**, F17–F21.

Loveday, C., (2001), International perspectives on antiretroviral resistance. Nucleoside reverse transcriptase inhibitor resistance. *J Acquir Immune Defic Syndr*, **26** (Suppl 1), S10–S24.

Loya, S., Bakhanashvili, M., Tal, R., Hughes, S.H., Boyer, P.L. and Hizi, A., (1994), Enzymatic properties of two mutants of reverse transcriptase of human immunodeficiency virus type 1 (tyrosine 181→isoleucine and tyrosine 188→leucine), resistant to nonnucleoside inhibitors. *AIDS Res Hum Retroviruses*, **10**, 939–946.

Lu, J., Sista, P., Giguel, F., Greenberg, M. and Kuritzkes, D.R., (2004), Relative replicative fitness of human immunodeficiency virus Type 1 mutants resistant to enfuvirtide (T-20). *J Virol*, **78**, 4628–4637.

Mangeat, B., Turelli, P., Caron, G., Friedli, M., Perrin, L. and Trono, D., (2003), Broad antiretroviral defence by human APOBEC3G through lethal editing of nascent reverse transcripts. *Nature*, **424**, 99–103.

Mansky, L.M., (2002), HIV mutagenesis and the evolution of antiretroviral drug resistance. *Drug Resist Updat*, **5**, 219–223.

Mansky, L.M., Le Rouzic, E., Benichou, S., Gajary, L.C., (2003), Influence of reverse transcriptase variants, drugs, and Vpr on human immunodeficiency virus type 1 mutant frequencies. *J Virol*, **77**, 2071–2080.

Marcelin, A.G., Lamotte, C., Delaugerre, C., Ktorza, N., Ait Mohand, H., Cacace, R., Bonmarchand, M., Wirden, M., Simon, A., Bossi, P., Bricaire, F., Costagliola, D., Katlama, C., Peytavin, G. and Calvez, V., (2003), Genotypic inhibitory quotient as predictor of virological response to ritonavir–amprenavir in human immunodeficiency virus type 1 protease inhibitor-experienced patients. *Antimicrob Agents Chemother*, **47**, 594–600.

Mariani, R., Chen, D., Schrofelbauer, B., Navarro, F., Konig, R., Bollman, B., Munk, C., Nymark-McMahon, H. and Landau, N.R., (2003), Species-specific exclusion of APOBEC3G from HIV-1 virions by Vif. *Cell*, **114**, 21–31.

Markham, R.B., Wang, W.-C., Weisstein, A.E., Wang, Z., Munoz, A., Templeton, A., Margolick, J., Vlahov, D., Quinn, T., Farzadegan, H. and Yu, X.-F., (1998), Patterns of HIV-1 evolution in individuals with differing rates of CD4 T cell decline. *Proc Nat Acad Sci USA*, **95**, 12568–12573.

Marozsan, A.J., Kuhmann, S.E., Morgan, T., Herrera, C., Rivera-Troche, E., Xu, S., Baroudy, B.M., Strizki, J. and Moore, J.P., (2005), Generation and properties of a human immunodeficiency virus type 1 isolate resistant to the small molecule CCR5 inhibitor, SCH-417690 (SCH-D). *Virology*, **338**, 182.

Martin, P., Vickrey, J.F., Proteasa, G., Jimenez, Y.L., Wawrzak, Z., Winters, M.A., Merigan, T.C. and Kovari, L.C., (2005), "Wide-Open" 1.3 A Structure of a Multidrug-Resistant HIV-1 Protease as a Drug Target. Structure. **13**, 1887–1895.

Martinez-Picado, J., Savara, A.V., Sutton, L. and D'Aquila, R.T., (1999), Replicative fitness of protease inhibitor-resistant mutants of human immunodeficiency virus type 1. *J Virol*, **73**, 3744–3752.

Martinez, J.L., McArthur, R., VIckrey, J., Martin, P., Proteasa, G., Kondapalli, K., Jimenez, Y., Wawrzak, W., Winters, M., Merigan, T. and Kovari, L., (2004), The Multi-Drug Resistance Protease Represents a Novel Drug Target 44th Annual Interscience Conference on Antimicrobial Agents and Chemotherapy. American Society of Microbiology, Washington, DC, USA.

Mayers, D.L., (1997), Prevalence and incidence of resistance to zidovudine and other antiretroviral drugs. *Am J Med*, **102**, 70–75.

Mazzotta, F., Lo Caputo, S., Torti, C., Tinelli, C., Pierotti, P., Castelli, F., Lazzarin, A., Angarano, G., Maserati, R., Gianotti, N., Ladisa, N., Quiros-Roldan, E., Rinehart, A.R. and Carosi, G., (2003), Real versus virtual phenotype to guide treatment in heavily pretreated patients: 48-week follow-up of the Genotipo-Fenotipo di Resistenza (GenPheRex) trial. *J Acquir Immune Defic Syndr*, **32**, 268–280.

Menendez-Arias, L., (2002), Molecular basis of fidelity of DNA synthesis and nucleotide specificity of retroviral reverse transcriptases. *Prog Nucleic Acid Res Mol Biol*, **71**, 91–147.

Menzo, S., Castagna, A., Monachetti, A., Hasson, H., Danise, A., Carini, E., Bagnarelli, P., Lazzarin, A. and Clementi, M., (2004), Resistance and replicative capacity of HIV-1 strains selected in vivo by long-term enfuvirtide treatment. *New Microbiol*, **27**, 51–61.

Mink, M., Greenberg, M., Mosier, S., Janumpalli, S., Davison, D., Jin, L., Melby, T., Sista, P., Lambert, D., Cammack, N., Salgo, M. and Matthews, T., (2002), Impact of HIV-1 gp41 amino acid substitutions (positions 36–45) on susceptibility to T-20 (enfuvirtide) in vitro: analysis of primary virus isolates recovered from patients during chronic enfuvirtide treatment and site directed mutants in NL4-3. In XI International HIV Drug Resistance Workshop, Seville, Spain, 2–5 July 2002. Abstract 22. *Antivir Ther*, **7**, S17.

Molla, A., Korneyeva, M., Gao, Q., Vasavanonda, S., Schipper, P.J., Mo, H.M., Markowitz, M., Chernyavskiy, T., Niu, P., Lyons, N., Hsu, A., Granneman, G.R., Ho, D.D., Boucher, C.A., Leonard, J.M., Norbeck, D.W. and Kempf, D.J., (1996), Ordered accumulation of mutations in HIV protease confers resistance to ritonavir. *Nat Med*, **2**, 760–766.

Murphy, R.L., (1999), New antiretroviral drugs part I: PIs. *AIDS Clin Care*, **11**, 35–37.

Myint, L., Matsuda, M., Matsuda, Z., Yokomaku, Y., Chiba, T., Okano, A., Yamada, K. and Sugiura, W., (2004), Gag non-cleavage site mutations contribute to full recovery of viral fitness in protease inhibitor-resistant human immunodeficiency virus type 1. *Antimicrob Agents Chemother*, **48**, 444–452.

Nijhuis, M., Deeks, S. and Boucher, C., (2001), Implications of antiretroviral resistance on viral fitness. *Curr Opin Infect Dis*, **14**, 23–28.

Nunberg, J.H., Schleif, W.A., Boots, E.J., O'Brien, J.A., Quintero, J.C., Hoffman, J.M., Emini, E.A. and Goldman, M.E., (1991), Viral resistance to human immunodeficiency virus type 1-specific pyridinone reverse transcriptase inhibitors. *J Virol*, **65**, 4887–4892.

Ohtaka, H., Velazquez-Campoy, A., Xie, D. and Freire, E., (2002), Overcoming drug resistance in HIV-1 chemotherapy: the binding thermodynamics of Amprenavir and TMC-126 to wild-type and drug-resistant mutants of the HIV-1 protease. *Protein Sci*, **11**, 1908–1916.

Ormaasen, V., Sandvik, L., Asjo, B., Holberg-Petersen, M., Gaarder, P.I. and Bruun, J.N., (2004), An algorithm-based genotypic resistance score is associated with clinical outcome in HIV-1-infected adults on antiretroviral therapy. *HIV Med*, **5**, 400–406.

Overbaugh, J. and Bangham, C.R., (2001), Selection forces and constraints on retroviral sequence variation. *Science*, **292**, 1106–1109.

Palella, F.J., Jr, Delaney, K.M., Moorman, A.C., Loveless, M.O., Fuhrer, J., Satten, G.A., Aschman, D.J. and Holmberg, S.D., (1998), Declining morbidity and mortality among patients with advanced human immunodeficiency virus infection. HIV Outpatient Study Investigators. *N Engl J Med*, **338**, 853–860.

Palmer, S., Alaeus, A., Albert, J. and Cox, S., (1998), Drug susceptibility of subtypes A,B,C,D, and E human immunodeficiency virus type 1 primary isolates. *AIDS Res Hum Retroviruses*, **14**, 157–162.

Palmer, S., Shafer, R.W. and Merigan, T.C., (1999), Highly drug-resistant HIV-1 clinical isolates are cross-resistant to many antiretroviral compounds in current clinical development. *AIDS*, **13**, 661–667.

Panidou, E.T., Trikalinos, T.A., Ioannidis, J.P., (2004), Limited benefit of antiretroviral resistance testing in treatment-experienced patients: a meta-analysis. *AIDS*, **18**, 2153–2161.

Patick, A.K., Duran, M., Cao, Y., Shugarts, D., Keller, M.R., Mazabel, E., Knowles, M., Chapman, S., Kuritzkes, D.R. and Markowitz, M., (1998), Genotypic and phenotypic characterization of human immunodeficiency virus Type 1 variants isolated from patients treated with the protease inhibitor nelfinavir. *Antimicrob Agents Chemother*, **42**, 2637–2644.

Perez-Alvarez, L., Carmona, R., Ocampo, A., Asorey, A., Miralles, C., Perez de Castro, S., Pinilla, M., Contreras, G., Taboada, J.A. and Najera, R., (2006), Long-term monitoring of genotypic and phenotypic resistance to T20 in treated patients infected with HIV-1. *J Med Virol*, **78**, 141–147.

Petrella, M., Brenner, B., Loemba, H. and Wainberg, M.A., (2001), HIV drug resistance and implications for the introduction of antiretroviral therapy in resource-poor countries. *Drug Resist Updat*, **4**, 339–346.

Petrella, M., Wainberg, M.A., (2002), Might the M184V substitution in HIV-1 RT confer clinical benefit? AIDS Rev 4:224–232.

Phillips, A., (2001), Will the drugs still work? Transmission of resistant HIV. *Nat Med*, **7**, 993–994.

Poveda, E., Rodes, B., Labernardiere, J.L., Benito, J.M., Toro, C., Gonzalez-Lahoz, J., Faudon, J.L., Clavel, F., Schapiro, J. and Soriano, V., (2004), Evolution of genotypic and phenotypic resistance to Enfuvirtide in HIV-infected patients experiencing prolonged virologic failure. *J Med Virol*, **74**, 21–28.

Poveda, E., Rodes, B., Lebel-Binay, S., Faudon, J.-L., Jimenez, V. and Soriano, V., (2005), Dynamics of enfuvirtide resistance in HIV-infected patients during and after long-term enfuvirtide salvage therapy. *Focus on HIV*, **34**, 295.

Preston, B.D. and Dougherty, J.P., (1996), Mechanisms of retroviral mutation. *Trends Microbiol*, **4**, 16–21.

Quinn, T.C., Wawer, M.J., Sewankambo, N., Serwadda, D., Li, C., Wabwire-Mangen, F., Meehan, M.O., Lutalo, T. and Gray, R.H., (2000), Viral load and heterosexual transmission of human immunodeficiency virus type 1. Rakai Project Study Group. *N Engl J Med*, **342**, 921–929.

Quiros-Roldan, E., Signorini, S., Castelli, F., Torti, C., Patroni, A., Airoldi, M. and Carosi, G., (2001), Analysis of HIV-1 mutation patterns in patients failing antiretroviral therapy. *J Clin Lab Anal*, **15**, 43–46.

Ramos, A., Hu, D.J., Nguyen, L., Phan, K.O., Vanichseni, S., Promadej, N., Choopanya, K., Callahan, M., Young, N.L., McNicholl, J., Mastro, T.D., Folks, T.M. and Subbarao, S., (2002), Intersubtype human immunodeficiency virus type 1 superinfection following seroconversion to primary infection in two injection drug users. *J Virol*, **76**, 7444–7452.

Reeves, J.D., Gallo, S.A., Ahmad, N., Miamidian, J.L., Harvey, P.E., Sharron, M., Pohlmann, S., Sfakianos, J.N., Derdeyn, C.A., Blumenthal, R., Hunter, E. and Doms, R.W., (2002), Sensitivity of HIV-1 to entry inhibitors correlates with envelope/coreceptor affinity, receptor density, and fusion kinetics. *Proc Natl Acad Sci*, **99**, 16249–16254.

Reeves, J.D., Lee, F.-H., Miamidian, J.L., Jabara, C.B., Juntilla, M.M. and Doms, R.W., (2005), Enfuvirtide resistance mutations: impact on human immunodeficiency virus envelope function, entry inhibitor sensitivity, and virus neutralization. *J Virol*, **79**, 4991–4999.

Resch, W., Ziermann, R., Parkin, N., Gamarnik, A. and Swanstrom, R., (2002), Nelfinavir-resistant, amprenavir-hypersusceptible strains of human immunodeficiency virus Type 1 carrying an N88S mutation in protease have reduced infectivity, reduced replication capacity, and reduced fitness and process the gag polyprotein precursor aberrantly. *J Virol*, **76**, 8659–8666.

Rezende, L.F., Drosopoulos, W.C. and Prasad, V.R., (1998), The influence of 3TC resistance mutation M184I on the fidelity and error specificity of human immunodeficiency virus type 1 reverse transcriptase. *Nucleic Acids Res*, **26**, 3066–3072.

Richman, D., Shih, C.K., Lowy, I., Rose, J., Prodanovich, P., Goff, S. and Griffin, J., (1991), Human immunodeficiency virus type 1 mutants resistant to nonnucleoside inhibitors of reverse transcriptase arise in tissue culture. *Proc Natl Acad Sci USA*, **88**, 11241–11245.

Rimsky, L.T., Shugars, D.C. and Matthews, T.J., (1998), Determinants of human immunodeficiency virus Type 1 resistance to gp41-derived inhibitory peptides. *J Virol*, **72**, 986–993.

Roberts, J.D., Bebenek, K. and Kunkel, T.A., (1988), The accuracy of reverse transcriptase from HIV-1. *Science*, **242**, 1171–1173.

Rousseau, M.N., Vergne, L., Montes, B., Peeters, M., Reynes, J., Delaporte, E. and Segondy, M., (2001), Patterns of resistance mutations to antiretroviral drugs in extensively treated HIV-1-infected patients with failure of highly active antiretroviral therapy. *J Acquir Immune Defic Syndr*, **26**, 36–43.

Salomon, H., Wainberg, M.A., Brenner, B., Quan, Y., Rouleau, D., Cote, P., LeBlanc, R., Lefebvre, E., Spira, B., Tsoukas, C., Sekaly, R.P., Conway, B., Mayers, D. and Routy, J.P., (2000), Prevalence of HIV-1 resistant to antiretroviral drugs in 81 individuals newly infected by sexual contact or injecting drug use. Investigators of the Quebec Primary Infection Study. *AIDS*, **14**, F17–F23.

Saracino, A., Monno, L., Locaputo, S., Torti, C., Scudeller, L., Ladisa, N., Antinori, A., Sighinolfi, L., Chirianni, A., Mazzotta, F., Carosi, G. and Angarano, G., (2004), Selection of antiretroviral therapy guided by genotypic or phenotypic resistance testing: an open-label, randomized, multicenter study (PhenGen). *J Acquir Immune Defic Syndr*, **37**, 1587–1598.

Sardana, V.V., Emini, E.A., Gotlib, L., Graham, D.J., Lineberger, D.W., Long, W.J., Schlabach, A.J., Wolfgang, J.A. and Condra, J.H., (1992), Functional analysis of HIV-1 reverse transcriptase amino acids involved in resistance to multiple nonnucleoside inhibitors. *J Biol Chem*, **267**, 17526–17530.

Sax, P.E., Islam, R., Walensky, R.P., Losina, E., Weinstein, M.C., Goldie, S.J., Sadownik, S.N. and Freedberg, K.A., (2005), Should resistance testing be performed for treatment-naive HIV-infected patients? A cost-effectiveness analysis. *Clin Infect Dis*, **41**, 1316–1323.

Schinazi, R.F., Larder, B.A. and Mellors, A.W., (2000), Mutations in retroviral genes associated in drug resistance: 2000–2001 update. *Int Antiviral News*, **8**, 65–91.

Schols, D., Este, J.A., Cabrera, C. and De Clercq, E., (1998), T-Cell-line-tropic human immunodeficiency virus Type 1 that is made resistant to stromal cell-derived factor 1alpha contains mutations in the envelope gp120 but does not show a switch in coreceptor use. *J Virol*, **72**, 4032–4037.

Shafer, R.W., Eisen, J.A., Merigan, T.C. and Katzenstein, D.A., (1997), Sequence and drug susceptibility of subtype C reverse transcriptase from human immunodeficiency virus type 1 seroconverters in Zimbabwe. *J Virol*, 71:5441–5448.

Shafer, R.W., Jung, D.R., Betts, B.J., Xi, Y. and Gonzales, M.J., (2000), Human immunodeficiency virus reverse transcriptase and protease sequence database. *Nucl Acids Res*, **28**, 346–348.

Shafer, R.W., Winters, M.A., Palmer, S. and Merigan, T.C., (1998), Multiple Concurrent reverse transcriptase and protease mutations and multidrug resistance of HIV-1 isolates from heavily treated patients. *Ann Intern Med*, **128**, 906–911.

Shulman, N., Zolopa, A., Havlir, D., Hsu, A., Renz, C., Boller, S., Jiang, P., Rode, R., Gallant, J., Race, E., Kempf, D.J. and Sun, E., (2002), Virtual inhibitory quotient predicts response to ritonavir boosting of indinavir-based therapy in human immunodeficiency virus-infected patients with ongoing viremia. *Antimicrob Agents Chemother*, **46**, 3907–3916.

Sista, P., Melby, T., Greenberg, M., Davison, D., Jin, L., Mosier, S., Mink, M., Nelson, E., Fang, L., Cammack, N., Salgo, M. and Matthews, T., (2002), Characterization of Baseline and Treatment-Emergent Resistance to T-20 (Enfuvirtide) Observed in Phase II Clinical Trials: Substitutions in Gp41 Amino Acids 36–45 and Enfuvirtide Susceptibility of Virus Isolates. In XI International HIV Drug Resistance Workshop, Seville, Spain, 2–5 July 2002. Abstract 21.

Sista, P., Melby, T., Greenberg, M., DeMasi, R., Kuritzkes, D., Nelson, M., Petropoulos, C., Salgo, M., Cammack, N. and Matthews, T., (2003), Subgroup analysis of baseline (BL) susceptibility and early virological response to enfuvirtide in the combined TORO studies. In XII International HIV Drug Resistance Workshop, Los Cabos, Mexico, 10–14 June 2003. Abstract 55.

Sista, P.R., Melby, T., Davison, D., Jin, L., Mosier, S., Mink, M., Nelson, E.L., DeMasi, R., Cammack, N., Salgo, M.P., Matthews, T.J. and Greenberg, M.L., (2004), Characterization of determinants of genotypic and phenotypic resistance to enfuvirtide in baseline and on-treatment HIV-1 isolates. *AIDS*, **18**:, 1787–1794.

Smith, D.M., Wong, J.K., Hightower, G.K., Ignacio, C.C., Koelsch, K.K., Daar, E.S., Richman, D.D. and Little, S.J., (2004), Incidence of HIV superinfection following primary infection. *JAMA*, **292**, 1177–1178.

Spence, R.A., Kati, W.M., Anderson, K.S. and Johnson, K.A., (1995), Mechanism of inhibition of HIV-1 reverse transcriptase by nonnucleoside inhibitors. *Science*, **267**, 988–993.

Su, C., Heilek-Snyder, G., Fenger, D., Ravindran, P., Tsai, K., Cammack, N., Sista, P. and Chiu, S., (2003), The relationship between susceptibility to enfuvirtide of baseline viral recombinants and polymorphisms in the env region of R5-tropic HIV-1. *Antivir Ther*, **8**, S59.

Tamiya, S., Mardy, S., Kavlick, M.F., Yoshimura, K. and Mistuya, H., (2004), Amino acid insertions near gag cleavage sites restore the otherwise compromised replication of human immunodeficiency virus Type 1 variants resistant to protease inhibitors. *J Virol*, **78**, 12030–12040.

Tantillo, C., Ding, J., Jacobo-Molina, A., Nanni, R.G., Boyer, P.L., Hughes, S.H., Pauwels, R., Andries, K., Janssen, P.A. and Arnold, E., (1994), Locations of anti-AIDS drug binding sites and resistance mutations in the three-dimensional structure of HIV-1 reverse transcriptase. Implications for mechanisms of drug inhibition and resistance. *J Mol Biol*, **243**, 369–387.

Torre, D. and Tambini, R., (2002), Antiretroviral drug resistance testing in patients with HIV-1 infection: a meta-analysis study. *HIV Clin Trials*, **3**, 1–8.

Trkola, A., Kuhmann, S.E., Strizki, J.M., Maxwell, E., Ketas, T., Morgan, T., Pugach, P., Xu, S., Wojcik, L., Tagat, J., Palani, A., Shapiro, S., Clader, J.W., McCombie, S., Reyes, G.R., Baroudy, B.M. and Moore, J.P., (2002), HIV-1 escape from a small molecule, CCR5-specific entry inhibitor does not involve CXCR4 use. *Proc Natl Acad Sci*, **99**, 395–400.

Tsui, R., Herring, B.L., Barbour, J.D., Grant, R.M., Bacchetti, P., Kral, A., Edlin, B.R. and Delwart, E.L., (2004), Human immunodeficiency virus type 1 superinfection was not detected following 215 years of injection drug user exposure. *J Virol*, **78**, 94–103.

Tural, C., Ruiz, L., Holtzer, C., Schapiro, J., Viciana, P., Gonzalez, J., Domingo, P., Boucher, C., Rey-Joly, C. and Clotet, B., (2002), Clinical utility of HIV-1 genotyping and expert advice: the Havana trial. *AIDS*, **16**, 209–218.

Turner, D., Brenner, B., Routy, J.P., Moisi, D., Rosberger, Z., Roger, M. and Wainberg, M.A., (2004), Diminished representation of HIV-1 variants containing select drug resistance-conferring mutations in primary HIV-1 infection. *J Acquir Immune Defic Syndr*, **37**, 1627–1631.

Vandamme, A.M., Debyser, Z., Pauwels, R., De Vreese, K., Goubau, P., Youle, M., Gazzard, B., Stoffels, P.A., Cauwenbergh, G.F. and Anne, J., (1994), Characterization of HIV-1 strains isolated from patients treated with TIBO R82913. *AIDS Res Hum Retroviruses*, **10**, 39–46.

Vandamme, A.M., Sonnerborg, A., Ait-Khaled, M., Albert, J., Asjo, B., Bacheler, L., Banhegyi, D., Boucher, C., Brun-Vezinet, F., Camacho, R., Clevenbergh, P., Clumeck, N., Dedes, N., De Luca, A., Doerr, H.W., Faudon, J.L., Gatti, G., Gerstoft, J., Hall, W.W., Hatzakis, A., Hellmann, N., Horban, A., Lundgren, J.D., Kempf, D., Miller, M., Miller, V., Myers, T.W., Nielsen, C., Opravil, M., Palmisano, L., Perno, C.F., Phillips, A., Pillay, D., Pumarola, T., Ruiz, L., Salminen, M., Schapiro, J., Schmidt, B., Schmit, J.C., Schuurman, R., Shulse, E., Soriano, V., Staszewski, S., Vella, S., Youle, M., Ziermann, R. and Perrin, L., (2004), Updated European recommendations for the clinical use of HIV drug resistance testing. *Antivir Ther*, **9**, 829–848.

Vanden Haesevelde, M., Decourt, J.L., De Leys, R.J., Vanderborght, B., van der Groen, G., van Heuverswijn, H. and Saman, E., (1994), Genomic cloning and complete sequence analysis of a highly divergent African human immunodeficiency virus isolate. *J Virol*, **68**, 1586–1596.

Velazquez-Campoy, A., Muzammil, S., Ohtaka, H., Schon, A., Vega, S. and Freire, E., (2003), Structural and thermodynamic basis of resistance to HIV-1 protease inhibition: implications for inhibitor design. *Curr Drug Targets Infect Disord*, **3**, 311–328.

Verhofstede, C., Wanzeele, F.V., Van Der Gucht, B., De Cabooter, N. and Plum, J., (1999), Interruption of reverse transcriptase inhibitors or a switch from reverse transcriptase to protease inhibitors resulted in a fast reappearance of virus strains with a reverse transcriptase inhibitor-sensitive genotype. *AIDS*, **13**, 2541–2546.

Wainberg, M.A. and Friedland, G., (1998), Public health implications of antiretroviral therapy and HIV drug resistance. *JAMA*, **279**, 1977–1983.

Wainberg, M.A., Hsu, M., Gu, Z., Borkow, G. and Parniak, M.A., (1996), Effectiveness of 3TC in HIV clinical trials may be due in part to the M184V substitution in 3TC-resistant HIV-1 reverse transcriptase. *AIDS*, **10** (Suppl 5), S3–S10.

Wei, X., Decker, J.M., Liu, H., Zhang, Z., Arani, R.B., Kilby, J.M., Saag, M.S., Wu, X., Shaw, G.M. and Kappes, J.C., (2002), Emergence of resistant human immunodeficiency virus Type 1 in patients receiving fusion inhibitor (T-20) monotherapy. *Antimicrob Agents Chemother*, **46**, 1896–1905.

Weinstock, H.S., Zaidi, I., Heneine, W., Bennett, D., Garcia-Lerma, J.G., Douglas, J.M., Jr, LaLota, M., Dickinson, G., Schwarcz, S., Torian, L., Wendell, D., Paul, S., Goza, G.A.,

Ruiz, J., Boyett, B. and Kaplan, J.E., (2004), The epidemiology of antiretroviral drug resistance among drug-naive HIV-1-infected persons in 10 US cities. *J Infect Dis*, **189**, 2174–2180.

Wensing, A.M., van de Vijver, D.A., Angarano, G., Asjo, B., Balotta, C., Boeri, E., Camacho, R., Chaix, M.L., Costagliola, D., De Luca, A., Derdelinckx, I., Grossman, Z., Hamouda, O., Hatzakis, A., Hemmer, R., Hoepelman, A., Horban, A., Korn, K., Kucherer, C., Leitner, T., Loveday, C., MacRae, E., Maljkovic, I., de Mendoza, C., Meyer, L., Nielsen, C., Op de Coul, E.L., Ormaasen, V., Paraskevis, D., Perrin, L., Puchhammer-Stockl, E., Ruiz, L., Salminen, M., Schmit, J.C., Schneider, F., Schuurman, R., Soriano, V., Stanczak, G., Stanojevic, M., Vandamme, A.M., Van Laethem, K., Violin, M., Wilbe, K., Yerly, S., Zazzi, M. and Boucher, C.A., (2005), Prevalence of drug-resistant HIV-1 variants in untreated individuals in Europe: implications for clinical management. *J Infect Dis*, **192**, 958–966.

Westby, M., Lewis, M., Whitcomb, J., Youle, M., Pozniak, A.L., James, I.T., Jenkins, T.M., Perros, M. and van der Ryst, E., (2006), Emergence of CXCR4-using human immunodeficiency virus Type 1 (HIV-1) variants in a minority of HIV-1-infected patients following treatment with the CCR5 antagonist Maraviroc is from a pretreatment CXCR4-using virus reservoir. *J Virol*, **80**, 4909–4920.

Westby, M., Mori, J., Smith-Burchnell, C., Lewis, M., Mosley, M., Perruccio, F., Mansfield, R., Dorr, P. and Perros, M., (2005), Maraviroc (UK-427,857)-resistant HIV-1 variants, selected by serial passage, are sensitive to CCR5 antagonists and T-20. *Antivir Ther*, **10**, S72.

Westby, M., Smith-Burchnell, C., Mori, J., Lewis, M., Mansfield, R., Whitcomb, J., Petropoulos, C. and Perros, M., (2004a), In vitro escape of R5 primary isolates from the CCR5 antagonist, UK-427,857, is difficult and involves continued use of the CCR5 receptor. In XIII Int. HIV Drug Resist Workshop. Tenerife, Spain. Abstract 6. *Antivir Ther*, **9**, S10.

Westby, M., Whitcomb, J., Huang, W., James, I., Abel, S., Petropoulos, C., Perros, M. and Ryst, E., (2004a), Reversible predominance of CXCR4 using variants in a non-responsive dual tropic patient receiving the CCR5 antagonist UK-427,857. In Eleventh Conference on Retroviruses and Opportunistic Infections, San Francisco, CA, 8–11 February 2004, USA. Abstract 538.

Wheat, L., Lalezari, J., Kilby, M., Wheeler, D., Salgo, M., DeMasi, R. and Delehanty, J., (2002), A week 48 assessment of high strength T-20 formulations in multi-class experienced patients. In 9th Conference on Retroviruses and Opportunistic Infections, Seattle, USA, 24–28 February 2002. Abstract 417-W.

Whitcomb, J., Huang, W., Fransen, S., Wrin, T., Paxind, E., Toma, J., Greenberg, M., Sista, P., Melby, T., Matthews, T., Cammack, N., Hellman, N. and Petropoulus C., (2003), Analysis of baseline enfuvirtide (T-20) susceptibility and co-receptor tropism in two phase III study populations. Tenth Conference on Retroviruses and Opportunistic Infections, Boston, MA, 10–14 February 2003, USA. Abstract 557.

Winston, A., Hales, G., Amin, J., van Schaick, E., Cooper, D.A. and Emery, S., (2005), The normalized inhibitory quotient of boosted protease inhibitors is predictive of viral load response in treatment-experienced HIV-1-infected individuals. *AIDS*, **19**, 1393–1399.

Winston, A., Mandalia, S., Pillay, D., Gazzard, B. and Pozniak, A., (2002), The prevalence and determinants of the K65R mutation in HIV-1 reverse transcriptase in tenofovir-naive patients. *AIDS*, **16**, 2087–2089.

Winters, M.A., Baxter, J.D., Mayers, D.L., Wentworth, D.N., Hoover, M.L., Neaton, J.D. and Merigan, T.C., (2000), Frequency of antiretroviral drug resistance mutations in HIV-1 strains from patients failing triple drug regimens. The Terry Beirn Community Programs for Clinical Research on AIDS. *Antivir Ther*, **5**, 57–63.

Wong, J.K., Hezareh, M., Gunthard, H.F., Havlir, D.V., Ignacio, C.C., Spina, C.A. and Richman, D.D., (1997), Recovery of replication-competent HIV despite prolonged suppression of plasma viremia. *Science*, **278**, 1291–1295.

Wu, J.C., Warren, T.C., Adams, J., Proudfoot, J., Skiles, J., Raghavan, P., Perry, C., Potocki, I., Farina, P.R. and Grob, P.M., (1991), A novel dipyridodiazepinone inhibitor of HIV-1 reverse transcriptase acts through a nonsubstrate binding site. *Biochemistry*, **30**, 2022–2026.

Xu, L., Pozniak, A., Wildfire, A., Stanfield-Oakley, S.A., Mosier, S.M., Ratcliffe, D., Workman, J., Joall, A., Myers, R., Smit, E., Cane, P.A., Greenberg, M.L. and Pillay, D., (2005), Emergence and evolution of enfuvirtide resistance following long-term therapy involves Heptad Repeat 2 mutations within gp41. *Antimicrob Agents Chemother*, **49**, 1113–1119.

Yeni, P.G., Hammer, S.M., Carpenter, C.C., Cooper, D.A., Fischl, M.A., Gatell, J.M., Gazzard, B.G., Hirsch, M.S., Jacobsen, D.M., Katzenstein, D.A., Montaner, J.S., Richman, D.D., Saag, M.S., Schechter, M., Schooley, R.T., Thompson, M.A., Vella, S. and Volberding, P.A., (2002), Antiretroviral treatment for adult HIV infection in 2002: updated recommendations of the International AIDS Society-USA Panel. *JAMA*, **288**, 222–235.

Yerly, S., Kaiser, L., Race, E., Bru, J.P., Clavel, F. and Perrin, L., (1999), Transmission of antiretroviral-drug-resistant HIV-1 variants. *Lancet*, **354**, 729–733.

Yerly, S., Vora, S., Rizzardi, P., Chave, J.P., Vernazza, P.L., Flepp, M., Telenti, A., Battegay, M., Veuthey, A.L., Bru, J.P., Rickenbach, M., Hirschel, B. and Perrin, L., (2001), Acute HIV infection: impact on the spread of HIV and transmission of drug resistance. *AIDS*, **15**, 2287–2292.

Zaccarelli, M., Tozzi, V., Lorenzini, P., Trotta, M.P., Forbici, F., Visco-Comandini, U., Gori, C., Narciso, P., Perno, C.F. and Antinori, A., (2005), Multiple drug class-wide resistance associated with poorer survival after treatment failure in a cohort of HIV-infected patients. *AIDS*, **19**, 1081–1089.

Chapter 8
Resistance of Herpesviruses to Antiviral Agents

G. Boivin and W. L. Drew

Acyclovir is the prototype of a series of antivirals which are effective against herpesviruses. However, resistance to this class of drugs can occur and is being seen mainly in immunocompromised patients with herpes simplex virus (HSV) and varicella-zoster virus (VZV) infections. Clinical use of intravenous (IV) ganciclovir began in 1984 for the treatment of life-threatening and sight-threatening human cytomegalovirus (HCMV) infections in immunocompromised patients. A few years later, description of ganciclovir-resistant HCMV strains were reported in AIDS patients with HCMV retinitis and, more recently, in organ or bone marrow transplant recipients. Foscarnet and cidofovir became available subsequently and resistance to these agents has also been described.

In this chapter we review the available antiviral drugs for HCMV, HSV and VZV, the methods for detecting antiviral resistance, the clinical significance of resistant strains and their treatment.

8.1 Antiviral Agents for Herpesvirus Infections

Three antiviral agents and a prodrug are currently available for the systemic treatment of HCMV infections. Ganciclovir (GCV, Cytovene, Hoffmann LaRoche) is a deoxyguanosine analog and was the first drug to be approved in 1988. Since then, it has remained the first line treatment for HCMV infections in immunocompromised patients. Upon entry in HCMV-infected cells, GCV is selectively phosphorylated by a viral protein kinase homolog (the product of the UL97 gene, pUL97). Subsequently, cellular kinases convert GCV-monophosphate into GCV-triphosphate, which acts as a potent inhibitor of the HCMV DNA polymerase (pol) by competing with deoxyguanosine triphosphate on the enzyme-binding site (Figure 8.1). Ganciclovir is also incorporated into the viral DNA where it slows down and eventually stops chain elongation (Balfour, 1999; Biron et al., 1985; Sullivan et al., 1992). Ganciclovir formulations are available for intravenous (IV) or oral administration and as ocular implants for the local treatment of HCMV retinitis. Due to its poor bioavailability (∼6%), efforts were made to develop prodrugs of GCV. Valganciclovir (VGCV, Valcyte, Hoffmann LaRoche) is a new valyl ester formulation of

I. W. Fong and K. Drlica (eds.), *Antimicrobial Resistance and Implications for the Twenty-First Century.* © Springer 2008

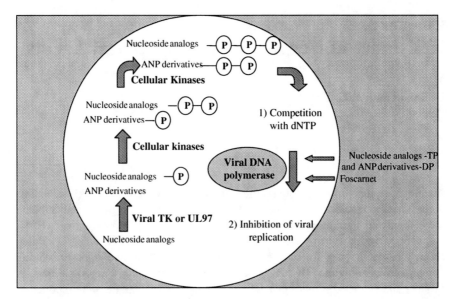

Fig. 8.1 Mechanisms of action of antiviral agents used in the treatment of herpesvirus infections. TK, thymidine kinase; ANP, acyclic nucleoside phosphonate.

GCV exhibiting about 10 times the bioavailability of GCV following oral administration (Pescovitz et al., 2000).

The other two compounds approved for systemic treatment of HCMV infections are also potent inhibitors of the viral DNA pol. However, due to their toxicity profiles, they are usually reserved for patients failing or not tolerating GCV therapy. Cidofovir (CDV, Vistide, Gilead Sciences) is a nucleotide analog of cytidine (also called acyclic nucleoside phosphonate) that only requires activation (phosphorylation) by cellular enzymes to exert its antiviral activity (Cihlar and Chen, 1996). Once in its diphosphate form, CDV inhibits the HCMV DNA pol by a mechanism similar to that of GCV (Figure 8.1). Foscarnet (FOS, Foscavir, Astra-Zeneca), a pyrophosphate analog, differs from the two previous antivirals both by its mechanism of action and by the fact that it does not require any activation step to exert its antiviral activity. Foscarnet binds to and blocks the pyrophosphate binding site on the viral polymerase, thus preventing incorporation of incoming dNTPs into viral DNA (Figure 8.1) (Chrisp and Clissold, 1991). Finally, formivirsen (Vitravene, Novartis) is a 21-nt long antisense oligonucleotide with sequence complementary to the HCMV immediate–early 2 mRNA that interferes with HCMV replication at an early stage during the replication cycle (Mulamba et al., 1998). Its only current indication is for the local treatment of HCMV retinitis in AIDS patients.

In addition to the treatment of established HCMV disease, antivirals have also been used to prevent such symptomatic episodes, especially in transplant recipients. The first strategy, defined as prophylaxis, consists of administering an antiviral to patients during the first 3 months or so after transplantation. The second strategy, referred to as "preemptive therapy," consists of using short courses of antivirals only for high-risk patients based on evidence of active viral replication (e.g., detection of early HCMV antigens such as the pp65 protein or sufficient amounts of viral DNA/mRNA) (Boeckh and Boivin, 1998). These preventive strategies have shown efficacy in preventing HCMV disease in both solid organ transplant (SOT) and hematopoietic stem cell transplant (HSCT) patients (Boeckh et al., 2003; Humar et al., 2005; Lowance et al., 1999; Paya et al., 2002, 2004). However, some studies suggest that "prevention" may sometimes only delay, rather than truly prevent, the onset of HCMV disease in predisposed patients (Limaye et al., 2002; Lowance et al., 1999; Nguyen et al., 1999; Razonable et al., 2001; Zaia et al., 1997).

Antiviral agents currently licensed for the treatment of HSV and VZV infections include acyclovir (ACV, Zovirax, GlaxoSmithKline) and its valyl-ester prodrug valacyclovir (VACV, Valtrex, GlaxoSmithKline), famciclovir (FCV, Famvir, Novartis) which is the valyl-ester prodrug of penciclovir (PCV), and FOS. Acyclovir and PCV are deoxyguanosine analogs that must be phosphorylated by the thymidine kinase (TK) of HSV (UL23) and VZV (ORF36) and then by cellular kinases to exert their antiviral activity (Biron et al., 1985; Fyfe et al., 1978). The triphosphate forms are competitive inhibitors of the viral DNA pol (Figure 8.1). In addition, incorporation of ACV triphosphate into the replicating viral DNA chain stops synthesis (Balfour, 1999). Brivudin or bromovinyldeoxyuridine (BVDU) is a nucleoside analog marketed in Europe for treatment of varicella-zoster infections (De Clercq, 2004). The compound is also effective against HSV-1 and its activity depends on phosphorylation by the viral TK and cellular kinases before interaction with the viral DNA polymerase. The pyrophosphate analog FOS is usually indicated for ACV- or PCV-resistant HSV or VZV infections (Safrin et al., 1991a–c). Both episodic (short term) or suppressive (continuous) therapy with the deoxyguanosine analogs can be used in the management of recurrent HSV infections (Balfour, 1999).

8.2 Human Cytomegalovirus Resistance

8.2.1 Phenotypic and Genotypic Assays to Evaluate HCMV Drug Susceptibility

Two different albeit complementary approaches have been developed to assess HCMV drug resistance. In the phenotypic method, the virus is grown in the presence of various concentrations of an antiviral in order to determine the concentration of drug that will inhibit a percentage (more commonly 50%) of viral growth in cell culture. In this assay, a standardized viral inoculum is inoculated in different wells. The virus is then allowed to grow for a few days (typically 7–10 days) in the presence of serial drug dilutions before staining the cells. The number of viral plaques per concentration is first determined. Then, the percentage of viral growth, as compared to a control well without antiviral, is plotted against drug concentrations to determine the concentration that will inhibit the growth of 50% of viral plaques (50% inhibitory concentration or IC_{50}). Even though recent efforts have been made to standardize this assay (Landry et al., 2000), the inter-assay and inter-laboratory variability is still problematic. In addition to the relative subjectivity of this method, there are some differences in the cut-off values defining drug resistance depending on the laboratory. Similar assays, either based on detection of HCMV DNA by hybridization (Dankner et al., 1990) or quantitative PCR, or detection of specific HCMV antigens by ELISA (Tatarowicz et al., 1991), flow cytometry (Kesson et al., 1998; McSharry et al., 1998), immunofluorescence (Telenti and Smith, 1989), or immunoperoxidase (Gerna et al., 1992), have also been developed. Among those, a commercial DNA:DNA hybridization assay (Hybriwix probe system/cytomegalovirus susceptibility test kit) developed by Diagnostic Hybrids (Athens, OH) has shown good correlation with the plaque reduction assay (Jabs et al., 1998a,b; Weinberg et al., 2003). Even if the readout method is more objective with the hybridization assay, IC_{50} cut-off values defining resistance are still a matter of debate although a value of 6 μM is most often used for GCV. Altogether, phenotypic assays are time consuming, subject to possible selection bias introduced during viral growth of mixed viral populations in cell culture (Gilbert and Boivin, 2003; Hamprecht et al., 2003), and may lack sensitivity to detect low-level resistance or minor resistant subpopulations (Chou et al., 2002; Gilbert and Boivin, 2003).

In contrast to phenotypic assays, which directly measure drug susceptibility of viral isolates, genotypic assays detect the presence of viral mutations known to be associated with drug resistance. Those assays are based on restriction fragment length polymorphism (RFLP) of PCR-amplified DNA fragments or on DNA sequencing of viral genes (UL97 and UL54) that are the sites of HCMV resistance to antivirals. One of the advantages of these assays is that they can be performed directly on clinical specimens (Boivin et al., 1996; Wolf et al., 1995a,b) thus reducing considerably the time required to obtain results. By omitting the need to grow the virus, such methods also minimize the risks of introducing a selection bias. The limited number of UL97 mutations responsible for GCV resistance has allowed the development of rapid RFLP assays to detect their presence in clinical samples (Chou et al., 1995; Hanson et al., 1995). Indeed, approximately 70% of GCV-resistant clinical isolates contain mutations in one of three UL97 codons (460, 594 and 595) (Boivin et al., 2001). Typically, the presence of a given mutation will either obliterate an existing restriction site or create a new one. The difference in RFLP patterns can then be visualized following gel electrophoresis. The major advantages of this assay include its short turnaround time (2–4 days) and its ability to detect as little as 20% of a mutant virus in a background of wild-type viruses (Chou et al., 1995). However, due to reports of resistance mutations at other codons, DNA sequence of the entire UL97 region involved in GCV resistance should be determined for a comprehensive analysis. For genotypic analysis of UL54 DNA polymerase mutations, DNA sequencing is required due to the large number of mutations reported within all conserved regions of this gene (Gilbert et al., 2002). Genotypic approaches are fast but their interpretation is not always straightforward, i.e., discriminating between mutations associated with natural polymorphisms (Boivin et al., 2005a; Chou et al., 1999; Fillet et al., 2004; Lurain et al., 2001) from those related to drug resistance. In order to prove that a new mutation is associated with drug resistance, recombinant viruses need to be generated using either overlapping cosmid/plasmid inserts (Cihlar et al., 1998a) or by marker transfer experiments of the mutated gene in a wild-type (Baldanti et al., 1995, 1996; Chou et al., 1997) or genetically engineered (Chou et al., 2002, 2003) virus prior to testing of this mutant virus in phenotypic assays. The introduction of a reporter gene (luciferase) in a permissive cell line (Gilbert and Boivin, 2005a) or directly in a recombinant virus (Chou et al., 2005) should accelerate the phenotypic testing of new mutants by allowing an automated and more objective evaluation of viral replication.

8.2.2 Clinical Significance, Incidence and Risk Factors for Drug-Resistant HCMV Infections

Drug-resistant HCMV strains first emerged as a significant problem in patients with AIDS. Numerous studies have documented the emergence of drug-resistant HCMV strains (detected by phenotypic or genotypic methods) and their correlation with progressive or recurrent HCMV disease (mainly retinitis) during therapy (Baldanti et al., 1995; Chou et al., 1997, 1998; Drew et al., 1999; Smith et al., 1996, 1998; Wolf et al., 1995a,b). The first study to evaluate the prevalence of GCV resistance in AIDS patients was conducted by evaluating the excretion of GCV-resistant strains in the urine of 31 patients with AIDS treated with IV GCV for HCMV retinitis. In this study, no resistant isolates were recovered in patients treated for ≤ 3 months whereas 38% of those excreting the virus in their urine after >3 months of GCV had a resistant isolate, representing 8% of the entire cohort of patients (Drew et al., 1991). Since then, larger studies have evaluated the temporal emergence of GCV-resistant strains using either phenotypic (Jabs et al., 1998a) or genotypic (Boivin et al., 2001) assays. In all studies, GCV resistance (defined by an IC_{50} value ≥ 6 μM) at the initiation of treatment was a rare event ($\leq 2.7\%$ of tested strains). Phenotypic evaluation of blood or urine isolates from 95 patients treated with GCV (mostly intravenous) for HCMV retinitis revealed that 7, 12, 27, and 27% of patients excreted a GCV-resistant strain after respectively 3, 6, 9, and 12 months of drug exposure (Jabs et al., 1998a). On the other hand, a more recent study of 148 AIDS patients treated for HCMV retinitis with oral VGCV has identified the presence of GCV resistance mutations in 2, 7, 9, and 13% of patients after 3, 6, 9, and 12 months of therapy, respectively (Boivin et al., 2001). The lower incidence of GCV resistance in the latter study despite the use of sensitive genotypic methods might be explained by differences in the treatment of the underlying HIV disease in the study population, notably improved antiretroviral therapy. Due to their less frequent use in clinic, fewer data have been reported on the temporal emergence of FOS- and CDV-resistant HCMV strains in HIV-infected individuals. One small study found an incidence of phenotypic resistance to FOS of 9, 26, 37, and 37% after 3, 6, 9, and 12 months of therapy using an IC_{50} cut-off value of 400 μM (Jabs et al., 1998b) whereas another one reported rates of 13, 24, and 37% after 6, 9, and 12 months using an IC_{50} cut-off value of 600 μM (Weinberg et al., 2003). The data on CDV resistance (IC_{50} value ≥ 2–4 μM) are even more limited but they seem to indicate a resistance rate similar to what has been observed with GCV and FOS (Jabs et al., 1998b).

Proposed risk factors for the development of HCMV resistance in this patient population include inadequate tissue drug concentrations due to poor tissue penetration (e.g., the eyes) or poor bioavailability (e.g., oral GCV), a sustained and profound immunosuppression status (CD4 counts <50 cells/μl), frequent discontinuation of treatment due to toxicity, and a high pre-therapy HCMV load (Drew, 2003; Nichols and Boeckh, 2001).

HCMV resistance to GCV appears to be an emerging problem in SOT recipients and has been associated with an increased number of asymptomatic and symptomatic viremic episodes, earlier onset of HCMV disease, graft loss, and an increased risk of death (Bhorade et al., 2002; Limaye et al., 2000). Due to the different HCMV preventive strategies and immunosuppressive regimens in use at different centers and considering the heterogeneity of the transplant populations, it has been difficult to precisely evaluate the temporal emergence of HCMV resistance in that setting. Lung transplant recipients appear to have the highest incidence of HCMV resistance development with rates of 3.6–9% after median cumulative GCV exposures ranging from 79 to 100 days (Kruger et al., 1999; Limaye et al., 2002; Lurain et al., 2002). In two of those studies, the incidence of resistance increased from 15.8 to 27% in D+/R– (HCMV antibody-positive donor with HCMV-negative recipient) patients (Limaye et al., 2002; Lurain et al., 2002) and occurred as a late complication, i.e., a median of 4.4 months after transplantation (Limaye et al., 2002). As opposed to what has been reported in lung transplant recipients, the incidence of GCV resistance in other SOT populations has been much lower and almost exclusively seen in D+/R– patients (Limaye et al., 2000; Lurain et al., 2002). More specifically, Lurain and colleagues studied two cohorts of SOT patients including heart, liver, and kidney recipients at two US centers (Lurain et al., 2002). Phenotypic evaluation for HCMV resistance prompted by either clinical suspicion or positive blood cultures indicated that rates of resistance were generally low (e.g., <0.5%) at one center and varied from 2.2 to 5.6 at another center depending on the transplanted organ. Another retrospective study by Limaye and colleagues evaluated 240 SOT patients including 67 D+/R– patients but excluded lung transplant recipients (Limaye et al., 2000). In their cohort, GCV-resistant HCMV disease developed only in D+/R– patients, with resistance rates of 7% in these patients. HCMV resistance was more frequently seen among recipients of kidney/pancreas or pancreas alone (21%) than among kidney (5%) or liver (0%) recipients. Of note, cases of GCV-resistant HCMV infections occurred at a median of 10 months after transplantation with a median total drug exposure of 194 days (129 days of oral GCV) including two to three treatment courses for HCMV disease per patient. Importantly, GCV-resistant HCMV infections accounted for 20% of HCMV diseases that occurred during the first year after transplantation (Limaye et al., 2000).

The first prospective study evaluating the emergence of GCV resistance in SOT recipients was reported by Boivin et al. (2004). In this study, molecular methods were used to assess the emergence of UL97 and UL54 mutations associated with GCV resistance in D+/R– patients (175 liver, 120 kidney, 56 heart, 11 kidney/pancreas, and 2 liver/kidney recipients) receiving HCMV prophylaxis with either oral GCV (1 g TID) or oral VGCV (900 mg OD). Among 301 evaluable patients, the incidence of GCV resistance at the end of the prophylactic period (day

100 post-transplant) was very low in both arms (0 and 1.9% for the VGCV and oral GCV arms, respectively). During the first year following transplantation, GCV resistance-associated mutations were found in none compared to 6.1% of patients at the time of suspected HCMV disease after receiving VGCV and oral GCV prophylaxis, respectively. Of note, however, no lung transplant and a small number of kidney/pancreas recipients were included in this trial, which might explain at least partly the low emergence of GCV resistance in this study as compared to previous ones. Interestingly, recent studies have shown that detection of known GCV resistance mutations is not always associated with adverse clinical consequences in non-lung transplants in contrast to more immunosuppressed lung transplants (Boivin et al., 2004, 2005b). Documented risk factors for the emergence of GCV resistance in SOT patients include the lack of HCMV-specific immunity (as encountered in the D+/R− group) (Baldanti et al., 2004; Benz et al., 2003), lung or kidney/pancreas transplantation, longer drug exposure (prophylaxis > preemptive therapy), suboptimal plasma or tissue drug concentrations (as seen with oral GCV), potent immunosuppressive regimens, a high HCMV viral load, and frequent episodes of HCMV disease (Bhorade et al., 2002; Limaye et al., 2000, 2002; Limaye, 2002).

Limited data from small-scale studies suggest that the incidence of GCV resistance in the BMT/HSCT population might not be as high as observed in SOT recipients and AIDS patients, perhaps because of the more limited immunosuppression exposure and the greater use of preemptive antiviral strategies. In a study published by Gilbert et al. (2001), molecular methods were used to detect the presence of the most common UL97 mutations associated with GCV resistance in blood samples of HSCT patients selected on the basis of having a positive HCMV PCR despite ≥14 days of preemptive IV GCV or a second viremic episode within the first 98 days after transplantation. No UL97 mutations associated with GCV resistance were detected in this cohort of 50 patients (10 of them fulfilling the above criteria for genotypic testing) (Gilbert et al., 2001). However, this was a small study and resistance would be unlikely after just a short period of preemptive treatment. In another study designed to evaluate risk factors and outcomes associated with rising HCMV antigenemia levels during the first 2–4 weeks of preemptive therapy, Nichols et al. (2001) prospectively evaluated 119 HSCT patients receiving preemptive GCV or FOS therapy following a positive pp65 antigenemia test. Among these subjects, 47 (39%) exhibited a significant rise in antigenemia levels despite antiviral administration and 15 had at least one isolate available for susceptibility testing. Only one GCV-resistant isolate was identified in a patient who received 4 weeks of GCV therapy (Nichols et al., 2001). In contrast, Erice et al. (1998) reported genotypic or phenotypic evidences of infection with a GCV-resistant HCMV strain in two of five selected patients who had received GCV for a median of 58 days. However, all five patients had also received ACV prophylaxis for a median of 47 days which could have predisposed to the selection of a GCV-resistant HCMV strain (Michel et al., 2001). Of note, the impact of prior ACV in selecting for GCV resistance has not been confirmed by another group (Drew et al., 1995). Springer et al. (2005) also reported two HSCT patients who developed persistent and severe drug-resistant HCMV infections, including one virus with a DNA polymerase mutation conferring multidrug resistance. Even though short courses of GCV therapy appear to be relatively safe in adult BMT patients, the situation might differ in pediatric patients receiving T-cell depleted unrelated transplants as reported by Eckle et al. (2000). In their study of 42 such patients, three showed genotypic evidences of GCV resistance, followed by the excretion of a resistant strain after 30–93 days of GCV exposure. Of note, in the same study, none of the 37 patients who underwent a similar procedure, but who received their transplant from a mismatched related donor, developed GCV resistance (Eckle et al., 2000). Rapid emergence of GCV resistance was also documented in four of five children with congenital immunodeficiency disorders who underwent T-cell-depleted BMT (Wolf et al., 1998). In those patients, genotypic evidence of GCV resistance was demonstrated after only 7–24 days (median 10 days) of cumulative GCV therapy. Finally, the emergence of GCV-resistant strains has been recently associated with previously uncommon central nervous system HCMV disease and retinitis occurring late after HSCT (Hamprecht et al., 2003; Wolf et al., 2003).

8.2.3 Role of HCMV UL97 and UL54 Mutations in Drug-Resistant Clinical Strains

Cumulative results obtained in three recent studies that have documented the emergence of UL97 mutations in clinical isolates (Jabs et al., 2001b; Lurain et al., 2002) or in blood samples (Boivin et al., 2001) from 61 AIDS and SOT patients are in general agreement with the proposed frequency of UL97 mutations based on characterization of 76 independent UL97 mutants gathered in a single laboratory over years (Chou et al., 2002). Those data suggest that mutations A594V (30–34.5%), L595S (20–24%), M460V (11.5–14.5%) and H520Q (5–11.5%) represent the most frequent UL97 mutations present in GCV-resistant mutants (Figure 8.2) (Drew et al., 2001; Gilbert and Boivin, 2005b). Other frequent UL97 mutations associated with resistance include C592G and C603W. Based on marker transfer experiments, mutations M460V (7× increase in resistance) (Chou et al., 1995), C603W (8×) (Chou et al., 1997), deletion of codons 595–603 (8.4×) (Chou and Meichsner, 2000), H520Q (10×) (Hanson et al., 1995), L595S (4.9–11.5×) (Chou et al., 1995, 2002),

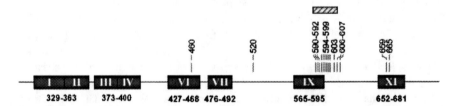

Fig. 8.2 HCMV UL97 mutations conferring ganciclovir (GCV) resistance. HCMV UL97 conserved regions are represented by *shaded boxes*. *Numbers* under the boxes indicate the positions (codons no.) of these conserved regions. *Vertical bars* indicate the presence of amino acids substitutions while the *hatched box* indicates a region (codons 590–607) in which diverse deletions (from 1 to 17 codons) have been reported.

A594V (10.7×) (Chou et al., 1995), C607Y (12.5×) (Baldanti et al., 1998), and deletion of codon 595 (13.3×) (Baldanti et al., 1995) appear to be associated with the highest increase in GCV resistance over the parental strain whereas mutations C592G, A594T, E596G, and deletion of codon 600 seem to confer only modest decreases in susceptibility (Chou et al., 2002). Interestingly, analysis of GCV-phosphorylating activity of mutated UL97 genes expressed in a recombinant vaccinia virus expression system would have predicted mutations H520Q and M460V to confer the greatest decrease in GCV susceptibility (Baldanti et al., 2002).

Among the most frequent DNA pol mutations associated with drug resistance, there are V715M, V781I, and L802M conferring FOS resistance and F412C, L501I/F, and P522S conferring GCV/CDV resistance (Figure 8.3) (Drew et al., 2001; Gilbert and Boivin, 2005b). Mutation A809V conferring GCV/FOS resistance has also been reported with some frequency. Importantly, some mutations (E756K, V812L, and del981–982) have been associated with resistance to all three antivirals (Chou et al., 2003; Cihlar et al., 1998a). With regard to the levels of resistance, mutations L501I, K513N, and deletion of codons 981–982 have been associated with a six to eightfold decrease in GCV susceptibility (Chou et al., 2003; Cihlar et al., 1998a,b), mutations F412C/V, K513N, and A987G with a 10 to 18-fold decrease in CDV susceptibility (Chou et al., 1997; Cihlar et al., 1998a,b) whereas mutations D588N, V715M, E756K, L802M, and T821I seem to confer a 5.5 to 21-fold resistance to FOS (Baldanti et al., 1996; Chou et al., 1997, 2003; Cihlar et al., 1998a; Mousavi-Jazi et al., 2001). A few UL54 mutations have been studied in marker transfer experiments for their effect on viral fitness. Among those, mutations T700A and V715M (conserved region II) (Baldanti et al., 1996), K513N (δ-region C) (Cihlar et al., 1998b), and D301N (Exo I motif) (Chou et al.,

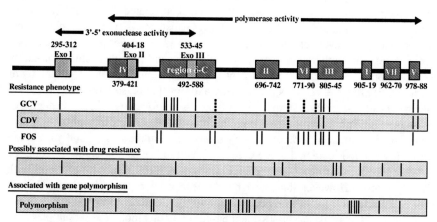

Fig. 8.3 HCMV DNA polymerase mutations conferring drug resistance. HCMV DNA polymerase conserved regions are represented by *shaded boxes*. *Numbers* under/over the boxes indicate the positions (codons no.) of these conserved regions. Note that resistant strains containing reported mutations were not necessarily tested for all antivirals listed. *Dashed lines* indicate that independent marker transfer experiments or substitution by different amino acids have resulted in discrepant results.

2003) were shown to significantly reduce the yield of progeny virus in cell culture supernatants whereas some others (D413E, T503I, L516R, and E756K/D) were only associated with a modest attenuation of viral replication (Chou et al., 2003). In the case of HCMV DNA pol mutants selected during GCV therapy, it should be noted that UL97 mutations have been generally shown to emerge first and to confer a low-level of resistance ($IC_{50} < 30$ µM) whereas subsequent emergence of UL54 mutations usually leads to a high-level of drug resistance ($IC_{50} > 30$ µM) (Erice et al., 1997; Jabs et al., 2001a; Smith et al., 1997). However, isolated UL54 mutations have been reported occasionally in clinical HCMV strains (Boivin et al., 2005a).

8.2.4 Management of Infections Caused by Resistant HCMV Strains

HCMV resistance to antivirals should be suspected in patients failing treatment who have been exposed to an antiviral for substantial periods of time (typically >3–4 months of treatment in AIDS patients and after long-term prophylaxis in transplant recipients), especially if some risk factors are present (i.e., D+/R– SOT, lung or kidney/pancreas transplant, AIDS patients with CD4 counts <50 cells/µl). Resistance should be suspected in pediatric patients with shorter periods of drug exposure if they had T-cell depletion. Clinical resistance is more likely if active viral replication (high or increasing levels of DNAemia/antigenemia or viremia) persists or recurs despite maximum IV doses of the antivirals (Limaye, 2002; Nichols and Boeckh, 2001). On the other hand, rising antigenemia levels during the first 2 weeks of antiviral therapy in HSCT recipients have not been associated with antiviral resistance, but rather with host and other transplant-related factors (Gerna et al., 2003; Nichols et al., 2001). Whenever antiviral resistance is suspected, phenotypic and/or genotypic investigation for resistance should be undertaken. As discussed above, genotypic methods are fast, more convenient, and provide useful information for selection of an alternative treatment. However, identification of mutations of unknown significance remains problematic and, for that reason, phenotypic assays may still be necessary. Furthermore, genotypic assays do not quantitate the degree of resistance while phenotypic assays do. The choice of the sample to analyze may also have some importance. Some studies have reported that there is a good correlation between genotypes detected in the eyes and the blood (93.5%) (Hu et al., 2002) or between blood and urine isolates (87.5%) (Jabs et al., 2001a) of AIDS patients with HCMV retinitis. However, there have been at least some reports of resistant HCMV strains restricted to specific body compartments (Eckle et al., 2000; Liu et al., 1998). This suggests that resistance assessment based solely on blood or urine samples may be suboptimal in some cases.

As mentioned, resistance is more likely when stable or rising viral loads (especially DNAemia levels) or persistence of clinical symptoms are observed more than 2 weeks after initiating appropriate full-dose IV antiviral therapy. In this context, clinical decisions on disease management should be based on genotypic analysis of UL97 and UL54 genes (when available), the patient's immune status (e.g., high risk D+/R– recipients, lung transplant recipients), and disease severity (i.e., sight or life-threatening conditions) (Drew, 2003; Limaye, 2002). Despite the limitation mentioned above, genotypic resistance testing is more practical and rapid (results in 72–96 h) than phenotypic assays. Thus, rescue therapy should be ideally based on results of the genotypic assays. In centers where genotypic testing is unavailable or performed infrequently, initial management should avoid the use of drugs with similar pathways of resistance. For instance, patients failing GCV should be given FOS in the absence of any sequencing data because of the possibility of UL54 mutations that could confer resistance to both GCV and CDV. On the other hand, if UL97 and UL54 sequencing data are available and indicate that only UL97 mutations are present, then CDV therapy can be attempted. Other empiric options for patients failing GCV therapy could consist in re-inducing the patient with higher than normal doses of GCV (up to 10 mg/kg IV BID) or combination therapy with reduced doses of GCV and FOS (Mylonakis et al., 2002; SOCA, 1996) although these strategies are associated with significant toxicity and can be clinically risky in patients with life or sight-threatening diseases. Whenever possible, improvement of the patient's immune status (i.e., reduction of immunosuppressive regimen in transplant patients or aggressive antiretroviral therapy in AIDS patients) should also be considered. HCMV viral load should be carefully monitored (once weekly) to quantitate a response to the change in therapy.

8.3 Herpes Simplex Virus and Varicella-Zoster Virus Resistance

8.3.1 Phenotypic and Genotypic Assays to Evaluate HSV and VZV Drug Susceptibility

Phenotypically, HSV resistance to ACV is related to one of the following mechanisms: (1) a complete deficiency in viral TK activity (TK deficient), (2) a decreased production of viral TK (TK low producer), (3) a viral TK protein with altered substrate specificity (TK altered), i.e., the enzyme is able to phosphorylate thymidine, the natural substrate, but does not phosphorylate ACV, and finally (4) a viral DNA pol with altered substrate specificity (DNA pol altered). Alteration or absence of the TK protein is the most frequent mechanism seen in the clinic, probably because TK is not essential for viral replication in most tissues and cultured cells (Gaudreau et al., 1998; Morfin et al., 2000a). However, several reports have demonstrated that at least some TK activity is needed for HSV reactivation from latency in neuronal ganglia (Coen et al., 1989; Tenser et al., 1989; Wilcox et al., 1992). The TK phenotype can be determined by the selective incorporation of iododeoxycytidine (IdC) and thymidine into infected cells using plaque autoradiography (Martin et al., 1985). However, it can be difficult to evaluate residual TK activity in HSV isolates of immunocompromised patients in whom heterogenous populations (TK-competent/TK-deficient) may coexist (Gaudreau et al., 1998; Sasadeusz and Sacks,

1996). Recently, the HSV TK gene was expressed in the protozoan parasite *Leish-mania*, normally devoid of TK activity, in order to evaluate the role of specific mutations in conferring resistance to nucleoside analogs (Bestman-Smith et al., 2001). Although such expression systems do not allow the determination of the resistance levels conferred by a specific TK mutation, they may facilitate the discrimination between resistance-associated mutations and polymorphic alterations. Only approximately 10% of ACV-resistant HSV strains have polymerase mutations (Chibo et al., 2004). Resistance to FOS and to CDV is conferred by specific mutations within the viral DNA pol which is the ultimate target of all current antiviral drugs. Depending upon the locus of the pol mutation, there may or may not be cross-resistance between ACV, FOS, and CDV (Bestman-Smith and Boivin, 2003). At the present time, no simple enzymatic assay has been described to rapidly assess the DNA polymerase activity of alphaherpesviruses.

Levels of drug resistance (IC_{50} values) are best measured by cell-based (phenotypic) assays. Such assays are more practical in the case of HSV (and to some extent VZV) than for HCMV considering the more rapid replication kinetics of the former viruses. HSV resistance cut-offs to ACV have varied in the literature from 4.4 to 13.2 μM according to the method selected, i.e., plaque reduction or dye uptake assays and various other factors (although a cut-off of 2 μg/ml or 8.8 μM is mostly used with the plaque reduction assay) (Chatis and Crumpacker, 1991; Collins and Ellis, 1993; Englund et al., 1990; Erlich et al., 1989a; Safrin et al., 1994). Due to this variability, a susceptibility index is said to be a better measure of viral resistance. The ratio of the IC_{50} of the patient's isolate should be at least $3\times$ or greater than the IC_{50} of a known, sensitive HSV control (Sarisky et al., 2002). An alternative to phenotypic assays is genotyping by sequence analysis. For a comprehensive genotypic analysis, the whole TK gene as well as the conserved regions of the HSV DNA pol gene should be determined because of the large number of TK mutations (substitutions, deletions, and additions) as well as DNA pol mutations associated with drug resistance (Morfin and Thouvenot, 2003). Different systems can be used to generate HSV recombinant viruses and evaluate specific HSV mutations such as transfection of a set of overlapping viral cosmids and plasmids allowing rapid site-directed mutagenesis (Bestman-Smith and Boivin, 2003) and the cloning of the viral genome into bacterial artificial chromosomes.

Acyclovir-resistant VZV infections are defined from a clinical point of view by the persistence of lesions after 7–10 days of ACV therapy. Susceptibility of VZV to acyclovir can be tested in plaque reduction assays using fibroblastic cell lines such as MRC-5 (Fillet et al., 1998). The end point for detecting resistance is a susceptibility index of greater than or equal to four, i.e., the test strain has an IC_{50} greater than four times that of a control, known sensitive strain, e.g., the OKA strain. As regards absolute values, three resistant strains from a single series had mean ACV IC_{50} values of 85 vs. 3.3 μM for the OKA strain (Saint-Leger et al., 2001). Recently, Sahli et al. (2000) have developed a rapid phenotypic assay for assessing ACV-resistant VZV strains following expression of the viral TK gene in a TK-deficient bacteria (*tdk* mutant). VZV TK activity can thus be indirectly evaluated by reduction of colony formation in the presence of the nucleoside analogue 5-fluorodeoxyruidine (FudR).

8.3.2 Clinical Significance, Incidence, and Risk Factors for Drug-Resistant HSV and VZV Infections

Antiviral drugs against herpesviruses provide some of the best examples of effective and selective antiviral therapy. However, drug-resistant viruses have been rapidly selected in the laboratory and also identified in the clinic. Contrasting with HCMV resistance data, no extensive survey has been performed to evaluate the rate of emergence of drug-resistant HSV isolates according to the duration of antiviral therapy. Such study would be a difficult task considering that oral and topical ACV formulations are widely used. Moreover, no uniform agreement has been reached on a standardized method to measure HSV susceptibility to antiviral agents. Thus, it is difficult to compare data from different laboratories and define the true incidence of HSV resistance in specific patient populations.

In immunocompetent hosts, HSV resistance to ACV is not a clinically important problem. Studies have shown that 0.1–0.6% of HSV isolates recovered from untreated, prophylaxed, or treated immunocompetent subjects harbor a resistant phenotype to ACV ($IC_{50} \geq 8.8$ μM) as assessed by a plaque reduction assay, and this seems to reflect the natural occurrence of TK-deficient mutants in a viral population (Boon et al., 2000; Christophers et al., 1998; Danve-Szatanek et al., 2004; Fife et al., 1994; Mertz et al., 1988; Whitley and Gnann, 1992). Except for a few notable cases (Kost et al., 1993; Kriesel et al., 2005; Swetter et al., 1998), the occasional recovery of ACV-resistant HSV-2 from immunocompetent hosts has not been associated with clinical failure and proved to be transient (Gupta et al., 2005; Whitley and Gnann, 1992). However, ACV-resistant HSV strains are more often isolated in immunocompromised hosts and such isolates have been associated with persistent and/or disseminated diseases (Boivin et al., 1993; Chen et al., 2000; Christophers et al., 1998; Englund et al., 1990; Erlich et al., 1989a; Hill et al., 1991; Morfin et al., 2000b; Safrin et al., 1990; Wade et al., 1982). In the few clinical surveys reported, the rate of ACV-resistant HSV isolates has varied from 4.3 to 14% among all immunocompromised groups (Christophers et al., 1998; Danve-Szatanek et al., 2004; DeJesus et al., 2003; Englund et al., 1990; Levin et al., 2004; Reyes et al., 2003). More specifically, Christopher et al. (1998) reported that 6.5% of HSV isolates obtained from patients with cancer were resistant to ACV compared to 10% from heart or lung transplant recipients, and 6% from AIDS patients. Similarly, Englund et al. (1990) showed that 7% of HSV isolates recovered from AIDS patients were resistant to ACV compared to 5–14% from diverse SOT and BMT recipients. In another study, 8% of allogeneic cell transplant recipients demonstrated persistent HSV excretion despite ACV therapy whereas 5% of HSV isolates showed significant level of ACV resistance in vitro (Chakrabarti et al., 2000). Morfin and Thouvenot (2003) reported that patients receiving either autologous or allogenic bone marrow have a similar incidence, i.e., 9%, of HSV infection but resistance only occurred in allogenic transplants, reaching a prevalence of 30%. The severity of immunosuppression and the prolonged use of ACV are considered two important factors for the development of drug resistance. The importance of the severity of immunosuppression is underscored by Langston et al. (2002) who studied adult patients undergoing lymphocyte depleted hematopoietic progenitor cell transplant from HLA-matched family donors. All seven evaluable, HSV-1 or -2 seropositive patients reactivated at a median of 40 days post-transplant and the five strains tested were all resistant to ACV. Furthermore, FOS resistance developed rapidly in the three patients treated with this drug (Langston et al., 2002). Importantly, the prevalence of ACV-resistant HSV isolates has remained stable in immunocompromised patients over the past two decades (Christophers et al., 1998; Danve-Szatanek et al., 2004) and there has been no unequivocal evidence of transmission of a resistant HSV strain from person to person.

Sparse data are available regarding the incidence of ACV-resistant VZV isolates in the clinic but some studies have shown that such resistant viruses have been mainly found in AIDS patients with low CD4 cell counts who presented with atypical, disseminated, or relapsing zoster lesions (Boivin et al., 1994; Morfin et al., 1999; Saint-Leger et al., 2001).

Only a few FOS-resistant HSV ($IC_{50} \geq 330$ μM or ≥ 100 μg/ml and at least a threefold increase in IC_{50} value compared to the parental susceptible strain) and VZV isolates have been reported in the clinic (Bendel et al., 1993; Collins et al., 1989; Fillet et al., 1995; Hwang et al., 1992; Safrin et al., 1991b; Schmit and Boivin, 1999). Nine FOS-resistant HSV clinical isolates from HIV-infected subjects in whom ACV and FOS therapy sequentially failed have been described (Bestman-Smith and Boivin, 2002; Schmit and Boivin, 1999).

8.3.3 Role of Viral TK and DNA Polymerase Mutations in Drug-Resistant Clinical Strains

Mutations in the HSV TK gene leading to ACV resistance consist of either additions or deletions in homopolymer runs of G's and C's associated with a premature stop codon or single nucleotide substitutions in conserved and non-conserved regions of the gene (Figure 8.4) (Gilbert et al., 2002). A few studies have shown that each mechanism accounts for approximately 50% of ACV-resistant phenotypes in the

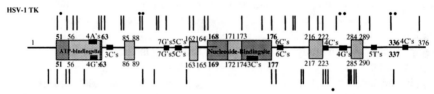

Fig. 8.4 HSV thymidine kinase mutations conferring acyclovir (ACV) resistance. HSV TK active and conserved regions are represented by *dark and light shaded boxes*, respectively. *Black boxes* indicate the position of homopolymer runs. *Numbers* under/over the boxes indicate the position (codons no.) of these regions. *Bars* indicate amino acids substitutions whereas *dots* indicate nucleotide substitutions leading to premature stop codons. Addition/deletions in homopolymer runs are not shown.

clinic (Gaudreau et al., 1998; Morfin et al., 2000a). Nucleotide substitutions are scattered within the TK gene including the three catalytic sites of the enzyme (ATP-binding site, nucleoside binding site, and aa 336) (Darby et al., 1986). Albeit rarely seen in clinic, most HSV DNA polymerase mutations conferring drug resistance are located in the conserved regions of the enzyme, most specifically in regions II, VI, and III (Figure 8.5) (Gilbert et al., 2002). Because the viral DNA pol is the ultimate target of all currently available anti-HSV agents, mutations within this gene can confer cross-resistance to many classes of drugs. A recent study by Bestman-Smith and Boivin (2003) has reported the drug resistance patterns of several HSV-1 recombinant viruses with mutations in the viral DNA polymerase gene and has shown that some mutations in region II (S724N, L778M) could confer resistance to ACV, FOS, and CDV. The general pattern of VZV TK or DNA polymerase mutations conferring drug resistance is generally similar to that seen with HSV although much less studies have been reported (Boivin et al., 1994; Morfin et al., 1999; Visse et al., 1999).

8.3.4 Management of Infections Caused by Resistant HSV and VZV Strains

With the emergence of ACV-resistant HSV infections observed in patients with AIDS and other immunocompromised hosts, several studies have examined the utility of alternative antiviral agents and treatment regimens. Standard doses of oral ACV have no clinical benefit if the HSV isolate is resistant to ACV in vitro.

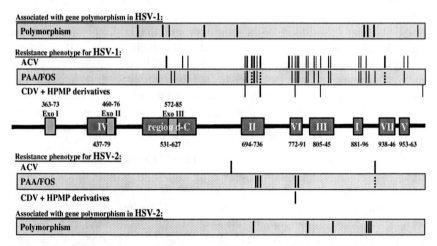

Fig. 8.5 HSV DNA polymerase mutations conferring drug resistance. HSV DNA polymerase conserved regions are represented by *shaded boxes*. *Numbers* under/over the boxes indicate the position (codons no.) of these conserved regions for HSV-1. Note that resistant strains containing reported mutations were not necessarily tested for all antivirals listed. *Bars in bold* indicates clinical isolates. *Dashed lines* indicate that independent marker transfer experiments or substitution by different amino acids have resulted in discrepant results.

Most ACV-resistant strains isolated from immunocompromised patients are TK deficient and are therefore also resistant to PCV and its prodrug FCV. In the presence of suspected or confirmed resistance to ACV, the options are either to switch to high-dose oral VACV or intravenous ACV or to use second-line agents such as FOS or CDV. If lesions do not begin to respond to high-dose oral VACV or intravenous ACV within 5–7 days, one of the other options should be chosen.

Acyclovir-resistant HSV strains remain susceptible in vitro to vidarabine, which is phosphorylated without TK, and to FOS, which does not require phosphorylation for activity. Studies have confirmed that FOS is superior to vidarabine in the treatment of these TK-deficient, drug-resistant HSV infections (Chatis et al., 1989; Erlich et al., 1989b; Safrin et al., 1991c). The dosage of FOS used for the treatment of ACV-resistant HSV infections is 40 mg/kg every 8 h (with reduction in dose for renal dysfunction). Continuous-infusion ACV therapy has been effective in a few patients with severe ACV-resistant HSV infection. Acyclovir has been administered at a dosage of 1.5–2.0 mg/kg/h for 6 weeks, and complete resolution of ACV-resistant HSV proctitis has been reported (Engel et al., 1990).

Cidofovir, in a 1–3% ointment, is effective in about 50% of patients with ACV-unresponsive HSV infections, when applied daily for 5 days (Lalezari et al., 1997). Intravenous CDV 5 mg/kg once weekly has also been effective (Kopp et al., 2002; Lalezari et al., 1994). A final option for topical therapy is trifluorothymidine (TFT) as an ophthalmic solution which may be applied to the affected area three to four times a day until the lesion is completely healed (Chilukuri and Rosen, 2003; Kessler et al., 1996).

As with many opportunistic infections in AIDS patients, there is a high incidence of recurrent HSV disease after successful treatment of drug-resistant HSV. Some (but not all) relapses in this setting have been due to drug-resistant strains, suggesting that these mutant viruses are capable of causing latency in the immunocompromised host. Chronic prophylaxis with daily ACV, VACV, FCV, or FOS can be considered in patients who have been treated successfully for drug-resistant HSV, although there are no data to confirm efficacy in this setting. Foscarnet-resistant strains of HSV have been reported, raising concerns over the possible selection for multidrug-resistant HSV with suppressive therapy (Schmit and Boivin, 1999; Snoeck et al., 1994).

Drug-resistant VZV strains have been identified in patients with AIDS. These patients may present with atypical-appearing cutaneous lesions that shed VZV intermittently despite ongoing high-dose antiviral therapy. Such strains have been isolated from patients previously treated with ACV for recurrent VZV or HSV infection, and these strains may be resistant to ACV, VACV, and FCV by deficiency of the TK enzyme (Boivin et al., 1994; Jacobson et al., 1990; Morfin et al., 1999). Foscarnet has been shown to be effective in small studies but, as with HSV, cross-resistance between ACV and FOS may occur due to viral DNA polymerase mutations (Safrin et al., 1991b). The intravenous dosage used has been 40 mg/kg three times daily.

8.4 Conclusions and Future Directions

With the increasing number of immunocompromised subjects and the prolonged administration of antiviral agents, the problem of drug resistance among herpesviruses is not expected to fade. Clearly, some drug-resistant mutants of HCMV and HSV are pathogenic and can result in significant morbidity and mortality among severely immunocompromised patients. Considerable advances have been made during recent years in our understanding of the mechanisms of herpesvirus resistance although significant work needs to be done to evaluate the impact of specific mutations on the virus replicating capacities (fitness) both in vitro and in vivo. It is anticipated that a better knowledge of the molecular mechanisms of drug resistance as well as rapid laboratory methods for detecting viral mutant sequences directly in clinical samples will result in more rational therapeutic strategies. Further work is also needed to compare prophylactic and preemptive antiviral strategies for their potential to select for HCMV resistance and on the true efficacy of alternatives when resistance does occur. On the other hand, since all currently available anti-herpetic agents target the viral DNA pol, there is an urgent need to develop and evaluate new antivirals with different mechanisms of action.

In that regard, a few alternative compounds with anti-herpetic activity (for example, ribonucleotide reductase inhibitors) have been recently described although none are in late stage of development (Duan et al., 1998; Liuzzi et al., 1994). In addition to efforts that are being made to develop oral formulation of CDV (Bidanset et al., 2004), new compounds with anti-HCMV activity are currently being developed. These drugs belong to different classes that include benzimidazoles derivatives, 4-sulfonamides substituted naphthalene derivatives, benzathidazine-modified acyclonucleosides, tricyclic inhibitors, indolocarbazoles, and an experimental immunosuppressive agent (Emery and Hassan-Walker, 2002; Michel and Mertens, 2004). Among promising candidates, maribavir (1263W94, GlaxoSmithKline) is an L-ribofuranosyl derivative of BDCRB (benzimidazole derivative) that has been shown to have good bioavailability and toxicity profile in humans (Koszalka et al., 2002; Wang et al., 2003) associated with a potent inhibitory effect on HCMV replication in vitro (Biron et al., 2002; Koszalka et al., 2002) and in humans (Lalezari et al., 2002). Maribavir is thought to prevent exit of nucleocapsids from the nucleus (nuclear egress) and DNA replication by direct inhibition of pUL97 (Biron et al., 2002; Krosky et al., 2003) and mutations conferring resistance to this drug have been mapped to this gene (Biron et al., 2002) as well as gene UL27 (Chou et al., 2004; Komazin et al., 2003). Of note, UL97 mutations conferring resistance to maribavir are different than those associated with GCV resistance. Other promising compounds include another derivative of the prototype BDCRB, the D-ribopyranosyl derivative GW275175X (GlaxoSmithKline), as well as the non-nucleosidic 4-sulfonamides-substituted naphthalene derivative tomeglovir (BAY-384766, Bayer). Both compounds showed good inhibitory activity against HCMV replication (McSharry et al., 2001; Reefschlaeger et al., 2001; Underwood et al., 2004) and seem to interfere with cleavage of concatameric DNA molecules and encapsidation which involve pUL89 and pUL56, respectively (Underwood et al., 2004).

References

Baldanti, F., Lilleri, D., Campanini, G., Comolli, G., Ridolfo, A.L., Rusconi, S. and Gerna, G., (2004), Human cytomegalovirus double resistance in a donor-positive/recipient-negative lung transplant patient with an impaired CD4-mediated specific immune response. *J Antimicrob Chemother*, **53**, 536–539.

Baldanti, F., Michel, D., Simoncini, L., Heuschmid, M., Zimmermann, A., Minisini, R., Schaarschmidt, P., Schmid, T., Gerna, G. and Mertens, T., (2002), Mutations in the UL97 orf of ganciclovir-resistant clinical cytomegalovirus isolates differentially affect GCV phosphorylation as determined in a recombinant vaccinia virus system. *Antiviral Res*, **54**, 59–67.

Baldanti, F., Silini, E., Sarasini, A., Talarico, C.L., Stanat, S.C., Biron, K.K., Furione, M., Bono, F., Palu, G. and Gerna, G., (1995), A three-nucleotide deletion in the UL97 open reading frame is responsible for the ganciclovir resistance of a human cytomegalovirus clinical isolate. *J Virol*, **69**, 796–800.

Baldanti, F., Underwood, M.R., Stanat, S.C., Biron, K.K., Chou, S., Sarasini, A., Silini, E. and Gerna, G., (1996), Single amino acid changes in the DNA polymerase confer foscarnet resistance and slow-growth phenotype, while mutations in the UL97-encoded phosphotransferase confer ganciclovir resistance in three double-resistant human cytomegalovirus strains recovered from patients with AIDS. *J Virol*, **70**, 1390–1395.

Baldanti, F., Underwood, M.R., Talarico, C.L., Simoncini, L., Sarasini, A., Biron, K.K. and Gerna, G., (1998), The cys607->tyr change in the UL97 phosphotransferase confers ganciclovir resistance to two human cytomegalovirus strains recovered from two immunocompromised patients. *Antimicrob Agents Chemother*, **42**, 444–446.

Balfour, H.H., Jr., (1999), Antiviral drugs. *N Engl J Med*, **340**, 1255–1268.

Bendel, A.E., Gross, T.G., Woods, W.G., Edelman, C.K. and Balfour, H.H., Jr., (1993), Failure of foscarnet in disseminated herpes zoster. *Lancet*, **341**, 1342.

Benz, C., Holz, G., Michel, D., Awerkiew, S., Dries, V., Stippel, D., Goeser, T. and Busch, D.H., (2003), Viral escape and T-cell immunity during ganciclovir treatment of cytomegalovirus infection: case report of a pancreatico-renal transplant recipient. *Transplantation*, **75**, 724–727.

Bestman-Smith, J. and Boivin, G., (2002), Herpes simplex virus isolates with reduced adefovir susceptibility selected in vivo by foscarnet therapy. *J Med Virol*, **67**, 88–91.

Bestman-Smith, J. and Boivin, G., (2003), Drug resistance patterns of recombinant herpes simplex virus DNA polymerase mutants generated with a set of overlapping cosmids and plasmids. *J Virol*, **77**, 7820–7829.

Bestman-Smith, J., Schmit, I., Papadopoulou, B. and Boivin, G., (2001), Highly reliable heterologous system for evaluating resistance of clinical herpes simplex virus isolates to nucleoside analogues. *J Virol*, **75**, 3105–3110.

Bhorade, S.M., Lurain, N.S., Jordan, A., Leischner, J., Villanueva, J., Durazo, R., Creech, S., Vigneswaran, W.T. and Garrity, E.R., (2002), Emergence of ganciclovir-resistant cytomegalovirus in lung transplant recipients. *J Heart Lung Transplant*, **21**, 1274–1282.

Bidanset, D.J., Beadle, J.R., Wan, W.B., Hostetler, K.Y. and Kern, E.R., (2004), Oral activity of ether lipid ester prodrugs of cidofovir against experimental human cytomegalovirus infection. *J Infect Dis*, **190**, 499–503.

Biron, K.K., Harvey, R.J., Chamberlain, S.C., Good, S.S., Smith, A.A., 3rd, Davis, M.G., Talarico, C.L., Miller, W.H., Ferris, R., Dornsife, R.E., Stanat, S.C., Drach, J.C., Townsend, L.B. and Koszalka, G.W., (2002), Potent and selective inhibition of human cytomegalovirus replication by 1263w94, a benzimidazole l-riboside with a unique mode of action. *Antimicrob Agents Chemother*, **46**, 2365–2372.

Biron, K.K., Stanat, S.C., Sorrell, J.B., Fyfe, J.A., Keller, P.M., Lambe, C.U. and Nelson, D.J., (1985), Metabolic activation of the nucleoside analog 9-[(2-hydroxy-1-(hydroxymethyl)ethoxy]methyl)guanine in human diploid fibroblasts infected with human cytomegalovirus. *Proc Natl Acad Sci USA*, **82**, 2473–2477.

Boeckh, M. and Boivin, G., (1998), Quantitation of cytomegalovirus: methodologic aspects and clinical applications. *Clin Microbiol Rev*, **11**, 533–554.

Boeckh, M., Leisenring, W., Riddell, S.R., Bowden, R.A., Huang, M.L., Myerson, D., Stevens-Ayers, T., Flowers, M.E., Cunningham, T. and Corey, L., (2003), Late cytomegalovirus disease and mortality in recipients of allogeneic hematopoietic stem cell transplants: importance of viral load and T-cell immunity. *Blood*, **101**, 407–414.

Boivin, G., Chou, S., Quirk, M.R., Erice, A. and Jordan, M.C., (1996), Detection of ganciclovir resistance mutations and quantitation of cytomegalovirus (CMV) DNA in leukocytes of patients with fatal disseminated CMV disease. *J Infect Dis*, **173**, 523–528.

Boivin, G., Edelman, C.K., Pedneault, L., Talarico, C.L., Biron, K.K. and Balfour, H.H.J., (1994), Phenotypic and genotypic characterization of acyclovir-resistant varicella-zoster viruses isolated from persons with AIDS. *J Infect Dis*, **170**, 68–75.

Boivin, G., Erice, A., Crane, D.D., Dunn, D.L. and Balfour, H.H., Jr., (1993), Acyclovir susceptibilities of herpes simplex virus strains isolated from solid organ transplant recipients after acyclovir or ganciclovir prophylaxis. *Antimicrob Agents Chemother*, **37**, 357–359.

Boivin, G., Gilbert, C., Gaudreau, A., Greenfield, I., Sudlow, R. and Roberts, N.A., (2001), Rate of emergence of cytomegalovirus (CMV) mutations in leukocytes of patients with acquired immunodeficiency syndrome who are receiving valganciclovir as induction and maintenance therapy for CMV retinitis. *J Infect Dis*, **184**, 1598–1602.

Boivin, G., Goyette, N., Gilbert, C. and Covington, E., (2005a), Analysis of cytomegalovirus DNA polymerase (UL54) mutations in solid organ transplant patients receiving valganciclovir or ganciclovir prophylaxis. *J Med Virol*, **77**, 425–429.

Boivin, G., Goyette, N., Gilbert, C., Humar, A. and Covington, E., (2005b), Clinical impact of ganciclovir-resistant cytomegalovirus infections in solid organ transplant patients. *Transpl Infect Dis*, **6**, 1–5.

Boivin, G., Goyette, N., Gilbert, C., Roberts, N., Macey, K., Paya, C., Pescovitz, M.D., Humar, A., Dominguez, E., Washburn, K., Blumberg, E., Alexander, B., Freeman, R., Heaton, N. and Covington, E., (2004), Absence of cytomegalovirus-resistance mutations after valganciclovir prophylaxis, in a prospective multicenter study of solid-organ transplant recipients. *J Infect Dis*, **189**, 1615–1618.

Boon, R.J., Bacon, T.H., Robey, H.L., Coleman, T.J., Connolly, A., Crosson, P. and Sacks, S.L., (2000), Antiviral susceptibilities of herpes simplex virus from immunocompetent subjects with recurrent herpes labialis: a UK-based survey. *J Antimicrob Chemother*, **46**, 1051.

Chakrabarti, S., Pillay, D., Ratcliffe, D., Cane, P.A., Collingham, K.E. and Milligan, D.W., (2000), Resistance to antiviral drugs in herpes simplex virus infections among allogeneic stem cell transplant recipients: risk factors and prognostic significance. *J Infect Dis*, **181**, 2055–2058.

Chatis, P.A. and Crumpacker, C.S., (1991), Analysis of the thymidine kinase gene from clinically isolated acyclovir-resistant herpes simplex viruses. *Virology* 180, 793–797.

Chatis, P.A., Miller, C.H., Schrager, L.E. and Crumpacker, C.S., (1989), Successful treatment with foscarnet of an acyclovir-resistant mucocutaneous infection with herpes simplex virus in a patient with acquired immunodeficiency syndrome. *N Engl J Med*, **320**, 297–300.

Chen, Y., Scieux, C., Garrait, V., Socie, G., Rocha, V., Molina, J.M., Thouvenot, D., Morfin, F., Hocqueloux, L., Garderet, L., Esperou, H., Selimi, F., Devergie, A., Leleu, G., Aymard, M., Morinet, F., Gluckman, E. and Ribaud, P., (2000), Resistant herpes simplex virus type 1 infection: an emerging concern after allogeneic stem cell transplantation. *Clin Infect Dis*, **31**, 927–935.

Chibo, D., Druce, J., Sasadeusz, J. and Birch, C., (2004), Molecular analysis of clinical isolates of acyclovir resistant herpes simplex virus. *Antiviral Res*, **61**, 83–91.

Chilukuri, S. and Rosen, T., (2003), Management of acyclovir-resistant herpes simplex virus. *Dermatol Clin*, **21**, 311–320.

Chou, S., Erice, A., Jordan, M.C., Vercellotti, G.M., Michels, K.R., Talarico, C.L., Stanat, S.C. and Biron, K.K., (1995), Analysis of the UL97 phosphotransferase coding sequence in clinical cytomegalovirus isolates and identification of mutations conferring ganciclovir resistance. *J Infect Dis*, **171**, 576–583.

Chou, S., Lurain, N.S., Thompson, K.D., Miner, R.C. and Drew, W.L., (2003), Viral DNA polymerase mutations associated with drug resistance in human cytomegalovirus. *J Infect Dis*, **188**, 32–39.

Chou, S., Lurain, N.S., Weinberg, A., Cai, G.Y., Sharma, P.L. and Crumpacker, C.S., (1999), Interstrain variation in the human cytomegalovirus DNA polymerase sequence and its effect on genotypic diagnosis of antiviral drug resistance. *Antimicrob Agents Chemother*, **43**, 1500–1502.

Chou, S., Marousek, G., Guentzel, S., Follansbee, S.E., Poscher, M.E., Lalezari, J.P., Miner, R.C. and Drew, W.L., (1997), Evolution of mutations conferring multidrug resistance during prophylaxis and therapy for cytomegalovirus disease. *J Infect Dis*, **176**, 786–789.

Chou, S., Marousek, G., Parenti, D.M., Gordon, S.M., LaVoy, A.G., Ross, J.G., Miner, R.C. and Drew, W.L., (1998), Mutation in region III of the DNA polymerase gene conferring foscarnet resistance in cytomegalovirus isolates from 3 subjects receiving prolonged antiviral therapy. *J Infect Dis*, **178**, 526–530.

Chou, S., Marousek, G.I., Senters, A.E., Davis, M.G. and Biron, K.K., (2004), Mutations in the human cytomegalovirus UL27 gene that confer resistance to maribavir. *J Virol*, **78**, 7124–7130.

Chou, S. and Meichsner, C.L., (2000), A nine-codon deletion mutation in the cytomegalovirus UL97 phosphotransferase gene confers resistance to ganciclovir. *Antimicrob Agents Chemother*, **44**, 183–185.

Chou, S., Waldemer, R.H., Senters, A.E., Michels, K.S., Kemble, G.W., Miner, R.C. and Drew, W.L., (2002), Cytomegalovirus UL97 phosphotransferase mutations that affect susceptibility to ganciclovir. *J Infect Dis*, **185**, 162–169.

Chou, S., Van Wechel, L.C., Lichy, H.M. and Marousek, G.I., (2005), Phenotyping of cytomegalovirus drug resistance mutations by using recombinant viruses incorporating a reporter gene. *Antimicrob Agents Chemother*, **49**, 2710–2715.

Chrisp, P. and Clissold, S.P., (1991), Foscarnet. A review of its antiviral activity, pharmacokinetic properties and therapeutic use in immunocompromised patients with cytomegalovirus retinitis. *Drugs*, **41**, 104–129.

Christophers, J., Clayton, J., Craske, J., Ward, R., Collins, P., Trowbridge, M. and Darby, G., (1998), Survey of resistance of herpes simplex virus to acyclovir in Northwest England. *Antimicrob Agents Chemother*, **42**, 868–872.

Cihlar, T. and Chen, M.S., (1996), Identification of enzymes catalyzing two-step phosphorylation of cidofovir and the effect of cytomegalovirus infection on their activities in host cells. *Mol Pharmacol*, **50**, 1502–1510.

Cihlar, T., Fuller, M. and Cherrington, J., (1998a), Characterization of drug resistance-associated mutations in the human cytomegalovirus DNA polymerase gene by using recombinant mutant viruses generated from overlapping DNA fragments. *J Virol*, **72**, 5927–5936.

Cihlar, T., Fuller, M.D., Mulato, A.S. and Cherrington, J.M., (1998b), A point mutation in the human cytomegalovirus DNA polymerase gene selected in vitro by cidofovir confers a slow replication phenotype in cell culture. *Virology* 248, 382–393.

Coen, D.M., Kosz Vnenchak, M., Jacobson, J.G., Leib, D.A., Bogard, C.L., Schaffer, P.A., Tyler, K.L. and Knipe, D.M., (1989), Thymidine kinase-negative herpes simplex virus mutants establish latency in mouse trigeminal ganglia but do not reactivate. *Proc Natl Acad Sci USA*, **86**, 4736–4740.

Collins, P. and Ellis, M.N., (1993), Sensitivity monitoring of clinical isolates of herpes simplex virus to acyclovir. *J Med Virol*, Supplement 1, 58–66.

Collins, P., Larder, B.A., Oliver, N.M., Kemp, S., Smith, I.W. and Darby, G., (1989), Characterization of a DNA polymerase mutant of herpes simplex virus from a severely immunocompromised patient receiving acyclovir. *J Gen Virol*, **70**, 375–382.

Dankner, W.M., Scholl, D., Stanat, S.C., Martin, M., Sonke, R.L. and Spector, S.A., (1990), Rapid antiviral DNA–DNA hybridization assay for human cytomegalovirus. *J Virol Meth*, **28**, 293–298.

Danve-Szatanek, C., Aymard, M., Thouvenot, D., Morfin, F., Agius, G., Bertin, I., Bil-laudel, S., Chanzy, B., Coste-Burel, M., Finkielsztejn, L., Fleury, H., Hadou, T., Henquell, C., Lafeuille, H., Lafon, M.E., Le Faou, A., Legrand, M.C., Maille, L., Mengelle, C., Morand, P., Morinet, F., Nicand, E., Omar, S., Picard, B., Pozzetto, B., Puel, J., Raoult, D., Scieux, C., Segondy, M., Seigneurin, J.M., Teyssou, R. and Zandotti, C., (2004), Surveillance network for herpes simplex virus resistance to antiviral drugs: 3-year follow-up. *J Clin Microbiol*, **42**, 242–249.

Darby, G., Larder, B.A. and Inglis, M.M., (1986), Evidence that the "active centre" of the herpes simplex virus thymidine kinase involves an interaction between three distinct regions of the polypeptide. *J Gen Virol*, **67**, 753–758.

De Clercq, E., (2004), Discovery and development of BVDU (brivudin) as a therapeutic for the treatment of herpes zoster. *Biochem Pharmacol*, **68**, 2301–2315.

DeJesus, E., Wald, A., Warren, T., Schacker, T.W., Trottier, S., Shahmanesh, M., Hill, J.L. and Brennan, C.A., (2003), Valacyclovir for the suppression of recurrent genital herpes in human immunodeficiency virus-infected subjects. *J Infect Dis*, **188**, 1009–1016.

Drew, W.L., (2003), Cytomegalovirus disease in the highly active antiretroviral therapy era. *Curr Infect Dis Rep*, **5**, 257–265.

Drew, W.L., Anderson, R., Lang, W., Miner, R.C., Davis, G. and Lalezari, J., (1995), Fail-ure of high-dose oral acyclovir to suppress CMV viruria or induce ganciclovir-resistant CMV in HIV antibody positive patients. *J Acquir Immune Defic Syndr Hum Retrovirol*, **8**, 289–291.

Drew, W.L., Miner, R.C., Busch, D.F., Follansbee, S.E., Gullett, J., Mehalko, S.G., Gordon, S.M., Owen, W.F., Jr, Matthews, T.R. and Buhles, W.C., (1991), Prevalence of resistance in patients receiving ganciclovir for serious cytomegalovirus infection. *J Infect Dis*, **163**, 716–719.

Drew, W.L., Paya, C.V. and Emery, V., (2001), Cytomegalovirus (CMV) resistance to antivirals. *Am J Transplant*, **1**, 307–312.

Drew, W.L., Stempien, M.J., Andrews, J., Shadman, A., Tan, S.J., Miner, R. and Buhles, W., (1999), Cytomegalovirus (CMV) resistance in patients with CMV retinitis and AIDS treated with oral or intravenous ganciclovir. *J Infect Dis*, **179**, 1352–1355.

Duan, J., Liuzzi, M., Paris, W., Lambert, M., Lawetz, C., Moss, N., Jaramillo, J., Gauthier, J., Deziel, R. and Cordingley, M.G., (1998), Antiviral activity of a selective ribonucleotide reduc-tase inhibitor against acyclovir-resistant herpes simplex virus type 1 in vivo. *Antimicrob Agents Chemother*, **42**, 1629–1635.

Eckle, T., Prix, L., Jahn, G., Klingebiel, T., Handgretinger, R., Selle, B. and Hamprecht, K., (2000), Drug-resistant human cytomegalovirus infection in children after allogeneic stem cell trans-plantation may have different clinical outcomes. *Blood*, **96**, 3286–3289.

Emery, V.C. and Hassan-Walker, A.F., (2002), Focus on new drugs in development against human cytomegalovirus. *Drugs*, **62**, 1853–1858.

Engel, J.P., Englund, J.A., Fletcher, C.V. and Hill, E.L., (1990), Treatment of resistant herpes sim-plex virus with continuous-infusion acyclovir. *JAMA*, 263, 1662–1664.

Englund, J.A., Zimmerman, M.E., Swierkosz, E.M., Goodman, J.L., Scholl, D.R. and Balfour, H.H., (1990), Herpes simplex virus resistant to acyclovir: a study in a tertiary care center. *Ann Intern Med*, **112**, 416–422.

Erice, A., Borrell, N., Li, W., Miller, W.J. and Balfour, H.H., Jr., (1998), Ganciclovir susceptibilities and analysis of UL97 region in cytomegalovirus (CMV) isolates from bone marrow recipients with CMV disease after antiviral prophylaxis. *J Infect Dis*, **178**, 531–534.

Erice, A., Gil-Roda, C., Perez, J.L., Balfour, H.H., Sannerud, K.J., Hanson, M.N., Boivin, G. and Chou, S., (1997), Antiviral susceptibilities and analysis of UL97 and DNA polymerase sequences of clinical cytomegalovirus isolates from immunocompromised patients. *J Infect Dis*, **175**, 1087–1092.

Erlich, K.S., Jacobson, M.A., Koehler, J.E., Follansbee, S.E., Drennan, D.P., Gooze, L., Safrin, S. and Mills, J., (1989b), Foscarnet therapy for severe acyclovir-resistant herpes simplex virus type-2 infections in patients with the acquired immunodeficiency syndrome (AIDS). An uncon-trolled trial. *Ann Intern Med*, **110**, 710–713.

Erlich, K.E., Mills, J., Chatis, P., Mertz, G.J., Busch, D.F., Follansbee, S.E., Grant, R.M. and Crumpacker, C.S., (1989a), Acyclovir-resistant herpes simplex virus infections in patients with the acquired immunodeficiency syndrome. *N Engl J Med*, **320**, 293–296.

Fife, K.H., Crumpacker, C.S., Mertz, G.J., Hill, E.L. and Boone, G.S., (1994), Recurrence and resistance patterns of herpes simplex virus following cessation of > or = 6 years of chronic suppression with acyclovir. *J Infect Dis*, **169**, 1338–1341.

Fillet, A.M., Auray, L., Alain, S., Gourlain, K., Imbert, B.M., Najioullah, F., Champier, G., Gouarin, S., Carquin, J., Houhou, N., Garrigue, I., Ducancelle, A., Thouvenot, D. and Mazeron, M.C., (2004), Natural polymorphism of cytomegalovirus DNA polymerase lies in two nonconserved regions located between domains delta-C and II and between domains III and I. *Antimicrob Agents Chemother*, **48**, 1865–1868.

Fillet, A.M., Dumont, B., Caumes, E., Visse, B., Agut, H., Bricaire, F. and Huraux, J.M., (1998), Acyclovir-resistant varicella-zoster virus: phenotypic and genetic characterization. *J Med Virol*, **55**, 250–254.

Fillet, A.M., Visse, B., Caumes, E., Dumont, B., Gentilini, M. and Huraux, J.M., (1995), Foscarnet-resistant multidermatomal zoster in a patient with AIDS. *Clin Infect Dis*, **21**, 1348–1349.

Fyfe, J.A., Keller, P.M., Furman, P.A., Miller, R.L. and Elion, G.B., (1978), Thymidine kinase from herpes simplex virus phosphorylates the new antiviral compound, 9-(2-hydroxyethoxymethyl)guanine. *J Biol Chem*, **253**, 8721–8727.

Gaudreau, A., Hill, E., Balfour, H.H., Jr, Erice, A. and Boivin, G., (1998), Phenotypic and genotypic characterization of acyclovir-resistant herpes simplex viruses from immunocompromised patients. *J Infect Dis*, **178**, 297–303.

Gerna, G., Baldanti, F., Zavattoni, M., Sarasini, A., Percivalle, E. and Revello, M.G., (1992), Monitoring of ganciclovir sensitivity of multiple human cytomegalovirus strains coinfecting blood of an AIDS patient by an immediate–early antigen plaque assay. *Antiviral Res*, **19**, 333–345.

Gerna, G., Sarasini, A., Lilleri, D., Percivalle, E., Torsellini, M., Baldanti, F. and Revello, M.G., (2003), In vitro model for the study of the dissociation of increasing antigenemia and decreasing DNAemia and viremia during treatment of human cytomegalovirus infection with ganciclovir in transplant recipients. *J Infect Dis*, **188**, 1639–1647.

Gilbert, C., Bestman-Smith, J. and Boivin, G., (2002), Resistance of herpesviruses to antiviral drugs: clinical impacts and molecular mechanisms. *Drug Resist Updat*, **5**, 88–114.

Gilbert, C. and Boivin, G., (2003), Discordant phenotypes and genotypes of cytomegalovirus (CMV) in patients with AIDS and relapsing CMV retinitis. *AIDS*, **17**, 337–341.

Gilbert, C. and Boivin, G., (2005a), New reporter cell line to evaluate the sequential emergence of multiple human cytomegalovirus mutations during in vitro drug exposure. *Antimicrob Agents Chemother*, **49**, 4860–4866.

Gilbert, C. and Boivin, G., (2005b), Human cytomegalovirus resistance to antiviral drugs. *Antimicrob Agents Chemother*, **49**, 873–883.

Gilbert, C., Roy, J., Belanger, R., Delage, R., Beliveau, C., Demers, C. and Boivin, G., (2001), Lack of emergence of cytomegalovirus UL97 mutations conferring ganciclovir (GCV) resistance following preemptive GCV therapy in allogeneic stem cell transplant recipients. *Antimicrob Agents Chemother*, **45**, 3669–3671.

Gupta, R., Hill, E.L., McClernon, D., Davis, G., Selke, S., Corey, L. and Wald, A., (2005), Acyclovir sensitivity of sequential herpes simplex virus type 2 isolates from the genital mucosa of immunocompetent women. *J Infect Dis*, **192**, 1102–1107.

Hamprecht, K., Eckle, T., Prix, L., Faul, C., Einsele, H. and Jahn, G., (2003), Ganciclovir-resistant cytomegalovirus disease after allogeneic stem cell transplantation: pitfalls of phenotypic diagnosis by in vitro selection of an UL97 mutant strain. *J Infect Dis*, **187**, 139–143.

Hanson, M.N., Preheim, L.C., Chou, S., Talarico, C.L., Biron, K.K. and Erice, A., (1995), Novel mutation in the UL97 gene of a clinical cytomegalovirus strain conferring resistance to ganciclovir. *Antimicrob Agents Chemother*, **39**, 1204–1205.

Hill, E.L., Hunter, G.A. and Ellis, M.N., (1991), In vitro and in vivo characterization of herpes simplex virus clinical isolates recovered from patients infected with human immunodeficiency virus. *Antimicrob Agents Chemother*, **35**, 2322–2328.

Hu, H., Jabs, D.A., Forman, M.S., Martin, B.K., Dunn, J.P., Weinberg, D.V. and Davis, J.L., (2002), Comparison of cytomegalovirus (CMV) UL97 gene sequences in the blood and vitreous of patients with acquired immunodeficiency syndrome and CMV retinitis. *J Infect Dis*, **185**, 861–867.

Humar, A., Kumar, D., Preiksaitis, J., Boivin, G., Siegal, D., Fenton, J., Jackson, K., Nia, S. and Lien, D., (2005), A trial of valganciclovir prophylaxis for cytomegalovirus prevention in lung transplant recipients. *Am J Transplant*, **5**, 1462–1468.

Hwang, C.B., Ruffner, K.L. and Coen, D.M., (1992), A point mutation within a distinct conserved region of the herpes simplex virus DNA polymerase gene confers drug resistance. *J Virol*, **66**, 1774–1776.

Jabs, D.A., Enger, C., Dunn, J.P. and Forman, M., (1998a), Cytomegalovirus retinitis and viral resistance: ganciclovir resistance. *J Infect Dis*, **177**, 770–773.

Jabs, D.A., Enger, C., Forman, M. and Dunn, J.P., (1998b), Incidence of foscarnet resistance and cidofovir resistance in patients treated for cytomegalovirus retinitis. The Cytomegalovirus Retinitis and Viral Resistance Study Group. *Antimicrob Agents Chemother*, **42**, 2240–2244.

Jabs, D.A., Martin, B.K., Forman, M.S., Dunn, J.P., Davis, J.L., Weinberg, D.V., Biron, K.K. and Baldanti, F., (2001a), Mutations conferring ganciclovir resistance in a cohort of patients with acquired immunodeficiency syndrome and cytomegalovirus retinitis. *J Infect Dis*, **183**, 333–337.

Jabs, D.A., Martin, B.K., Forman, M.S., Dunn, J.P., Davis, J.L., Weinberg, D.V., Biron, K.K., Baldanti, F. and Hu, H., (2001b), Longitudinal observations on mutations conferring ganciclovir resistance in patients with acquired immunodeficiency syndrome and cytomegalovirus retinitis: The Cytomegalovirus and Viral Resistance Study Group report number 8. *Am J Ophthalmol*, **132**, 700–710.

Jacobson, M.A., Berger, T.G., Fikrig, S., Becherer, P., Moohr, J.W., Stanat, S.C. and Biron, K.K., (1990), Acyclovir-resistant varicella zoster virus infection after chronic oral acyclovir therapy in patients with the acquired immunodeficiency syndrome (AIDS). *Ann Intern Med*, **112**, 187–191.

Kessler, H.A., Hurwitz, S., Farthing, C., Benson, C.A., Feinberg, J., Kuritzkes, D.R., Bailey, T.C., Safrin, S., Steigbigel, R.T., Cheeseman, S.H., McKinley, G.F., Wettlaufer, B., Owens, S., Nevin, T. and Korvick, J.A., (1996), Pilot study of topical trifluridine for the treatment of acyclovir-resistant mucocutaneous herpes simplex disease in patients with AIDS (ACTG 172). aIDS Clinical Trials Group. *J Acquir Immune Defic Syndr Hum Retrovirol*, **12**, 147–152.

Kesson, A., Zeng, F., Cunningham, A. and Rawlinson, W., (1998), The use of flow cytometry to detect antiviral resistance in human cytomegalovirus. *J Virol Meth*, **71**, 177–186.

Komazin, G., Ptak, R.G., Emmer, B.T., Townsend, L.B. and Drach, J.C., (2003), Resistance of human cytomegalovirus to the benzimidazole l-ribonucleoside maribavir maps to UL27. *J Virol*, **77**, 11499–11506.

Kopp, T., Geusau, A., Rieger, A. and Stingl, G., (2002), Successful treatment of an aciclovir-resistant herpes simplex type 2 infection with cidofovir in an AIDS patient. *Br J Dermatol*, **147**, 134–138.

Kost, R.G., Hill, E.L., Tigges, M. and Straus, S.E., (1993), Brief report: recurrent acyclovir-resistant genital herpes in an immunocompetent patient. *N Engl J Med*, **329**, 1777–1782.

Koszalka, G.W., Johnson, N.W., Good, S.S., Boyd, L., Chamberlain, S.C., Townsend, L.B., Drach, J.C. and Biron, K.K., (2002), Preclinical and toxicology studies of 1263w94, a potent and selective inhibitor of human cytomegalovirus replication. *Antimicrob Agents Chemother*, **46**, 2373–2380.

Kriesel, J.D., Spruance, S.L., Prichard, M., Parker, J.N. and Kern, E.R., (2005), Recurrent antiviral-resistant genital herpes in an immunocompetent patient. *J Infect Dis*, **192**, 156–161.

Krosky, P.M., Baek, M.C. and Coen, D.M., (2003), The human cytomegalovirus UL97 protein kinase, an antiviral drug target, is required at the stage of nuclear egress. *J Virol*, **77**, 905–914.

Kruger, R.M., Shannon, W.D., Arens, M.Q., Lynch, J.P., Storch, G.A. and Trulock, E.P., (1999), The impact of ganciclovir-resistant cytomegalovirus infection after lung transplantation. *Transplantation*, **68**, 1272–1279.

Lalezari, J., Schacker, T., Feinberg, J., Gathe, J., Lee, S., Cheung, T., Kramer, F., Kessler, H., Corey, L., Drew, W.L., Boggs, J., McGuire, B., Jaffe, H.S. and Safrin, S., (1997), A randomized, double-blind, placebo-controlled trial of cidofovir gel for the treatment of acyclovir-unresponsive mucocutaneous herpes simplex virus infection in patients with AIDS. *J Infect Dis*, **176**, 892–898.

Lalezari, J.P., Aberg, J.A., Wang, L.H., Wire, M.B., Miner, R., Snowden, W., Talarico, C.L., Shaw, S., Jacobson, M.A. and Drew, W.L., (2002), Phase I dose escalation trial evaluating the pharmacokinetics, anti-human cytomegalovirus (HCMV) activity, and safety of 1263w94 in human immunodeficiency virus-infected men with asymptomatic HCMV shedding. *Antimicrob Agents Chemother*, **46**, 2969–2976.

Lalezari, J.P., Drew, W.L., Glutzer, E., Miner, D., Safrin, S., Owen, W.F., Jr, Davidson, J.M., Fisher, P.E. and Jaffe, H.S., (1994), Treatment with intravenous (s)-1-[3-hydroxy-2-(phosphonylmethoxy)propyl]-cytosine of acyclovir-resistant mucocutaneous infection with herpes simplex virus in a patient with AIDS. *J Infect Dis*, **170**, 570–572.

Landry, M.L., Stanat, S., Biron, K., Brambilla, D., Britt, W., Jokela, J., Chou, S., Drew, W.L., Erice, A., Gilliam, B., Lurain, N., Manischewitz, J., Miner, R., Nokta, M., Reichelderfer, P., Spector, S., Weinberg, A., Yen-Lieberman, B. and Crumpacker, C., (2000), A standardized plaque reduction assay for determination of drug susceptibilities of cytomegalovirus clinical isolates. *Antimicrob Agents Chemother*, **44**, 688–692.

Langston, A.A., Redei, I., Caliendo, A.M., Somani, J., Hutcherson, D., Lonial, S., Bucur, S., Cherry, J., Allen, A. and Waller, E.K., (2002), Development of drug-resistant herpes simplex virus infection after haploidentical hematopoietic progenitor cell transplantation. *Blood*, **99**, 1085–1088.

Levin, M.J., Bacon, T.H. and Leary, J.J., (2004), Resistance of herpes simplex virus infections to nucleoside analogues in HIV-infected patients. *Clin Infect Dis*, **39** (Suppl 5), S248–S257.

Limaye, A.P., (2002), Ganciclovir-resistant cytomegalovirus in organ transplant recipients. *Clin Infect Dis*, **35**, 866–872.

Limaye, A.P., Corey, L., Koelle, D.M., Davis, C.L. and Boeckh, M., (2000), Emergence of ganciclovir-resistant cytomegalovirus disease among recipients of solid-organ transplants. *Lancet*, **356**, 645–649.

Limaye, A.P., Raghu, G., Koelle, D.M., Ferrenberg, J., Huang, M.L. and Boeckh, M., (2002), High incidence of ganciclovir-resistant cytomegalovirus infection among lung transplant recipients receiving preemptive therapy. *J Infect Dis*, **185**, 20–27.

Liu, W., Kuppermann, B.D., Martin, D.F., Wolitz, R.A. and Margolis, T.P., (1998), Mutations in the cytomegalovirus UL97 gene associated with ganciclovir-resistant retinitis. *J Infect Dis*, **177**, 1176–1181.

Liuzzi, M., Deziel, R., Moss, N., Beaulieu, P., Bonneau, A.M., Bousquet, C., Chafouleas, J.G., Garneau, M., Jaramillo, J., Krogsrud, R.L., Lagacé, L., McCollum, R.S., Nawoot, S. and Guindon, Y., (1994), A potent peptidomimetic inhibitor of HSV ribonucleotide reductase with antiviral activity in vivo. *Nature*, 372, 695–698.

Lowance, D., Neumayer, H.H., Legendre, C.M., Squifflet, J.P., Kovarik, J., Brennan, P.J., Norman, D., Mendez, R., Keating, M.R., Coggon, G.L., Crisp, A. and Lee, I.C., (1999), Valacyclovir for the prevention of cytomegalovirus disease after renal transplantation. *N Engl J Med*, **340**, 1462–1470.

Lurain, N.S., Bhorade, S.M., Pursell, K.J., Avery, R.K., Yeldandi, V.V., Isada, C.M., Robert, E.S., Kohn, D.J., Arens, M.Q., Garrity, E.R., Taege, A.J., Mullen, M.G., Todd, K.M., Bremer, J.W. and Yen-Lieberman, B., (2002), Analysis and characterization of antiviral drug-resistant cytomegalovirus isolates from solid organ transplant recipients. *J Infect Dis*, **186**, 760–768.

Lurain, N.S., Weinberg, A., Crumpacker, C.S. and Chou, S., (2001), Sequencing of cytomegalovirus UL97 gene for genotypic antiviral resistance testing. *Antimicrob Agents Chemother*, **45**, 2775–2780.

Martin, J.L., Ellis, M.N., Keller, P.M., Biron, K.K., Lehrman, S.N., Barry, D.W. and Furman, P.A., (1985), Plaque autoradiography assay for the detection and quantification of thymidine kinase-deficient and thymidine kinase-altered mutants of herpes simplex virus in clinical isolates. *Antimicrob Agents Chemother*, **28**, 181–187.

McSharry, J.J., McDonough, A., Olson, B., Hallenberger, S., Reefschlaeger, J., Bender, W. and Drusano, G.L., (2001), Susceptibilities of human cytomegalovirus clinical isolates to BAY38-4766, BAY43-9695, and ganciclovir. *Antimicrob Agents Chemother*, **45**, 2925–2927.

McSharry, J.M., Lurain, N.S., Drusano, G.L., Landay, A., Manischewitz, J., Nokta, M., O'Gorman, M., Shapiro, H.M., Weinberg, A., Reichelderfer, P. and Crumpacker, C., (1998), Flow cytometric determination of ganciclovir susceptibilities of human cytomegalovirus clinical isolates. *J Clin Microbiol*, **36**, 958–964.

Mertz, G.J., Jones, C.C., Mills, J., Fife, K.H., Lemon, S.M., Stapleton, J.T., Hill, E.L. and Davis, L.G., (1988), Long-term acyclovir suppression of frequently recurring genital herpes simplex virus infection. A multicenter double-blind trial. *JAMA*, **260**, 201–206.

Michel, D., Hohn, S., Haller, T., Jun, D. and Mertens, T., (2001), Aciclovir selects for ganciclovir-cross-resistance of human cytomegalovirus in vitro that is only in part explained by known mutations in the UL97 protein. *J Med Virol*, **65**, 70–76.

Michel, D. and Mertens, T., (2004), The UL97 protein kinase of human cytomegalovirus and homologues in other herpesviruses: impact on virus and host. *Biochim Biophys Acta*, **1697**, 169–180.

Morfin, F., Souillet, G., Bilger, K., Ooka, T., Aymard, M. and Thouvenot, D., (2000a), Genetic characterization of thymidine kinase from acyclovir-resistant and susceptible herpes simplex virus type 1 isolated from bone marrow transplant recipients. *J Infect Dis*, **182**, 290–293.

Morfin, F. and Thouvenot, D., (2003), Herpes simplex virus resistance to antiviral drugs. *J Clin Virol*, **26**, 29–37.

Morfin, F., Thouvenot, D., Aymard, M. and Souillet, G., (2000b), Reactivation of acyclovir-resistant thymidine kinase-deficient herpes simplex virus harbouring single base insertion within a 7 Gs homopolymer repeat of the thymidine kinase gene. *J Med Virol*, **62**, 247–250.

Morfin, F., Thouvenot, D., De Turenne-Tessier, M., Lina, B., Aymard, M. and Ooka, T., (1999), Phenotypic and genetic characterization of thymidine kinase from clinical strains of varicella-zoster virus resistant to acyclovir. *Antimicrob Agents Chemother*, **43**, 2412–2416.

Mousavi-Jazi, M., Schloss, L., Drew, W.L., Linde, A., Miner, R.C., Harmenberg, J., Wahren, B. and Brytting, M., (2001), Variations in the cytomegalovirus DNA polymerase and phosphotransferase genes in relation to foscarnet and ganciclovir sensitivity. *J Clin Virol*, **23**, 1–15.

Mulamba, G.B., Hu, A., Azad, R.F., Anderson, K.P. and Coen, D.M., (1998), Human cytomegalovirus mutant with sequence-dependent resistance to the phosphorothioate oligonucleotide fomivirsen (ISIS 2922). *Antimicrob Agents Chemother*, **42**, 971–973.

Mylonakis, E., Kallas, W.M. and Fishman, J.A., (2002), Combination antiviral therapy for ganciclovir-resistant cytomegalovirus infection in solid-organ transplant recipients. *Clin Infect Dis*, **34**, 1337–1341.

Nguyen, Q., Champlin, R., Giralt, S., Rolston, K., Raad, I., Jacobson, K., Ippoliti, C., Hecht, D., Tarrand, J., Luna, M. and Whimbey, E., (1999), Late cytomegalovirus pneumonia in adult allogeneic blood and marrow transplant recipients. *Clin Infect Dis*, **28**, 618–623.

Nichols, W. and Boeckh, M., (2001), Cytomegalovirus infections. *Curr Treat Options Infect Dis*, **3**, 78–91.

Nichols, W.G., Corey, L., Gooley, T., Drew, W.L., Miner, R., Huang, M., Davis, C. and Boeckh, M., (2001), Rising pp65 antigenemia during preemptive anticytomegalovirus therapy after allogeneic hematopoietic stem cell transplantation: risk factors, correlation with DNA load, and outcomes. *Blood*, **97**, 867–874.

Paya, C., Humar, A., Dominguez, E., Washburn, K., Blumberg, E., Alexander, B., Freeman, R., Heaton, N. and Pescovitz, M.D., (2004), Efficacy and safety of valganciclovir vs. oral ganciclovir for prevention of cytomegalovirus disease in solid organ transplant recipients. *Am J Transplant*, **4**, 611–620.

Paya, C.V., Wilson, J.A., Espy, M.J., Sia, I.G., DeBernardi, M.J., Smith, T.F., Patel, R., Jenkins, G., Harmsen, W.S., Vanness, D.J. and Wiesner, R.H., (2002), Preemptive use of oral ganciclovir to prevent cytomegalovirus infection in liver transplant patients: a randomized, placebo-controlled trial. *J Infect Dis*, **185**, 854–860.

Pescovitz, M.D., Rabkin, J., Merion, R.M., Paya, C.V., Pirsch, J., Freeman, R.B., O'Grady, J., Robinson, C., To, Z., Wren, K., Banken, L., Buhles, W. and Brown, F., (2000), Valganciclovir results in improved oral absorption of ganciclovir in liver transplant recipients. *Antimicrob Agents Chemother*, **44**, 2811–2815.

Razonable, R.R., Rivero, A., Rodriguez, A., Wilson, J., Daniels, J., Jenkins, G., Larson, T., Hellinger, W.C., Spivey, J.R. and Paya, C.V., (2001), Allograft rejection predicts the occurrence of late-onset cytomegalovirus (CMV) disease among CMV-mismatched solid organ transplant patients receiving prophylaxis with oral ganciclovir. *J Infect Dis*, **184**, 1461–1464.

Reefschlaeger, J., Bender, W., Hallenberger, S., Weber, O., Eckenberg, P., Goldmann, S., Haerter, M., Buerger, I., Trappe, J., Herrington, J.A., Haebich, D. and Ruebsamen-Waigmann, H., (2001), Novel non-nucleoside inhibitors of cytomegaloviruses (BAY 38-4766): in vitro and in vivo antiviral activity and mechanism of action. *J Antimicrob Chemother*, **48**, 757–767.

Reyes, M., Shaik, N.S., Graber, J.M., Nisenbaum, R., Wetherall, N.T., Fukuda, K. and Reeves, W.C., (2003), Acyclovir-resistant genital herpes among persons attending sexually transmitted disease and human immunodeficiency virus clinics. *Arch Intern Med*, **163**, 76–80.

Safrin, S., Ashley, R., Houlihan, C., Cusick, P.S. and Mills, J., (1991a), Clinical and serologic features of herpes simplex virus infection in patients with AIDS. *AIDS*, **5**, 1107–1110.

Safrin, S., Assaykeen, T., Follansbee, S. and Mills, J., (1990), Foscarnet therapy for acyclovir-resistant mucocutaneous herpes simplex virus infection in 26 AIDS patients: preliminary data. *J Infect Dis*, **161**, 1078–1084.

Safrin, S., Berger, T.G., Gilson, I., Wolfe, P.R., Wofsy, C.B., Mills, J. and Biron, K.K., (1991b), Foscarnet therapy in five patients with AIDS and acyclovir-resistant varicella-zoster virus infection. *Ann Intern Med*, **115**, 19–21.

Safrin, S., Crumpacker, C., Chatis, P., Davis, R., Hafner, R., Rush, J., Kessler, H.A., Landry, B. and Mills, J., (1991c), A controlled trial comparing foscarnet with vidarabine for acyclovir-resistant mucocutaneous herpes simplex in the acquired immunodeficiency syndrome. *N Engl J Med*, **325**, 551–555.

Safrin, S., Kemmerly, S., Plotkin, B., Smith, T., Weissbach, N., De Veranez, D., Phan, L.D. and Cohn, D., (1994), Foscarnet-resistant herpes simplex virus infection in patients with AIDS. *J Infect Dis*, **169**, 193–196.

Sahli, R., Andrei, G., Estrade, C., Snoeck, R. and Meylan, P.R., (2000), A rapid phenotypic assay for detection of acyclovir-resistant varicella-zoster virus with mutations in the thymidine kinase open reading frame. *Antimicrob Agents Chemother*, **44**, 873–878.

Saint-Leger, E., Caumes, E., Breton, G., Douard, D., Saiag, P., Huraux, J.M., Bricaire, F., Agut, H. and Fillet, A.M., (2001), Clinical and virologic characterization of acyclovir-resistant varicella-zoster viruses isolated from 11 patients with acquired immunodeficiency syndrome. *Clin Infect Dis*, **33**, 2061–2067.

Sarisky, R.T., Crosson, P., Cano, R., Quail, M.R., Nguyen, T.T., Wittrock, R.J., Bacon, T.H., Sacks, S.L., Caspers-Velu, L., Hodinka, R.L. and Leary, J.J., (2002), Comparison of methods for identifying resistant herpes simplex virus and measuring antiviral susceptibility. *J Clin Virol*, **23**, 191–200.

Sasadeusz, J.J. and Sacks, S.L., (1996), Spontaneous reactivation of thymidine kinase-deficient, acyclovir-resistant type 2 herpes simplex virus: masked heterogeneity or reversion? *J Infect Dis*, **174**, 476–482.

Schmit, I. and Boivin, G., (1999), Characterization of the DNA polymerase and thymidine kinase genes of herpes simplex virus isolates from AIDS patients in whom acyclovir and foscarnet therapy sequentially failed. *J Infect Dis*, **180**, 487–490.

Smith, I.L., Cherrington, J.M., Jiles, R.E., Fuller, M.D., Freeman, W.R. and Spector, S.A., (1997), High-level resistance of cytomegalovirus to ganciclovir is associated with alterations in both the UL97 and DNA polymerase genes. *J Infect Dis*, **176**, 69–77.

Smith, I.L., Shinkai, M., Freeman, W.R. and Spector, S.A., (1996), Polyradiculopathy associated with ganciclovir-resistant cytomegalovirus in an AIDS patient: phenotypic and genotypic characterization of sequential virus isolates. *J Infect Dis*, **173**, 1481–1484.

Smith, I.L., Taskintuna, I., Rahhal, F.M., Powell, H.C., Ai, E., Mueller, A.J., Spector, S.A. and Freeman, W.R., (1998), Clinical failure of CMV retinitis with intravitreal cidofovir is associated with antiviral resistance. *Arch Ophthalmol*, **116**, 178–185.

Snoeck, R, Andrei, G., Gerard, M., Silverman, A., Hedderman, A., Balzarini, J., Sadzot-Delvaux, C., Tricot, G., Clumeck, N. and De Clercq, E., (1994), Successful treatment of progressive mucocutaneous infection due to acyclovir- and foscarnet-resistant herpes simplex virus with (s)-1-(3-hydroxy-2-phosphonylmethoxypropyl)cytosine (HPMPC). *Clin Infect Dis*, **18**, 570–578.

SOCA (Studies of Ocular Complications of AIDS Research Group and AIDS Clinical Trials Group), (1996), Combination foscarnet and ganciclovir therapy vs monotherapy for the treatment of relapsed cytomegalovirus retinitis in patients with AIDS. *Arch Ophtalmol*, **114**, 23–33.

Springer, K.L., Chou, S., Li, S., Giller, R.H., Quinones, R., Shira, J.E. and Weinberg, A., (2005), How evolution of mutations conferring drug resistance affects viral dynamics and clinical outcomes of cytomegalovirus-infected hematopoietic cell transplant recipients. *J Clin Microbiol*, **43**, 208–213.

Sullivan, V., Talarico, C.L., Stanat, S.C., Davis, M., Coen, D.M. and Biron, K.K., (1992), A protein kinase homologue controls phosphorylation of ganciclovir in human cytomegalovirus-infected cells. *Nature*, **359**, 85.

Swetter, S.M., Hill, E.L., Kern, E.R., Koelle, D.M., Posavard, C.M., Lawrence, W. and Safrin, S., (1998), Chronic vulvar ulceration in an immunocompetent woman due to acyclovir-resistant, thymidine kinase-deficient herpes simplex virus. *J Infect Dis*, **177**, 543–550.

Tatarowicz, W.A., Lurain, N.S. and Thompson, K.D., (1991), In situ ELISA for the evaluation of antiviral compounds effective against human cytomegalovirus. *J Virol Meth*, **35**, 207–215.

Telenti, A. and Smith, T.F., (1989), Screening with a shell vial assay for antiviral activity against cytomegalovirus. *Diagn Microbiol Infect Dis*, **12**, 5–8.

Tenser, R.B., Hay, K.A. and Edris, W.A., (1989), Latency-associated transcript but not reactivatable virus is present in sensory ganglion neurons after inoculation of thymidine kinase-negative mutants of herpes simplex virus type 1. *J Virol*, **63**, 2861–2865.

Underwood, M.R., Ferris, R.G., Selleseth, D.W., Davis, M.G., Drach, J.C., Townsend, L.B., Biron, K.K. and Boyd, F.L., (2004), Mechanism of action of the ribopyranoside benzimidazole GW275175X against human cytomegalovirus. *Antimicrob Agents Chemother*, **48**, 1647–1651.

Visse, B., Huraux, J.M. and Fillet, A.M., (1999), Point mutations in the varicella-zoster virus DNA polymerase gene confers resistance to foscarnet and slow growth phenotype. *J Med Virol*, **59**, 84–90.

Wade, J.C., Newton, B., McLaren, C., Flournoy, N., Keeney, R.E. and Meyers, J.D., (1982), Intravenous acyclovir to treat mucocutaneous herpes simplex virus infection after marrow transplantation: a double-blind trial. *Ann Intern Med*, **96**, 265–269.

Wang, L.H., Peck, R.W., Yin, Y., Allanson, J., Wiggs, R. and Wire, M.B., (2003), Phase I safety and pharmacokinetic trials of 1263w94, a novel oral anti-human cytomegalovirus agent, in healthy and human immunodeficiency virus-infected subjects. *Antimicrob Agents Chemother*, **47**, 1334–1342.

Weinberg, A., Jabs, D.A., Chou, S., Martin, B.K., Lurain, N.S., Forman, M.S. and Crumpacker, C., (2003), Mutations conferring foscarnet resistance in a cohort of patients with acquired immunodeficiency syndrome and cytomegalovirus retinitis. *J Infect Dis*, **187**, 777–784.

Whitley, R.J. and Gnann, J.W., Jr., (1992), Acyclovir: a decade later. *N Engl J Med*, **327**, 782–789.

Wilcox, C.L., Crnic, L.S. and Pizer, L.I., (1992), Replication, latent infection, and reactivation in neuronal culture with a herpes simplex virus thymidine kinase-negative mutant. *Virology*, **187**, 348–352.

Wolf, D.G., Lee, D.J. and Spector, S.A., (1995a), Detection of human cytomegalovirus mutations associated with ganciclovir resistance in cerebrospinal fluid of AIDS patients with central nervous system disease. *Antimicrob Agents Chemother*, **39**, 2552–2554.

Wolf, D.G., Lurain, N.S., Zuckerman, T., Hoffman, R., Satinger, J., Honigman, A., Saleh, N., Robert, E.S., Rowe, J.M. and Kra-Oz, Z., (2003), Emergence of late cytomegalovirus central nervous system disease in hematopoietic stem cell transplant recipients. *Blood*, **101**, 463–465.

Wolf, D.G., Smith, I.L., Lee, D.J., Freeman, W.R., Flores Aguilar, M. and Spector, S.A., (1995b), Mutations in human cytomegalovirus UL97 gene confer clinical resistance to ganciclovir and can be detected directly in patient plasma. *J Clin Invest*, **95**, 257–263.

Wolf, D.G., Yaniv, I., Honigman, A., Kassis, I., Schonfeld, T. and Ashkenazi, S., (1998), Early emergence of ganciclovir-resistant human cytomegalovirus strains in children with primary combined immunodeficiency. *J Infect Dis*, **178**, 535–538.

Zaia, J.A., GallezHawkins, G.M., Tegtmeier, B.R., terVeer, A., Li, X.L., Niland, J.C. and Forman, S.J., (1997), Late cytomegalovirus disease in marrow transplantation is predicted by virus load in plasma. *J Infect Dis*, **176**, 782–785.

Chapter 9
Hepatitis Virus Resistance

Jean-Michel Pawlotsky

9.1 Introduction

In antiviral therapy, an antiviral effect is defined as a minimum reduction in viral replication from the pretreatment baseline. Failure to achieve this decrease constitutes primary treatment failure. A confirmed increase in viral replication from the nadir following initially effective treatment constitutes secondary treatment failure (Locarnini et al., 2004). Viral resistance can be defined as a primary or secondary treatment failure that has a viral cause. The mechanisms underlying viral resistance to antiviral drugs depend on the antiviral mechanisms of the drugs. In hepatitis B virus (HBV) therapy, where specific inhibitors of viral replication are used, resistance is related to the selection of HBV variants with amino acid substitutions in the drug target, i.e. the reverse transcriptase domain of HBV DNA polymerase, that confer reduced susceptibility to the inhibitory action of the drug. In HCV therapy, which is based on the use of a combination of pegylated interferon (IFN) alpha and ribavirin that exerts complex and intricate non-virus specific antiviral and immunomodulatory effects, treatment failure appears to be multifactorial and the mechanisms underlying the viral component of treatment failure remain poorly understood. This chapter describes our current knowledge of HBV and HCV treatment failures, the role of viral resistance and the ways to prevent and manage resistance when it occurs.

9.2 Hepatitis B Virus Resistance

Hepatitis B virus (HBV) infection is a major public health problem, with approximately 350 million individuals chronically infected worldwide (Lee, 1997). HBV is highly endemic in sub-Saharan Africa, China, and South-East Asia. It is also highly endemic in the Mediterranean basin and it is present at significant levels in most industrialized countries (Lee, 1997). Two forms of chronic HBV infection can be individualized according to the presence or the absence of HBe antigen, but transitional forms exist (Ganem and Prince, 2004; Lee, 1997). Chronic HBV carriers are exposed to increased risk of complications such as chronic hepatitis, cirrhosis,

I. W. Fong and K. Drlica (eds.), *Antimicrobial Resistance and Implications for the Twenty-First Century.* © Springer 2008

and hepatocellular carcinoma, of which HBV is currently the most frequent cause (Ganem and Prince, 2004). Up to one million people die every year from the complications of HBV infection (Lee, 1997).

9.2.1 HBV Virological Characteristics

HBV is a member of the *Hepadnaviridae* family. Its partially double-stranded circular DNA genome is contained in an icosahedral capsid, itself enveloped by a lipid bilayer bearing three different surface proteins. The HBV lifecycle starts with virion attachment to an unknown specific receptor complex (reviewed in (Ganem and Prince, 2004)). The viral envelope then fuses with the cell membrane, releasing the nucleocapsid into the cytoplasm. The virus is decapsidated, and the genomic HBV DNA and HBV DNA polymerase are transferred to the nucleus. Viral DNA is repaired in the nucleus to yield a fully double-stranded DNA molecule, which is subsequently supercoiled to form covalently closed circular proviral DNA (cccDNA). cccDNA is the form under which HBV persists in host cells (Tuttleman et al., 1986). cccDNA has a very long half-life, despite the fact that it is not integrated into the cellular genome, and can probably be transmitted to progeny cells. It serves as a reservoir for viral reactivation when antiviral therapy is interrupted. cccDNA is transcribed by cellular RNA polymerases into messenger RNAs for viral protein synthesis, and into a pregenomic RNA which is subsequently encapsidated in the cell cytoplasm together with a molecule of HBV DNA polymerase. The latter has a reverse transcriptase function that catalyzes synthesis of the negatively stranded genomic DNA, while the pregenomic RNA is gradually degraded by the RNAse H activity of the polymerase in the nucleocapsid. A positive DNA strand is then synthesized by the polymerase, using the negative strand as template. Newly generated nucleocapsids can be recycled to yield additional cccDNA molecules in the nucleus, but most of them bud into the endoplasmic reticulum to form mature virions that are subsequently released into the pericellular space by exocytosis (Ganem and Prince, 2004).

HBV infection is characterized by high levels of virus production and turnover (Nowak et al., 1996; Whalley et al., 2001), whereas the HBV reverse transcriptase, like the HIV reverse transcriptase, is an error-prone enzyme lacking 3'-5' exonuclease proofreading capacity (Cane et al., 1999; Gunther et al., 1999; Pallier et al., 2006a). This results in a large number of nucleotide substitutions during replication, which principally accumulate at the reverse transcription step. The misincorporation rate has been estimated to be of the order of 10^{10} incorrect nucleotide incorporations per day in a given patient (Zoulim, 2004). As a result, HBV, like other viruses with error-prone polymerases, such as HIV, hepatitis C virus (HCV) and poliovirus, has a "quasispecies" distribution in infected individuals (Cane et al., 1999; Gunther et al., 1999; Pallier et al., 2006a). This means that HBV circulates as a complex mixture of genetically distinct but closely related variants that are in equilibrium at a given time point of infection in a given replicative environment. The fact that the four main HBV open reading frames overlap, i.e. that each nucleotide is part of the coding sequence of more than one functional viral protein or regulatory element, explains why HBV quasispecies variability is subject to stronger conservatory constraints than RNA viruses with non-overlapping reading frames, such as HCV or poliovirus. The quasispecies distribution of HBV implies that any newly generated mutation conferring a selective advantage to the virus in a given replicative environment will allow the corresponding viral population to overtake the other variants, following a classical Darwinian evolutionary process (Duarte et al., 1994). HBV quasispecies distribution constitutes the mechanistic background for selection of resistant HBV variants by antiviral drugs.

9.2.2 Current Treatment of Chronic HBV Infection

Treatment of chronic hepatitis B is aimed at driving viral replication to the lowest possible level, and thereby to halt the progression of liver disease and prevent the onset of complications. However, HBV infection cannot be fully eradicated because of cccDNA persistence. Two categories of drugs are used in this setting, namely IFN alpha (now pegylated IFN alpha, which can be administered every week with sustained efficacy over the week following administration) and specific HBV inhibitors that primarily target the reverse transcriptase function of HBV DNA polymerase. The current HBV inhibitors split into three structural categories: (i) L-nucleoside analogues, including lamivudine (3TC), its 5-fluoro-derivative emtricitabine (FTC), telbivudine (LdT), valtorcitabine (LdC), and clevudine (L-FMAU); (ii) acyclic nucleoside phosphonates, represented by the dAMP analogues adefovir and tenofovir and other molecules in development; (iii) deoxyguanosine analogues in which the deoxyribose is replaced by a cyclopentane derivative, including entecavir and abacavir, or otherwise modified (Shaw et al., 2006).

Four HBV inhibitors are currently approved for HBV therapy in the United States and Europe (Table 9.1): lamivudine, adefovir dipivoxil, entecavir, and telbivudine. In addition, tenofovir disoproxil fumarate and the combination of tenofovir and emtricitabine are approved for use in the treatment of human immunodeficiency virus (HIV) infection and clinical trials are ongoing to assess their utility in the treatment of HBV infection. The goal of treatment with specific HBV inhibitors

Table 9.1 Approved drugs for chronic hepatitis B therapy in the United States and Europe

Interferons	Specific HBV inhibitors
Standard IFN alpha	Lamivudine (3TC)
Pegylated IFN alpha (2a and 2b)	Entecavir
	Telbivudine (LdT)
	Adefovir dipivoxil
	Tenofovir disoproxil fumarate[a]
	Tenofovir disoproxil fumarate + Emtricitabine[a]

HBV, hepatitis B virus; IFN, interferon.
[a] approved for human immunodeficiency virus therapy.

is to induce an antiviral effect that is as profound as possible and as sustained as possible in order to efficiently prevent the complications of chronic HBV infection on the long-term. The principal target of HBV inhibitors is the YMDD catalytic motif of HBV reverse transcriptase, located in domain C of the polymerase molecule (Figure 9.1) (Das et al., 2001). They exert their antipolymerase/reverse transcriptase activity by inhibiting the elongation of the HBV DNA minus strand through competition with the natural polymerase substrates, and by acting as chain terminators through their incorporation in the nascent DNA strand (Severini et al., 1995; Zoulim et al., 1996).

There is currently no clear consensus on the indications of HBV therapy (de Franchis et al., 2003; Lok et al., 2001). Schematically, short-term IFN alpha-based therapy is indicated for HBeAg-positive patients with a high likelihood of HBe seroconversion (such as patients with low viral load and high serum amino-transferase levels), whereas HBV polymerase inhibitor-based therapy is indicated as first-line treatment for HBeAg-positive patients with a low likelihood of HBe seroconversion and for HBeAg-negative patients, and as second-line treatment for HBeAg-positive patients who failed to seroconvert during IFN therapy. In most cases, HBV inhibitors will need to be administered for life, because these drugs are only virustatic, and viral replication restarts soon after their withdrawal, whereas liver disease again begins to progress.

9.2.3 HBV Resistance to Specific Inhibitors

Experience with highly active antiretroviral therapy (HAART) in HIV-infected patients shows that resistance to HIV reverse transcriptase inhibitors is acquired

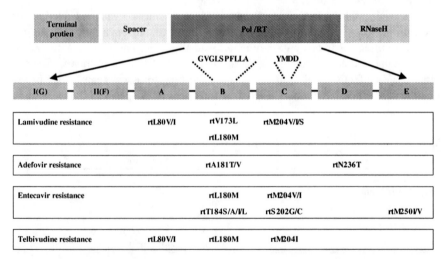

Fig. 9.1 Amino acid substitutions known to be associated with resistance in patients with chronic hepatitis B receiving lamivudine, adefovir dipivoxil, entecavir and telbivudine.

gradually, through the selection of preexisting resistant variants and gradual accumulation of new amino acid substitutions that confer stepwise increases in the level
of drug resistance (reviewed in (Clavel and Hance, 2004)). Schematically, partial
resistance conferred by a substitution in a preexisting viral population impairs drug
efficacy sufficiently to restore a level of replication compatible with the accumulation of further mutations. The latter may restore the in vivo fitness of the resistant
virus, allowing viral replication to return to pretreatment levels.

9.2.3.1 Lamivudine and L-Nucleoside Analogues Resistance

Long-term lamivudine administration frequently elicits viral resistance, characterized by a rebound of viral replication in an adherent patient. The incidence of
lamivudine resistance is 14–32% after 1 year of treatment, 38% after 2 years and
53–76% after 3 years (Lai et al., 2003). Lamivudine resistance leads to the loss of
benefits of therapy and to a gradual aggravation of liver disease (Dienstag et al.,
2003; Leung et al., 2001; Liaw et al., 1999, 2004; Lok et al., 2003; Papatheodoridis
et al., 2002; Yuen et al., 2003). Lamivudine resistance appears to arise from the
selection, under lamivudine pressure, of preexisting HBV variants bearing amino
acid substitutions that confer improved replicative capacities in the presence of
lamivudine relative to wild-type lamivudine-sensitive variants. The principal mutations associated with lamivudine resistance are located in domain C, in the YMDD
catalytic motif of the HBV reverse transcriptase (Figure 9.1). They include rtM204V
(YVDD sequence) and rtM204I (YIDD) (Allen et al., 1998), and the more recently
identified rtM204S (YSDD) (Bozdayi et al., 2003; Niesters et al., 2002), according to the genotype-independent numbering convention for the reverse transcriptase (rt) domain of human HBV polymerase proposed by Stuyver et al. (2001).
Recent data based on molecular modeling have suggested that lamivudine resistance conferred by amino acid substitutions at position rtM204 is due to both steric
hindrance and electrostatic repulsion, because the substituted amino acids decrease
the size of the dNTP binding pocket and change the surrounding charge distribution
(Chong et al., 2003).

Lamivudine-resistant mutants with amino acid substitutions in the YMDD motif
appear to replicate less efficiently than the wild-type virus in vitro. Additional
substitutions that are often co-selected with the resistance substitutions at reverse
transcriptase position 204 of domain C, such as rtL180M and rtV173L located
in the B domain, can compensate for this loss of replication efficiency in vitro
(Figure 9.1) (Allen et al., 1998; Delaney et al., 2003; Fu and Cheng, 1998; Seigneres
et al., 2002). Molecular modeling has suggested that rtL180M decreases affinity for
the oxothiolane ring of lamivudine-triphosphate by increasing local electronega-
tivity around the deoxyribose-binding site, allowing better discrimination between
lamivudine triphosphate and dCTP, and that rtV173L may alter either the alignment
of the nucleic acid template and/or the environment around the catalytic site in such
a way as to increase polymerization efficiency (Shaw et al., 2006). Other amino acid
substitutions have been reported to be associated with YMDD variations. We have
shown recently that, in vivo, the fluctuating environment in which these variants

replicate during lamivudine administration to a given patient potentially favors one variant population over another, regardless of intrinsic replication capacity of the viral populations in vitro (Pallier et al., 2006a).

Primary resistance to any individual drug appears to confer at least some degree of cross-resistance to other members of its group, and may also diminish sensitivity to HBV inhibitors from other groups (Shaw et al., 2006). In vitro cross-resistance has indeed been reported between lamivudine and other L-nucleoside analogs, such as telbivudine, emtricitabine, valtorcitabine, clevudine, or elvucitabine which all showed significantly reduced antiviral potency on HBV variants bearing the V, I, or S substitutions at position rtM204 (Table 9.2) (Cane et al., 1999; Chin et al., 2001; Delaney et al., 2001; Dent et al., 2000; Fu and Cheng, 2000; Gish et al., 2002; Ono et al., 2001; Yamamoto et al., 2002). Resistance to telbivudine has been reported to be of 22 and 9% in HBe antigen-positive and -negative patients, respectively, after 1 year of administration. All resistant variants selected so far carried an rtM204I (YIDD) substitution (Figure 9.1) (Lai et al., 2006a).

9.2.3.2 Entecavir Resistance

Mutations at reverse transcriptase position 204, which confer resistance to lamivudine, also confer reduced susceptibility to entecavir in vitro (Levine et al., 2002), showing cross-resistance between the two drugs (Table 9.2). In vivo, entecavir is able to significantly reduce replication of lamivudine-resistant variants, but to a lesser extent than wild-type HBV variants (Sherman et al., 2006). Additional substitutions were recently reported to restore baseline viral replication capacities when associated with substitutions at reverse transcriptase position 204 (with or without the rtL180M substitution). They include rtT184S/A/I/L, rtS202G/C, and rtM250I/V (Figure 9.1) (Tenney et al., 2004, 2007). Some of them further reduce the entecavir susceptibility of lamivudine-resistant variants in vitro (Tenney et al., 2004), whereas they all substantially improve lamivudine-resistant variants replicative fitness in the presence of entecavir. Modeling has suggested that rtM250 substitutions could interact with either the nucleic acid primer or template, whereas substitutions for rtT184 or rtS202 could alter the geometry of the nearby dNTP-binding pocket (Shaw et al., 2006). Full resistance to entecavir thus results of a three-step selection-mutation accumulation process, subsequently involving lamivudine-resistance substitutions at position rt204, the rtL180M substitution that improves lamivudine-resistant variants fitness, and one or more of the entecavir-specific substitutions that restore full-replication fitness in the presence of the drug (it should be noted, however, that some of these substitutions may be selected by lamivudine monotherapy (Tenney et al., 2007)). This three-step process, together with the strong antiviral potency of entecavir, results in the high genetic barrier of the drug to HBV resistance and explains the low incidence of resistance in treatment-naïve patients. The cumulative probability of entecavir resistance emergence at 4 years of monotherapy has indeed been estimated to be 1.2% only (Colonno et al., 2006b). In contrast, in lamivudine-refractory patients, the substitutions at positions 204 and 180 are already present at the start of entecavir therapy. Thus, full resistance occurs as soon as one or more

Table 9.2 In vitro cross-resistance to HBV inhibitors conferred by HBV reverse transcriptase amino acid substitutions (phenotypic assays)

	Lamivudine (3TC)	Emtricitabine (FTC)	Telbivudine (LdT)	Valtorcitabine (LdC)	Clevudine (L-FMAU)	Elvucitabine (L-Fd4C)	Entecavir	Adefovir	Tenofovir
Wild-type	S	S	S	S	S	S	S	S	S
M204I	R	NDA	RS	RS	RS	NDA	RS	S	S
M204I+L180M	R	NDA	NDA	NDA	NDA	NDA	RS	S	S
M204V+180M	R	RS	RS	RS	R	RS	RS	S	S
M204V+L180M+V173L	R	NDA	NDA	NDA	NDA	NDA	RS	S	S
N236T	S	S	S	NDA	RS	NDA	S	RS	RS
A181V	RS	RS	R	R	R	NDA	RS	RS	S

S, sensitive; RS, reduced susceptibility; R, fully resistant; NDA, no data available.

of the entecavir resistance substitutions is added. In these patients, the cumulative incidence has been reported to be of the order of 40% after 4 years of entecavir administration and is most often associated with a viral breakthrough (Colonno et al., 2006b).

9.2.3.3 Adefovir and Tenofovir Resistance

HBV resistance to adefovir dipivoxil is complex. Long-term administration has been shown to select variants bearing rtN236T or rtA181V, both of which confer low-level resistance (5–10-fold increase) to adefovir in vitro (Figure 9.1). The emergence of adefovir resistance is slow, but after 5 years of monotherapy, adefovir-resistant variants have been selected in 29% of cases (Hadziyannis et al., 2006). However, their presence may be associated with different patterns of HBV DNA kinetics, including a typical escape with restoration of baseline replication levels, a gradual increase in viral replication preceding the emergence of the mutant virus by several months, or persistence of full adefovir antiviral efficacy (Angus et al., 2003; Hadziyannis et al., 2006; Villeneuve et al., 2003; Westland et al., 2003; Yang et al., 2002). The mechanisms underlying adefovir resistance are not completely understood. Molecular modeling has suggested that rtN236T increases preference for the natural dATP over its analogue adefovir diphosphate, whereas rtA181V/T (and probably other substitutions in the same region) causes an allosteric change in conformation of the catalytic site by forcing the repositioning of rtM204 (Shaw et al., 2006). Preliminary data from our group based on extensive quasispecies analyses of serial blood samples taken during adefovir administration suggest that rtN236T and rtA181V/T are present alone or together in the same variants, and that the dynamics of the different resistant populations over time vary from one patient to the next (Pallier et al., 2006b). Other amino acid substitutions in the reverse transcriptase sequence have been suggested to be associated with reduced susceptibility to adefovir antiviral action, including rtL80V/I, rtV84M, rtV214A, rtS85A, rt Q215S, rtP237H, and rtN238T/D. They may appear alone or in conjunction with rtN236T and/or rtA181V/T (Shaw et al., 2006). A substitution at reverse transcriptase position rt233 has been recently suggested to be associated with primary resistance to adefovir in three patients (Schildgen et al., 2006). However, more recent phenotypic analysis has ruled out a role for rt233 substitutions in adefovir resistance (William E. Delaney, 4th, personal communication, November 2006). Interestingly, mutations that confer resistance to adefovir do not confer significant cross-resistance to L-nucleosides or entecavir, at least in vitro (Table 9.2) (Shaw et al., 2006). However, the rtA181T substitution has been reported in patients undergoing long-term lamivudine therapy and appears to confer reduced susceptibility to this drug in vitro, raising the issue of its selection as a resistance mutation in patients receiving the combination of lamivudine and adefovir (Table 9.2) (Shaw et al., 2006). It is unclear whether adefovir resistance substitutions can confer resistance to other nucleotide analogs such as tenofovir.

The amino acid changes that confer resistance to tenofovir administration in vivo are not known. This is principally due to the fact that the HBV-infected patients who have received tenofovir so far were generally coinfected with HIV and also receiving lamivudine or emtricitabine, a combination that prevented tenofovir resistance. The rtA194T substitution has been suggested to be associated with tenofovir resistance (Sheldon et al., 2005), but more recent analysis has shown that it does not confer in vitro resistance to tenofovir as a single mutation or in a lamivudine-resistant viral background (Delaney et al., 2006). Tenofovir administration in monotherapy may favor the emergence of tenofovir resistance on the mid- to long-term in ongoing trials.

9.2.4 Assessment of HBV Resistance

9.2.4.1 Genotypic Analysis

Amino acid substitutions known to be associated with HBV resistance to specific reverse transcriptase inhibitors can be identified in blood samples taken from patients undergoing antiviral therapy by means of molecular methods based on direct sequence analysis, which provides the full sequence of the analyzed fragment, or on alternative techniques that identify specific sequences at given positions (Table 9.3) (Allen et al., 1998; Bozdayi et al., 2003; Colonno et al., 2006a,b; Hadziyannis et al., 2006; Lai et al., 2006a; Niesters et al., 2002; Tenney et al., 2004, 2007; Yeon et al., 2006).

Direct sequence analysis of PCR amplicons can be used to determine the exact nucleotide and deduced amino acid sequence of the analyzed fragment. The sequences must be analyzed and interpreted carefully, because the coexistence of variant viral populations in the same blood sample (quasispecies distribution of viral genomes) may lead to sequence ambiguities (Pawlotsky, 2005b). A commercial assay, Trugene® HBV Genotyping Kit (distributed by Siemens Medical Solutions Diagnostics, Tarry town, New York), has been developed that can determine the

Table 9.3 Methods for assessment of HBV resistance

Genotypic analysis
Direct sequence analysis of PCR amplicons (population sequencing)
Clonal (quasispecies) analysis of PCR amplicons
Reverse hybridization of PCR amplicons to membrane-bound probes
Fluorometric real-time PCR
Mini-sequencing (mixed hybridization-sequencing PCR technique)
Microchip-based assays
Mass spectrometry
Phenotypic analysis
Transient transfection assays
Site-directed mutagenesis of a prototype HBV strain
Analysis of full-length HBV genomes from clinical isolates
Viral vectors delivering replication competent genomes into hepatoma cells
Continuous cell lines expressing resistance-associated substitutions

PCR, polymerase chain reaction.

HBV genotype and identify HBV resistance mutations based on the determination of the HBV reverse transcriptase sequence (Gintowt et al., 2005; Roque-Afonso et al., 2003).

An alternative approach is the use of reverse hybridization of PCR amplicons to membrane-bound probes. Line probe assays (INNO-LiPA, Innogenetics, Gent, Belgium) use a series of short immobilized oligonucleotide probes to discriminate among different PCR fragments. INNO-LiPA HBV DR v2 is a line probe assay that detects wild-type HBV and HBV variants bearing substitutions associated with lamivudine and adefovir dipivoxil resistance (Hussain et al., 2006; Osiowy et al., 2006; Sertoz et al., 2005). Other hybridization-based techniques have been developed but are not yet commercially available as "ready-to-use" kits, including fluorometric real-time PCR detecting resistant HBV variants from differences in melting temperature, mixed hybridization-sequencing-PCR techniques (so-called "mini-sequencing") involving the differential extension of multiple oligonucleotide primers with fluorescent ddNTPs, or inhibition of primer extension, which relies on differences in efficiency of amplification of perfectly matched primer-templates and those with mismatched 3' termini (Shaw et al., 2006). New technologies are under study for the identification of drug-resistant mutations, such as mass spectrometry or microchip-based assays (Table 9.3) (Hong et al., 2004; Kim et al., 2005). These techniques allow the simultaneous identification of a large number of nucleotide substitutions but they remain expensive and are not yet applicable in many settings.

In clinical trials with already approved and new drugs, systematic genotypic analysis coupled with the study of any primary or secondary treatment failure by means of the molecular techniques described above is necessary in order to characterize the incidence and profile of resistance to the specific drug and to identify new resistance mutations. With only one available specific antiviral drug for many years in clinical practice (lamivudine), there was no utility to identify HBV resistance mutations in order to guide therapy. With more drugs on the market and more and more described resistance mutations, with more and more complex profiles, such a determination will be useful. However, its implementation in clinical practice will need that clear recommendations and decision algorithms be issued to tailor therapy according to HBV resistance profiles. Such recommendations do not exist so far, but international initiatives have been launched that could reach this goal in the near future.

9.2.4.2 Phenotypic Analysis

Phenotypic analysis allows the determination of inhibitory concentrations (IC) of specific HBV inhibitors in in vitro experiments, i.e. allows to test the susceptibility of a given reverse transcriptase sequence to the antiviral action of a given drug (Table 9.3). Two methods based on transient transfection have been described. The first one uses site-directed mutagenesis to insert amino acid substitutions in the sequence of a prototype HBV strain (Chin et al., 2001; Fu and Cheng, 1998). This method allows to study the effect of the mutation on drug susceptibility, but not the intrinsic susceptibility of a "natural" HBV sequence, recovered from a patient's blood. The other method studies mutations in their natural genetic background after amplification of full-length HBV genomes from clinical isolates (Chin et al., 2001; Durantel et al., 2004). Another approach uses viral vectors to deliver replication competent HBV genomes to hepatoma cell lines, ensuring better levels of HBV replication than in transient transfection assays (Delaney et al., 2002, 2001). Continuous cell lines expressing amino acid substitutions known to be associated with resistance to HBV reverse transcriptase inhibitors have also been used (Fu and Cheng, 2000; Walters et al., 2003; Ying et al., 2000).

Phenotypic analyses need to be performed every time a new mutation is shown to have emerged during therapy with one or several antiviral drugs, including when this occurs during clinical trials where systematic genotypic analyses are generally performed or when it is reported in patients treated in clinical practice with approved drugs. Phenotypic analyses allow establishment of the drug susceptibility associated with a given amino acid substitution and eventual cross-resistance with other drugs from the same or other families. These results are necessary to better understand the mechanisms underlying viral breakthrough and to derive treatment strategies to avoid resistance to occur in the clinic. However, one must be careful in interpreting the results of in vitro phenotypic analysis, since the replication properties observed in vitro may not always translate in vivo where the virus is under the influence of a much more complex replicative environment (Pallier et al., 2006a).

9.2.5 Prevention of HBV Resistance

Selection of resistant HBV variants cannot be avoided forever, because chronic HBV infection is not curable and viral genetic information persists in cells for the patients' life under the form of cccDNA. Improving treatment strategies will allow at best to delay the emergence of resistance for several years so that viral replication remains profoundly inhibited and the immune response has a chance to take over control in a substantial proportion of patients.

There are different strategies to delay the development of resistance to anti-HBV drugs, most of which are inspired by HIV treatment with HAART (Richman, 2006). These strategies are currently being validated in HBV therapy. It should be stressed however that the target enzyme, the HBV reverse transcriptase, is very similar to the HIV reverse transcriptase in that HIV and HBV share the same virological properties that favor the emergence and selection of mutant viruses, and that many drugs used for both infections are the same or closely related. Therefore, it can be anticipated that many of the concepts developed for HIV therapy will also apply to HBV treatment.

9.2.5.1 Improving Potency of the Antiviral Treatment

As shown by Richman, the relationship between the antiviral activity of a treatment regimen and the probability of developing resistance during therapy is a bell-shaped curve (Richman, 2000, 2006). A compound with limited antiviral activity does not exert great selective pressure on the virus and the likelihood of selecting drug-resistant mutants is low. A compound with intermediate antiviral activity exerts more selective pressure, but insufficient potency to completely suppress viral replication. Thus the probability of selecting for drug resistance is high. Finally, a treatment regimen that suppresses viral replication to below the limit of detection of sensitive assays is usually associated with a low likelihood of resistance (Richman, 2000, 2006). In this respect, using the most potent antiviral drug regimen is an excellent way to avoid the rapid emergence of resistance. This can be achieved by using the most potent drug and/or by combining antiviral drugs with additive or synergistic effects. Entecavir, telbivudine and tenofovir have been shown to be more potent than lamivudine and the low dose of adefovir (10 mg per day) used in HBV treatment (Chang et al., 2006; Lai et al., 2005, 2006b; Lampertico, 2006; van Bommel et al., 2004). No additive or synergistic effect of two HBV drugs has been observed so far in vivo as compared to the drugs in monotherapy. However, such an effect could have been masked by the poor sensitivity of the HBV DNA assays, and future drugs may exhibit additive or synergistic effects.

9.2.5.2 Raising the "Pharmacologic Barrier" to Viral Escape

It is crucial that high drug trough levels be achieved and that the drug efficacy can tolerate suboptimal pharmacokinetic parameters in order to allow the patients to miss doses. Drugs should be administered at the maximum effective dose if they are not toxic, at the maximum tolerated dose if they have side-effects (Richman, 2006). For instance, adefovir is used at the suboptimal dose of 10 mg/day, whereas 30 mg/day has been shown to provide better antiviral efficacy but to increase the risk of renal laboratory abnormalities (Hadziyannis et al., 2003; Marcellin et al., 2003). However, the most important parameter is adherence to therapy, as drug interruptions and poor adherence are the principal causes of treatment failure.

9.2.5.3 Raising the "Genetic Barrier" to Viral Escape

The optimal antiviral regimen should combine two drugs with a different mechanism of action and that lack cross-resistance. Drug combinations substantially reduce the risk of selection of resistant HBV mutants, because the likelihood that preexisting HBV variants bear mutations that confer resistance to both drugs is low and, if these variants exist, their replicative fitness is extremely low so they will take much more time than variants that are resistant to only one drug to take over and yield a virologic breakthrough (Pawlotsky, 2005b; Richman, 1996, 2000). It has indeed been shown that adefovir prevents the development of lamivudine resistance when both drugs are given in combination. On the other hand, adefovir resistance has been observed only in patients treated with adefovir monotherapy in the pivotal trials, whereas no patient treated with a combination of adefovir plus lamivudine or adefovir plus emtricitabine developed resistance to adefovir (Hadziyannis et al., 2006; Lampertico et al., 2006). Selection of multiresistant HBV variants can occur after several years of administration (or if the patients are nor fully adherent)

(Villet et al., 2006), but it takes much longer for this to occur on combination therapy compared with sequential therapy. Therefore, *de novo* combination therapy gives the patients a better chance of a sustained inhibition of viral replication. This is particularly important in those patients who will need life-long therapy.

Overall, first-line treatment of chronic hepatitis B should be based on the most potent antiviral regimen. The use of combination therapy as first-line treatment remains debated in spite of its many advantages. It indeed raises cost and reimbursement issues. The combination of two potent drugs is however the best guarantee that resistance will be considerably delayed if the patients are fully adherent. Two drugs with a good intrinsic resistance profile and with no cross-resistance among them should be used. Currently, the combination options are entecavir or telbivudine plus adefovir. In the future, the utility of tenofovir, the resistance profile of which is unknown but which is active on lamivudine-, telbivudine- and entecavir-resistant variants, instead of adefovir in combination regimens will need to be assessed. On the basis of current knowledge, the combination of entecavir and tenofovir appears as potentially the most potent and the least likely to expose to rapid emergence of HBV resistance.

9.2.6 Management of HBV Resistance

HBV resistance is characterized by a relapse of HBV replication during treatment of more than $1 \, \mathrm{Log}_{10}$ IU/mL above the nadir in a patient who is still fully adherent. HBV resistance, when it occurs, exposes patients to liver disease progression and the onset of complications. Resistance mutations can be identified routinely by means of virological assays based on direct sequence analysis or reverse hybridization of polymerase chain reaction products (Pawlotsky, 2002). However, consensual decisional algorithms will need to be established before resistance detection becomes systematic, in order to adapt treatment strategy to the resistance profile of the infecting viral strain.

When a patient develops resistance to an HBV inhibitor in monotherapy, add-on therapy should always be preferred to a switch, because the initial drug will retain its efficacy on sensitive viruses that can harbor mutations conferring resistance to the second drug, while the second drug will control viruses resistant to the first one. Based on the low probability of selecting variants bearing mutations that confer resistance to both drugs and are replication fit, combination therapy will ensure a sustained inhibition of viral replication in the vast majority of patients. Practically, in patients who develop lamivudine resistance, it is reasonable to switch to entecavir that will potently suppress lamivudine-sensitive variants and will also inhibit, although to a lower extent, lamivudine-resistant variants. Given the risk of selecting additional "secondary" entecavir resistance mutations in lamivudine-resistant strains in these patients and loosing the benefit of the switch, adefovir should be added as a second drug. Tenofovir will probably offer a good alternative to adefovir because it will avoid the suboptimal response to adefovir which is seen in a number of patients receiving 10 mg of adefovir daily. In patients who develop adefovir resistance, it is reasonable to continue on adefovir and add-on entecavir. Again, tenofovir will constitute a good alternative to adefovir in the combination when it is approved for HBV treatment. Finally, if entecavir resistance occurs in naïve or lamivudine-resistant patients treated with entecavir monotherapy, adefovir or, ideally, tenofovir (when approved) should be added. Ongoing clinical trials will soon confirm these options.

9.3 Hepatitis C Virus Resistance

Chronic hepatitis C virus (HCV) infection is another major public health problem with more than 170 million people infected worldwide. HCV infection prevalence is 1–2% in industrialized countries. An estimated 20% of HCV-infected patients have or will develop cirrhosis, with an annual risk of developing hepatocellular carcinoma of 1–4% among patients with cirrhosis (NIH Consensus Development Program, 2002). Up to 250,000 people die every year from the complications of HCV infection (*source*: World Health Organization, 2000) and chronic HCV infection has become the most common indication for liver transplantation. In contrast with other chronic viral infections in man, HCV infection is curable. Thus its complications can be prevented by antiviral therapy. Treatment of chronic hepatitis C is currently based on a combination of pegylated IFN alpha and ribavirin (NIH Consensus Development Program, 2002), but a number of new anti-HCV therapies are in development (Pawlotsky, 2006b; Pawlotsky and Gish, 2006).

9.3.1 HCV Virological Characteristics

HCV is a member of the *Flaviviridae* family, genus Hepacivirus. Six HCV genotypes (1–6) and a large number of subtypes (1a, 1b, 1c, etc.) have been identified (Simmonds et al., 2005). HCV only natural host is man. All HCV genotypes have a common ancestor virus. However, HCV genotypes 1, 2, and 4 emerged and diversified in Central and Western Africa, genotype 5 in South Africa, and genotypes 3 and 6 in China, South-East Asia and the Indian subcontinent. In these areas, a large number of subtypes of these genotypes are found (Simmonds et al., 2005). The rest of the world, in particular industrialized areas, harbor a small number of HCV subtypes that could widely spread because they met an efficient route of transmission, such as blood transfusion or the intravenous use of drugs. They include genotypes 1a, 1b, 2a, 2b, 2c, 3a, 4a, and 5a (Simmonds et al., 2005).

The HCV virion is made of a single-stranded positive RNA genome, contained into an icosahedral capsid, itself enveloped by a lipid bilayer into which two different glycoproteins are anchored (Penin, 2003). The genome contains three distinct regions: a short 5' non-coding region that contains the internal ribosome entry site (IRES), the RNA structure responsible for attachment of the ribosome and polyprotein translation; a long, unique open reading frame of more than 9000 nucleotides which is translated into a precursor polyprotein, secondarily cleaved to give birth to the structural proteins (the capsid protein C and the two envelope glycoproteins E1 and E2) and to the non-structural proteins (p7, NS2, NS3, NS4A, NS4B, NS5A, and NS5B); a short 3′ non-coding region principally involved in the priming of HCV replication (Penin, 2003).

The HCV lifecycle starts with virion attachment to its specific receptor. Several candidate molecules have been suggested to play a role in the receptor complex, including tetraspanin CD81, the scavenger receptor B I (SR-BI), the adhesion molecules DC-SIGN and L-SIGN, the low-density lipoprotein (LDL) receptor and, more recently, claudin-I (Bartenschlager et al., 2004; von Hahn et al., 2006). The HCV lifecyle is poorly known because of the lack, until very recently, of a productive culture system. It is supposed, by analogy with other *Flaviviridae*, that the virus linked to its receptor complex is internalized and that the nucleocapsid is released into the cytoplasm. The virus is decapsidated, and the genomic HCV RNA is used both for polyprotein translation and replication in the cytoplasm. Replication and posttranslational processing appear to take place in a membranous web made of the non-structural proteins and host cell proteins called "replication complex", located in close contact with perinuclear membranes. Genome encapsidation appears to take place in the endoplasmic reticulum and nucleocapsids are enveloped and matured into the Golgi apparatus before newly produced virions are released in the pericellular space by exocytosis (Bartenschlager et al., 2004; Penin, 2003).

HCV infection, like HBV infection, is characterized by high levels of virus production and turnover (Neumann et al., 1998), whereas the HCV RNA-dependent RNA polymerase, like other viral RNA polymerases, is an error-prone enzyme lacking 3'-5' exonuclease proofreading capacity that result in the "quasispecies" distribution of HCV populations in infected individuals (Martell et al., 1992; Pawlotsky, 2006a). This means that HCV circulates as a complex mixture of genetically distinct but closely related variants that are in equilibrium at a given time point of infection in a given replicative environment. The quasispecies distribution of HCV implies that any newly generated mutation conferring a selective advantage to the virus in a given replicative environment will allow the corresponding viral population to overtake the other variants, following a classical Darwinian evolutionary process (Duarte et al., 1994). HCV quasispecies distribution constitutes the mechanistic background for selection of resistant HCV variants by antiviral drugs. Its role in IFN resistance is unclear.

9.3.2 Current Treatment of Chronic HCV Infection

The current standard treatment for chronic hepatitis C is the combination of pegylated IFN alpha and ribavirin (NIH Consensus Development Program, 2002). The efficacy endpoint of hepatitis C treatment is the "sustained virological response", defined by the absence of detectable HCV RNA in serum 24 weeks after the end of treatment, which corresponds to a cure of infection in more than 99% of cases, as recently shown (NIH Consensus Development Program, 2002; McHutchison et al., 2006). Only patients with detectable HCV RNA should be considered for pegylated IFN alpha and ribavirin combination therapy (NIH consensus Development

Program, 2002). The decision to treat patients with chronic hepatitis C however depends on multiple parameters, including a precise assessment of the severity of liver disease and of its foreseeable outcome, the presence of absolute or relative contra-indications to therapy, and the patient's willingness to be treated. The HCV genotype should be systematically determined before treatment, as it determines the indication, the duration of treatment, the dose of ribavirin and the virological monitoring procedure (Hadziyannis et al., 2004). Given the likelihood of a sustained virological response, a precise assessment of liver disease prognosis by means of a liver biopsy or a non-invasive method based on serological markers of fibrosis or ultrasound-based testing (Castera and Pawlotsky, 2005; Castera et al., 2005; Poynard et al., 2004, 2005) must be performed in order to help with the treatment decision in patients infected with HCV genotypes 1, 4, 5, and 6, but not in patients infected with HCV genotypes 2 and 3.

The approved dose of pegylated IFN alpha-2a is 180 μg per week, independent of body weight, whereas that of pegylated IFN alpha-2b is weight-adjusted at 1.5 μg/kg per week, identical for all HCV genotypes. Patients infected with HCV genotype 1 should receive a high dose of ribavirin, i.e. 1,000–1,200 mg daily, based on body weight less than or greater than 75 kg. The heaviest patients could benefit from a higher ribavirin dose, up to 1,600 mg daily. Genotype 1-infected patients theoretically require 48 weeks of treatment (NIH Consensus Development Program, 2002). Monitoring of HCV RNA load decrease during therapy must be performed in order to avoid treating patients with no likelihood of a subsequent sustained virological response (Davis et al., 2003; Ferenci et al., 2005; Fried et al., 2002). In patients infected with HCV genotypes 2 and 3, the recommended dose of pegylated IFN alpha-2a or alpha-2b is the same as for HCV genotype 1. The fixed recommended dose of ribavirin is 800 mg per day (NIH Consensus Development Program, 2002). No monitoring of HCV RNA level changes during therapy is recommended because the vast majority of the patients with genotype 2 or 3 infection become HCV RNA-negative early during treatment. Finally, it is recommended to treat patients infected with HCV genotypes 4, 5, and 6 like those infected with HCV genotype 1, i.e. with pegylated IFN alpha at the usual dose, combined with a high dose of ribavirin (1,000–1,200 mg per day, according to body weight less or greater than 75 kg). In the absence of published data, no stopping rules have been defined and it is recommended to treat these patients for a total of 48 weeks. In all instances, the sustained virological response must be assessed 24 weeks after treatment withdrawal (NIH Consensus Development Program, 2002). Earlier assessment of the virological response to therapy, such as 4 weeks after treatment initiation, may be useful whatever the HCV genotype and tailoring rules will be derived from ongoing trials.

9.3.3 HCV Resistance to Interferon alpha-Ribavirin Therapy

9.3.3.1 Antiviral Mechanisms of IFN alpha and Ribavirin

IFN alpha directly inhibits HCV replication (Castet et al., 2002; Frese et al., 2001; Guo et al., 2001; Lanford et al., 2003; Lindenbach et al., 2005; Neumann et al.,

1998). The IFN-induced proteins and enzymatic pathways that establish an antiviral state in infected and uninfected cells are not yet fully identified. In addition, IFN alpha binding to its receptors at the surface of immune cells triggers complex and intricate effects, such as class I major histocompatibility complex antigen expression, activation of effector cells, and complex interactions with the cytokine cascade (Peters, 1996; Tilg, 1997). Mathematical modeling of the second slope of viral decay during therapy points to gradual clearance of infected cells. It is unclear whether IFN alpha, through its immunomodulatory properties, accelerates the clearance of infected cells at the same time as it inhibits viral replication. Thus IFN alpha could act mainly, or solely, as an antiviral agent.

Ribavirin is a synthetic guanosine analog that is transformed into ribavirin triphosphate, the active form of the drug, by intracellular phosphorylation. Ribavirin has only a moderate and transient dose-dependent inhibitory effect on HCV replication in vivo (Pawlotsky et al., 2004). The principal clinical effect of ribavirin is to prevent relapses in patients who respond to the antiviral effect of IFN alpha (Bronowicki et al., 2006). The precise mechanisms underlying the action of ribavirin in chronic hepatitis C are unknown. Its main properties do not appear to be related to inhibition of inosine monophosphate dehydrogenase (IMPDH), an enzyme that transiently depletes intracellular GTP pools (Hezode et al., 2006; Lau et al., 2002). Ribavirin has been suggested to tilt the T-helper 1 (Th1)/Th2 balance towards Th1 (Hultgren et al., 1998), but these findings have not been confirmed. Finally, a recent mathematical model of HCV kinetics during IFN-alpha-ribavirin therapy suggests that, during combination therapy, ribavirin acts mainly by making HCV virions less infectious (Dixit et al., 2004). It was recently shown in vitro and with other viruses that ribavirin, an RNA mutagen, can drive viral quasispecies to "error catastrophe", i.e. loss of fitness by lethal accumulation of nucleotide mutations during replication (Crotty et al., 2001, 2000; Lanford et al., 2003). Nevertheless, studies of human HCV infection have shown no acceleration of mutagenesis during ribavirin therapy (Chevaliez et al., 2005; Lutchman et al., 2004). Thus, the mechanism by which ribavirin promotes sustained viral eradication when combined with IFN alpha remains unknown.

9.3.3.2 Mechanisms of Pegylated IFN alpha-Ribavirin Treatment Failure

Pegylated IFN alpha-ribavirin treatment failure is defined as a failure to achieve a sustained virological response. Several factors have been shown to play a role in treatment failure, such as the treatment regimen, host factors (age, gender, race, body weight, genetic background, alcohol or intravenous drug use, adherence to therapy), disease-related factors (advanced fibrosis and cirrhosis, HIV coinfection, extra-hepatic manifestations of HCV infection), and finally viral factors that account for so-called "HCV resistance" (Pawlotsky, 2003, 2005a). Indeed, HCV genotypes 1 and 4 are intrinsically more resistant to the antiviral action of IFN alpha than are genotypes 2 and 3 (Pawlotsky et al., 2002). In addition, within each genotype, different strains may display very different intrinsic sensitivities to IFN.

Various explanations have been forwarded for differences in HCV sensitivity to IFN alpha antiviral properties. The role of specific HCV proteins is strongly suggested by the fact that patients infected by different genotypes respond differently to therapy (Fried et al., 2002; Hadziyannis et al., 2004; Manns et al., 2001). Differences in the treatment response among patients infected by the same genotype could be related to differences in the sequence and function of certain viral proteins interacting with IFN-alpha-induced pathways. Various interactions of HCV proteins with cellular proteins or pathways have been described in vitro, and some have been suggested to mediate intrinsic HCV resistance to IFN alpha. The full-length HCV polyprotein has been shown to strongly inhibit IFN alpha-induced signal transduction through the Jak-Stat pathway downstream of Stat tyrosine phosphorylation in both cell cultures and transgenic mice (Blindenbacher et al., 2003; Heim et al., 1999). Non-structural protein 5A (NS5A) has been reported to inhibit phosphorylation and activation of double-stranded RNA-activated, IFN-induced protein kinase (PKR), a major antiviral effector of IFN alpha (Gale et al., 1998, 1999), to activate transcription of interleukin-8, which would in turn inhibit activation of 2'–5' oligoadenylate synthetase, another antiviral effector of IFN alpha (Polyak et al., 2001a,b; Wagoner et al., 2007), and to inhibit extracellular mitogen-activated protein kinase (MAPK) in vitro, possibly via an interaction with Grb2, which could in turn reduce the level of Stat1 phosphorylation and impair the antiviral efficacy of IFN alpha (He et al., 2002; Macdonald et al., 2003; Tan et al., 1999). A short region of the E2 envelope glycoprotein, called PePHD (for PKR-eIF2 phosphorylation homology domain), has been reported in vitro to inhibit the kinase activity of PKR and block its inhibitory effect on protein synthesis and cell growth (Taylor et al., 1999). Finally, the HCV NS3/4A protease has been shown to act as an antagonist of virus-induced interferon regulatory factor (IRF)-3 activation and IFN-β expression through its ability to block retinoic-acid-inducible gene I (RIG-I) signaling and to ablate Toll-like receptor (TLR) 3 signaling by cleaving the Toll–interleukin-1 receptor-domain-containing (TRIF) adaptor protein (Foy et al., 2003, 2005; Gale and Foy, 2005; Li et al., 2005; Loo et al., 2006; Sumpter et al., 2005). These mechanisms underlying "IFN resistance", together with others that remain unknown, probably play a crucial role in antagonizing natural type I IFN responses in the initial stages of HCV infection, when viral replication triggers the host's innate response, thereby allowing the virus to become established. Some of these mechanisms could theoretically play a role in the difference in IFN sensitivity between the different HCV genotypes. However, this has never been demonstrated. In addition, any viral protein antagonism of the antiviral properties of IFN alpha would probably be overwhelmed by the amounts of exogenous IFN alpha administered during therapy. Finally, the involved viral sequences are generally highly conserved within each HCV genotype, and their functional properties have never been shown to differ, qualitatively or quantitatively, according to IFN antiviral efficacy in vivo. On the other hand, reports that ribavirin can select resistant variants in vivo (Young et al., 2003) have not been confirmed, a result not surprising given the very modest antiviral effect of ribavirin on HCV replication in vivo (Pawlotsky et al., 2004).

Overall, viral factors responsible for different HCV sensitivity to IFN-induced antiviral effectors are likely to play a partial role in the multifactorial failure of IFN therapy to clear infection in a substantial proportion of patients with the current molecules and dosages. The underlying mechanisms however remain unknown.

9.3.4 HCV Resistance to Specific HCV Inhibitors

Recent technological advances have helped identify new candidate HCV drug targets. Schematically, future drugs for HCV infection may belong to five main categories, namely new IFNs, alternatives to ribavirin, immune modulators, cyclophilin B inhibitors, and specific HCV inhibitors that may theoretically target any step of the HCV lifecycle (Pawlotsky, 2006b; Pawlotsky and Gish, 2006; Pawlotsky and McHutchison, 2004). If none of the first four categories of drugs is likely to select HCV variants bearing specific sequences or motifs conferring resistance to the corresponding drug(s), experience from the treatment of HIV and HBV infections with specific viral inhibitors has shown that together with non-adherence to therapy, the principal cause of treatment failure is viral resistance, characterized by the selection of minor viral populations bearing mutations that confer resistance to the specific drug (Buckheit, 2004). Among HCV specific inhibitors, the most advanced drugs target the NS3/4A protease and the NS5B RNA-dependent RNA polymerase. Numerous molecules are currently at the clinical and preclinical development stage. Resistance to these molecules has been described in vitro and in patients receiving monotherapy for the most advanced of them.

9.3.4.1 Resistance to NS5B RNA-Dependent RNA Polymerase Inhibitors

The HCV RNA-dependent RNA polymerase harbors several potential target sites for specific inhibitors, including the active enzymatic site, a target for nucleosidic/nucleotidic inhibitors, and four target sites (A–D) for non-nucleosidic, allosteric inhibitors (NNIs) (Carroll and Olsen, 2006; Koch and Narjes, 2006). A large number of families of HCV polymerase inhibitors targeting these different sites is at the developmental stage (Carroll and Olsen, 2006; Koch and Narjes, 2006). All of these drugs have been shown to be able to select for specific resistant variants in vitro in the HCV replicon system (Eldrup et al., 2004; Klumpp et al., 2006; Kukolj et al., 2005; Le Pogam et al., 2006; Migliaccio et al., 2003; Mo et al., 2005; Murakami et al., 2006; Stuyver et al., 2006; Tomei et al., 2003, 2004). These findings are summarized in Table 9.4.

Valopicitabine (NM283, Idenix Pharmaceuticals/Novartis) is a prodrug of 2'-C-methyl-cytidine that targets the HCV polymerase catalytic site. Valopicitabine administration has been reported to induce a moderate (less than one log) and sustained dose-related viral load reduction. This effect appears to be additive to the antiviral effect of pegylated IFN, in both naïve patients and non-sustained virological responders to IFN-based therapy (Lawitz et al., 2006; O'Brien et al., 2005). Cases were recently reported where a variant virus bearing an S282T mutation has been selected after several weeks of administration of the drug in monotherapy (Idenix Pharmaceuticals, personal communication, November 2006). The development of Valopicitabine has been recently halted. Another drug belonging to the same family (Merck Sharp & Dohme) (Tomassini et al., 2005) has induced a sharp decrease of approximately 5 logs of the HCV RNA level after intravenous

Table 9.4 Amino acid substitutions reported to be selected by and confer resistance to specific HCV inhibitors, including RNA-dependent RNA polymerase inhibitors and NS3/4A protease inhibitors, in the subgenomic HCV replicon system in vitro

NS5BRNA-dependent RNA polymeraseinhibitors	NS3/4A protease inhibitors
2' methyl nucleosides	BILN 2061
S282T	D168V/A/E
	A156T/V
4' azido-cytidine	Telaprevir (VX-950)
S96T	A156T/S
S96T/N142T	
Benzimidazoles (NNI site A)	SCH503034
P495S/L/A/T	T54A
P496S/A	A156T/S
V499A	V170A
Thiophenes (NNI site B)	ITMN 191
L419M	D168V/A/E
M423T/I	A156S/V
I482L	
L419M/M423T	
Benzothiadiazines (NNI site C)	
H95Q	
N411S	
M414L/T	
Y448H	
Benzofurans	
C316Y	

administration to HCV-infected chimpanzees. Selection of the same S282T variant occurred after only a few days of administration, as expected in the context of potent inhibition of sensitive virus replication (Olsen D, et al., 1st International Workshop on hepatitis C resistance and new compounds, Boston, Massachusetts, October 25–26, 2006).

HCV 796 (Wyeth Pharmaceuticals) is a non-nucleosidic inhibitor of HCV polymerase belonging to the benzofuran family and targeting NNI site D (Gopalsamy et al., 2006). Its administration to man induced a dose-dependent reduction of HCV replication of up to 1.5 log at the highest administered doses (1,000–1,500 mg daily). However, a relapse was observed upon therapy in most of the patients after 3–4 days of administration, with a return of HCV RNA to nearly baseline levels at the end of the 14 days treatment period in most cases. This relapse was related to the selection of variants bearing the C316Y mutation, an amino acid substitution identified to confer resistance to this class of drugs in vitro. Additional substitutions were found to be associated with the C316Y mutation, essentially in patients infected with HCV genotype 1b (Villano S, et al., 1st International Workshop on hepatitis C resistance and new compounds, Boston, Massachusetts, October 25–26, 2006).

9.3.4.2 Resistance to NS3/4A Serine Protease Inhibitors

The three-dimensional structures of NS3 combined with its cofactor NS4A have been resolved (Kim et al., 1996; Love et al., 1996; Yan et al., 1998). Various classes of NS3/4A protease inhibitors have been developed that potently inhibit the protease function in cell-free systems and HCV replication in various models. Three drugs belonging to different families but all targeting the active site of the NS3/4A protease have been administered to HCV genotype 1-infected patients so far. All three induced significant reductions (2–4 log) in HCV RNA levels. BILN 2061 (Boehringer-Ingelheim) will not be further developed because of myocardial toxicity in animals (Hinrichsen et al., 2004; Reiser et al., 2005). Telaprevir (VX-950, Vertex Pharmaceuticals) and SCH 503034 (Schering-Plough Pharmaceuticals) appear to be safe and are currently undergoing clinical evaluation in combination with pegylated IFN alpha and ribavirin (Reesink et al., 2006b; Sarrazin et al., 2007b). Another drug also targeting the active site of the enzyme, ITMN 191 (Inter-Mune/Roche), will soon start clinical evaluation. Table 9.4 shows the amino acid substitutions reported to be selected by these drugs and to confer resistance in the replicon system in vitro (Lin et al., 2004, 2005; Lu et al., 2004; Tong et al., 2006; Yi et al., 2006).

The administration of telaprevir in monotherapy induced a sharp decline of viral replication of the order of 3–4 logs in patients infected with HCV genotype 1 treated for 14 days (Reesink et al., 2006b). The patients with a high exposure to the drug generally developed a biphasic decline of viral replication and sometimes became HCV RNA undetectable after several days of administration. In contrast, the patients who had a lower exposure to the drug experienced either a true relapse or a plateau phase after the initial viral level reduction (Tong et al., 2006). A careful analysis of the dynamics of viral variant populations during the 14 days treatment period in the latter patients has shown the selection of HCV variants exhibiting different levels of resistance to telaprevir. The principal mutations were V36M/A, T54A, R155K/T, and A156S/T/V. Some patients harbored double mutant populations (Sarrazin et al., 2007a). Amino acids 155 and 156 are directly in contact with the drug within the active site of the enzyme and the changes could therefore directly influence the interaction, whereas changes at positions 36 and 54 could induce conformational changes that would indirectly affect the affinity of the enzyme for the telaprevir molecule. Interestingly, the most resistant virus, i.e. the virus bearing amino acid substitutions at position 156, was also the least fit in vivo. The other variants, that were intrinsically less resistant to telaprevir, had better replication fitness and were predominant in the patients who recovered high-level replication upon telaprevir administration (Sarrazin et al., 2007a).

SCH 503034 has been shown in vitro in the replicon system to select for three amino acid substitutions that confer low to moderate levels of resistance, T54A, V170A and A156S, and for one substitution that confers high-level resistance, A156T (Table 9.4) (Tong et al., 2006). Interestingly, several of these substitutions have been shown to confer resistance to telaprevir both in vitro and in vivo, pointing to cross-resistance between the two molecules (Lin et al., 2005; Lin et al., 2004; Sarrazin et al., 2007a). Like with telaprevir in vivo, the most resistant variant (A156T) also appeared to be the least fit in vitro, whereas the most fit variant (V170A) was also the least resistant (Tong et al., 2006). In vivo resistance data are currently awaited in patients treated with SCH 503034.

9.3.5 Prevention of HCV Resistance to Specific HCV Inhibitors

In contrast with HBV infection, HCV infection is curable. Thus strategies that prevent the development of resistance to anti-HCV drugs may allow for infection cure. These strategies are, in theory, the same as for HBV and HIV drugs. They include the use of the most potent drugs and/or combinations of antivirals with additive or synergistic effects, the use of drugs with a high "pharmacologic barrier" to viral escape, high adherence to therapy, and strategies that raise the "genetic barrier" to viral escape, such as the combination of several drugs with a different mechanism of action and that lack cross-resistance. Currently, the new HCV antiviral drugs are being developed by different companies which do not collaborate. In addition, toxicity and drug–drug-interaction data are lacking, and there are not yet potent drugs in every drug class in clinical development. Thus, no trials have been initiated where several specific inhibitors of HCV replication are used in combination. Instead, HCV inhibitors, such as telaprevir, SCH503034, valopicitabine or R1626 are used in combination with pegylated IFN alpha, with or without ribavirin (Lawitz et al., 2006; O'Brien et al., 2005; Reesink et al., 2006a; Rodriguez-Torres et al., 2006). The action of IFN alpha has been shown to be at least additive in vivo, and to ensure undetectable HCV RNA in a majority of patients, when potent antivirals, such as telaprevir, are used (Reesink et al., 2006a; Rodriguez-Torres et al., 2006) Ongoing trials where such combinations are being administered for 12–48 weeks will tell whether the sustained virological response rates will be improved relative to standard therapy, whether ribavirin is needed and whether resistance is efficiently prevented. Preliminary data however suggest that not all patients are able to achieve HCV RNA clearance during pegylated IFN alpha-telaprevir combination therapy (Vertex Pharmaceuticals' Press Release, Cambridge, Massachusetts, December 12, 2006). This finding is not surprising if one thinks that patients who are null or poor responders to IFN alpha are, in fact, receiving a monotherapy of the HCV inhibitor. Future trials with the currently developed and future specific HCV inhibitors will determine whether IFN-ribavirin combination will remain the basis for HCV therapy as well as which is (are) the best combination(s) of drugs for each subgroup of HCV patients in order to achieve the highest cure rate.

9.3.6 Management of HCV Treatment Failure

Unlike HBV, there are few options for the patients who fail on standard pegylated IFN alpha-ribavirin therapy, because no other therapies are approved so far. Ongoing trials are assessing the utility of higher doses and/or more frequent injections of pegylated IFNs, higher doses of ribavirin, and drug combinations with novel antiviral drugs. Due to the complexity of the mechanisms underlying HCV treatment failure, the best strategy in the non-responder to pegylated IFN alpha-ribavirin therapy remains to be established.

9.4 Conclusion

HBV resistance to specific inhibitor molecules is a frequent cause of HBV treatment failure. HBV resistance incidence when the drugs are used in monotherapy is known and the underlying mechanisms are well understood. Prevention of HBV resistance is based on the use of combinations of highly potent antivirals and improvement of patients' adherence to therapy. Pegylated IFN alpha-ribavirin treament failure in HCV therapy is multifactorial, and its viral components are poorly understood. Currently, the only option is to increase the exposure to IFN and/or ribavirin. A number of new HCV drugs are at the developmental stage, including several classes of specific HCV inhibitors. The latter have been shown to select for resistant HCV variants both in vitro and in vivo. The frequency and rapidity of resistance emergence makes it unlikely that these drugs will be used in monotherapy for the treatment of chronic hepatitis C. Ongoing trials are assessing the efficacy of combinations of specific HCV inhibitors with pegylated IFN alpha, with or without ribavirin. Combinations of potent orally administered HCV inhibitors will need to be tested for their ability to achieve high cure rates in the future.

References

NIH Consensus Development Program, (2002), National Institutes of Health Consensus Development Conference Statement: management of hepatitis C: 2002-June 10–12, 2002. *Hepatology*, 36, S3–S20.

Allen, M.I., Deslauriers, M., Andrews, C.W., Tipples, G.A., Walters, K.A., Tyrrell, D.L., Brown, N. and Condreay, L.D., (1998), Identification and characterization of mutations in hepatitis B virus resistant to lamivudine. *Hepatology*, 27, 1670–1677.

Angus, P., Vaughan, R., Xiong, S., Yang, H., Delaney, W., Gibbs, C., Brosgart, C., Colledge, D., Edwards, R., Ayres, A., Bartholomeusz, A. and Locarnini, S., (2003), Resistance to adefovir dipivoxil therapy associated with the selection of a novel mutation in the HBV polymerase. *Gastroenterology*, 125, 292–297.

Bartenschlager, R., Frese, M. and Pietschmann, T., (2004), Novel insights into hepatitis C virus replication and persistence. *Adv Virus Res*, 63, 71–180.

Blindenbacher, A., Duong, F.H., Hunziker, L., Stutvoet, S.T., Wang, X., Terracciano, L., Moradpour, D., Blum, H.E., Alonzi, T., Tripodi, M., La Monica, N. and Heim, M.H., (2003), Expression of hepatitis C virus proteins inhibits interferon alpha signaling in the liver of transgenic mice. *Gastroenterology*, 124, 1465–1475.

Bozdayi, A.M., Uzunalimoglu, O., Turkyilmaz, A.R., Aslan, N., Sezgin, O., Sahin, T., Bozdayi, G., Cinar, K., Pai, S.B., Pai, R., Bozkaya, H., Karayalcin, S., Yurdaydin, C. and Schinazi, R.F., (2003), YSDD: a novel mutation in HBV DNA polymerase confers clinical resistance to lamivudine. *J Viral Hepat*, 10, 256–265.

Bronowicki, J.P., Ouzan, D., Asselah, T., Desmorat, H., Zarski, J.P., Foucher, J., Bourliere, M., Renou, C., Tran, A., Melin, P., Hezode, C., Chevalier, M., Bouvier-Alias, M., Chevaliez, S., Montestruc, F., Lonjon-Domanec, I. and Pawlotsky, J.M., (2006), Effect of ribavirin in genotype 1 patients with hepatitis C responding to pegylated interferon alfa-2a plus ribavirin. *Gastroenterology*, 131, 1040–1048.

Buckheit, R.W., Jr., (2004), Understanding HIV resistance, fitness, replication capacity and compensation: targeting viral fitness as a therapeutic strategy. *Expert Opin Investig Drugs*, 13, 933–958.

Cane, P.A., Mutimer, D., Ratcliffe, D., Cook, P., Beards, G., Elias, E. and Pillay, D., (1999), Analysis of hepatitis B virus quasispecies changes during emergence and reversion of lamivudine resistance in liver transplantation. *Antivir Ther*, **4**, 7–14.

Carroll, S.S. and Olsen, D.B., (2006), Nucleoside analog inhibitors of hepatitis C virus replication. *Infect Disord Drug Targets*, **6**, 17–29.

Castera, L. and Pawlotsky, J.M., (2005), Noninvasive diagnosis of liver fibrosis in patients with chronic hepatitis C. *MedGenMed*, **7**, 39.

Castera, L., Vergniol, J., Foucher, J., Le Bail, B., Chanteloup, E., Haaser, M., Darriet, M., Couzigou, P. and De Ledinghen, V., (2005), Prospective comparison of transient elastography, Fibrotest, APRI, and liver biopsy for the assessment of fibrosis in chronic hepatitis C. *Gastroenterology*, **128**, 343–350.

Castet, V., Fournier, C., Soulier, A., Brillet, R., Coste, J., Larrey, D., Dhumeaux, D., Maurel, P. and Pawlotsky, J.M., (2002), Alpha interferon inhibits hepatitis C virus replication in primary human hepatocytes infected in vitro. *J Virol*, **76**, 8189–8199.

Chang, T.T., Gish, R.G., de Man, R., Gadano, A., Sollano, J., Chao, Y.C., Lok, A.S., Han, K.H., Goodman, Z., Zhu, J., Cross, A., DeHertogh, D., Wilber, R., Colonno, R. and Apelian, D., (2006), A comparison of entecavir and lamivudine for HBeAg-positive chronic hepatitis B. *N Engl J Med*, **354**, 1001–1010.

Chevaliez, S., Brillet, R., Hezode, C., Dhumeaux, D. and Pawlotsky, J.M., (2005), Assessment of ribavirin mutagenic properties in vivo and the influence of interferon alpha-induced reduction of hepatitis C virus RNA load upon therapy. *Hepatology*, **42** (Suppl. 1), 679A.

Chin, R., Shaw, T., Torresi, J., Sozzi, V., Trautwein, C., Bock, T., Manns, M., Isom, H., Furman, P. and Locarnini, S., (2001), In vitro susceptibilities of wild-type or drug-resistant hepatitis B virus to (–)-beta-D-2,6-diaminopurine dioxolane and 2'-fluoro-5-methyl-beta-L-arabinofuranosyluracil. *Antimicrob Agents Chemother*, **45**, 2495–2501.

Chong, Y., Stuyver, L., Otto, M.J., Schinazi, R.F. and Chu, C.K., (2003), Mechanism of antiviral activities of 3'-substituted ʟ-nucleosides against 3TC-resistant HBV polymerase: a molecular modelling approach. *Antivir Chem Chemother*, **14**, 309–319.

Clavel, F. and Hance, A.J., (2004), HIV drug resistance. *N Engl J Med*, **350**, 1023–1035.

Colonno, R.J., Rose, R.E., Baldick, C.J., Levine, S.M., Klesczewski, K. and Tenney, D.J., (2006a), High barrier to resistance results in no emergence of entecavir resistance in nucleoside-naïve subjects during the first two years of therapy. *J Hepatol*, **44** (Suppl. 2), S182.

Colonno, R.J., Rose, R.E., Pokornowski, K., Baldick, C.J., Klesczewski, K. and Tenney, D.J., (2006b), Assessment at three years shows high barrier to resistance is maintained in entecavir-treated nucleoside naive patients while resistance emergence increases over time in lamivudine refractory patients. *Hepatology*, **44** (Suppl. 2), 229A.

Crotty, S., Cameron, C.E. and Andino, R., (2001), RNA virus error catastrophe: direct molecular test by using ribavirin. *Proc Natl Acad Sci USA*, **98**, 6895–6900.

Crotty, S., Maag, D., Arnold, J.J., Zhong, W., Lau, J.Y., Hong, Z., Andino, R. and Cameron, C.E., (2000), The broad-spectrum antiviral ribonucleoside ribavirin is an RNA virus mutagen. *Nat Med*, **6**, 1375–1379.

Das, K., Xiong, X., Yang, H., Westland, C.E., Gibbs, C.S., Sarafianos, S.G. and Arnold, E., (2001), Molecular modeling and biochemical characterization reveal the mechanism of hepatitis B virus polymerase resistance to lamivudine (3TC) and emtricitabine (FTC). *J Virol*, **75**, 4771–4779.

Davis, G.L., Wong, J.B., McHutchison, J.G., Manns, M.P., Harvey, J. and Albrecht, J., (2003), Early virologic response to treatment with peginterferon alfa-2b plus ribavirin in patients with chronic hepatitis C. *Hepatology*, **38**, 645–652.

de Franchis, R., Hadengue, A., Lau, G., Lavanchy, D., Lok, A., McIntyre, N., Mele, A., Paumgartner, G., Pietrangelo, A., Rodes, J., Rosenberg, W. and Valla, D., (2003), EASL International Consensus Conference on Hepatitis B. 13–14 September, 2002 Geneva, Switzerland. Consensus statement (long version). *J Hepatol*, **39** (Suppl. 1), S3–S25.

Delaney, W.E.t., Edwards, R., Colledge, D., Shaw, T., Furman, P., Painter, G. and Locarnini, S., (2002), Phenylpropenamide derivatives AT-61 and AT-130 inhibit replication of wild-type and lamivudine-resistant strains of hepatitis B virus in vitro. *Antimicrob Agents Chemother*, **46**, 3057–3060.

Delaney, W.E.t., Edwards, R., Colledge, D., Shaw, T., Torresi, J., Miller, T.G., Isom, H.C., Bock, C.T., Manns, M.P., Trautwein, C. and Locarnini, S., (2001), Cross-resistance testing of anti-hepadnaviral compounds using novel recombinant baculoviruses which encode drug-resistant strains of hepatitis B virus. *Antimicrob Agents Chemother*, **45**, 1705–1713.

Delaney, W.E.t., Ray, A.S., Yang, H., Qi, X., Xiong, S., Zhu, Y. and Miller, M.D., (2006), Intracellular metabolism and in vitro activity of tenofovir against hepatitis B virus. *Antimicrob Agents Chemother*, **50**, 2471–2477.

Delaney, W.E.t., Yang, H., Westland, C.E., Das, K., Arnold, E., Gibbs, C.S., Miller, M.D. and Xiong, S., (2003), The hepatitis B virus polymerase mutation rtV173L is selected during lamivudine therapy and enhances viral replication in vitro. *J Virol*, **77**, 11833–11841.

Dent, B., Chin, R., Trautwein, C., Bock, T.C., Manns, M.P., Cleary, D., Furman, P. and Locarnini, S.A., (2000), Antiviral cross-resistance profiles of two novel antiviral agents, DAPD and L-FMAU, against nucleoside analogue resistant HBV. *Hepatology*, **32** (Suppl. 1), 457A.

Dienstag, J.L., Goldin, R.D., Heathcote, E.J., Hann, H.W., Woessner, M., Stephenson, S.L., Gardner, S., Gray, D.F. and Schiff, E.R., (2003), Histological outcome during long-term lamivudine therapy. *Gastroenterology*, **124**, 105–117.

Dixit, N.M., Layden-Almer, J.E., Layden, T.J. and Perelson, A.S., (2004), Modelling how ribavirin improves interferon response rates in hepatitis C virus infection. *Nature*, **432**, 922–924.

Duarte, E.A., Novella, I.S., Weaver, S.C., Domingo, E., Wain-Hobson, S., Clarke, D.K., Moya, A., Elena, S.F., de la Torre, J.C. and Holland, J.J., (1994), RNA virus quasispecies: significance for viral disease and epidemiology. *Infect Agents Dis*, **3**, 201–214.

Durantel, D., Carrouee-Durantel, S., Werle-Lapostolle, B., Brunelle, M.N., Pichoud, C., Trepo, C. and Zoulim, F., (2004), A new strategy for studying in vitro the drug susceptibility of clinical isolates of human hepatitis B virus. *Hepatology*, **40**, 855–864.

Eldrup, A.B., Allerson, C.R., Bennett, C.F., Bera, S., Bhat, B., Bhat, N., Bosserman, M.R., Brooks, J., Burlein, C., Carroll, S.S., Cook, P.D., Getty, K.L., MacCoss, M., McMasters, D.R., Olsen, D.B., Prakash, T.P., Prhavc, M., Song, Q., Tomassini, J.E. and Xia, J., (2004), Structure-activity relationship of purine ribonucleosides for inhibition of hepatitis C virus RNA-dependent RNA polymerase. *J Med Chem*, **47**, 2283–2295.

Ferenci, P., Fried, M.W., Shiffman, M.L., Smith, C.I., Marinos, G., Goncales, F.L., Jr, Haussinger, D., Diago, M., Carosi, G., Dhumeaux, D., Craxi, A., Chaneac, M. and Reddy, K.R., (2005), Predicting sustained virological responses in chronic hepatitis C patients treated with peginterferon alfa-2a (40 KD)/ribavirin. *J Hepatol*, **43**, 425–433.

Foy, E., Li, K., Sumpter, R., Jr., Loo, Y.M., Johnson, C.L., Wang, C., Fish, P.M., Yoneyama, M., Fujita, T., Lemon, S.M. and Gale, M., Jr., (2005), Control of antiviral defenses through hepatitis C virus disruption of retinoic acid-inducible gene-I signaling. *Proc Natl Acad Sci USA*, **102**, 2986–2991.

Foy, E., Li, K., Wang, C., Sumpter, R., Jr., Ikeda, M., Lemon, S.M. and Gale, M., Jr., (2003), Regulation of interferon regulatory factor-3 by the hepatitis C virus serine protease. *Science*, **300**, 1145–1148.

Frese, M., Pietschmann, T., Moradpour, D., Haller, O. and Bartenschlager, R., (2001), Interferon-alpha inhibits hepatitis C virus subgenomic RNA replication by an MxA-independent pathway. *J Gen Virol*, **82**, 723–733.

Fried, M.W., Shiffman, M.L., Reddy, K.R., Smith, C., Marinos, G., Goncales, F.L., Jr, Haussinger, D., Diago, M., Carosi, G., Dhumeaux, D., Craxi, A., Lin, A., Hoffman, J. and Yu, J., (2002), Peginterferon alfa-2a plus ribavirin for chronic hepatitis C virus infection. *N Engl J Med*, **347**, 975–982.

Fu, L. and Cheng, Y.C., (1998), Role of additional mutations outside the YMDD motif of hepatitis B virus polymerase in ι(–)SddC (3TC) resistance. *Biochem Pharmacol*, **55**, 1567–1572.

Fu, L. and Cheng, Y.C., (2000), Characterization of novel human hepatoma cell lines with stable hepatitis B virus secretion for evaluating new compounds against lamivudine- and penciclovir-resistant virus. *Antimicrob Agents Chemother*, **44**, 3402–3407.

Gale, M., Jr, Blakely, C.M., Kwieciszewski, B., Tan, S.L., Dossett, M., Tang, N.M., Korth, M.J., Polyak, S.J., Gretch, D.R. and Katze, M.G., (1998), Control of PKR protein kinase by hepatitis

C virus nonstructural 5A protein: molecular mechanisms of kinase regulation. *Mol Cell Biol*, **18**, 5208–5218.

Gale, M., Jr and Foy, E.M., (2005), Evasion of intracellular host defence by hepatitis C virus. *Nature*, **436**, 939–945.

Gale, M., Jr, Kwieciszewski, B., Dossett, M., Nakao, H. and Katze, M.G., (1999), Antiapoptotic and oncogenic potentials of hepatitis C virus are linked to interferon resistance by viral repression of the PKR protein kinase. *J Virol*, **73**, 6506–6516.

Ganem, D. and Prince, A.M., (2004), Hepatitis B virus infection—natural history and clinical consequences. *N Engl J Med*, **350**, 1118–1129.

Gintowt, A.A., Germer, J.J., Mitchell, P.S. and Yao, J.D., (2005), Evaluation of the MagNA Pure LC used with the Trugene HBV Genotyping Kit. *J Clin Virol*, **34**, 155–157.

Gish, R.G., Leung, N.W., Wright, T.L., Trinh, H., Lang, W., Kessler, H.A., Fang, L., Wang, L.H., Delehanty, J., Rigney, A., Mondou, E., Snow, A. and Rousseau, F., (2002), Dose range study of pharmacokinetics, safety, and preliminary antiviral activity of emtricitabine in adults with hepatitis B virus infection. *Antimicrob Agents Chemother*, **46**, 1734–1740.

Gopalsamy, A., Aplasca, A., Ciszewski, G., Park, K., Ellingboe, J.W., Orlowski, M., Feld, B. and Howe, A.Y., (2006), Design and synthesis of 3,4-dihydro-1H-[1]-benzothieno[2,3-c]pyran and 3,4-dihydro-1H-pyrano[3,4-b]benzofuran derivatives as non-nucleoside inhibitors of HCV NS5B RNA dependent RNA polymerase. *Bioorg Med Chem Lett*, **16**, 457–460.

Gunther, S., Fischer, L., Pult, I., Sterneck, M. and Will, H., (1999), Naturally occurring variants of hepatitis B virus. *Adv Virus Res*, **52**, 25–137.

Guo, J.T., Bichko, V.V. and Seeger, C., (2001), Effect of alpha interferon on the hepatitis C virus replicon. *J Virol*, **75**, 8516–8523.

Hadziyannis, S.J., Sette, H., Jr, Morgan, T.R., Balan, V., Diago, M., Marcellin, P., Ramadori, G., Bodenheimer, H., Jr, Bernstein, D., Rizzetto, M., Zeuzem, S., Pockros, P.J., Lin, A. and Ackrill, A.M., (2004), Peginterferon-alpha2a and ribavirin combination therapy in chronic hepatitis C: a randomized study of treatment duration and ribavirin dose. *Ann Intern Med*, **140**, 346–355.

Hadziyannis, S.J., Tassopoulos, N.C., Heathcote, E.J., Chang, T.T., Kitis, G., Rizzetto, M., Marcellin, P., Lim, S.G., Goodman, Z., Ma, J., Brosgart, C.L., Borroto-Esoda, K., Arterburn, S. and Chuck, S.L., (2006), Long-term therapy with Adefovir Dipivoxil for HBeAg-negative chronic Hepatitis B for up to 5 years. *Gastroenterology*, **131**, 1743–1751.

Hadziyannis, S.J., Tassopoulos, N.C., Heathcote, E.J., Chang, T.T., Kitis, G., Rizzetto, M., Marcellin, P., Lim, S.G., Goodman, Z., Wulfsohn, M.S., Xiong, S., Fry, J. and Brosgart, C.L., (2003), Adefovir dipivoxil for the treatment of hepatitis B e antigen-negative chronic hepatitis B. *N Engl J Med*, **348**, 800–807.

He, Y., Nakao, H., Tan, S.L., Polyak, S.J., Neddermann, P., Vijaysri, S., Jacobs, B.L. and Katze, M.G., (2002), Subversion of cell signaling pathways by hepatitis C virus nonstructural 5A protein via interaction with Grb2 and P85 phosphatidylinositol 3-kinase. *J Virol*, **76**, 9207–9217.

Heim, M.H., Moradpour, D. and Blum, H.E., (1999), Expression of hepatitis C virus proteins inhibits signal transduction through the Jak-STAT pathway. *J Virol*, **73**, 8469–8475.

Hezode, C., Bouvier-Alias, M., Costentin, C., Medkour, F., Franc-Poole, E., Aalyson, M., Alam, J., Dhumeaux, D. and Pawlotsky, J.M., (2006), Effect of an IMPDH inhibitor, merimepodib (MMPD), assessed alone and in combination with ribavirin, on HCV replication: implications regarding ribavirin's mechanism of action. *Hepatology*, **44** (Suppl. 2), 615A.

Hinrichsen, H., Benhamou, Y., Wedemeyer, H., Reiser, M., Sentjens, R.E., Calleja, J.L., Forns, X., Erhardt, A., Cronlein, J., Chaves, R.L., Yong, C.L., Nehmiz, G. and Steinmann, G.G., (2004), Short-term antiviral efficacy of BILN 2061, a hepatitis C virus serine protease inhibitor, in hepatitis C genotype 1 patients. *Gastroenterology*, **127**, 1347–1355.

Hong, S.P., Kim, N.K., Hwang, S.G., Chung, H.J., Kim, S., Han, J.H., Kim, H.T., Rim, K.S., Kang, M.S., Yoo, W. and Kim, S.O., (2004), Detection of hepatitis B virus YMDD variants using mass spectrometric analysis of oligonucleotide fragments. *J Hepatol*, **40**, 837–844.

Hultgren, C., Milich, D.R., Weiland, O. and Sallberg, M., (1998), The antiviral compound ribavirin modulates the T helper (Th) 1/Th2 subset balance in hepatitis B and C virus-specific immune responses. *J Gen Virol*, **79**, 2381–2391.

Hussain, M., Fung, S., Libbrecht, E., Sablon, E., Cursaro, C., Andreone, P. and Lok, A.S., (2006), Sensitive line probe assay that simultaneously detects mutations conveying resistance to lamivudine and adefovir. *J Clin Microbiol*, **44**, 1094–1097.

Kim, H.S., Han, K.H., Ahn, S.H., Kim, E.O., Chang, H.Y., Moon, M.S., Chung, H.J., Yoo, W., Kim, S.O. and Hong, S.P., (2005), Evaluation of methods for monitoring drug resistance in chronic hepatitis B patients during lamivudine therapy based on mass spectrometry and reverse hybridization. *Antivir Ther*, **10**, 441–449.

Kim, J.L., Morgenstern, K.A., Lin, C., Fox, T., Dwyer, M.D., Landro, J.A., Chambers, S.P., Markland, W., Lepre, C.A., O'Malley, E.T., Harbeson, S.L., Rice, C.M., Murcko, M.A., Caron, P.R. and Thomson, J.A., (1996), Crystal structure of the hepatitis C virus NS3 protease domain complexed with a synthetic NS4A cofactor peptide. *Cell*, **87**, 343–355.

Klumpp, K., Leveque, V., Le Pogam, S., Ma, H., Jiang, W.R., Kang, H., Granycome, C., Singer, M., Laxton, C., Hang, J.Q., Sarma, K., Smith, D.B., Heindl, D., Hobbs, C.J., Merrett, J.H., Symons, J., Cammack, N., Martin, J.A., Devos, R. and Najera, I., (2006), The novel nucleoside analog R1479 (4'-azidocytidine) is a potent inhibitor of NS5B-dependent RNA synthesis and hepatitis C virus replication in cell culture. *J Biol Chem*, **281**, 3793–3799.

Koch, U. and Narjes, F., (2006), Allosteric inhibition of the hepatitis C virus NS5B RNA dependent RNA polymerase. *Infect Disord Drug Targets*, **6**, 31–41.

Kukolj, G., McGibbon, G.A., McKercher, G., Marquis, M., Lefebvre, S., Thauvette, L., Gauthier, J., Goulet, S., Poupart, M.A. and Beaulieu, P.L., (2005), Binding site characterization and resistance to a class of non-nucleoside inhibitors of the hepatitis C virus NS5B polymerase. *J Biol Chem*, **280**, 39260–39267.

Lai, C.L., Dienstag, J., Schiff, E., Leung, N.W., Atkins, M., Hunt, C., Brown, N., Woessner, M., Boehme, R. and Condreay, L., (2003), Prevalence and clinical correlates of YMDD variants during lamivudine therapy for patients with chronic hepatitis B. *Clin Infect Dis*, **36**, 687–696.

Lai, C.L., Gane, E., Hsu, C.W., Thongsawat, S., Wang, Y., Chen, Y., Heathcote, E.J., Rasenack, J., Bzowej, N., Naoumov, N., Zeuzem, S., Di Bisceglie, A., Chao, G.C., Fielman-Constance, B.A. and Brown, N.A., (2006a), Two-year results from the GLOBE trial in patients with hepatitis B: greater clinical and antiviral efficacy for telbivudine (LdT) versus lamivudine. *Hepatology*, **44** (Suppl. 1), 222A.

Lai, C.L., Gane, E., Liaw, Y.F., Thongsawat, S., Wang, Y., Chen, Y., Heathcote, E.J., Rasenack, J., Bzowej, N., Naoumov, N., Chao, G., Constance, B.F. and Brown, N.A., (2005), Maximal early HBV suppression is predictive of optimal virologic and clinical efficacy in nucleoside-treated hepatitis B patients: scientific observations from a large multinational trial (the GLOBE study). *Hepatology*, **42** (Suppl. 1), 232A.

Lai, C.L., Shouval, D., Lok, A.S., Chang, T.T., Cheinquer, H., Goodman, Z., DeHertogh, D., Wilber, R., Zink, R.C., Cross, A., Colonno, R. and Fernandes, L., (2006b), Entecavir versus lamivudine for patients with HBeAg-negative chronic hepatitis B. *N Engl J Med*, **354**, 1011–1020.

Lampertico, P., (2006), Entecavir versus lamivudine for HBeAg positive and negative chronic hepatitis B. *J Hepatol*, **45**, 457–460.

Lampertico, P., Marzano, A., Levrero, M., Santantonio, T., Di Marco, V., Brunetto, M., Andreone, P., Sagnelli, E., Fagiuoli, S., Mazzella, G., Raimondo, G., Gaeta, G. and Ascione, A., (2006), Adefovir and lamivudine combination therapy is superior to adefovir monotherapy for lamivudine-resistant patients with HBeAg-negative chronic hepatitis B. *Hepatology*, **44** (Suppl. 2), 693A.

Lanford, R.E., Guerra, B., Lee, H., Averett, D.R., Pfeiffer, B., Chavez, D., Notvall, L. and Bigger, C., (2003), Antiviral effect and virus-host interactions in response to alpha interferon, gamma interferon, poly(I)-poly(C), tumor necrosis factor alpha, and ribavirin in hepatitis C virus subgenomic replicons. *J Virol*, **77**, 1092–1104.

Lau, J.Y., Tam, R.C., Liang, T.J. and Hong, Z., (2002), Mechanism of action of ribavirin in the combination treatment of chronic HCV infection. *Hepatology*, **35**, 1002–1009.

Lawitz, E., Nguyen, T., Younes, Z., Santoro, J., Gitlin, N., McEniry, D., Chasen, R., Goff, J., Knox, S., Kleber, K., Belanger, B., Brown, N.A. and Dieterich, D.T., (2006), Valopicitabine (NM283) plus peg-interferon in treatment-naive hepatitis C patients with HCV genotype 1 infection: HCV RNA clearance during 24 weeks of treatment. *Hepatology*, **44** (Suppl. 2), 223A.

Le Pogam, S., Jiang, W.R., Leveque, V., Rajyaguru, S., Ma, H., Kang, H., Jiang, S., Singer, M., Ali, S., Klumpp, K., Smith, D., Symons, J., Cammack, N. and Najera, I., (2006), In vitro selected Con1 subgenomic replicons resistant to 2'-C-methyl-cytidine or to R1479 show lack of cross resistance. *Virology*, **351**, 349–359.

Lee, W.M., (1997), Hepatitis B virus infection. *N Engl J Med*, **337**, 1733–1745.

Leung, N.W., Lai, C.L., Chang, T.T., Guan, R., Lee, C.M. and Ng, K.Y., (2001), Extended lamivudine treatment inpatients with chronic hepatitis B enhances hepatitis B e antigen seroconversion rates: results after 3 years of therapy. *Hepatology*, **33**, 1527–1532.

Levine, S., Hernandez, D., Yamanaka, G., Zhang, S., Rose, R., Weinheimer, S. and Colonno, R.J., (2002), Efficacies of entecavir against lamivudine-resistant hepatitis B virus replication and recombinant polymerases in vitro. *Antimicrob Agents Chemother*, **46**, 2525–2532.

Li, K., Foy, E., Ferreon, J.C., Nakamura, M., Ferreon, A.C., Ikeda, M., Ray, S.C., Gale, M., Jr and Lemon, S.M., (2005), Immune evasion by hepatitis C virus NS3/4A protease-mediated cleavage of the Toll-like receptor 3 adaptor protein TRIF. *Proc Natl Acad Sci USA*, **102**, 2992–2997.

Liaw, Y.F., Chien, R.N., Yeh, C.T., Tsai, S.L. and Chu, C.M., (1999), Acute exacerbation and hepatitis B virus clearance after emergence of YMDD motif mutation during lamivudine therapy. *Hepatology*, **30**, 567–572.

Liaw, Y.F., Sung, J.J., Chow, W.C., Farrell, G., Lee, C.Z., Yuen, H., Tanwandee, T., Tao, Q.M., Shue, K., Keene, O.N., Dixon, J.S., Gray, D.F. and Sabbat, J., (2004), Lamivudine for patients with chronic hepatitis B and advanced liver disease. *N Engl J Med*, **351**, 1521–1531.

Lin, C., Gates, C.A., Rao, B.G., Brennan, D.L., Fulghum, J.R., Luong, Y.P., Frantz, J.D., Lin, K., Ma, S., Wei, Y.Y., Perni, R.B. and Kwong, A.D., (2005), In vitro studies of cross-resistance mutations against two hepatitis C virus serine protease inhibitors, VX-950 and BILN 2061. *J Biol Chem*, **280**, 36784–36791.

Lin, C., Lin, K., Luong, Y.P., Rao, B.G., Wei, Y.Y., Brennan, D.L., Fulghum, J.R., Hsiao, H.M., Ma, S., Maxwell, J.P., Cottrell, K.M., Perni, R.B., Gates, C.A. and Kwong, A.D., (2004), In vitro resistance studies of hepatitis C virus serine protease inhibitors, VX-950 and BILN 2061: structural analysis indicates different resistance mechanisms. *J Biol Chem*, **279**, 17508–17514.

Lindenbach, B.D., Evans, M.J., Syder, A.J., Wolk, B., Tellinghuisen, T.L., Liu, C.C., Maruyama, T., Hynes, R.O., Burton, D.R., McKeating, J.A. and Rice, C.M., (2005), Complete replication of hepatitis C virus in cell culture. *Science*, **309**, 623–626.

Locarnini, S., Hatzakis, A., Heathcote, J., Keeffe, E.B., Liang, T.J., Mutimer, D., Pawlotsky, J.M. and Zoulim, F., (2004), Management of antiviral resistance in patients with chronic hepatitis B. *Antivir Ther*, **9**, 679–693.

Lok, A.S., Heathcote, E.J. and Hoofnagle, J.H., (2001), Management of hepatitis B: 2000— summary of a workshop. *Gastroenterology*, **120**, 1828–1853.

Lok, A.S., Lai, C.L., Leung, N., Yao, G.B., Cui, Z.Y., Schiff, E.R., Dienstag, J.L., Heathcote, E.J., Little, N.R., Griffiths, D.A., Gardner, S.D. and Castiglia, M., (2003), Long-term safety of lamivudine treatment in patients with chronic hepatitis B. *Gastroenterology*, **125**, 1714–1722.

Loo, Y.M., Owen, D.M., Li, K., Erickson, A.K., Johnson, C.L., Fish, P.M., Carney, D.S., Wang, T., Ishida, H., Yoneyama, M., Fujita, T., Saito, T., Lee, W.M., Hagedorn, C.H., Lau, D.T., Weinman, S.A., Lemon, S.M. and Gale, M., Jr., (2006), Viral and therapeutic control of IFN-beta promoter stimulator 1 during hepatitis C virus infection. *Proc Natl Acad Sci USA*, **103**, 6001–6006.

Love, R.A., Parge, H.E., Wickersham, J.A., Hostomsky, Z., Habuka, N., Moomaw, E.W., Adachi, T. and Hostomska, Z., (1996), The crystal structure of hepatitis C virus NS3 proteinase reveals a trypsin-like fold and a structural zinc binding site. *Cell*, **87**, 331–342.

Lu, L., Pilot-Matias, T.J., Stewart, K.D., Randolph, J.T., Pithawalla, R., He, W., Huang, P.P., Klein, L.L., Mo, H. and Molla, A., (2004), Mutations conferring resistance to a potent hepatitis C virus serine protease inhibitor in vitro. *Antimicrob Agents Chemother*, **48**, 2260–2266.

Lutchman, G.A., Danehower, S., Park, Y., Ward, C., Liang, T.J., Hoofnagle, J.H., Thomson, M.Z. and Ghany, M.G., (2004), Mutation rate of hepatitis C virus in patients during ribavirin monotherapy. *Hepatology*, **40** (Suppl. 1), 385A.

Macdonald, A., Crowder, K., Street, A., McCormick, C., Saksela, K. and Harris, M., (2003), The hepatitis C virus non-structural NS5A protein inhibits activating protein-1 function by perturbing ras-ERK pathway signaling. *J Biol Chem*, **278**, 17775–17784.

Manns, M.P., McHutchison, J.G., Gordon, S.C., Rustgi, V.K., Shiffman, M., Reindollar, R., Goodman, Z.D., Koury, K., Ling, M. and Albrecht, J.K., (2001), Peginterferon alfa-2b plus ribavirin compared with interferon alfa-2b plus ribavirin for initial treatment of chronic hepatitis C: a randomised trial. *Lancet* **358**, 958–965.

Marcellin, P., Chang, T.T., Lim, S.G., Tong, M.J., Sievert, W., Shiffman, M.L., Jeffers, L., Goodman, Z., Wulfsohn, M.S., Xiong, S., Fry, J. and Brosgart, C.L., (2003), Adefovir dipivoxil for the treatment of hepatitis B e antigen-positive chronic hepatitis B. *N Engl J Med*, **348**, 808–816.

Martell, M., Esteban, J.I., Quer, J., Genesca, J., Weiner, A., Esteban, R., Guardia, J. and Gomez, J., (1992), Hepatitis C virus (HCV) circulates as a population of different but closely related genomes: quasispecies nature of HCV genome distribution. *J Virol*, **66**, 3225–3229.

McHutchison, J.G., Shiffman, M.L., Gordon, S.C., Lindsay, K.L., Morgan, T., Norkrans, G., Esteban-Mur, R., Poynard, T., Pockros, P.J., Albrecht, J.K. and Brass, C., (2006), Sustained virologic response (SVR) to interferon-alpha-2b +/– ribavirin therapy at 6 months reliably predicts long-term clearance of HCV at 5-year follow-up. *J Hepatol*, **44** (Suppl. 2), S275.

Migliaccio, G., Tomassini, J.E., Carroll, S.S., Tomei, L., Altamura, S., Bhat, B., Bartholomew, L., Bosserman, M.R., Ceccacci, A., Colwell, L.F., Cortese, R., De Francesco, R., Eldrup, A.B., Getty, K.L., Hou, X.S., LaFemina, R.L., Ludmerer, S.W., MacCoss, M., McMasters, D.R., Stahlhut, M.W., Olsen, D.B., Hazuda, D.J. and Flores, O.A., (2003), Characterization of resistance to non-obligate chain-terminating ribonucleoside analogs that inhibit hepatitis C virus replication in vitro. *J Biol Chem*, **278**, 49164–49170.

Mo, H., Lu, L., Pilot-Matias, T., Pithawalla, R., Mondal, R., Masse, S., Dekhtyar, T., Ng, T., Koev, G., Stoll, V., Stewart, K.D., Pratt, J., Donner, P., Rockway, T., Maring, C. and Molla, A., (2005), Mutations conferring resistance to a hepatitis C virus (HCV) RNA-dependent RNA polymerase inhibitor alone or in combination with an HCV serine protease inhibitor in vitro. *Antimicrob Agents Chemother*, **49**, 4305–4314.

Murakami, E., Bao, H., Ramesh, M., McBrayer, T.R., Whitaker, T., Micolochick Steuer, H.M., Schinazi, R.F., Stuyver, L.J., Obikhod, A., Otto, M.J. and Furman, P.A., (2006), Mechanism of activation of beta-D-2'-deoxy-2'-fluoro-2'-C-methylcytidine and inhibition of hepatitis C virus NS5B RNA polymerase. *Antimicrob Agents Chemother*, **51**, 503–509.

Neumann, A.U., Lam, N.P., Dahari, H., Gretch, D.R., Wiley, T.E., Layden, T.J. and Perelson, A.S., (1998), Hepatitis C viral dynamics in vivo and the antiviral efficacy of interferon-alpha therapy. *Science*, **282**, 103–107.

Niesters, H.G., De Man, R.A., Pas, S.D., Fries, E. and Osterhaus, A.D., (2002), Identification of a new variant in the YMDD motif of the hepatitis B virus polymerase gene selected during lamivudine therapy. *J Med Microbiol*, **51**, 695–699.

Nowak, M.A., Bonhoeffer, S., Hill, A.M., Boehme, R., Thomas, H.C. and McDade, H., (1996), Viral dynamics in hepatitis B virus infection. *Proc Natl Acad Sci USA*, **93**, 4398–4402.

O'Brien, C., Godofsky, E., Rodriguez-Torres, M., Afdhal, N.H., Pappas, S.C., Pockros, P., Lawitz, E., Bzowej, N., Rustgi, V., Sulkowski, M.S., Sherman, K.E., Jacobson, I.M., Chao, G., Knox, S., Pietropaolo, K. and Brown, N.A., (2005), Randomized trial of valopicitabine (NM283), alone or with peginterferon, *vs* retreatment with peginterferon plus ribavirin in hepatitis C patients with previous non response to peginterferon-ribavirin: first interim results. *Hepatology*, **42** (Suppl. 1), 234A.

Ono, S.K., Kato, N., Shiratori, Y., Kato, J., Goto, T., Schinazi, R.F., Carrilho, F.J. and Omata, M., (2001), The polymerase L528M mutation cooperates with nucleotide binding-site mutations, increasing hepatitis B virus replication and drug resistance. *J Clin Investig*, **107**, 449–455.

Osiowy, C., Villeneuve, J.P., Heathcote, E.J., Giles, E. and Borlang, J., (2006), Detection of rtN236T and rtA181V/T mutations associated with resistance to adefovir dipivoxil in samples from patients with chronic hepatitis B virus infection by the INNO-LiPA HBV DR line probe assay (version 2). *J Clin Microbiol*, **44**, 1994–1997.

Pallier, C., Castera, L., Soulier, A., Hezode, C., Nordmann, P., Dhumeaux, D. and Pawlotsky, J.M., (2006a), Dynamics of hepatitis B virus resistance to lamivudine. *J Virol*, **80**, 643–653.

Pallier, C., Rodriguez, C., Brillet, R. and Pawlotsky, J.M., (2006b), Dynamics of hepatitis B virus resistance to adefovir dipivoxil unraveled by a thorough quasispecies analysis. *Hepatology*, **44** (Suppl. 2), 565A.

Papatheodoridis, G.V., Dimou, E., Laras, A., Papadimitropoulos, V. and Hadziyannis, S.J., (2002), Course of virologic breakthroughs under long-term lamivudine in HBeAg-negative precore mutant HBV liver disease. *Hepatology*, **36**, 219–226.

Pawlotsky, J.M., (2002), Molecular diagnosis of viral hepatitis. *Gastroenterology*, **122**, 1554–1568.

Pawlotsky, J.M., (2003), Mechanisms of antiviral treatment efficacy and failure in chronic hepatitis C. *Antiviral Res*, **59**, 1–11.

Pawlotsky, J.M., (2005a), Current and future concepts in hepatitis C therapy. *Semin Liver Dis*, **25**, 72–83.

Pawlotsky, J.M., (2005b), The concept of hepatitis B virus mutant escape. *J Clin Virol*, **34** (Suppl. 1), S125–S129.

Pawlotsky, J.M., (2006a), Hepatitis C virus population dynamics during infection. *Curr Top Microbiol Immunol*, **299**, 261–284.

Pawlotsky, J.M., (2006b), Therapy of hepatitis C: from empiricism to eradication. *Hepatology*, **43**, S207–S220.

Pawlotsky, J.M., Dahari, H., Neumann, A.U., Hezode, C., Germanidis, G., Lonjon, I., Castera, L. and Dhumeaux, D., (2004), Antiviral action of ribavirin in chronic hepatitis C. *Gastroenterology*, **126**, 703–714.

Pawlotsky, J.M. and Gish, R.G., (2006), Future therapies for hepatitis C. *Antivir Ther*, **11**, 397–408.

Pawlotsky, J.M., Hezode, C., Pellegrin, B., Soulier, A., von Wagner, M., Brouwer, J.T., Missale, G., Germanidis, G., Lurie, Y., Negro, F., Esteban, J., Hellstrand, K., Ferrari, C., Zeuzem, S., Schalm, S.W. and Neumann, A.U., (2002), Early HCV genotype 4 replication kinetics during treatment with peginterferon alpha-2a (Pegasys)-ribavirin combination: a comparison with HCV genotypes 1 and 3 kinetics. *Hepatology*, **36**, 291A.

Pawlotsky, J.M. and McHutchison, J.G., (2004), Hepatitis C. Development of new drugs and clinical trials: promises and pitfalls. Summary of an AASLD hepatitis single topic conference, Chicago, IL, 27 February–1 March 2003. *Hepatology*, **39**, 554–567.

Penin, F., (2003), Structural biology of hepatitis C virus. *Clin Liver Dis*, **7**, 1–21.

Peters, M., (1996), Actions of cytokines on the immune response and viral interactions: an overview. *Hepatology*, **23**, 909–916.

Polyak, S.J., Khabar, K.S., Paschal, D.M., Ezelle, H.J., Duverlie, G., Barber, G.N., Levy, D.E., Mukaida, N. and Gretch, D.R., (2001a), Hepatitis C virus nonstructural 5A protein induces interleukin-8, leading to partial inhibition of the interferon-induced antiviral response. *J Virol*, **75**, 6095–6106.

Polyak, S.J., Khabar, K.S., Rezeiq, M. and Gretch, D.R., (2001b), Elevated levels of interleukin-8 in serum are associated with hepatitis C virus infection and resistance to interferon therapy. *J Virol*, **75**, 6209–6211.

Poynard, T., Imbert-Bismut, F., Munteanu, M., Messous, D., Myers, R.P., Thabut, D., Ratziu, V., Mercadier, A., Benhamou, Y. and Hainque, B., (2004), Overview of the diagnostic value of biochemical markers of liver fibrosis (FibroTest, HCV FibroSure) and necrosis (ActiTest) in patients with chronic hepatitis C. *Comp Hepatol*, **3**, 8.

Poynard, T., Imbert-Bismut, F., Munteanu, M. and Ratziu, V., (2005), FibroTest-FibroSURE: towards a universal biomarker of liver fibrosis? *Expert Rev Mol Diagn*, **5**, 15–21.

Reesink, H., Forestier, N., Weegink, C., Zeuzem, S., McNair, L., Purdy, M., Chu, H.M. and Jansen, P.L.M., (2006a), Initial results of a 14-day study of the hepatitis C virus inhibitor protease VX-950 in combination with peginterferon alpha-2a. *J Hepatol*, **44** (Suppl. 2), S272.

Reesink, H.W., Zeuzem, S., Weegink, C.J., Forestier, N., van Vliet, A., van de Wetering de Rooij, J., McNair, L., Purdy, S., Kauffman, R., Alam, J. and Jansen, P.L., (2006b), Rapid decline of viral

RNA in hepatitis C patients treated with VX-950: a phase Ib, placebo-controlled, randomized study. *Gastroenterology*, **131**, 997–1002.

Reiser, M., Hinrichsen, H., Benhamou, Y., Reesink, H.W., Wedemeyer, H., Avendano, C., Riba, N., Yong, C.L., Nehmiz, G. and Steinmann, G.G., (2005), Antiviral efficacy of NS3-serine protease inhibitor BILN-2061 in patients with chronic genotype 2 and 3 hepatitis C. *Hepatology*, **41**, 832–835.

Richman, D.D., (1996), The implications of drug resistance for strategies of combination antiviral chemotherapy. *Antiviral Res*, **29**, 31–33.

Richman, D.D., (2000), The impact of drug resistance on the effectiveness of chemotherapy for chronic hepatitis B. *Hepatology*, **32**, 866–867.

Richman, D.D., (2006), Contending with HBV resistance: lessons learned from HIV resistance and its management. Clinical Care Options, LLC, Atlanta, Georgia.

Rodriguez-Torres, M., Lawitz, E., Muir, A.J., Keane, J., Kieffer, T.L., McNair, L. and McHutchison, J., (2006), Current status of subjects receiving peginterferon alfa-2a (PEG-IFN) and ribavirin (RBV) follow-on therapy after 28-day treatment with the hepatitis C protease inhibitor telaprevir (VX-950), PEG-IFN and RBV. *Hepatology*, **44** (Suppl. 2), 532A.

Roque-Afonso, A.M., Ferey, M.P., Mackiewicz, V., Fki, L. and Dussaix, E., (2003), Monitoring the emergence of hepatitis B virus polymerase gene variants during lamivudine therapy in human immunodeficiency virus coinfected patients: performance of CLIP sequencing and line probe assay. *Antivir Ther*, **8**, 627–634.

Sarrazin, C., Kieffer, T.L., Bartels, D., Hanzelka, B., Muh, U., Welker, M., Wincheringer, D., Zhou, Y., Chu, H.M., Lin, C., Weegink, C., Reesink, H., Zeuzem, S. and Kwong, A.D., (2007a), Dynamic HCV genotypic and phenotypic changes in patients treated with the protease inhibitor telaprevir (VX-950). *Gastroenterology*, **132**, 1767–1777.

Sarrazin, C., Rouzier, R., Wagner, F., Forestier, N., Larrey, D., Gupta, S., Hussain, M., Shah, A., Cutler, D., Zhang, J. and Zeuzem, S., (2007b), SCH 503034, a novel hepatitis C virus protease inhibitor, plus pegylated interferon-a2b for genotype 1 non-responders. *Gastroenterology*, **132**, 1270–1278.

Schildgen, O., Sirma, H., Funk, A., Olotu, C., Wend, U.C., Hartmann, H., Helm, M., Rockstroh, J.K., Willems, W.R., Will, H. and Gerlich, W.H., (2006), Variant of hepatitis B virus with primary resistance to adefovir. *N Engl J Med*, **354**, 1807–1812.

Seigneres, B., Pichoud, C., Martin, P., Furman, P., Trepo, C. and Zoulim, F., (2002), Inhibitory activity of dioxolane purine analogs on wild-type and lamivudine-resistant mutants of hepadnaviruses. *Hepatology*, **36**, 710–722.

Sertoz, R.Y., Erensoy, S., Pas, S., Akarca, U.S., Ozgenc, F., Yamazhan, T., Ozacar, T. and Niesters, H.G., (2005), Comparison of sequence analysis and INNO-LiPA HBV DR line probe assay in patients with chronic hepatitis B. *J Chemother*, **17**, 514–520.

Severini, A., Liu, X.Y., Wilson, J.S. and Tyrrell, D.L., (1995), Mechanism of inhibition of duck hepatitis B virus polymerase by (–)-beta-L-2',3'-dideoxy-3'-thiacytidine. *Antimicrob Agents Chemother*, **39**, 1430–1435.

Shaw, T., Bartholomeusz, A. and Locarnini, S., (2006), HBV drug resistance: mechanisms, detection and interpretation. *J Hepatol*, **44**, 593–606.

Sheldon, J., Camino, N., Rodes, B., Bartholomeusz, A., Kuiper, M., Tacke, F., Nunez, M., Mauss, S., Lutz, T., Klausen, G., Locarnini, S. and Soriano, V., (2005), Selection of hepatitis B virus polymerase mutations in HIV-coinfected patients treated with tenofovir. *Antivir Ther*, **10**, 727–734.

Sherman, M., Yurdaydin, C., Sollano, J., Silva, M., Liaw, Y.F., Cianciara, J., Boron-Kaczmarska, A., Martin, P., Goodman, Z., Colonno, R., Cross, A., Denisky, G., Kreter, B. and Hindes, R., (2006), Entecavir for treatment of lamivudine-refractory, HBeAg-positive chronic hepatitis B. *Gastroenterology*, **130**, 2039–2049.

Simmonds, P., Bukh, J., Combet, C., Deleage, G., Enomoto, N., Feinstone, S., Halfon, P., Inchauspe, G., Kuiken, C., Maertens, G., Mizokami, M., Murphy, D.G., Okamoto, H., Pawlotsky, J.M., Penin, F., Sablon, E., Shin, I.T., Stuyver, L.J., Thiel, H.J., Viazov, S., Weiner, A.J. and Widell, A., (2005), Consensus proposals for a unified system of nomenclature of hepatitis C virus genotypes. *Hepatology*, **42**, 962–973.

Stuyver, L.J., Locarnini, S.A., Lok, A., Richman, D.D., Carman, W.F., Dienstag, J.L. and Schinazi, R.F., (2001), Nomenclature for antiviral-resistant human hepatitis B virus mutations in the polymerase region. *Hepatology*, **33**, 751–757.

Stuyver, L.J., McBrayer, T.R., Tharnish, P.M., Clark, J., Hollecker, L., Lostia, S., Nachman, T., Grier, J., Bennett, M.A., Xie, M.Y., Schinazi, R.F., Morrey, J.D., Julander, J.L., Furman, P.A. and Otto, M.J., (2006), Inhibition of hepatitis C replicon RNA synthesis by beta-D-2'-deoxy-2'-fluoro-2'-C-methylcytidine: a specific inhibitor of hepatitis C virus replication. *Antivir Chem Chemother*, **17**, 79–87.

Sumpter, R., Jr., Loo, Y.M., Foy, E., Li, K., Yoneyama, M., Fujita, T., Lemon, S.M. and Gale, M., Jr., (2005), Regulating intracellular antiviral defense and permissiveness to hepatitis C virus RNA replication through a cellular RNA helicase, RIG-I. *J Virol*, **79**, 2689–2699.

Tan, S.L., Nakao, H., He, Y., Vijaysri, S., Neddermann, P., Jacobs, B.L., Mayer, B.J. and Katze, M.G., (1999), NS5A, a nonstructural protein of hepatitis C virus, binds growth factor receptor-bound protein 2 adaptor protein in a Src homology 3 domain/ligand-dependent manner and perturbs mitogenic signaling. *Proc Natl Acad Sci USA*, **96**, 5533–5538.

Taylor, D.R., Shi, S.T., Romano, P.R., Barber, G.N. and Lai, M.M., (1999), Inhibition of the interferon-inducible protein kinase PKR by HCV E2 protein. *Science*, **285**, 107–110.

Tenney, D.J., Levine, S.M., Rose, R.E., Walsh, A.W., Weinheimer, S.P., Discotto, L., Plym, M., Pokornowski, K., Yu, C.F., Angus, P., Ayres, A., Bartholomeusz, A., Sievert, W., Thompson, G., Warner, N., Locarnini, S. and Colonno, R.J., (2004), Clinical emergence of entecavir-resistant hepatitis B virus requires additional substitutions in virus already resistant to Lamivudine. *Antimicrob Agents Chemother*, **48**, 3498–3507.

Tenney, D.J., Rose, R.E., Baldick, C.J., Levine, S.M., Pokornowski, K.A., Walsh, A.W., Fang, J., Yu, C.F., Zhang, S., Mazzucco, C.M., Eggers, B., Hsu, M., Plym, M.J., Poundstone, P., Yang, J. and Colonno, R.J., (2007), Two-year entecavir resistance assessment in lamivudine refractory HBV patients reveals different clinical outcomes depending on the resistance substitutions present. *Antimicrob Agents Chemother*, **51**, 902–911.

Tilg, H., (1997), New insights into the mechanisms of interferon alfa: an immunoregulatory and anti-inflammatory cytokine. *Gastroenterology*, **112**, 1017–1021.

Tomassini, J.E., Getty, K., Stahlhut, M.W., Shim, S., Bhat, B., Eldrup, A.B., Prakash, T.P., Carroll, S.S., Flores, O., MacCoss, M., McMasters, D.R., Migliaccio, G. and Olsen, D.B., (2005), Inhibitory effect of 2'-substituted nucleosides on hepatitis C virus replication correlates with metabolic properties in replicon cells. *Antimicrob Agents Chemother*, **49**, 2050–2058.

Tomei, L., Altamura, S., Bartholomew, L., Biroccio, A., Ceccacci, A., Pacini, L., Narjes, F., Gennari, N., Bisbocci, M., Incitti, I., Orsatti, L., Harper, S., Stansfield, I., Rowley, M., De Francesco, R. and Migliaccio, G., (2003), Mechanism of action and antiviral activity of benzimidazole-based allosteric inhibitors of the hepatitis C virus RNA-dependent RNA polymerase. *J Virol*, **77**, 13225–13231.

Tomei, L., Altamura, S., Bartholomew, L., Bisbocci, M., Bailey, C., Bosserman, M., Cellucci, A., Forte, E., Incitti, I., Orsatti, L., Koch, U., De Francesco, R., Olsen, D.B., Carroll, S.S. and Migliaccio, G., (2004), Characterization of the inhibition of hepatitis C virus RNA replication by nonnucleosides. *J Virol*, **78**, 938–946.

Tong, X., Chase, R., Skelton, A., Chen, T., Wright-Minogue, J. and Malcolm, B.A., (2006), Identification and analysis of fitness of resistance mutations against the HCV protease inhibitor SCH 503034. *Antiviral Res*, **70**, 28–38.

Tuttleman, J.S., Pourcel, C. and Summers, J., (1986), Formation of the pool of covalently closed circular viral DNA in hepadnavirus-infected cells. *Cell*, **47**, 451–460.

van Bommel, F., Wunsche, T., Mauss, S., Reinke, P., Bergk, A., Schurmann, D., Wiedenmann, B. and Berg, T., (2004), Comparison of adefovir and tenofovir in the treatment of lamivudine-resistant hepatitis B virus infection. *Hepatology*, **40**, 1421–1425.

Villeneuve, J.P., Durantel, D., Durantel, S., Westland, C., Xiong, S., Brosgart, C.L., Gibbs, C.S., Parvaz, P., Werle, B., Trepo, C. and Zoulim, F., (2003), Selection of a hepatitis B virus strain resistant to adefovir in a liver transplantation patient. *J Hepatol*, **39**, 1085–1089.

Villet, S., Pichoud, C., Villeneuve, J.P., Trepo, C. and Zoulim, F., (2006), Selection of a multiple drug-resistant hepatitis B virus strain in a liver-transplanted patient. *Gastroenterology*, **131**, 1253–1261.

von Hahn, T., Evans, M.J., Syder, A.J., Hatziioannou, T., McKeating, J.A., Bieniasz, P.D. and Rice, C.M., (2006), Identification of claudin-1 as an essential cellular cell entry factor for hepatitis C virus. *Hepatology*, **44** (Suppl. 2), 197A.

Wagoner, J., Austin, M., Green, J., Imaizumi, T., Casola, A., Brasier, A., Khabar, K.S., Wakita, T., Gale, M., Jr and Polyak, S.J., (2007), Regulation of CXCL-8 (interleukin-8) induction by double-stranded RNA signaling pathways during hepatitis C virus infection. *J Virol*, **81**, 309–318.

Walters, K.A., Tipples, G.A., Allen, M.I., Condreay, L.D., Addison, W.R. and Tyrrell, L., (2003), Generation of stable cell lines expressing Lamivudine-resistant hepatitis B virus for antiviral-compound screening. *Antimicrob Agents Chemother*, **47**, 1936–1942.

Westland, C., Delaney, W.t., Yang, H., Chen, S.S., Marcellin, P., Hadziyannis, S., Gish, R., Fry, J., Brosgart, C., Gibbs, C., Miller, M. and Xiong, S., (2003), Hepatitis B virus genotypes and virologic response in 694 patients in phase III studies of adefovir dipivoxil1. *Gastroenterology*, **125**, 107–116.

Whalley, S.A., Murray, J.M., Brown, D., Webster, G.J., Emery, V.C., Dusheiko, G.M. and Perelson, A.S., (2001), Kinetics of acute hepatitis B virus infection in humans. *J Exp Med*, **193**, 847–854.

Yamamoto, T., Litwin, S., Zhou, T., Zhu, Y., Condreay, L., Furman, P. and Mason, W.S., (2002), Mutations of the woodchuck hepatitis virus polymerase gene that confer resistance to lamivudine and 2'-fluoro-5-methyl-beta-L-arabinofuranosyluracil. *J Virol*, **76**, 1213–1223.

Yan, Y., Li, Y., Munshi, S., Sardana, V., Cole, J.L., Sardana, M., Steinkuehler, C., Tomei, L., De Francesco, R., Kuo, L.C. and Chen, Z., (1998), Complex of NS3 protease and NS4A peptide of BK strain hepatitis C virus: a 2.2 A resolution structure in a hexagonal crystal form. *Protein Sci*, **7**, 837–847.

Yang, H., Westland, C.E., Delaney, W.E.t., Heathcote, E.J., Ho, V., Fry, J., Brosgart, C., Gibbs, C.S., Miller, M.D. and Xiong, S., (2002), Resistance surveillance in chronic hepatitis B patients treated with adefovir dipivoxil for up to 60 weeks. *Hepatology*, **36**, 464–473.

Yeon, J.E., Yoo, W., Hong, S.P., Chang, Y.J., Yu, S.K., Kim, J.H., Seo, Y.S., Chung, H.J., Moon, M.S., Kim, S.O., Byun, K.S. and Lee, C.H., (2006), Resistance to adefovir dipivoxil (ADV) in lamivudine-resistant chronic hepatitis B patients treated with ADV. *Gut*, **55**, 1488–1495.

Yi, M., Tong, X., Skelton, A., Chase, R., Chen, T., Prongay, A., Bogen, S.L., Saksena, A.K., Njoroge, F.G., Veselenak, R.L., Pyles, R.B., Bourne, N., Malcolm, B.A. and Lemon, S.M., (2006), Mutations conferring resistance to SCH6, a novel hepatitis C virus NS3/4A protease inhibitor. Reduced RNA replication fitness and partial rescue by second-site mutations. *J Biol Chem*, **281**, 8205–8215.

Ying, C., De Clercq, E. and Neyts, J., (2000), Lamivudine, adefovir and tenofovir exhibit long-lasting anti-hepatitis B virus activity in cell culture. *J Viral Hepat*, **7**, 79–83.

Young, K.C., Lindsay, K.L., Lee, K.J., Liu, W.C., He, J.W., Milstein, S.L. and Lai, M.M., (2003), Identification of a ribavirin-resistant NS5B mutation of hepatitis C virus during ribavirin monotherapy. *Hepatology*, **38**, 869–878.

Yuen, M.F., Kato, T., Mizokami, M., Chan, A.O., Yuen, J.C., Yuan, H.J., Wong, D.K., Sum, S.M., Ng, I.O., Fan, S.T. and Lai, C.L., (2003), Clinical outcome and virologic profiles of severe hepatitis B exacerbation due to YMDD mutations. *J Hepatol*, **39**, 850–855.

Zoulim, F., (2004), Mechanism of viral persistence and resistance to nucleoside and nucleotide analogs in chronic hepatitis B virus infection. *Antiviral Res*, **64**, 1–15.

Zoulim, F., Dannaoui, E., Borel, C., Hantz, O., Lin, T.S., Liu, S.H., Trepo, C. and Cheng, Y.C., (1996), 2',3'-dideoxy-beta-L-5-fluorocytidine inhibits duck hepatitis B virus reverse transcription and suppresses viral DNA synthesis in hepatocytes, both in vitro and in vivo. *Antimicrob Agents Chemother*, **40**, 448–453.

Chapter 10
Resistance to Antifungal Agents

Beth A. Arthington-Skaggs and John H. Rex

10.1 Introduction

10.1.1 Incidence of Fungal Infections

The twenty-first century is an exciting time for antifungal therapy with the number of therapeutic agents available for the treatment of mycotic diseases greater than ever. Unfortunately, the number of life-threatening fungal infections is also on the rise (Golan, 2005; McNeil et al., 2001). Multiple factors have contributed to the growing incidence of invasive fungal infections such as an increasing number of patients with severe immunosuppression, including HIV-infected patients, cancer patients with chemotherapy-induced neutropenia, and organ transplant recipients who are receiving immunosuppressive therapy. Other contributing factors are associated more generally with severely ill patients and include the frequent use of increasingly more invasive medical procedures, treatment with broad-spectrum antibiotics and glucocorticoids, receipt of parenteral nutrition, and receipt of peritoneal dialysis or hemodialysis (Lortholary and Dupont, 1997).

Not all forms of immunosuppression and risk are equal and the incidence of invasive fungal infections varies by patient group. Among solid organ transplant recipients, the incidence varies from about 5% after renal transplantation to 35% in lung and heart–lung transplant recipients and up to 50% after small bowel transplants (Singh, 2004). Most of these infections are due to either *Candida* (35–91%) or *Aspergillus* (9–52%) species (Singh, 2004). Among neutropenic adults with acute leukemia, 4–8% will develop an invasive fungal infection whereas 16–18% of patients undergoing hematopoietic stem cell transplantation will develop invasive fungal disease (Menichetti, 2004; Wingard, 2004). In patients with AIDS, oropharyngeal candidiasis occurs in as many as 90% of patients at some time during the progression of their disease and cryptococcal meningitis, one of the most prevalent AIDS-defining infections, will develop in up to 10% of patients not receiving highly active antiretroviral therapy (HAART) (Leigh et al., 2004; Loeffler and Stevens, 2003). In light of this clinical picture, the therapeutic and prophylactic use of antifungal compounds has increased and likewise, so has the concern about emerging antifungal drug resistance (Brown, 2004; Lortholary and Dupont, 1997; Marr, 2004; Perfect, 2004a).

I. W. Fong and K. Drlica (eds.), *Antimicrobial Resistance and Implications for the Twenty-First Century.* © Springer 2008

10.1.2 Microbiological Resistance Versus Clinical Resistance

The term "resistance" can often be used to describe two distinctly different phenomena: (1) the relative insensitivity of a fungus to an antifungal drug as determined in vitro and compared with other isolates of the same species and (2) persistence of an infection despite adequate therapy (Loeffler and Stevens, 2003). For this review, microbiological resistance will refer to the former and clinical resistance will be used to describe the latter. Clinical resistance is often multifactorial with microbiological resistance being just one of several contributing factors. Other factors include impaired host immune function, insufficient access of agent to the infected site, accelerated metabolism of the drug, presence of contaminated implanted medical device, as well as other reasons (Rex et al., 1997). On the other hand, microbiological resistance is more straightforward and can be objectively and reproducibly measured in the laboratory (National Committee for Clinical Laboratory Standards, 2002a,b, 2004). Furthermore, molecular mechanisms associated with a resistant phenotype can be identified and characterized in individual fungal isolates (Ghannoum and Rice, 1999; Marichal and Vanden Bossche, 1995; Sanglard and Odds, 2002; Vanden Bossche et al., 1994; White et al., 1998).

10.2 Mechanisms of Resistance

10.2.1 Antifungal Drugs

The majority of clinically relevant antifungal agents used today can be placed into four distinct categories based on their fungal target. These targets include (1) ergosterol in the plasma membrane, (2) ergosterol synthesis, (3) fungal nucleic acid synthesis, and (4) β-1,3 glucan synthesis. Each class of antifungal is discussed below with regard to its mechanism of action, spectrum of activity, scope of resistance, and molecular mechanisms of resistance (Table 10.1).

10.2.2 Polyenes

The polyene class of antifungal drugs, of which amphotericin B and nystatin are the most commonly used, are natural products produced by *Streptomyces* spp. (Hamilton-Miller, 1973; Ryley et al., 1981). These compounds are relatively large and similar in size and shape to plasma membrane phospholipids. Polyene antibiotics exert their fungicidal effect by binding to sterol molecules in the plasma membrane of sensitive organisms causing an impairment of barrier function, leakage of cellular constituents, and ultimately cell death (Kerridge, 1986). Although sterols are major components in the plasma membranes of lower and higher eukaryotes alike, it is the increased affinity of the polyene antifungal agents

Table 10.1 Antifungal agents: spectrum of activity, principal modes of action, and mechanisms of fungal resistance.

Antifungal agent	Spectrum of activity	Mode of action	Mechanisms of resistance
Polyenes			
Amphotericin B, nystatin	Broad spectrum of activity against *Candida* spp. (except *C. lusitaniae*), *C. neoformans*, the dimorphic fungi, and the filamentous fungi (except *Aspergillus terreus*, *Fusarium* spp., *Scedosporium* spp., and *Trichosporon beigelii*)	Binding to ergosterol and disruption of cell membrane function	Alteration or decrease in membrane ergosterol
Pyrimidine analogs			
5-Fluorocytosine (5-FC)	Active against *Candida* spp., *C. neoformans*, and dematiaceous fungi causing chromoblastomycosis. Resistance can rapidly emerge when 5-FC is used as monotherapy	Impairment of nucleic acid and protein synthesis by formation of toxic fluorinated pyrimidine antimetabolites	Decreased uptake of 5-FC; decreased metabolism to form toxic antimetabolites
Azoles			
Fluconazole	Active against *Candida* spp. (except *C. krusei* and reduced activity against *C. glabrata*) and *C. neoformans*; no activity against filamentous fungi; active against *Coccidioides immitis* but limited activity against *Histoplasma capsulatum* and *Blastomyces dermatitidis*	Inhibition of cytochrome P-450 14α-lanosterol demethylase (encoded by *ERG11*) required for ergosterol biosynthesis	Enhanced drug efflux by membrane-bound efflux pumps; mutation(s) in *ERG11* leading to alterations of the azole–target interaction; overexpression of *ERG11*; alteration in other ergosterol biosynthetic genes
Voriconazole	Like fluconazole but enhanced activity against filamentous fungi, including *Fusarium* spp., *H. capsulatum*, *B. dermatitidis*, and dermatophytes; limited activity against zygomycetes		

(continued)

Table 10.1 (continued)

Antifungal agent	Spectrum of activity	Mode of action	Mechanisms of resistance
Ravuconazole	Like fluconazole but enhanced activity against filamentous fungi. Limited activity against *Fusarium* spp. and zygomycetes		
Itraconazole	Like fluconazole but enhanced activity against filamentous fungi, *H. capsulatum, B. dermatitidis,* and dermatophytes; limited activity against *Fusarium* spp. and zygomycetes		
Posaconazole	Like itraconazole but enhanced activity against *Fusarium* spp.		
Allylamines			
Terbinafine	Active against most dermatophytes; limited activity against *Candida* spp.	Inhibition of squalene epoxidase (encoded by *ERG1*) required for ergosterol biosynthesis	Alteration of *ERG1* or factor essential for its activity
Echinocandins			
Caspofungin, micafungin, anidulafungin	Active against most *Candida* spp. (higher MICs observed among *C. parapsilosis, C. krusei, C. guilliermondii,* and *C. lusitaniae*) and *Aspergillus* spp.; no activity against *C. neoformans, Fusarium* spp., or zygomycetes	Inhibition of β-1,3 glucan synthase required for cell wall biosynthesis	Reduced echinocandin susceptibility of β-1,3 glucan synthase activity

for ergosterol-containing fungal membranes relative to cholesterol-containing mammalian membranes that mediates the specificity of these drugs and enables their therapeutic use (Archer and Gale, 1975). A less well-characterized mechanism of action of amphotericin B, mediated via the host's immune system, involves a stimulatory effect of low concentrations of this drug on macrophages (Martin et al., 1994b). Unfortunately, the severe side effects of fever and chills associated with amphotericin B treatment may also be mediated through this mechanism of amphotericin B action. Recent reports have suggested that amphotericin B exposure causes increased production of the pro-inflammatory cytokines, interleukin-1β (IL1-β) and tumor necrosis factor-α (TNFα), known to mediate many adverse pathophysiological events (Medoff et al., 1983; Perfect et al., 1987; Vonk et al., 1998). Results of an in vitro study demonstrated that amphotericin B slightly increased the production of these pro-inflammatory cytokines by human mononuclear cells whereas the production of the anti-inflammatory cytokine, IL1-receptor antagonist, was significantly inhibited (Vonk et al., 1998). The net result is a shift towards pro-inflammatory cytokine production.

Despite the specificity of polyene antifungals for fungal membranes, these compounds are also associated with significant nephrotoxicity when administered intravenously (Brezis et al., 1984). However, because of poor oral absorption, amphotericin B is limited to systemic administration. In an effort to decrease the toxicity of amphotericin B therapy, a variety of lipid-associated formulations have been developed that package amphotericin B in liposomes (AmBisome) or in ribbon-like (Abelcet) and disc-like (Amphotec and Amphocil) lipid complexes (Dupont, 2002). Lipid-associated preparations of amphotericin B presumably are delivered to the kidneys at a slower rate compared to conventional amphotericin B resulting in reduced nephrotoxicity and increased tolerable doses (Odds et al., 2003). Decreased production of IL1-β and TNFα following treatment with lipid-associated amphotericin B compared to conventional amphotericin B may also explain the observed decrease in host toxicity. In a murine model, investigators showed that treatment with conventional amphotericin B caused a significantly greater production of IL1-β and TNFα than treatment with a lipid formulation of amphotericin B, suggesting that the superior host tolerance for lipid-associated amphotericin B might derive from the reduced expression of these pro-inflammatory cytokines (Martin et al., 1994b). While nystatin is most commonly used as oral and topical therapy for superficial mycoses (Meade, 1979), liposomal formulations of nystatin have been studied for systemic use. Although these formulations have been shown to be associated with reduced side effects while maintaining antifungal activity in vitro and in vivo (Denning and Warn, 1999; Groll et al., 1999; Mehta et al., 1987), they do not appear to be high-priority candidates for further development.

Despite these limitations, the polyene antifungals have been a mainstay in the antifungal armamentarium since the 1960s and amphotericin B is still considered one of the more reliable options for treatment of severe invasive disease. This is primarily due to broad-spectrum fungicidal activity against both yeasts and moulds coupled with a low incidence of resistance in clinical isolates (Odds et al., 2003). Although resistance to polyene antifungals is rare, mechanisms associated with decreased susceptibility have been studied and are almost always associated with an absence or decrease of membrane ergosterol, in favor of non-ergosterol sterols (Hamilton-Miller, 1973; Kelly et al., 1994; Kim et al., 1974; Vanden Bossche et al., 1994). An exception to this was reported by Capek et al. (1970), who isolated nystatin-resistant mutants of *Trichophyton mentagrophytes*, *Trichophyton rubrum*, and *Microsporum gypseum* shown to produce an inducible enzyme that degrades nystatin.

The incidence of primary or intrinsic resistance to amphotericin B is relatively limited and includes yeasts such as *Trichosporon beigelii*, *Candida lusitaniae*, *and Candida guilliermondii*, as well as filamentous fungi such as *Aspergillus terreus*, *Scedosporium apiospermum*, *Malassezia furfur*, and *Fusarium* spp. (Berenguer et al., 1997; Blumberg and Reboli, 1996; Boutati and Anaissie, 1997; Dick et al., 1985; Powderly et al., 1988; Tritz and Woods, 1993; Viudes et al., 2002). Secondary, or acquired, resistance to amphotericin B during or following amphotericin B therapy appears to be uncommon as "breakthrough" candidemias in patients treated with amphotericin B have not been noted in all surveys (Hoban et al., 1999; Pfaller et al., 1995b). Isolated cases of secondary resistance to amphotericin B have been described in *C. albicans*, *C. glabrata*, *C. krusei*, *C. tropicalis*, and *C. parapsilosis* isolates causing disseminated infections in patients with cancer, patients undergoing myelosuppressive chemotherapy and/or bone marrow transplantation, and neonates receiving amphotericin B therapy (Blumberg and Reboli, 1996; Dick et al., 1980, 1985; Ellis, 2002; Fan-Havard et al., 1991; Kovacicova et al., 2001; Nolte et al., 1997; Powderly et al., 1988; Safe et al., 1977; Zaoutis et al., 2005).

A recent publication suggested that amphotericin B resistance among *C. parapsilosis* isolates causing candidemia in children may represent an emerging threat. Authors reported that nearly 20% of *C. parapsilosis* isolates obtained from pediatric patients in four US children's hospitals were resistant to amphotericin B, as defined by a minimum inhibition concentration (MIC) $\geq$ 2 µg/ml (Zaoutis et al., 2005). Another report described a case of amphotericin B-resistant candidemia caused by *C. parapsilosis* in a pediatric patient following neurosurgery for a brain tumor (Kovacicova et al., 2001). A study that investigated phenotypic differences among clinical isolates of *C. parapsilosis* from the three genetically distinct subgroups, I, II, and III, found no difference in amphotericin B susceptibility among isolates from any of the three groups (Lin et al., 1995).

Resistance to amphotericin B has also been associated with the formation of *Candida* biofilms, as for example on indwelling catheters (Hawser and Douglas, 1994). *C. albicans* as well as *C. parapsilosis* isolates have been shown to form prominent biofilms in in vitro model systems (Kojic and Darouiche, 2003) and amphotericin B MICs for biofilm-grown cells can be up to 100× higher than the same isolates grown as planktonic cells (Mukherjee and Chandra, 2004). The exact mechanism(s) underlying increased resistance of *Candida* biofilms is unknown but may be related to insufficient penetration of the drug into the biofilm due to the production of extracelluar matrix material (Baillie and Douglas, 2000) and/or reduced growth rate of cells within a mature biofilm (Baillie and Douglas, 1998).

Yet another mechanism of amphotericin B resistance has been described in a study of two *Cryptococcus neoformans* isolates obtained from a patient with AIDS before and after treatment failure with amphotericin B. Results demonstrated that the resistant strain had a defect in the ergosterol biosynthetic enzyme, Δ8-7 sterol isomerase, leading to a depletion of ergosterol and a concomitant increase in non-ergosterol sterols (Kelly et al., 1994).

Fungal resistance to amphotericin B following previous azole antifungal treatment has been described in vitro (Cosgrove et al., 1978; Johnson et al., 2000; Martin et al., 1994a) and in vivo (Polak et al., 1982; Schaffner and Bohler, 1993; Schaffner and Frick, 1985). The basis for antagonism is thought to be depletion of membrane ergosterol as a result of azole-induced inhibition of ergosterol biosynthesis. However, antagonism between these two agents may not be universal, but instead limited to (1) certain fungal–drug combinations, (2) specific models of testing, and (3) the timing of the interaction between the compounds. For example, in vitro studies of combination amphotericin B and fluconazole, ketoconazole, miconazole, or clotrimazole therapy demonstrated antagonism for *Candida* spp. and *Aspergillus fumigatus* (Cosgrove et al., 1978; Johnson et al., 2000; Martin et al., 1994a). In contrast, in vivo studies using both immunocompetent and immunosuppressed mice or rats demonstrated polyene–azole antagonism in cases of invasive aspergillosis (Polak et al., 1982; Schaffner and Bohler, 1993; Schaffner and Frick, 1985) whereas indifference or synergy was observed when the agents were combined to treat invasive candidiasis or meningeal cryptococcosis (Polak et al., 1982; Schmitt et al., 1991; Sugar et al., 1995). Finally, a clinical study of the combination of amphotericin B and fluconazole as therapy for candidemia failed to demonstrate antagonism when the two agents were given simultaneously (Rex et al., 1995b). In broad terms, antagonism seems easiest to show when azole therapy precedes polyene therapy. While the majority of amphotericin B resistance can be predicted based on the species of the organism causing disease, these examples highlight the need to consider the possibility of acquired amphotericin B resistance when managing patients infected with these fungal species (Collin et al., 1999; Rex et al., 2000).

10.2.3 Flucytosine (5-FC)

Flucytosine, also know as 5-FC, is a synthetic molecule with fungistatic activity against *Candida* spp., *C. neoformans*, and some dematiaceous fungi. The antifungal action of flucytosine is based on the disturbance of protein synthesis and/or DNA synthesis. Inhibition of protein synthesis occurs when uracil in fungal RNA is replaced by 5-fluorouracil. Flucytosine-induced inhibition of DNA synthesis occurs via inhibition of thymidylate synthase, an enzyme responsible for the de novo synthesis of thymidine required for DNA synthesis (Scholer, 1980). To be susceptible to flucytosine, target cells must possess cytosine permease to internalize flucytosine, cytosine deaminase to convert it to 5-fluorouracil, and uridine monophosphate pyrophosphorylase to convert 5-fluorouracil into 5-fluorouridylate which becomes a substrate for RNA synthesis.

Resistance to flucytosine is produced by lack or deficiency of an enzyme at any step of the pathway or by a surplus of de novo synthesis of normal compounds competing (again at any step) with the fluorinated antimetabolites (Polak and Scholer,

1975; Whelan, 1987). In *C. albicans*, the most common enzyme defect associated with resistance is a deficiency in the activity of uridine monophosphate pyrophosphorylase (Whelan and Kerridge, 1984). Most filamentous fungi naturally lack the enzymes necessary to internalize and metabolize flucytosine, explaining the absence of activity against these organisms.

Despite its limited spectrum of activity, flucytosine offers the advantages of being well tolerated, available in both oral and parenteral formulations, and providing good oral absorption and tissue distribution. Unfortunately, the use of flucytosine for primary therapy is restricted by the high prevalence of resistance among clinically important yeast species and by the speed with which yeast isolates develop resistance during treatment (Sanglard and Odds, 2002). A review by Scholer in 1980 (Scholer, 1980) showed primary flucytosine resistance among *Candida* spp. isolates from United States, Europe, Africa, and the Middle East ranging from 7.9% for *C. albicans* to 20.7% for *Candida* non-*albicans* spp. These data may be compromised by the absence, at that time, of a standardized method for susceptibility testing and interpretive MIC breakpoints. Surveys of flucytosine resistance based on the NCCLS M27-A2 reference method for susceptibility testing of *Candida* spp. and *C. neoformans* do not corroborate the impression of high rates of resistance. Recent surveys conducted in North America have shown flucytosine resistance ranging from 3–4.3% for *C. albicans* (Hajjeh et al., 2004; Ostrosky-Zeichner et al., 2003; St-Germain et al., 2001) to 2–3.5% for *Candida* non-*albicans* (Hajjeh et al., 2004; Ostrosky-Zeichner et al., 2003; St-Germain et al., 2001). European surveys show flucytosine resistance in *C. albicans* to be only 0–0.2% (Barchiesi et al., 2000a; Cuenca-Estrella et al., 2001). It remains unclear whether the drop in prevalence of flucytosine resistance is due to biological reasons occurring in the organism or to the implementation of a standardized method for in vitro susceptibility testing. However, flucytosine use in the clinic remains limited to adjunct therapy and specifically in combination with amphotericin B for the treatment of cryptococcal meningitis where combination therapy has been shown to have superior efficacy over amphotericin B monotherapy (Saag et al., 2000). The combined use of these antifungal drugs suppresses the development of flucytosine resistance and lowers the required dose of amphotericin B.

10.2.4 Ergosterol Biosynthesis Inhibitors

The ergosterol biosynthesis inhibitors include both the azole derivative and the allylamine antifungals. The azoles are by far the largest class of antifungal agents in clinical use (reviewed in Sheehan et al., 1999). The first azole antifungal, chlormidazole, became available for topical use in 1958 (Herrling et al., 1959), long before the mechanism of action was understood. Nearly five decades later, the azole molecule continues to be modified to increase potency and spectrum of antifungal activity. The azole antifungals currently in clinical use contain either two or three nitrogens in the azole ring and are thereby classified as imidazoles (e.g. clotrimazole, miconazole, econazole, ketoconazole) or triazoles (e.g. fluconazole, itraconazole, voriconazole, posaconazole, ravuconazole), respectively. With the exception of ketoconazole, the imidazoles are restricted to the treatment of superficial mycoses whereas the triazoles are used to treat both superficial and systemic infections (Sheehan et al., 1999).

The fungistatic action of azoles in susceptible cells is produced by inhibition of ergosterol biosynthesis, the major sterol component of fungal plasma membranes. Azoles accomplish this by blocking the activity of a key ergosterol biosynthetic enzyme, cytochrome P-450-dependent 14α-lanosterol demethylase (encoded by *ERG11*), which catalyses the oxidative removal of the 14α-methyl group of lanosterol and/or eburicol in fungi (Vanden Bossche et al., 1995). It is the increased affinity of the azole molecule for the fungal P-450 over mammalian P-450 enzymes that creates the antifungal specificity of the azoles (Vanden Bossche et al., 1995). Just as fungal and mammalian P-450 enzymes differ in conformation, P-450 enzymes differ among fungal species. It is the precise nature of the interaction between a specific azole molecule and a specific P-450 that determines the extent of azole activity in different fungal species and thus accounts for the varying degrees of susceptibility observed for specific azole–microbe combinations (Vanden Bossche et al., 1995). This was further demonstrated in a study of three-dimensional models of wild-type and mutated forms of cytochrome P-450 14α-demethylase from *A. fumigatus* and *C. albicans* which revealed distinct differences in the way specific azole antifungals bind to specific P-450 enzymes. Results indicated that fluconazole and voriconazole binding were impacted most dramatically by mutation of the P-450 enzyme in the heme-binding domain whereas itraconazole and posaconazole utilized their long side chains, unique to these compounds, to occupy a specific channel of the P-450 enzyme. Therefore, itraconazole and posaconazole possessed increased binding affinity for P-450 enzymes mutated in the heme-binding domain and thus more antifungal activity relative to fluconazole and voriconazole (Xiao et al., 2004). Based on these data, investigators predicted that mutations in the P-450 enzyme that disrupt binding of the long side chains would result in an enzyme with decreased susceptibility specifically to itraconazole and posaconazole (Xiao et al., 2004).

Exposure of fungal cells to azole antifungals causes depletion of ergosterol and accumulation of 14α-methylated sterols in fungal membranes (Kelly et al., 1995). The substitution of non-ergosterol sterols disrupts the normal permeability and fluidity of the plasma membrane and leads to secondary consequences for membrane-bound enzymes involved in nutrient transport and cell wall biosynthesis (Georgopapadakou and Walsh, 1996; Marichal et al., 1985). In addition to disruption of membrane function, azole-induced accumulation of the 14α-methyl sterol intermediate, 14α-methylergosta-8,24(28)-dien-3β,6α-diol, or more commonly 14α-methyl-3,6-diol has been shown to be toxic for *Saccharomyces cerevisiae* and *C. albicans* (Joseph-Horne et al., 1995b; Kelly, 1993; Kelly et al., 1995).

To date, the majority of studies on azole resistance have focused on *C. albicans* and a number of detailed investigations and reviews have been published (Denning et al., 1997; Ghannoum and Rice, 1999; Rex et al., 1995c; Venkateswarlu et al., 1997; Wheat et al., 1997; White et al., 1998). Molecular mechanisms contributing to azole resistance have been well described in *C. albicans* isolates causing oropharyngeal candidiasis in human immunodeficiency virus-infected patients (Albertson et al., 1996; Clark et al., 1996; de Micheli et al., 2002; Fling et al., 1991; Ghannoum and Rice, 1999; Henry et al., 2000; Hernaez et al., 1998; Hitchcock, 1993; Kakeya et al., 2000; Kohli et al., 2002; Loeffler et al., 2000; Lopez-Ribot et al., 1998, 1999; Mago and Khuller, 1989; Michaelis and Berkower, 1995; Perea et al., 2001; Prasad et al., 1995; Sanglard et al., 1995, 1997, 1998; White, 1997; White et al., 1997, 1998, 2002) with a smaller number of studies focused on isolates causing disseminated disease in non-HIV-infected patients, particularly those receiving chemotherapy for treatment of hematological disorders (Marr et al., 1997, 1998; Mori et al., 1997; Nolte et al., 1997). Molecular mechanisms of azole resistance have also been studied in *Candida* species other than *C. albicans*, including, *C. glabrata* (Bennett et al., 2004; Borst et al., 2005; Geber et al., 1995; Hitchcock et al., 1993; Sanglard et al., 1999), *C. dubliniensis* (Moran et al., 1998; Perea et al., 2002), *C. krusei* (Fukuoka et al., 2003; Katiyar and Edlind, 2001; Orozco et al., 1998; Venkateswarlu et al., 1996), *C. tropicalis* (Barchiesi et al., 2000b), as well as *C. neoformans* (Joseph-Horne et al., 1995a; Lamb et al., 1995; Posteraro et al., 2003; Venkateswarlu et al., 1997) and *A. fumigatus* (da Silva Ferreira et al., 2004; Denning et al., 1997; Nascimento et al., 2003) (Table 10.2).

Overall, the molecular mechanisms of azole resistance fall into four general categories. First, alteration in the quality or quantity of the target enzyme, cytochrome P-450 lanosterol demethylase, either by overexpression or by point mutations in the

Table 10.2 Molecular mechanisms of azole resistance in non-*albicans Candida* spp., *Cryptococcus* neoformans, and *Aspergillus fumigatus*.

Organism	Mechanism of resistance	Ref.
Candida glabrata	Drug efflux mediated by Cg*CDR* and *PDH1*; reduced permeability of drug; *ERG3* mutation	Bennett et al. (2004), Geber et al. (1995), Hitchcock et al. (1993), Sanglard et al. (1999)
Candida dubliniensis	Drug efflux mediated by Cd*CDR1* and Cd*MDR1*; overexpression of *ERG11*; *ERG11* point mutations	Moran et al. (1998), Perea et al. (2002)
Candida tropicalis	Drug efflux mediated by *MDR1* and *CDR1*	Barchiesi et al. (2000b)
Candida krusei	Drug efflux mediated by *ABC1* (*CDR1* homolog); reduced susceptibility of 14α-demethylase	Fukuoka et al. (2003), Katiyar and Edlind (2001), Orozco et al. (1998), Venkateswarlu et al. (1996)
Cryptococcus neoformans	Drug efflux mediated by Cn*AFR1* (*CDR1* homolog); overexpression of *ERG11*; reduced susceptibility of 14α-demethylase	Joseph-Horne et al. (1995b), Lamb et al. (1995), Posteraro et al. (2003), Venkateswarlu et al. (1997)
Aspergillus fumigatus	Drug efflux mediated by Afu*MDR3* (*MDR1* homolog) and Afu*MDR4* (*CDR* homolog); reduced susceptibility of 14α-demethylase	da Silva Ferreira et al. (2004), Denning et al. (1997), Nascimento et al. (2003)

structural gene (White, 1997) produces moderate levels of resistance. Overexpression of *ERG11* results in a requirement for higher intracellular concentrations of azole to complex with and inhibit the excess lanosterol demethylase molecules in the cell (Henry et al., 2000; Kohli et al., 2002; White, 1997). *ERG11* point mutations can lead to amino acid substitutions, changing the conformation of the enzyme and decreasing the affinity for azole derivatives, while still maintaining essential 14α-demethylase activity (Kakeya et al., 2000; Sanglard et al., 1998).

The second, reduced intracellular accumulation of azole antifungal agents as a consequence of enhanced drug efflux (Albertson et al., 1996; Clark et al., 1996; Sanglard et al., 1995) tends to be associated with the highest levels of microbiological resistance to the azoles. This mechanism is mediated by two types of multidrug efflux transporters: the major facilitators, encoded by Ca*MDR1* in *C. albicans* (previously designated *BEN*r) (Fling et al., 1991; Hitchcock, 1993; White, 1997), and the ABC transporters belonging to the ATP-binding cassette superfamily and encoded by the *Candida Drug Resistance* genes *CDR1* and *CDR2* (de Micheli et al., 2002; Hernaez et al., 1998; Michaelis and Berkower, 1995; Moran et al., 1998; Perea and Patterson, 2002; Prasad et al., 1995; Sanglard et al., 1997, 1999; White, 1997). Overexpression of *CDR1/2* is known to confer cross-resistance to multiple azole derivatives, whereas overexpression of Ca*MDR1* appears to be specific for fluconazole resistance (Albertson et al., 1996; Sanglard et al., 1995; White, 1997).

A third mechanism of azole resistance involves mutation of a second ergosterol biosynthetic gene, *ERG3*, which encodes the C5–6 sterol desaturase enzyme catalyzing the conversion of sterol intermediates downstream of *ERG11* (Kelly et al., 1996b, 1997; Lupetti et al., 2002; Nolte et al., 1997; Sanglard et al., 2003). The role of *ERG3* in azole resistance originates from the observation that production of the toxic sterol intermediate, 14α-methyl-3,6-diol, in response to azole exposure, requires C5–6 sterol desaturase activity. Disruption of *ERG3* prevents the formation of the toxic sterol intermediate following azole exposure and renders the cell resistant (Lupetti et al., 2002). Azole resistance due to *ERG3* inactivation has also been shown to be associated with cross-resistance to amphotericin B (Kelly et al., 1996b, 1997; Nolte et al., 1997). In the absence of C5–6 desaturase activity, ergosterol biosynthesis is blocked leaving the plasma membrane devoid of ergosterol and thus resistant to amphotericin B. However, *ERG3* disruption does not always result in azole resistance as demonstrated in *C. glabrata* where a null mutation in *ERG3* did not result in resistance (Geber et al., 1995).

A fourth mechanism of azole resistance involves changes in plasma membrane fluidity and asymmetry leading to reduced azole permeability (Kohli et al., 2002; Loeffler et al., 2000). Previous studies using fluconazole-resistant strains of *C. albicans* from AIDS patients receiving long-term fluconazole therapy (Loeffler et al., 2000) or in vitro fluconazole-adapted strains of *C. albicans* (Kohli et al., 2002) have demonstrated an association between reduced azole susceptibility and altered sterol to phospholipid ratios. Another study suggested that altered phospholipid and fatty acid composition of the *C. albicans* plasma membrane may lead to miconazole resistance (Mago and Khuller, 1989).

While one or more of these mechanisms can describe the molecular basis of azole resistance for the majority of fungal isolates studied, resistant isolates that do not express any of the above-described mechanism have been identified, strongly suggesting the existence of yet unknown mechanisms of azole resistance (White et al., 2002). Multiple mechanisms of resistance may be active in an individual isolate at the same time and the stepwise accumulation of multiple resistance mechanisms during the course of azole exposure has been described and will be discussed later in the context of acquired azole resistance.

Primary or intrinsic resistance to azole antifungals is limited to a few fungal species but is well known for *C. krusei*. Intrinsic fluconazole resistance in this species is due, at least in part, to reduced ability of the azoles to inhibit the *C. krusei* 14α-lanosterol demethylase enzyme (Marichal and Vanden Bossche, 1995; Orozco et al., 1998). Evidence for this resistance mechanism is further supported by the observation that the increased activity of voriconazole against *C. krusei* is due to increased sensitivity of the *C. krusei* 14α-lanosterol demethylase to this specific azole derivative (Fukuoka et al., 2003). A role for drug efflux in *C. krusei* has also been suggested following the identification of two *C. krusei* genes, *ABC1* and *ABC2*, encoding close homologs of the *C. albicans* ABC transporters, *CDR1* and *CDR2* (Katiyar and Edlind, 2001). Analysis of *C. krusei ABC1* and *ABC2* expressions did not demonstrate rapid or pronounced upregulation following fluconazole exposure suggesting that drug efflux may not be the primary mechanism of intrinsic resistance in *C. krusei*. In another report, high-level itraconazole resistance in a clinical isolate of *C. krusei* was associated with a marked decrease in intracellular itraconazole following exposure to this azole (Venkateswarlu et al., 1996). Analysis of the cytochrome P-450 lanosterol demethylase in this isolate indicated no difference in affinity for itraconazole compared to the ATCC *C. krusei* 6258, excluding alteration in the azole target as a mechanism of resistance. However, the role of increased efflux versus decreased permeability of the azole drug in reducing the intracellular accumulation of itraconazole could not be definitively concluded (Venkateswarlu et al., 1996).

Aspergillus spp. are intrinsically resistant (or less susceptible) to fluconazole (Bodey, 1992; Denning et al., 1992) although they are generally susceptible to the later generation azoles, including itraconazole, voriconazole, and posaconazole (Marichal and Vanden Bossche, 1995; McGinnis et al., 1997; Oakley et al., 1997), due to increased affinity of these agents for the *Aspergillus* P-450 enzymes. *C. glabrata* is often misclassified as intrinsically resistant to azole antifungals. However, this species is not intrinsically resistant but does possess the ability to rapidly acquire resistance, both in vitro and in vivo, during azole exposure (Bennett et al., 2004; Borst et al., 2005; Hitchcock et al., 1993; Warnock et al., 1988). Azole susceptibility of moulds varies by species and azole derivative. In general, fluconazole demonstrates the narrowest spectrum of activity while the newer azole antifungals, voriconazole, posaconazole, and ravuconazole, have shown activity against a broader range of species including *Aspergillus* spp., the dimorphic fungi, *Penicillium marneffei*, and *Fusarium* species (Sheehan et al., 1999). Azole antifungals, new or old, appear to have no meaningful activity against the zygomycetes including *Rhizopus* spp., *Mucor* spp., and *Rhizomucor* spp. (Sheehan et al., 1999).

Secondary or acquired azole resistance in response to drug exposure is characterized by the emergence of resistance in an organism that is naturally susceptible.

Studies to characterize the molecular mechanisms underlying the emergence of secondary azole resistance have focused primarily on sequentially obtained *C. albicans* isolates from AIDS patients receiving long-term fluconazole therapy for the treatment or prevention of recurrent oropharyngeal candidiasis (Fichtenbaum et al., 2000; Johnson et al., 1995; Laguna et al., 1997; Lopez-Ribot et al., 1999; Ng and Denning, 1993; Reef and Mayer, 1995; Rex et al., 1995c; White et al., 1997). In this setting, isolates from the same patient before and after the emergence of resistance can be compared with regard to the presence or absence of individual resistance mechanisms. Molecular subtyping of serial isolates allows for differentiation between resistance emerging in the same strain versus strain replacement (Bart-Delabesse et al., 1993; Diaz-Guerra et al., 1998; Lasker et al., 2001; Le Guennec et al., 1995; Le Monte et al., 2001; Lischewski et al., 1995; McCullough and Hume, 1995; Redding et al., 1994, 1997) and has facilitated molecular epidemiologic studies of drug-resistant oropharyngeal candidiasis in AIDS patients. In the best-studied cases using sets of susceptible and resistant isolates from the same patient, genotyping results have shown that the emergence of resistance is most often due to sequential changes in a single strain. Furthermore, it appears that individual mechanisms of resistance in *C. albicans* are activated gradually, in a step-wise manner over time, and that high-level resistance results from the simultaneous expression of multiple resistance mechanisms in a single isolate (Lopez-Ribot et al., 1998, 1999; White, 1997). A recent study to determine the prevalence of molecular mechanisms of azole resistance in highly resistant strains of *C. albicans* demonstrated that multiple mechanisms were acting simultaneously in 75% of isolates (Perea et al., 2001). The most prevalent mechanism of azole resistance was overexpression of drug efflux pumps, observed in 85% of isolates. Mutation in the target gene, *ERG11*, was found in 65% of isolates while overexpression of *ERG11* occurred in 35% of isolates (Perea et al., 2001). Another study using a set of unmatched clinical isolates of *C. albicans* including both fluconazole-susceptible and -resistant strains, some expressing cross-resistance to multiple azoles, found that *CDR1* and *CDR2* overexpression correlated with resistance while Ca*MDR1* and *ERG11* overexpression as well as point mutations in *ERG11* showed little or no correlation with a phenotype of azole resistance (White et al., 2002). These data further emphasized the diversity of mechanisms that resulted in azole resistance and the presence of yet undiscovered resistance mechanisms.

In the setting of disseminated *Candida* infections, the incidence of azole resistance has remained quite low among *C. albicans* isolates, the most frequent cause of disease (Hajjeh et al., 2004). This is in spite of the increased incidence of candidemia as well as the increased use of fluconazole in the management and prevention of invasive disease (Hajjeh et al., 2004). A review of data reported from a number of longitudinal and cross-sectional surveys conducted worldwide over the past 15 years revealed that frequencies of fluconazole resistance among *C. albicans* bloodstream isolates, defined as an MIC $\geq$ 64 µg/ml, ranged from 0 to 9.6% (Antunes et al., 2004; Cuenca-Estrella et al., 2005; Diekema et al., 2002; Godoy et al., 2003; Hajjeh et al., 2004; Kao et al., 1999; Pfaller et al., 1998a, 2001, 2002; Takakura et al., 2004) (Table 10.3). Furthermore, there was no consistent trend suggesting that fluconazole MICs were creeping upwards over time. What

Table 10.3 In vitro fluconazole susceptibility of *Candida* spp. bloodstream isolates from different surveillance programs.

Surveillance program	Years	Ref.	No. of isolates tested	Fluconazole MIC90 (range; in µg/ml)					
				C. albicans	C. glabrata	C. parapsilosis	C. tropicalis	C. krusei	Non-C. albicans[a]
CDC	1992–1993	Kao et al. (1999)	394	0.5(0.12–256)	64(1–256)	2(0.25–4)	2(0.25–256)	64(8–64)	
CDC	1998–2000	Hajjeh et al. (2004)	935	0.5(0.12–>64)	16(0.12–>64)	1(0.12–4)	8(0.12–>64)	64(2–>64)	
Sweden	1994–1998	Chryssanthou (2001)	233	0.5(0.12–1)					>64(0.12–>64)
Quebec	1996–1998	St-Germain et al. (2001)	442	128(0.25–>256)	64(2–>256)	2(0.25–4)	>256(0.25–>256)	128(16–128)	
Barcelona	2002–2003	Cuenca-Estrella et al. (2005)	351	0.25(0.12–8)	16(2–>64)	1(0.12–>64)	0.5(0.12–>64)	64(32–>64)	
Brazil	2002–2003	Antunes et al. (2004)	120	<8(0.12–8)	<8(1–8)	<8(0.12–2)	<8(0.12–2)	16(16)	
Japan	2001–2002	Takakura et al. (2004)	535	1(0.12–>128)	32(0.25–>128)	2(0.12–>128)	>128(0.12–>128)	128(32–128)	
Global	1992–2001	Pfaller and Diekema (2004)	6082	0.5(0.12–>128)	16(1–>128)	2(0.25–64)	2(0.25–32)	8(0.12–>128)	

[a]C. glabrata, C. tropicalis, C. parapsilosis, C. krusei, and C. lusitaniae

was apparent is a generally consistent upward trend in the number of bloodstream infections caused by *Candida* spp. other than *C. albicans*, particularly *C. glabrata* over the same time (Table 10.4). Prior to the mid-1990s, *C. albicans* accounted for more than 50% of the candidemias whereas surveys conducted within the past 5–7 years found *C. albicans* to be responsible for less than half of all *Candida* bloodstream infections (Antunes et al., 2004; Cuenca-Estrella et al., 2005; Diekema et al., 2002; Edmond et al., 1999; Godoy et al., 2003; Hajjeh et al., 2004; Kao et al., 1999; Pfaller et al., 1998a, 1999, 2000, 2001, 2002; Takakura et al., 2004). The specific proportion of disease caused by non-*albicans Candida* spp. differs by geographic location both within a given country (Diekema et al., 2002; Edmond et al., 1999; Pfaller et al., 1998a) and between countries (Antunes et al., 2004; Godoy et al., 2003; Pfaller et al., 1999, 2000, 2001; Takakura et al., 2004). Furthermore, the prevalence of individual *Candida* spp. causing invasive disease varies between patient groups with *C. parapsilosis* more common in neonatal intensive care units, *C. krusei* and *C. tropicalis* more commonly associated with hematological malignancy, and *C. albicans* and *C. glabrata* associated with solid tumors (Meunier et al., 1992; Pfaller et al., 2002). While the changing trend in species distribution parallels the introduction of widespread fluconazole use, it cannot be implicated to the exclusion of other host factors influencing species prevalence (Sanglard and Odds, 2002; White, 1998). For example, *C. glabrata* was seen to rise in prevalence in a French institution where amphotericin B, not fluconazole, is routinely used for antifungal prophylaxis (Sanglard and Odds, 2002). Other studies have shown that introduction of routine fluconazole prophylaxis was followed by a reduction in the prevalence of *C. glabrata* and *C. krusei* (Baran et al., 2001; Kunova et al., 1997). Taken together, these data suggest that multiple factors are driving the changing species distribution including (1) host factors such as underlying disease and immunocompetence, (2) antifungal prophylaxis practices, (3) antibacterial use, and (4) immunosuppressive regimens (Sanglard and Odds, 2002).

10.2.5 Allylamines

The allylamines antifungals, like the azole derivatives, exert their antifungal affect by inhibiting ergosterol biosynthesis. They do so via inhibition of squalene expoxidase, encoded by *ERG1*, which catalyzes the formation of lanosterol, the first sterol intermediate in the ergosterol biosynthetic pathway. Terbinafine is the most common allylamine antifungal used clinically and is the treatment choice for dermatophytic infections caused by *Trichophyton* spp., *Microsporon* spp., and *Epidermophyton* spp. While the allylamines have limited activity against other fungi, including *Candida* spp., combination therapy with fluconazole and terbinafine has been shown to be effective for the treatment of fluconazole-resistant oropharyngeal candidiasis (Ghannoum and Elewski, 1999).

Terbinafine resistance among the dermatophytes is rare with no reports of resistance prior to 2003. Investigators reported clinical as well as microbial resistance in six *Trichophyton rubrum* isolates obtained sequentially from a single

Table 10.4 Trends in *Candida* spp. causing bloodstream infections as determined by different surveillance programs.

Surveillance program	Years	Ref.	No. of isolates tested	Percent of bloodstream infections				
				C. albicans	C. glabrata	C. parapsilosis	C. tropicalis	C. krusei
CDC	1992–1993	Kao et al. (1999)	837	52	12	21	10	4
CDC	1998–2000	Hajjeh et al. (2004)	1143	45	24	13	12	2
Sweden	1994–1998	Chryssanthou (2001)	233	53	22	14	4.7	3.9
Quebec	1996–1998	St-Germain et al. (2001)	442	54	15	12	9	3
SENTRY	1997–2000	Pfaller et al. (2002)	2047	54	16	15	10	2
Barcelona	2002–2003	Cuenca-Estrella et al. (2005)	351	51	9	23	10	4
Brazil	2002–2003	Antunes et al. (2004)	120	48	3	26	13	2
MSG/NIH	1995–1999	Ostrosky-Zeichner et al. (2003)	2000	37	23	20	15	3
Japan	2001–2002	Takakura et al. (2004)	535	41	18	23	12	2
Global	1992–2001	Pfaller and Diekema (2004)	6082	56	16	13	10	3

onychomycosis patient who failed oral terbinafine therapy (Mukherjee et al., 2003). Molecular characterization of the isolates demonstrated that resistance to terbinafine was due to alterations in the squalene epoxidase gene or a factor essential for its activity (Favre et al., 2004).

10.2.6 β-1,3 Glucan synthesis inhibitors

For more than four decades, the principal target of antifungal therapy has been the fungal sterol, ergosterol: inhibition of its biosynthesis by the azoles and allylamines or its function in the plasma membrane by the polyene agents. Introduction of the echinocandin antifungals represented a new class of agents with a novel mode of action, non-competitive inhibition of the synthesis of β-1,3 glucan by the β-1,3 glucan synthase enzyme complex. β-1,3 glucan is a major structural component of the fungal cell wall of most fungi (Denning, 2003). Higher eukaryotes lack β-1,3 glucans and therefore these molecules provide selective targets for antifungal drugs without target-associated toxicity in mammalian hosts (Denning, 1997).

The β-1,3 glucan synthase complex has been studied extensively in *S. cerevisiae* and genes encoding the major catalytic and regulatory subunits, *FKS1* and *FKS2*, have been identified (Douglas, 2001). *FKS1* and *FKS2* homologs have been found in *C. albicans* (Mio et al., 1997), *Aspergillus nidulans* (Kelly et al., 1996a), *Paracoccidioides brasiliensis* (Pereira et al., 2000), and *C. neoformans* (Thompson et al., 1999); however, specific mechanistic details of glucan synthesis and its inhibition by echinocandins still remain obscure. Previous studies have linked in vitro echinocandin resistance to mutations in *FKS1*, supporting it as a potential site of drug–target interaction, but the exact nature of the interaction is unknown (Douglas et al., 1994; Kurtz et al., 1996).

The echinocandins are fungicidal for most *Candida* spp. and higher MICs have been observed among *C. parapsilosis*, *C. krusei*, *C. guilliermondii*, and *C. lusitaniae* isolates compared to *C. albicans*, *C. tropicalis*, and *C. glabrata* (Bartizal et al., 1997; Douglas, 2001; Marco et al., 1998; Vazquez et al., 1997). Echinocandins are fungistatic for *Aspergillus* spp. but exhibit no meaningful activity against zygomycetes, *C. neoformans*, or *Fusarium* spp., presumably because these fungi lack the echinocandin target and possess glucan polysaccharides with α-linkages instead of β-linkages (Arikan et al., 2001; Del Poeta et al., 1997; Denning, 2003; Espinel-Ingroff, 1998; Franzot and Casadevall, 1997; Pfaller et al., 1998b).

Caspofungin acetate (Cancidas) was the first representative of this new class of antifungals to receive approval by the US Federal Drug Agency in 2001 and is licensed for the treatment of candidemia, other forms of invasive candidiasis, esophageal candidiasis, presumed fungal infections in neutropenic patients, and invasive apergillosis in patients who are refractory to or intolerant of other therapies (Arathoon et al., 2002; Dinubile et al., 2002; Kartsonis et al., 2003; Maertens et al., 2000). Based on limited data, micafungin was approved in 2005 for the treatment of esophageal candidiasis as well as for *Candida* infection prophylaxis in hematopoietic stem cell transplant patients. A third echinocandin agent, anidulafungin, is currently in phase III clinical trials.

Caspofungin has been a welcome addition to the list of treatment options for invasive disease and its use is continuing to increase. Thus far, development of secondary resistance or reduced susceptibility appears to be a rare event (Denning, 2003). Echinocandin resistance has been described and partially characterized in spontaneous *C. albicans* mutants selected on candin-containing medium and was shown to be specific for echinocandins without cross-resistance to other non-candin antifungals (Kurtz et al., 1996). Analysis of glucan synthase activity in crude membrane fractions from resistant strains demonstrated a reduced susceptibility to echinocandins compared to their susceptible parental strains (Kurtz et al., 1996). Furthermore, multidrug resistance mechanisms, such as drug efflux, did not seem to contribute to the echinocandin resistance phenotype in these strains. Evaluation of resistant strains in a murine model of disseminated candidiasis revealed a lack of in vitro–in vivo correlation as infected animals responded to lower dose of drug than would have been predicted based on the in vitro MIC result. This discrepancy may be attributed to in vitro susceptibility testing methods which have not been optimized for the echinocandin class of drugs (Kurtz et al., 1996).

Another report describes the development of secondary multidrug (echinocandin–azole) resistance in a patient receiving antifungal therapy for prosthetic valve endocarditis caused by *C. parapsilosis* (Moudgal et al., 2005). Prior to combination therapy with amphotericin B and flucytosine, the *C. parapsilosis* isolate obtained from the patient's blood was susceptible to fluconazole, voriconazole, caspofungin, anidulafungin, and amphotericin B and resistant to micafungin. Therapy was stopped due to renal insufficiency and the patient was subsequently treated successfully with caspofungin and fluconazole. Three months later, the patient was re-admitted and the *C. parapsilosis* isolate obtained from the blood was now resistant to fluconazole, voriconazole, caspofungin, and micafungin but remained susceptible to anidulafungin and amphotericin B. Molecular subtyping indicated that both isolates were the same strain, confirming that secondary cross-resistance to azole and echinocandin antifungals had emerged (Moudgal et al., 2005). The explanation for the relatively high micafungin MIC prior to the initiation of antifungal therapy as well as the absence of acquired anidulafungin resistance after therapy was not clear but suggests that, like the azole antifungals, specific echinocandin antifungal compounds may interact with their targets in a unique way that translates into compound-specific susceptibilities.

To date, only a few reports have proposed mechanisms of echinocandin resistance that are not *FKS1*-mediated. These include (1) overexpression of *CDR1* and *CDR2* (Schuetzer-Muehlbauer et al., 2003), (2) overexpression of *SBE2*, encoding a Golgi protein involved in transport of cell wall components (Osherov et al., 2002; Santos and Snyder, 2000), and (3) alteration in drug influx- and/or efflux-mediated membrane-bound translocators (Paderu et al., 2004). Schuetzer-Muehlbauer et al., found that azole-resistant clinical isolates of *C. albicans* overexpressing *CDR1* and *CDR2* were less susceptible to caspofungin when tested in a semi-quantitative agar plate resistance assay than a non-isogenic azole-susceptible control strain (Schuetzer-Muehlbauer et al., 2003). In addition, they showed that *CDR2* conferred caspofungin resistance when constitutively overexpressed in a drug-sensitive laboratory strain of *C. albicans* (Schuetzer-Muehlbauer et al., 2003). Osherov et al. showed that caspofungin resistance mediated by overexpression of a Golgi-resident protein Sbe2p, represents a novel mechanism of antifungal resistance never before described. In this study, regulated overexpression of Sbe2p resulted in resistance to caspofungin in *S. cerevisiae* (Osherov et al., 2002). Interestingly, *FKS1* and *FKS2* were not identified in their genetic screen for proteins that confer candin resistance when overexpressed. This may suggest that overexpression of these genes alone is not sufficient to confer a resistant phenotype and would require overexpression of other proteins that make up the large enzymatic complex of glucan synthase (Kurtz and Douglas, 1997; Osherov et al., 2002). Paderu et al. described the transport properties of caspofungin across the cell membrane of *C. albicans* and suggested that alterations in drug transport could be a potential source of reduced drug susceptibility (Paderu et al., 2004). Based on their results, a dual-uptake model for caspofungin transport, dependent upon the drug concentration in the surrounding environment, was suggested. When drug concentrations were low (at or below the MIC) a high-affinity, energy-independent, facilitated diffusion carrier was proposed to mediate drug uptake into the cell; at high drug concentrations, non-specific drug uptake through normal diffusion across the plasma membrane was suggested (Paderu et al., 2004). The identity of the drug transporter(s) and the resistance phenotype of strains lacking such transporter(s) will be important first steps in exploring the possible role of candin transport in resistance.

10.3 Clinical Implications

10.3.1 Detecting In Vitro Resistance

A growing interest and demand for in vitro antifungal susceptibility testing of fungal pathogens has been driven by several factors including (1) an increase in number of life-threatening fungal infections with a high mortality rate in immunocompromised patients, (2) a recognized shift in species causing disease, (3) a concern for emerging resistance, (4) an increased number of antifungal drug choices, (5) available standardized, reference methods as well as commercial test kits for determining in vitro antifungal susceptibilities, and (6) evidence for clinical correlation of in vitro MICs (reviewed in Rex et al., 2001). The Clinical Laboratory Standards Institute (CLSI; formerly National Committee for Clinical Laboratory Standards) published in 1997 the M27-A broth macro- and microdilution methods, now revised in 2002 to the M27-A2, for antifungal susceptibility testing of *Candida* spp. and *C. neoformans* (National Committee for Clinical Laboratory Standards, 2002b). In addition, CLSI has published an approved broth dilution method for antifungal susceptibility testing of filamentous fungi, M38-A (National Committee for Clinical Laboratory Standards, 2002a), as well as an agar-based disk diffusion method for determining in vitro antifungal susceptibilities of yeasts (National Committee for

Clinical Laboratory Standards, 2004). These methods were established through a consensus process whereby multiple laboratories collaborated together to evaluate several test variables such as test medium, inoculum size, incubation time and temperature, definitions of MIC endpoints, and identification of quality control strains. The availability of standardized, reference methods has dramatically improved the reproducibility and reliability of antifungal susceptibility testing and has provided a means by which inter-laboratory MIC studies can be conducted, novel MIC test methods can be evaluated, and in vitro activity of new antifungal agents can be assessed.

A number of commercial systems for antifungal susceptibility testing of yeasts and moulds are now under development, including the colorimetric broth dilution-based Sensititer YeastOne (TREK Diagnostics Systems Inc., Westlake, OH) and the agar dilution-based Etest (AB BIODISK North America Inc., Piscataway, NJ) methods. Both have been extensively tested, and agreement with the CLSI-approved reference methods for yeasts and moulds varies from acceptable to excellent depending upon the fungal species and antifungal agent combination tested (Castro et al., 2004; Meletiadis et al., 2002; Pfaller et al., 1998c, 2003, 2004; Warnock et al., 1998). The Sensititer YeastOne system has recently received approval for antifungal susceptibility testing of yeasts by the US Food and Drug Administration. The availability of commercial systems has led to an increase in the number of clinical microbiology laboratories willing and able to perform these tests. However, caution must be used in determining which isolates are appropriate for testing and how MIC results are interpreted to insure maximum clinical relevance and cost-effectiveness.

10.3.2 MIC Interpretation

The real value of an MIC comes from the corresponding ability to interpret its clinical meaning (Rex et al., 2001). For antimicrobials in general, treatment success rates can be predicted to fall as MIC values rise (Sanglard and Odds, 2002). Nevertheless, in vivo correlation of in vitro MICs is not perfect and microbial resistance, defined as an elevated MIC, is intended to convey a high, but not absolute, probability of treatment failure. Currently, CLSI has established interpretive MIC breakpoints for *Candida* spp. isolates tested against fluconazole, itraconazole, and 5-FC using CLSI-recommended guidelines for broth dilution testing. Based on these breakpoints, resistance is defined as an MIC ≥ 64 μg/ml for fluconazole, ≥ 1μg/ml for itraconazole, and ≥ 32 μg/ml for 5-FC (National Committee for Clinical Laboratory Standards, 2002b; Rex et al., 1997). Fluconazole breakpoints were based on an analysis of treatment outcomes in both mucosal and invasive diseases whereas itraconazole breakpoints were based exclusively on outcome data from cases of AIDS-associated oropharyngeal candidiasis (Rex et al., 1997). Resistance breakpoints for 5-FC are based exclusively on data from pharmacokinetic and animal model studies (Rex et al., 1997). Although it has been demonstrated that children have significantly different pharmacokinetic parameters from fluconazole than adults (Brammer and Coates, 1994; Saxen et al., 1993; Seay et al., 1995), none of the data are derived from pediatric cases and in vitro–in vivo correlations in this population have not been addressed (Rex et al., 2001).

Interpretive MIC breakpoints for other fungal pathogen–antifungal drug combinations covered in the CLSI documents have not been defined. In these cases, establishment of interpretive MIC breakpoints has been challenging for a number of reasons such as (1) the relative paucity of clinical outcome data for isolates with elevated MICs as such potentially resistant isolates are rarely encountered in most early clinical trials, (2) the relatively low incidence of most opportunistic mould pathogens which precludes large-scale prospective comparison of antifungal MICs for moulds with the clinical results of antifungal treatment, (3) the inability to accurately and consistently assess clinical outcome of antifungal therapy in patients with severe underlying disease, and (4) the absence of a clinically relevant method for determining in vitro susceptibility of fungi to the glucan synthesis inhibitors (Kurtz et al., 1996; Odds et al., 1998; Rex et al., 2001; Stone et al., 2002).

Alternative approaches to progress in this area have been undertaken with some success. The integration of an in vitro MIC with data from time-kill studies and pharmacodynamic analyses can increase the clinical predictive value of an MIC result and has been shown to provide important insight into how these factors are intertwined with regard to their role in clinical outcome (Klepser et al., 1997, 1998; Lee et al., 2000; Lewis et al., 1998; Rex et al., 2001).

However, these approaches do have limitations. At the simplest level, any given MIC determination contains an inherent potential for variation of at least one twofold dilution step (see below). This in turn means the simplistic analysis of exposure/MIC cut-off values is inappropriate. Further, animal models have provided a valuable approach to studying in vivo correlation of in vitro MICs, but have limitations as interpretable in vivo systems have not been successfully established for studying several mould–drug interactions (Diamond et al., 1998; Graybill, 2000; Johnson et al., 2000; Mukherjee et al., 2003; Odds et al., 1998; Rex et al., 1998). An extensive study of nine mould isolates in mouse and guinea pig models of disseminated disease found that amphotericin B and itraconazole MICs obtained by the NCCLS reference method provided few clues to the likelihood of in vivo response (Odds et al., 1998).

10.3.3 Limitations of In Vitro Susceptibility Tests

As a result of the knowledge generated over the past decade, in vitro antifungal susceptibility testing has become a valuable tool to investigators in this field. However, there are well-recognized limitations of the current methods that continue to be addressed through a variety of scientific studies.

Like all susceptibility testing methodologies, whether for bacteria or fungi, the MIC tests contain the potential for day-to-day variation of no less than one twofold dilution step. This is readily seen in studies where efforts are made to select reproducible quality control isolates for laboratory use. It can take considerable effort to find isolates that routinely produce the same MIC in different laboratories and on different days in the same laboratory (Barry et al., 2000; Pfaller et al., 1995a). The implications of this are significant. Any given MIC measurement should be viewed as being equally likely to be one dilution higher or lower, even when the testing has been conducted in the best of laboratories.

Beyond this basic limitation, testing of fungi has other problems for specific organism–drug combinations. The M27-A2 method, as well as other broth dilution methods using RPMI 1640 medium, is not optimal for detecting amphotericin B resistance (Rex et al., 1993). Under these test conditions, MICs cluster between two and three drug dilutions, precluding the ability to differentiate amphotericin B-susceptible from -resistant organisms. Use of an alternative test medium, such as Antibiotic Medium 3, has been shown to broaden the range of MICs and improve detection of amphoterin B-resistant isolates in broth dilution assays (Lozano-Chiu et al., 1997; Rex et al., 1995a). Agar dilution methods for amphotericin B susceptibility testing, such as the Etest, have also been shown to improve the reliability of distinguishing resistant isolates and increase correlation with clinical outcome (Clancy and Nguyen, 1999; Peyron et al., 2001).

Detection of in vitro resistance among *C. neoformans* isolates is also less than optimal using the M27-A2 method as this organism grows more slowly or not at all in the recommended RPMI 1640 test medium (Odds et al., 1995; Rex et al., 2001). Useful modifications including (1) use of RPMI medium containing 2% glucose (Rodriguez-Tudela and Martinez-Suarez, 1995), (2) use of yeast nitrogen base medium and increased inoculum size (Jessup et al., 1998), and (3) shaking of the microtiter plate during incubation (Rodriguez-Tudela et al., 2000) have been shown to improve the reproducibility and clinical correlation of MIC results.

Azole antifungal drug susceptibility testing of *Candida* spp., especially *C. albicans* and *C. tropicalis*, using the M27-A2 method, can be complicated by the presence of trailing growth at high drug concentrations. Trailing growth is an in vitro phenomenon characterized by reduced or persistent growth above the MIC that is significantly more apparent after 48 versus 24 h of incubation (Arthington-Skaggs et al., 2002; Revankar et al., 1998; Rex et al., 1998, 2001; Smith and Edlind, 2002; St-Germain, 2001). In fact, trailing growth after 48 h can be so great that azole-susceptible isolates can be mistaken as resistant. From previous reports, the incidence of trailing growth observed in fluconazole broth microdilution MIC tests can range from 11–18% for *C. albicans* isolates to 22–59% for *C. tropicalis* isolates (Arthington-Skaggs et al., 2002; Ostrosky-Zeichner et al., 2003; St-Germain, 2001). Consequences associated with misinterpreting trailing growth isolates as resistant can be seen at the individual patient level, whereby a cheaper, better tolerated antifungal therapy would be withheld and a more toxic, expensive, or broader spectrum agent used, as well as at the institutional or population levels where falsely inflated rates of azole resistance could cause unwarranted concern and a potential for unnecessarily restrictive guidelines for azole antifungal use.

Although the molecular mechanisms underlying the trailing growth phenotype are not fully understood, previous studies have demonstrated that trailing growth isolates of *C. albicans* respond to azole exposure by up-regulating genes associated with azole resistance, *ERG11*, *CDR1/2*, and *MDR1*, in a pattern distinct from that of fluconazole-susceptible and -resistant isolates (Lee et al., 2004; Smith and Edlind,

2002). To investigate the clinical significance of trailing growth, two independent studies using an immunocompetent mouse model of candidemia to evaluate the efficacy of fluconazole treatment found that trailing growth isolates were susceptible in vivo (Arthington-Skaggs et al., 2000; Rex et al., 1998). Furthermore, in a study of human immunodeficiency virus-positive patients with recurrent oropharyngeal candidiasis (3–12 episodes over 3- to 15-month periods), Revankar et al. detected trailing isolates mixed with non-trailers; the patients in all cases responded to 7-day courses of fluconazole (Revankar et al., 1998). While these data suggest that trailing growth isolates, unlike resistant isolates, do not appear to be associated with treatment failure, association with recurrent infection, i.e., by surviving treatment and seeding the next infection, is an important area for further investigation (Smith and Edlind, 2002).

For antifungal susceptibility testing of moulds, the M38-A reference method is targeted to the more common filamentous fungi that cause invasive disease and does not address testing of monilaceous and dematiaceous moulds or the dimorphic fungi, such as *Blastomyces dermatitidis*, *Coccidioides immitis*, and *Histoplasma capsulatum* (National Committee for Clinical Laboratory Standards, 2002a). In addition, M38-A specifies preparation of inocula using non-germinated conidial or sporangiospore suspensions versus hyphae, as conidia are easier to handle and can be more accurately standardized. However, it is the filamentous form of growth that is observed in vivo and it is not clear how the differences in susceptibility of conidia versus hyphae effect the in vitro–in vivo correlation of MICs. Most studies comparing the in vitro antifungal susceptibilities of conidia versus hyphae of *Aspergillus* spp. have found no difference (Espinel-Ingroff, 2001; Manavathu et al., 1999), although higher MICs for hyphae have been reported (Lass-Florl et al., 2002).

Finally, clinically relevant methods for in vitro susceptibility testing of yeasts or moulds and interpretive MIC breakpoints for the newer antifungals are unknown. Despite this limitation, quality control MIC ranges for yeasts using the M27-A2 broth microdilution method have been published for voriconazole, ravuconazole, posaconazole, caspofungin, and anidulafungin (Barry et al., 2000).

10.4 Antifungal Prophylaxis: A Breeding Ground For Resistance?

The steady introduction over the past 20 years of increasingly safe and well-tolerated alternatives to amphotericin B has made antifungal prophylaxis both feasible and attractive. Despite the increased use of fluconazole prophylaxis in non-HIV-infected patients, it does not appear to be a "breeding ground for resistance" as was observed in AIDS patients receiving fluconazole for the treatment and prevention of oropharyngeal candidiasis, prior to widespread use of HAART (Laguna et al., 1997; Lopez-Ribot et al., 1998; Perea et al., 2001; Redding et al., 1994). With respect to invasive fungal infections, the most prevalent being candidemia and aspergillosis, prophylactic therapy has been shown to be of significant value for neutropenic patients with acute leukemia and/or allogeneic stem cell transplants (Bohme et al., 1999; Goodman et al., 1992; Gotzsche and Johansen, 1997; Marr et al., 2000a,b; Rex et al., 2000; Slavin et al., 1995) as well as high-risk liver transplant recipients (Patel, 2000; Playford et al., 2004; Singh, 2004; Singh and Yu, 2000). Based on the success observed in these patients, the value of antifungal prophylaxis in other patient populations including lung transplant patients (Dummer et al., 2004; Minari et al., 2002), surgical intensive care patients (Lipsett, 2004; Ostrosky-Zeichner, 2004; Swoboda et al., 2003), and pre-term, low birth weight infants is being actively studied (Austin and Darlow, 2004; Kaufman et al., 2001).

The benefits of antifungal prophylaxis, however, must be balanced against the potential selection or induction of resistant isolates. While there are individual reports of antifungal resistance in patients receiving antifungal prophylaxis (Johnson et al., 1995; Kelly et al., 1997; Lopez-Ribot et al., 1999; Marr et al., 1997; Muller et al., 2000; Nolte et al., 1997; Redding et al., 1994), the data overwhelmingly indicate that widespread antifungal resistance is not emerging. Instead, resistance appears to be more specifically associated with factors related to the host, such as underlying disease and severity of immunosuppression and overall duration and cumulative dose of fluconazole received (Fan-Havard et al., 1991; Kicklighter et al., 2001; Pfaller and Diekema, 2004).

10.4.1 Vulvovaginal Candidiasis

Unlike oropharyngeal candidiasis, which is almost exclusively associated with immunosuppression of the host, chronic and/or recurrent vulvovaginal candidiasis affects immunosuppressed as well as immunocompetent, healthy women in all strata of society (Fidel and Sobel, 1996; Sobel, 1985). However, despite the many years of azole use in this clinical setting, there is no consistent evidence for emerging resistance (Sobel et al., 2004). Other work has shown an increased prevalence of non-*albicans Candida* spp., for which azole agents are less likely to be effective, as the cause of recurrent infections (Richter et al., 2005). While previous azole antifungal exposure is most certainly an important factor in the selection of these non-*albicans Candida* spp., it is not the sole factor.

10.4.2 Invasive Fungal Infection

10.4.2.1 Neutropenic Cancer Patients

In the neutropenic patient population, acquired fluconazole resistance is distinctly uncommon among *Candida* spp. bloodstream isolates of the azole-susceptible species, likely due to the shorter duration of exposure and lower cumulative dose used to treat or prevent candidemia. Population-based and sentinel surveillance data have demonstrated that azole resistance, as assessed by NCCLS reference testing methods, is rare among *C. albicans*, *C. parapsilosis*, and *C. tropicalis* bloodstream isolates with no trend towards increasing resistance over time (Chryssanthou, 2001; Diekema et al., 2002; Hajjeh et al., 2004; Kao et al., 1999; Pfaller et al., 2002; St-Germain et al., 2001; Takakura et al., 2004). Another study assessing the impact on resistance of 12 years of fluconazole use in clinical practice found very little variation in fluconazole susceptibility among *Candida* spp. isolates collected between 1992 and 2002 (Pfaller and Diekema, 2004).

Although mutation from azole susceptible to resistance remains uncommon among *Candida* spp. bloodstream isolates, a consistent trend towards increased rates of colonization and infection by the less azole-susceptible, non-*albicans Candida* spp., namely *C. glabrata* and *C. krusei*, has been observed among leukemic or bone marrow transplant recipients receiving fluconazole prophylaxis (Chryssanthou et al., 2004; Laverdiere et al., 2000; Marr, 2004; Marr et al., 2000a,b; Redding et al., 2004; Safdar et al., 2001; Wingard et al., 1991, 1993). A study describing the effect of fluconazole prophylaxis on *Candida* colonization and infection among 266 neutropenic cancer patients with acute leukemia or autologous bone marrow transplantation found that although *Candida* colonization and invasive disease were reduced in patients randomized to the fluconazole arm, compared to those receiving placebo, colonization with non-*albicans Candida* spp., particularly *C. glabrata*, was greater among patients receiving fluconazole prophylaxis and one definitive invasive *C. glabrata* infection was noted in the group receiving fluconazole (Laverdiere et al., 2000). Another study of 585 cancer patients receiving allogeneic blood and marrow transplantations and fluconazole prophylaxis found that more than half of the patients positive for *Candida* colonization were colonized with a non-*albicans Candida* spp. at some point in the study (Marr et al., 2000a,b). Furthermore, of the 27 patients that developed candidemia, 25 were infected with a non-*albicans Candida* spp. and the remaining 2 with fluconazole-resistant *C. albicans* (Marr et al., 2000a,b). Overall, however, the risk for selection of an azole-resistant strain or species is low compared to the benefit of prophylaxis in these patients at high risk for invasive fungal infection.

Although the overall number of invasive infections caused by filamentous fungi is lower than that of *Candida* spp., an increase in invasive infections caused by antifungal-resistant filamentous fungal strains and species has been observed in the hematology/oncology patient population receiving fluconazole prophylaxis (Castagnola et al., 2004). A recent study of cancer patients previously exposed to amphotericin B or triazole antifungals described the increased frequency of invasive aspergillosis caused by *A. terreus* (Lionakis et al., 2005). While *A. fumigatus* remains the most common cause of invasive aspergillosis, the increased incidence of *A. terreus*, which is intrinsically resistant to amphotericin B and less susceptible to itraconazole and voriconazole, is noteworthy.

The recent introduction of voriconazole prophylaxis against aspergillosis has been well received due to the drug's broad spectrum of activity, minimal host toxicity, and capacity for oral administration (Herbrecht et al., 2002). Thus far, emergence of voriconazole resistance among *Aspergillus* isolates has not occurred. However, breakthrough zygomycosis during voriconazole prophylaxis in both stem cell transplant (Imhof et al., 2004; Kontoyiannis et al., 2005; Marty et al., 2004; Siwek et al., 2004; Vigouroux et al., 2005) and lung transplant recipients (Mattner et al., 2004) has been described. Because the zygomycetes are intrinsically resistant to all azole antifungals, the magnitude of this consequence of voriconazole prophylaxis requires further investigation.

10.4.2.2 Pre-term, Low Birth Weight Neonates

Pre-term, low birth weight neonates, cared for in neonatal intensive care units, represent another group of patients at increased risk for invasive *Candida* infections who may benefit from antifungal prophylaxis (Austin and Darlow, 2004). Of the few studies that have investigated the impact of fluconazole prophylaxis to prevent candidemia in this patient population, the overall conclusion is that selection of resistant strains and species of *Candida* is not common (Kicklighter et al., 2001). However, the potential for selection for non-*albicans Candida* spp. as well as emergence of resistance in previously susceptible strains and species must be considered. A study of neonatal baboons receiving fluconazole prophylaxis to prevent candidemia found that the number of positive *C. parapsilosis* blood cultures increased from 1 between 1997 and 1999 to 11 between 2000 and 2002 with three fluconazole-resistant *C. parapsilosis* isolates collected in 2002 (Yoder et al., 2004).

10.4.2.3 Cryptococcosis in AIDS Patients

The widespread use of HAART has significantly decreased the incidence of opportunistic fungal infections in AIDS patients including cryptococcal meningitis. However, therapy is not available to all HIV-positive patients and antifungal prophylaxis to suppress recurrent cryptococcal disease is still common in some settings. Fortunately, overall resistance, as determined by NCCLS reference testing methods, remains uncommon based on data from surveys monitoring global trends in the antifungal susceptibility of *C. neoformans* (Pfaller et al., 2005). Isolated reports of increasing in vitro resistance to fluconazole among *C. neoformans* isolates have been published (Sar et al., 2004). For example, Sar et al. described an increase in fluconazole resistance among strains isolated from the cerebral spinal fluid of AIDS patients in Cambodia from 2.5 to 14% between 2000 and 2002 (Sar et al., 2004). Authors conclude that the increased resistance is probably linked to the recent widespread prescription of fluconazole for AIDS patients in Cambodia. However, it is essential to consider that correlation of in vitro resistance with treatment failure has not been determined in these patients and thus it is unclear whether or not fluconazole prophylaxis is associated with resistance in vivo among AIDS patients.

10.5 Future Considerations and Newer Agents

10.5.1 Echinocandins

The role of echinocandins for prophylaxis against invasive fungal disease remains to be determined. A few reports exist suggesting that echinocandins can effectively prevent invasive aspergillosis and candidiasis without the side effects of amphotericin B (Ifran et al., 2005; Perfect, 2004a,b; van Burik et al., 2004). Ifran et al. described the successful use of caspofungin for prophylaxis against invasive pulmonary aspergillosis in an allogeneic stem cell transplant patient (Ifran et al., 2005). Another report, based on data from a randomized, double-blind, multinational trial comparing caspofungin to liposomal amphotericin B as empirical therapy in neutropenic patients with persistent fever, found that caspofungin was as effective and generally better tolerated than liposomal amphotericin B (Walsh et al., 2004). Van Burik et al. studied 882 adult and pediatric hematopoetic stem cell transplant recipients and observed the overall efficacy of micafungin for the prevention of proven or probably invasive fungal infection to be superior to that of fluconazole during the neutropenic phase after transplant (van Burik et al., 2004). In this study, success was defined as the absence of suspected, proven, or probable invasive fungal infection through the end of therapy as well as the 4-week period following treatment (van Burik et al., 2004). Results from these studies are promising; however, at this early stage it is difficult to predict whether or not selection of resistant fungal strains and species will occur.

10.5.2 Combination Therapy

Combination antifungal therapy, particularly when combining drugs with distinct mechanisms of action, may offer improved protection and increased efficacy against invasive fungal infections along with a decreased likelihood for the emergence of resistance. Furthermore, when amphotericin B is used in combination with another antifungal, lower doses were possible in one case (therapy of cryptococcal meningitis in non-HIV-infected adults), thus reducing the nephrotoxic side effects of this agent (Dismukes et al., 1987). Combinations of caspofungin and amphotericin B (Castagnola et al., 2004; Ehrmann et al., 2005), voriconazole and amphotericin B (Ehrmann et al., 2005), and caspofungin and voriconazole (Castagnola et al., 2004; Damaj et al., 2004; Marr, 2004) have been shown to be effective for difficult-to-treat invasive fungal infections, particularly those caused by filamentous fungi. However, much of what we know about the efficacy of combination therapy in humans comes from individual case reports, as well-designed clinical trials of combination antifungal therapy are lacking (Cuenca-Estrella, 2004; Steinbach et al., 2003).

10.5.3 Cispentacin

Cispentacin, first described in 1989, is currently under development for the treatment of *Candida* infections. Also known as icofungipen, cispentacin has a distinct mechanism of action: inhibition of *Candida* isoleucyl-tRNA synthetase and thus inhibition of protein synthesis and cell growth (Konishi et al., 1989; Oki et al., 1989). In vitro studies have demonstrated good anti-*Candida* activity, although antifungal susceptibility testing methods have not been optimized. In vivo efficacy against animal models of candidemia (Oki et al., 1989; Petraitiene et al., 2005) has been demonstrated and notably, activity against azole-resistant *C. albicans* has been described (Petraitis et al., 2004).

10.5.4 Sordarins

The sordarins represent yet another class of antifungals in development. These compounds inhibit fungal growth by selectively inhibiting fungal protein synthesis through a highly specific interaction with the fungal elongation factor 2 (EF2) (Dominguez and Martin, 1998). Intrinsically resistant strains owe their resistance to molecular differences in EF2 (Dominguez and Martin, 1998; Dominguez et al., 1998). Originally discovered by routine screening in the early 1970s, interest in this class of compounds was revitalized as the result of a prospective screen for inhibitors of *C. albicans* protein synthesis (Dominguez and Martin, 1998). Different sordarin derivatives have different spectra of susceptible fungal spp. but generally include *Candida* spp., *C. neoformans*, the dimorphic fungi, and *Pneumocystis carinii* (Herreros et al., 1998). In a murine model of histoplasmosis, three sordarin derivatives were found to compare favorably with amphotericin B and fluconazole (Graybill et al., 1999). The sordarin antifungals, however, have only limited activity against filamentous fungi (Herreros et al., 1998).

10.6 Conclusions

As long as we experience continued success in developing advanced medical capabilities, the number of high-risk patients susceptible to severe fungal infections will continue to grow. As a result, antifungal use for both treatment and prophylaxis will increase along with the concern for the development of resistance. The launching of the echinocandin antifungals has, at last, broadened the diversity of antifungal drugs available for treating systemic fungal infections. Resistance to these agents remains to be seen at this point in time, but must be considered when deciding how and when to use these agents. Our knowledge of resistance mechanisms and methods for detecting in vitro resistance for the polyene and azole antifungals has become well established in the past 10 years and continues to be an area of intense research.

Likewise, we are gaining in our understanding of the complex interactions between host, microbe, and drug that ultimately dictate clinical outcome. As we continue to work in this area and translate relevant results into clinical practice, we should expect to maximize the utility of the current antifungal drugs and improve clinical outcomes of fungal infections for all patients.

References

Albertson, G.D., Niimi, M., Cannon, R.D. and Jenkinson, H.F., (1996), Multiple efflux mechanisms are involved in *Candida albicans* fluconazole resistance. *Antimicrob Agents Chemother*, **40**, 2835–2841.

Antunes, A.G., Pasqualotto, A.C., Diaz, M.C., d'Azevedo, P.A. and Severo, L.C., (2004), Candidemia in a Brazilian tertiary care hospital, species distribution and antifungal susceptibility patterns. *Rev Inst Med Trop Sao Paulo*, **46**, 239–241.

Arathoon, E.G., Gotuzzo, E., Noriega, L.M., Berman, R.S., DiNubile, M.J. and Sable, C.A., (2002), Randomized, double-blind, multicenter study of caspofungin versus amphotericin B for treatment of oropharyngeal and esophageal candidiases. *Antimicrob Agents Chemother*, **46**, 451–457.

Archer, D.B. and Gale, E.F., (1975), Antagonism by sterols of the action of amphotericin and filipin on the release of potassium ions from *Candida albicans* and *Mycoplasma mycoides* subsp. *capri*. *J Gen Microbiol*, **90**, 187–190.

Arikan, S., Lozano-Chiu, M., Paetznick, V. and Rex, J.H., (2001), *In vitro* susceptibility testing methods for caspofungin against *Aspergillus* and *Fusarium* isolates. *Antimicrob Agents Chemother*, **45**, 327–330.

Arthington-Skaggs, B.A., Lee-Yang, W., Ciblak, M.A., Frade, J.P., Brandt, M.E., Hajjeh, R.A., Harrison, L.H., Sofair, A.N. and Warnock, D.W., (2002), Comparison of visual and spectrophotometric methods of broth microdilution MIC end point determination and evaluation of a sterol quantitation method for *in vitro* susceptibility testing of fluconazole and itraconazole against trailing and nontrailing *Candida* isolates. *Antimicrob Agents Chemother*, **46**, 2477–2481.

Arthington-Skaggs, B.A., Warnock, D.W. and Morrison, C.J., (2000), Quantitation of *Candida albicans* ergosterol content improves the correlation between *in vitro* antifungal susceptibility test results and *in vivo* outcome after fluconazole treatment in a murine model of invasive candidiasis. *Antimicrob Agents Chemother*, **44**, 2081–2085.

Austin, N.C. and Darlow, B., (2004), Prophylactic oral antifungal agents to prevent systemic candida infection in preterm infants. *Cochrane Database Syst Rev*, CD003478.

Baillie, G.S. and Douglas, L.J., (1998), Effect of growth rate on resistance of *Candida albicans* biofilms to antifungal agents. *Antimicrob Agents Chemother*, **42**, 1900–1905.

Baillie, G.S. and Douglas, L.J., (2000), Matrix polymers of *Candida* biofilms and their possible role in biofilm resistance to antifungal agents. *J Antimicrob Chemother*, **46**, 397–403.

Baran, J., Jr, Muckatira, B. and Khatib, R., (2001), Candidemia before and during the fluconazole era: prevalence, type of species and approach to treatment in a tertiary care community hospital. *Scand J Infect Dis*, **33**, 137–139.

Barchiesi, F., Arzeni, D., Caselli, F. and Scalise, G., (2000a), Primary resistance to flucytosine among clinical isolates of *Candida* spp. *J Antimicrob Chemother*, **45**, 408–409.

Barchiesi, F., Calabrese, D., Sanglard, D., Falconi Di Francesco, L., Caselli, F., Giannini, D., Giacometti, A., Gavaudan, S. and Scalise, G., (2000b), Experimental induction of fluconazole resistance in *Candida tropicalis* ATCC 750. *Antimicrob Agents Chemother*, **44**, 1578–1584.

Barry, A.L., Pfaller, M.A., Brown, S.D., Espinel-Ingroff, A., Ghannoum, M.A., Knapp, C., Rennie, R.P., Rex, J.H. and Rinaldi, M.G., (2000), Quality control limits for broth microdilution susceptibility tests of ten antifungal agents. *J Clin Microbiol*, **38**, 3457–3459.

Bart-Delabesse, E., Boiron, P., Carlotti, A. and Dupont, B., (1993), *Candida albicans* genotyping in studies with patients with AIDS developing resistance to fluconazole. *J Clin Microbiol*, **31**, 2933–2937.

Bartizal, K., Gill, C.J., Abruzzo, G.K., Flattery, A.M., Kong, L., Scott, P.M., Smith, J.G., Leighton, C.E., Bouffard, A., Dropinski, J.F. and Balkovec, J., (1997), *In vitro* preclinical evaluation studies with the echinocandin antifungal MK-0991 (L-743,872). *Antimicrob Agents Chemother*, **41**, 2326–2332.

Bennett, J.E., Izumikawa, K. and Marr, K.A., (2004), Mechanism of increased fluconazole resistance in *Candida glabrata* during prophylaxis. *Antimicrob Agents Chemother*, **48**, 1773–1777.

Berenguer, J., Rodriguez-Tudela, J.L., Richard, C., Alvarez, M., Sanz, M.A., Gaztelurrutia, L., Ayats, J. and Martinez-Suarez, J.V., (1997), Deep infections caused by *Scedosporium prolificans*. A report on 16 cases in Spain and a review of the literature. Scedosporium Prolificans Spanish Study Group. *Medicine (Baltimore)*, **76**, 256–265.

Blumberg, E.A. and Reboli, A.C., (1996), Failure of systemic empirical treatment with amphotericin B to prevent candidemia in neutropenic patients with cancer. *Clin Infect Dis*, **22**, 462–466.

Bodey, G.P., (1992), Azole antifungal agents. *Clin Infect Dis*, **14** (Suppl 1), S161–S169.

Bohme, A., Karthaus, M. and Hoelzer, D., (1999), Antifungal prophylaxis in neutropenic patients with hematologic malignancies: is there a real benefit? *Chemotherapy*, **45**, 224–232.

Borst, A., Raimer, M.T., Warnock, D.W., Morrison, C.J. and Arthington-Skaggs, B.A., (2005), Rapid acquisition of stable azole resistance by *Candida glabrata* isolates obtained before the clinical introduction of fluconazole. *Antimicrob Agents Chemother*, **49**, 783–787.

Boutati, E.I. and Anaissie, E.J., (1997), *Fusarium*, a significant emerging pathogen in patients with hematologic malignancy: ten years' experience at a cancer center and implications for management. *Blood*, **90**, 999–1008.

Brammer, K.W. and Coates, P.E., (1994), Pharmacokinetics of fluconazole in pediatric patients. *Eur J Clin Microbiol Infect Dis*, **13**, 325–329.

Brezis, M., Rosen, S., Silva, P., Spokes, K. and Epstein, F.H., (1984), Polyene toxicity in renal medulla: injury mediated by transport activity. *Science*, **224**, 66–68.

Brown, J.M., (2004), Fungal infections in bone marrow transplant patients. *Curr Opin Infect Dis*, **17**, 347–352.

Capek, A., Simek, A., Leiner, J. and Weichet, J., (1970), Antimicrobial agents. VI. Antimycotic activity and problems of resistance. *Folia Microbiol (Praha)*, **15**, 314–317.

Castagnola, E., Machetti, M., Bucci, B. and Viscoli, C., (2004), Antifungal prophylaxis with azole derivatives. *Clin Microbiol Infect*, 10 (Suppl 1), 86–95.

Castro, C., Serrano, M.C., Flores, B., Espinel-Ingroff, A. and Martin-Mazuelos, E., (2004), Comparison of the Sensititre YeastOne colorimetric antifungal panel with a modified NCCLS M38-A method to determine the activity of voriconazole against clinical isolates of *Aspergillus* spp. *J Clin Microbiol*, **42**, 4358–4360.

Chryssanthou, E., (2001), Trends in antifungal susceptibility among Swedish *Candida* species bloodstream isolates from 1994 to 1998: comparison of the E-test and the Sensititre YeastOne Colorimetric Antifungal Panel with the NCCLS M27-A reference method. *J Clin Microbiol*, **39**, 4181–4183.

Chryssanthou, E., Cherif, H., Petrini, B., Kalin, M. and Bjorkholm, M., (2004), Surveillance of triazole susceptibility of colonizing yeasts in patients with haematological malignancies. *Scand J Infect Dis*, **36**, 855–859.

Clancy, C.J. and Nguyen, M.H., (1999), Correlation between *in vitro* susceptibility determined by E test and response to therapy with amphotericin B: results from a multicenter prospective study of candidemia. *Antimicrob Agents Chemother*, **43**, 1289–1290.

Clark, F.S., Parkinson, T., Hitchcock, C.A. and Gow, N.A., (1996), Correlation between rhodamine 123 accumulation and azole sensitivity in *Candida* species: possible role for drug efflux in drug resistance. *Antimicrob Agents Chemother*, **40**, 419–425.

Collin, B., Clancy, C.J. and Nguyen, M.H., (1999), Antifungal resistance in non-*albicans Candida* species. *Drug Resist Updat*, 2, 9–14.

Cosgrove, R.F., Beezer, A.E. and Miles, R.J., (1978), *In vitro* studies of amphotericin B in combination with the imidazole antifungal compounds clotrimazole and miconazole. *J Infect Dis*, **138**, 681–685.

Cuenca-Estrella, M., (2004), Combinations of antifungal agents in therapy—what value are they? *J Antimicrob Chemother*, **54**, 854–869.

Cuenca-Estrella, M., Diaz-Guerra, T.M., Mellado, E. and Rodriguez-Tudela, J.L., (2001), Flucytosine primary resistance in *Candida* species and *Cryptococcus neoformans*. *Eur J Clin Microbiol Infect Dis*, **20**, 276–279.

Cuenca-Estrella, M., Rodriguez, D., Almirante, B., Morgan, J., Planes, A.M., Almela, M., Mensa, J., Sanchez, F., Ayats, J., Gimenez, M., Salvado, M., Warnock, D.W., Pahissa, A. and Rodriguez-Tudela, J.L., (2005), *In vitro* susceptibilities of bloodstream isolates of *Candida* species to six antifungal agents: results from a population-based active surveillance programme, Barcelona, Spain, 2002–2003. *J Antimicrob Chemother*, **55**, 194–199.

da Silva Ferreira, M.E., Capellaro, J.L., dos Reis Marques, E., Malavazi, I., Perlin, D., Park, S., Anderson, J.B., Colombo, A.L., Arthington-Skaggs, B.A., Goldman, M.H. and Goldman, G.H., (2004), *In vitro* evolution of itraconazole resistance in *Aspergillus fumigatus* involves multiple mechanisms of resistance. *Antimicrob Agents Chemother*, **48**, 4405–4413.

Damaj, G., Ivanov, V., Le Brigand, B., D'Incan, E., Doglio, M.F., Bilger, K., Faucher, C., Vey, N. and Gastaut, J.A., (2004), Rapid improvement of disseminated aspergillosis with caspofungin/voriconazole combination in an adult leukemic patient. *Ann Hematol*, **83**, 390–393.

de Micheli, M., Bille, J., Schueller, C. and Sanglard, D., (2002), A common drug-responsive element mediates the upregulation of the *Candida albicans* ABC transporters *CDR1* and *CDR2*, two genes involved in antifungal drug resistance. *Mol Microbiol*, **43**, 1197–1214.

Del Poeta, M., Schell, W.A. and Perfect, J.R., (1997), In vitro antifungal activity of pneumocandin L-743,872 against a variety of clinically important molds. *Antimicrob Agents Chemother*, **41**, 1835–1836.

Denning, D.W., (1997), Echinocandins and pneumocandins—a new antifungal class with a novel mode of action. *J Antimicrob Chemother*, **40**, 611–614.

Denning, D.W., (2003), Echinocandin antifungal drugs. *Lancet*, **362**, 1142–1151.

Denning, D.W., Hanson, L.H., Perlman, A.M. and Stevens, D.A., (1992), *In vitro* susceptibility and synergy studies of *Aspergillus* species to conventional and new agents. *Diagn Microbiol Infect Dis*, **15**, 21–34.

Denning, D.W., Venkateswarlu, K., Oakley, K.L., Anderson, M.J., Manning, N.J., Stevens, D.A., Warnock, D.W. and Kelly, S.L., (1997), Itraconazole resistance in *Aspergillus fumigatus*. *Antimicrob Agents Chemother*, **41**, 1364–1368.

Denning, D.W. and Warn, P., (1999), Dose range evaluation of liposomal nystatin and comparisons with amphotericin B and amphotericin B lipid complex in temporarily neutropenic mice infected with an isolate of *Aspergillus fumigatus* with reduced susceptibility to amphotericin B. *Antimicrob Agents Chemother*, **43**, 2592–2599.

Diamond, D.M., Bauer, M., Daniel, B.E., Leal, M.A., Johnson, D., Williams, B.K., Thomas, A.M., Ding, J.C., Najvar, L., Graybill, J.R. and Larsen, R.A., (1998), Amphotericin B colloidal dispersion combined with flucytosine with or without fluconazole for treatment of murine cryptococcal meningitis. *Antimicrob Agents Chemother*, **42**, 528–533.

Diaz-Guerra, T.M., Martinez-Suarez, J.V., Laguna, F., Valencia, E. and Rodriguez-Tudela, J.L., (1998), Change in fluconazole susceptibility patterns and genetic relationship among oral *Candida albicans* isolates. *AIDS*, **12**, 1601–1610.

Dick, J.D., Merz, W.G. and Saral, R., (1980), Incidence of polyene-resistant yeasts recovered from clinical specimens. *Antimicrob Agents Chemother*, **18**, 158–163.

Dick, J.D., Rosengard, B.R., Merz, W.G., Stuart, R.K., Hutchins, G.M. and Saral, R., (1985), Fatal disseminated candidiasis due to amphotericin-B-resistant *Candida guilliermondii*. *Ann Intern Med*, **102**, 67–68.

Diekema, D.J., Messer, S.A., Brueggemann, A.B., Coffman, S.L., Doern, G.V., Herwaldt, L.A. and Pfaller, M.A., (2002), Epidemiology of candidemia: 3-year results from the emerging infections and the epidemiology of Iowa organisms study. *J Clin Microbiol*, **40**, 1298–1302.

Dinubile, M.J., Lupinacci, R.J., Berman, R.S. and Sable, C.A., (2002), Response and relapse rates of candidal esophagitis in HIV-infected patients treated with caspofungin. *AIDS Res Hum Retroviruses*, **18**, 903–908.

Dismukes, W.E., Cloud, G., Gallis, H.A., Kerkering, T.M., Medoff, G., Craven, P.C., Kaplowitz, L.G., Fisher, J.F., Gregg, C.R., Bowles, C.A., Shadomy, S., Stamm, A.M., Diasio, R.B., Kaufman, L., Soong, S.-J., Blackwelder, W. and the National Institute of Allergy and Infectious Diseases Mycoses Study Group, (1987), Treatment of cryptococcal meningitis with combination amphotericin B and flucytosine for four as compared with six weeks. *N Engl J Med*, **317**, 334–341.

Dominguez, J.M., Kelly, V.A., Kinsman, O.S., Marriott, M.S., Gomez de las Heras, F. and Martin, J.J., (1998), Sordarins: a new class of antifungals with selective inhibition of the protein synthesis elongation cycle in yeasts. *Antimicrob Agents Chemother*, **42**, 2274–2278.

Dominguez, J.M. and Martin, J.J., (1998), Identification of elongation factor 2 as the essential protein targeted by sordarins in *Candida albicans*. *Antimicrob Agents Chemother*, **42**, 2279–2283.

Douglas, C.M., (2001), Fungal beta(1,3)-d-glucan synthesis. *Med Mycol*, 39 (Suppl 1), 55–66.

Douglas, C.M., Foor, F., Marrinan, J.A., Morin, N., Nielsen, J.B., Dahl, A.M., Mazur, P., Baginsky, W., Li, W., el-Sherbeini, M., Clemas, J.A., Mandala, S.M., Frommer, B.R. and Kurtz, M.B., (1994), The *Saccharomyces cerevisiae FKS1* (*ETG1*) gene encodes an integral membrane protein which is a subunit of 1,3-beta-d-glucan synthase. *Proc Natl Acad Sci USA*, **91**, 12907–12911.

Dummer, J.S., Lazariashvilli, N., Barnes, J., Ninan, M. and Milstone, A.P., (2004), A survey of anti-fungal management in lung transplantation. *J Heart Lung Transplant*, **23**, 1376–1381.

Dupont, B., (2002), Overview of the lipid formulations of amphotericin B. *J Antimicrob Chemother*, 49 (Suppl 1), 31–36.

Edmond, M.B., Wallace, S.E., McClish, D.K., Pfaller, M.A., Jones, R.N. and Wenzel, R.P., (1999), Nosocomial bloodstream infections in United States hospitals: a three-year analysis. *Clin Infect Dis*, **29**, 239–244.

Ehrmann, S., Bastides, F., Gissot, V., Mercier, E., Magro, P., Bailly, E. and Legras, A., (2005), Cerebral aspergillosis in the critically ill: two cases of successful medical treatment. *Intensive Care Med*, **31**, 738–742.

Ellis, D., (2002), Amphotericin B: spectrum and resistance. *J Antimicrob Chemother*, 49 (Suppl 1), 7–10.

Espinel-Ingroff, A., (1998), Comparison of *in vitro* activities of the new triazole SCH56592 and the echinocandins MK-0991 (L-743,872) and LY303366 against opportunistic filamentous and dimorphic fungi and yeasts. *J Clin Microbiol*, **36**, 2950–2956.

Espinel-Ingroff, A., (2001), Germinated and nongerminated conidial suspensions for testing of susceptibilities of *Aspergillus* spp. to amphotericin B, itraconazole, posaconazole, ravuconazole, and voriconazole. *Antimicrob Agents Chemother*, **45**, 605–607.

Fan-Havard, P., Capano, D., Smith, S.M., Mangia, A. and Eng, R.H., (1991), Development of resistance in *Candida* isolates from patients receiving prolonged antifungal therapy. *Antimicrob Agents Chemother*, **35**, 2302–2305.

Favre, B., Ghannoum, M.A. and Ryder, N.S., (2004), Biochemical characterization of terbinafine-resistant *Trichophyton rubrum* isolates. *Med Mycol*, **42**, 525–529.

Fichtenbaum, C.J., Koletar, S., Yiannoutsos, C., Holland, F., Pottage, J., Cohn, S.E., Walawander, A., Frame, P., Feinberg, J., Saag, M., Van der Horst, C. and Powderly, W.G., (2000), Refractory mucosal candidiasis in advanced human immunodeficiency virus infection. *Clin Infect Dis*, **30**, 749–756.

Fidel, P.L., Jr. and Sobel, J.D., (1996), Immunopathogenesis of recurrent vulvovaginal candidiasis. *Clin Microbiol Rev*, 9, 335–348.

Fling, M.E., Kopf, J., Tamarkin, A., Gorman, J.A., Smith, H.A. and Koltin, Y., (1991), Analysis of a *Candida albicans* gene that encodes a novel mechanism for resistance to benomyl and methotrexate. *Mol Gen Genet*, **227**, 318–329.

Franzot, S.P. and Casadevall, A., (1997), Pneumocandin L-743,872 enhances the activities of amphotericin B and fluconazole against *Cryptococcus neoformans in vitro*. *Antimicrob Agents Chemother*, **41**, 331–336.

Fukuoka, T., Johnston, D.A., Winslow, C.A., de Groot, M.J., Burt, C., Hitchcock, C.A. and Filler, S.G., (2003), Genetic basis for differential activities of fluconazole and voriconazole against *Candida krusei*. *Antimicrob Agents Chemother*, **47**, 1213–1219.

Geber, A., Hitchcock, C.A., Swartz, J.E., Pullen, F.S., Marsden, K.E., Kwon-Chung, K.J. and Bennett, J.E., (1995), Deletion of the *Candida glabrata ERG3* and *ERG11* genes: effect on cell viability, cell growth, sterol composition, and antifungal susceptibility. *Antimicrob Agents Chemother*, **39**, 2708–2717.

Georgopapadakou, N.H. and Walsh, T.J., (1996), Antifungal agents: chemotherapeutic targets and immunologic strategies. *Antimicrob Agents Chemother*, **40**, 279–291.

Ghannoum, M.A. and Elewski, B., (1999), Successful treatment of fluconazole-resistant oropharyngeal candidiasis by a combination of fluconazole and terbinafine. *Clin Diagn Lab Immunol*, 6, 921–923.

Ghannoum, M.A. and Rice, L.B., (1999), Antifungal agents: mode of action, mechanisms of resistance and correlation of these mechanisms with bacterial resistance. *Clin Microbiol Rev*, **12**, 501–517.

Godoy, P., Tiraboschi, I.N., Severo, L.C., Bustamante, B., Calvo, B., Almeida, L.P., da Matta, D.A. and Colombo, A.L., (2003), Species distribution and antifungal susceptibility profile of *Candida* spp. bloodstream isolates from Latin American hospitals. *Mem Inst Oswaldo Cruz*, **98**, 401–405.

Golan, Y., (2005), Overview of transplant mycology. *Am J Health Syst Pharm*, **62**, S17–21.

Goodman, J.L., Winston, D.J., Greenfield, R.A., Chandrasekar, P.H., Fox, B., Kaizer, H., Shadduck, R.K., Shea, T.C., Stiff, P., Friedman, D.J., Powderly, W.G., Silber, J.L., Horowitz, H., Lichtin, A., Wolff, S.N., Mangan, K.F., Silver, S.M., Weisdorf, D., Ho, W.G., Gilbert, G. and Buell, D., (1992), A controlled trial of fluconazole to prevent fungal infections in patients undergoing bone marrow transplantation. *N Engl J Med*, **326**, 845–851.

Gotzsche, P.C. and Johansen, H.K., (1997), Meta-analysis of prophylactic or empirical antifungal treatment versus placebo or no treatment in patients with cancer complicated by neutropenia. *BMJ*, **314**, 1238–1244.

Graybill, J.R., (2000), The role of murine models in the development of antifungal therapy for systemic mycoses. *Drug Resist Updat*, 3, 364–383.

Graybill, J.R., Najvar, L., Fothergill, A., Bocanegra, R. and de las Heras, F.G., (1999), Activities of sordarins in murine histoplasmosis. *Antimicrob Agents Chemother*, **43**, 1716–1718.

Groll, A.H., Petraitis, V., Petraitiene, R., Field-Ridley, A., Calendario, M., Bacher, J., Piscitelli, S.C. and Walsh, T.J., (1999), Safety and efficacy of multilamellar liposomal nystatin against disseminated candidiasis in persistently neutropenic rabbits. *Antimicrob Agents Chemother*, **43**, 2463–2467.

Hajjeh, R.A., Sofair, A.N., Harrison, L.H., Lyon, G.M., Arthington-Skaggs, B.A., Mirza, S.A., Phelan, M., Morgan, J., Lee-Yang, W., Ciblak, M.A., Benjamin, L.E., Sanza, L.T., Huie, S., Yeo, S.F., Brandt, M.E. and Warnock, D.W., (2004), Incidence of bloodstream infections due to *Candida* species and *in vitro* susceptibilities of isolates collected from 1998 to 2000 in a population-based active surveillance program. *J Clin Microbiol*, **42**, 1519–1527.

Hamilton-Miller, J.M., (1973), Chemistry and biology of the polyene macrolide antibiotics. *Bacteriol Rev*, **37**, 166–196.

Hawser, S.P. and Douglas, L.J., (1994), Biofilm formation by *Candida* species on the surface of catheter materials *in vitro*. *Infect Immun*, **62**, 915–921.

Henry, K.W., Nickels, J.T. and Edlind, T.D., (2000), Upregulation of *ERG* genes in *Candida* species by azoles and other sterol biosynthesis inhibitors. *Antimicrob Agents Chemother*, **44**, 2693–2700.

Herbrecht, R., Denning, D.W., Patterson, T.F., Bennett, J.E., Greene, R.E., Oestmann, J.W., Kern, W.V., Marr, K.A., Ribaud, P., Lortholary, O., Sylvester, R., Rubin, R.H., Wingard, J.R., Stark, P., Durand, C., Caillot, D., Thiel, E., Chandrasekar, P.H., Hodges, M.R., Schlamm, H.T.,

Troke, P.F. and de Pauw, B., (2002), Voriconazole versus amphotericin B for primary therapy of invasive aspergillosis. *N Engl J Med*, **347**, 408–415.

Hernaez, M.L., Gil, C., Pla, J. and Nombela, C., (1998), Induced expression of the *Candida albicans* multidrug resistance gene *CDR1* in response to fluconazole and other antifungals. *Yeast*, **14**, 517–526.

Herreros, E., Martinez, C.M., Almela, M.J., Marriott, M.S., De Las Heras, F.G. and Gargallo-Viola, D., (1998), Sordarins: *in vitro* activities of new antifungal derivatives against pathogenic yeasts, *Pneumocystis carinii*, and filamentous fungi. *Antimicrob Agents Chemother*, **42**, 2863–2869.

Herrling, S., Sous, H., Kruepe, W., Osterloh, G. and Mueckter, H., (1959), Experimental studies on a new combination effective against fungi. *Arzneimittelforschung*, **9**, 489–494.

Hitchcock, C.A., (1993), Resistance of *Candida albicans* to azole antifungal agents. *Biochem Soc Trans*, **21**, 1039–1047.

Hitchcock, C.A., Pye, G.W., Troke, P.F., Johnson, E.M. and Warnock, D.W., (1993), Fluconazole resistance in *Candida glabrata*. *Antimicrob Agents Chemother*, **37**, 1962–1965.

Hoban, D.J., Zhanel, G.G. and Karlowsky, J.A., (1999), *In vitro* susceptibilities of *Candida* and *Cryptococcus neoformans* isolates from blood cultures of neutropenic patients. *Antimicrob Agents Chemother*, **43**, 1463–1464.

Ifran, A., Kaptan, K. and Beyan, C., (2005), Efficacy of caspofungin in prophylaxis and treatment of an adult leukemic patient with invasive pulmonary aspergillosis in allogeneic stem cell transplantation. *Mycoses*, **48**, 146–148.

Imhof, A., Balajee, S.A., Fredricks, D.N., Englund, J.A. and Marr, K.A., (2004), Breakthrough fungal infections in stem cell transplant recipients receiving voriconazole. *Clin Infect Dis*, **39**, 743–746.

Jessup, C.J., Pfaller, M.A., Messer, S.A., Zhang, J., Tumberland, M., Mbidde, E.K. and Ghannoum, M.A., (1998), Fluconazole susceptibility testing of *Cryptococcus neoformans*: comparison of two broth microdilution methods and clinical correlates among isolates from Ugandan AIDS patients. *J Clin Microbiol*, **36**, 2874–2876.

Johnson, E.M., Oakley, K.L., Radford, S.A., Moore, C.B., Warn, P., Warnock, D.W. and Denning, D.W., (2000), Lack of correlation of in vitro amphotericin B susceptibility testing with outcome in a murine model of *Aspergillus* infection. *J Antimicrob Chemother*, **45**, 85–93.

Johnson, E.M., Warnock, D.W., Luker, J., Porter, S.R. and Scully, C., (1995), Emergence of azole drug resistance in *Candida* species from HIV-infected patients receiving prolonged fluconazole therapy for oral candidosis. *J Antimicrob Chemother*, **35**, 103–114.

Joseph-Horne, T., Hollomon, D., Loeffler, R.S. and Kelly, S.L., (1995a), Cross-resistance to polyene and azole drugs in *Cryptococcus neoformans*. *Antimicrob Agents Chemother*, **39**, 1526–1529.

Joseph-Horne, T., Manning, N.J., Hollomon, D. and Kelly, S.L., (1995b), Defective sterol delta 5(6) desaturase as a cause of azole resistance in *Ustilago maydis*. *FEMS Microbiol Lett*, **127**, 29–34.

Kakeya, H., Miyazaki, Y., Miyazaki, H., Nyswaner, K., Grimberg, B. and Bennett, J.E., (2000), Genetic analysis of azole resistance in the Darlington strain of *Candida albicans*. *Antimicrob Agents Chemother*, **44**, 2985–2990.

Kao, A.S., Brandt, M.E., Pruitt, W.R., Conn, L.A., Perkins, B.A., Stephens, D.S., Baughman, W.S., Reingold, A.L., Rothrock, G.A., Pfaller, M.A., Pinner, R.W. and Hajjeh, R.A., (1999), The epidemiology of candidemia in two United States cities: results of a population-based active surveillance. *Clin Infect Dis*, **29**, 1164–1170.

Kartsonis, N.A., Nielsen, J. and Douglas, C.M., (2003), Caspofungin: the first in a new class of antifungal agents. *Drug Resist Updat*, 6, 197–218.

Katiyar, S.K. and Edlind, T.D., (2001), Identification and expression of multidrug resistance-related ABC transporter genes in *Candida krusei*. *Med Mycol*, **39**, 109–116.

Kaufman, D., Boyle, R., Hazen, K.C., Patrie, J.T., Robinson, M. and Donowitz, L.G., (2001), Fluconazole prophylaxis against fungal colonization and infection in preterm infants. *N Engl J Med*, **345**, 1660–1666.

Kelly, R., Register, E., Hsu, M.J., Kurtz, M. and Nielsen, J., (1996a), Isolation of a gene involved in 1,3-beta-glucan synthesis in *Aspergillus nidulans* and purification of the corresponding protein. *J Bacteriol*, **178**, 4381–4391.

Kelly, S.L., (1993), Molecular studies on azole sensitivity in fungi, pp. 199–213. In B.K. Maresca, G.S.; Yamaguchi, H. (Ed.): *Molecular Biology and its Application to Medical Mycology*, Springer-Verlag, Berlin, Heidelberg.

Kelly, S.L., Lamb, D.C., Corran, A.J., Baldwin, B.C. and Kelly, D.E., (1995), Mode of action and resistance to azole antifungals associated with the formation of 14 alpha-methylergosta-8,24(28)-dien-3 beta,6 alpha-diol. *Biochem Biophys Res Commun*, **207**, 910–915.

Kelly, S.L., Lamb, D.C., Kelly, D.E., Loeffler, J. and Einsele, H., (1996b), Resistance to fluconazole and amphotericin in *Candida albicans* from AIDS patients. *Lancet*, **348**, 1523–1524.

Kelly, S.L., Lamb, D.C., Kelly, D.E., Manning, N.J., Loeffler, J., Hebart, H., Schumacher, U. and Einsele, H., (1997), Resistance to fluconazole and cross-resistance to amphotericin B in *Candida albicans* from AIDS patients caused by defective sterol delta5,6-desaturation. *FEBS Lett*, **400**, 80–82.

Kelly, S.L., Lamb, D.C., Taylor, M., Corran, A.J., Baldwin, B.C. and Powderly, W.G., (1994), Resistance to amphotericin B associated with defective sterol delta 8–>7 isomerase in a *Cryptococcus neoformans* strain from an AIDS patient. *FEMS Microbiol Lett*, **122**, 39–42.

Kerridge, D., (1986), Mode of action of clinically important antifungal drugs. *Adv Microb Physiol*, **27**, 1–72.

Kicklighter, S.D., Springer, S.C., Cox, T., Hulsey, T.C. and Turner, R.B., (2001), Fluconazole for prophylaxis against candidal rectal colonization in the very low birth weight infant. *Pediatrics*, **107**, 293–298.

Kim, S.J., Kwon-Chung, K.J., Milne, G.W. and Prescott, B., (1974), Polyene-resistant mutants of *Aspergillus fennelliae*: identification of sterols. *Antimicrob Agents Chemother*, **6**, 405–410.

Klepser, M.E., Wolfe, E.J. and Pfaller, M.A., (1998), Antifungal pharmacodynamic characteristics of fluconazole and amphotericin B against *Cryptococcus neoformans*. *J Antimicrob Chemother*, **41**, 397–401.

Klepser, M.E., Wolfe, E.J., Jones, R.N., Nightingale, C.H. and Pfaller, M.A., (1997), Antifungal pharmacodynamic characteristics of fluconazole and amphotericin B tested against *Candida albicans*. *Antimicrob Agents Chemother*, **41**, 1392–1395.

Kohli, A., Smriti, Mukhopadhyay, K., Rattan, A. and Prasad, R., (2002), *In vitro* low-level resistance to azoles in *Candida albicans* is associated with changes in membrane lipid fluidity and asymmetry. *Antimicrob Agents Chemother*, **46**, 1046–1052.

Kojic, E.M. and Darouiche, R.O., (2003), Comparison of adherence of *Candida albicans* and *Candida parapsilosis* to silicone catheters *in vitro* and *in vivo*. *Clin Microbiol Infect*, **9**, 684–690.

Konishi, M., Nishio, M., Saitoh, K., Miyaki, T., Oki, T. and Kawaguchi, H., (1989), Cispentacin, a new antifungal antibiotic. I. Production, isolation, physico-chemical properties and structure. *J Antibiot (Tokyo)*, **42**, 1749–1755.

Kontoyiannis, D.P., Lionakis, M.S., Lewis, R.E., Chamilos, G., Healy, M., Perego, C., Safdar, A., Kantarjian, H., Champlin, R., Walsh, T.J. and Raad, I.I., (2005), Zygomycosis in a tertiary-care cancer center in the era of *Aspergillus*-active antifungal therapy: a case–control observational study of 27 recent cases. *J Infect Dis*, **191**, 1350–1360.

Kovacicova, G., Hanzen, J., Pisarcikova, M., Sejnova, D., Horn, J., Babela, R., Svetlansky, I., Lovaszova, M., Gogova, M. and Krcmery, V., (2001), Nosocomial fungemia due to amphotericin B-resistant *Candida* spp. in three pediatric patients after previous neurosurgery for brain tumors. *J Infect Chemother*, **7**, 45–48.

Kunova, A., Trupl, J., Demitrovicova, A., Jesenska, Z., Grausova, S., Grey, E., Pichna, P., Kralovicova, K., Sorkovska, D., Krupova, I., Spanik, S., Studena, M., Koren, P. and Krcmery, V., Jr., (1997), Eight-year surveillance of non-albicans *Candida* spp. in an oncology department prior to and after fluconazole had been introduced into antifungal prophylaxis. *Microb Drug Resist*, 3, 283–287.

Kurtz, M.B. and Douglas, C.M., (1997), Lipopeptide inhibitors of fungal glucan synthase. *J Med Vet Mycol*, **35**, 79–86.

Kurtz, M.B., Abruzzo, G., Flattery, A., Bartizal, K., Marrinan, J.A., Li, W., Milligan, J., Nollstadt, K. and Douglas, C.M., (1996), Characterization of echinocandin-resistant mutants of *Candida albicans*: genetic, biochemical, and virulence studies. *Infect Immun*, **64**, 3244–3251.

Laguna, F., Rodriguez-Tudela, J.L., Martinez-Suarez, J.V., Polo, R., Valencia, E., Diaz-Guerra, T.M., Dronda, F. and Pulido, F., (1997), Patterns of fluconazole susceptibility in isolates from human immunodeficiency virus-infected patients with oropharyngeal candidiasis due to *Candida albicans*. *Clin Infect Dis*, **24**, 124–130.

Lamb, D.C., Corran, A., Baldwin, B.C., Kwon-Chung, J. and Kelly, S.L., (1995), Resistant P45051A1 activity in azole antifungal tolerant *Cryptococcus neoformans* from AIDS patients. *FEBS Lett*, **368**, 326–330.

Lasker, B.A., Elie, C.M., Lott, T.J., Espinel-Ingroff, A., Gallagher, L., Kuykendall, R.J., Kellum, M.E., Pruitt, W.R., Warnock, D.W., Rimland, D., McNeil, M.M. and Reiss, E., (2001), Molecular epidemiology of *Candida albicans* strains isolated from the oropharynx of HIV-positive patients at successive clinic visits. *Med Mycol*, **39**, 341–352.

Lass-Florl, C., Nagl, M., Gunsilius, E., Speth, C., Ulmer, H. and Wurzner, R., (2002), *In vitro* studies on the activity of amphotericin B and lipid-based amphotericin B formulations against *Aspergillus* conidia and hyphae. *Mycoses*, **45**, 166–169.

Laverdiere, M., Rotstein, C., Bow, E.J., Roberts, R.S., Ioannou, S., Carr, D. and Moghaddam, N., (2000), Impact of fluconazole prophylaxis on fungal colonization and infection rates in neutropenic patients. The Canadian Fluconazole Study. *J Antimicrob Chemother*, **46**, 1001–1008.

Le Guennec, R., Reynes, J., Mallie, M., Pujol, C., Janbon, F. and Bastide, J.M., (1995), Fluconazole- and itraconazole-resistant *Candida albicans* strains from AIDS patients: multilocus enzyme electrophoresis analysis and antifungal susceptibilities. *J Clin Microbiol*, **33**, 2732–2737.

Le Monte, A.M., Goldman, M., Smedema, M.L., Connolly, P.A., McKinsey, D.S., Cloud, G.A., Kauffman, C.A. and Wheat, L.J., (2001), DNA fingerprinting of serial *Candida albicans* isolates obtained during itraconazole prophylaxis in patients with AIDS. *Med Mycol*, **39**, 207–213.

Lee, M.K., Williams, L.E., Warnock, D.W. and Arthington-Skaggs, B.A., (2004), Drug resistance genes and trailing growth in *Candida albicans* isolates. *J Antimicrob Chemother*, **53**, 217–224.

Lee, S.C., Fung, C.P., Huang, J.S., Tsai, C.J., Chen, K.S., Chen, H.Y., Lee, N., See, L.C. and Shieh, W.B., (2000), Clinical correlates of antifungal macrodilution susceptibility test results for non-AIDS patients with severe *Candida* infections treated with fluconazole. *Antimicrob Agents Chemother*, **44**, 2715–2718.

Leigh, J.E., Shetty, K. and Fidel, P.L., Jr., (2004), Oral opportunistic infections in HIV-positive individuals: review and role of mucosal immunity. *AIDS Patient Care STDS*, **18**, 443–456.

Lewis, R.E., Lund, B.C., Klepser, M.E., Ernst, E.J. and Pfaller, M.A., (1998), Assessment of antifungal activities of fluconazole and amphotericin B administered alone and in combination against *Candida albicans* by using a dynamic *in vitro* mycotic infection model. *Antimicrob Agents Chemother*, **42**, 1382–1386.

Lin, D., Wu, L.C., Rinaldi, M.G. and Lehmann, P.F., (1995), Three distinct genotypes within *Candida parapsilosis* from clinical sources. *J Clin Microbiol*, **33**, 1815–1821.

Lionakis, M.S., Lewis, R.E., Torres, H.A., Albert, N.D., Raad, II and Kontoyiannis, D.P., (2005), Increased frequency of non-*fumigatus Aspergillus* species in amphotericin B- or triazole-preexposed cancer patients with positive cultures for aspergilli. *Diagn Microbiol Infect Dis*, **52**, 15–20.

Lipsett, P.A., (2004), Clinical trials of antifungal prophylaxis among patients in surgical intensive care units: concepts and considerations. *Clin Infect Dis*, 39 (Suppl 4), S193–S199.

Lischewski, A., Ruhnke, M., Tennagen, I., Schonian, G., Morschhauser, J. and Hacker, J., (1995), Molecular epidemiology of *Candida* isolates from AIDS patients showing different fluconazole resistance profiles. *J Clin Microbiol*, **33**, 769–771.

Loeffler, J., Einsele, H., Hebart, H., Schumacher, U., Hrastnik, C. and Daum, G., (2000), Phospholipid and sterol analysis of plasma membranes of azole-resistant *Candida albicans* strains. *FEMS Microbiol Lett*, **185**, 59–63.

Loeffler, J. and Stevens, D.A., (2003), Antifungal drug resistance. *Clin Infect Dis*, **36**, S31–S41.

Lopez-Ribot, J.L., McAtee, R.K., Lee, L.N., Kirkpatrick, W.R., White, T.C., Sanglard, D. and Patterson, T.F., (1998), Distinct patterns of gene expression associated with development of fluconazole resistance in serial *Candida albicans* isolates from human immunodeficiency virus-infected patients with oropharyngeal candidiasis. *Antimicrob Agents Chemother*, **42**, 2932–2937.

Lopez-Ribot, J.L., McAtee, R.K., Perea, S., Kirkpatrick, W.R., Rinaldi, M.G. and Patterson, T.F., (1999), Multiple resistant phenotypes of *Candida albicans* coexist during episodes of oropharyngeal candidiasis in human immunodeficiency virus-infected patients. *Antimicrob Agents Chemother*, **43**, 1621–1630.

Lortholary, O. and Dupont, B., (1997), Antifungal prophylaxis during neutropenia and immunodeficiency. *Clin Microbiol Rev*, **10**, 477–504.

Lozano-Chiu, M., Nelson, P.W., Lancaster, M., Pfaller, M.A. and Rex, J.H., (1997), Lot-to-lot variability of antibiotic medium 3 used for testing susceptibility of *Candida* isolates to amphotericin B. *J Clin Microbiol*, **35**, 270–272.

Lupetti, A., Danesi, R., Campa, M., Del Tacca, M. and Kelly, S., (2002), Molecular basis of resistance to azole antifungals. *Trends Mol Med*, 8, 76–81.

Maertens, J., Raad, I., Sable, C.A., Ngai, A., Berman, R., Patterson, T.F., Denning, D. and Walsh, T., (2000), Multicenter, noncomparative study to evaluate safety and efficacy of caspofungin in adults with aspergillis refractory or intolerant to amphotericin B, amphotericin B lipid formulation, or azoles. *40th Interscience Conference on Antimicrobial Agents and Chemotherapy*, Toronto, Canada.

Mago, N. and Khuller, G.K., (1989), Influence of lipid composition on the sensitivity of *Candida albicans* to antifungal agents. *Indian J Biochem Biophys*, **26**, 30–33.

Manavathu, E.K., Cutright, J. and Chandrasekar, P.H., (1999), Comparative study of susceptibilities of germinated and ungerminated conidia of *Aspergillus fumigatus* to various antifungal agents. *J Clin Microbiol*, **37**, 858–861.

Marco, F., Pfaller, M.A., Messer, S.A. and Jones, R.N., (1998), Activity of MK-0991 (L-743,872), a new echinocandin, compared with those of LY303366 and four other antifungal agents tested against blood stream isolates of *Candida* spp. *Diagn Microbiol Infect Dis*, **32**, 33–37.

Marichal, P., Gorrens, J. and Vanden Bossche, H., (1985), The action of itraconazole and ketoconazole on growth and sterol synthesis in *Aspergillus fumigatus* and *Aspergillus niger*. *Sabouraudia*, **23**, 13–21.

Marichal, P. and Vanden Bossche, H., (1995), Mechanisms of resistance to azole antifungals. *Acta Biochim Pol*, **42**, 509–516.

Marr, K.A., (2004), Invasive *Candida* infections: the changing epidemiology. *Oncology (Huntingt)*, **18**, 9-14.

Marr, K.A., Lyons, C.N., Rustad, T.R., Bowden, R.A. and White, T.C., (1998), Rapid, transient fluconazole resistance in *Candida albicans* is associated with increased mRNA levels of *CDR*. *Antimicrob Agents Chemother*, **42**, 2584–2589.

Marr, K.A., Seidel, K., Slavin, M.A., Bowden, R.A., Schoch, H.G., Flowers, M.E., Corey, L. and Boeckh, M., (2000a), Prolonged fluconazole prophylaxis is associated with persistent protection against candidiasis-related death in allogeneic marrow transplant recipients: long-term follow-up of a randomized, placebo-controlled trial. *Blood*, **96**, 2055–2061.

Marr, K.A., Seidel, K., White, T.C. and Bowden, R.A., (2000b), Candidemia in allogeneic blood and marrow transplant recipients: evolution of risk factors after the adoption of prophylactic fluconazole. *J Infect Dis*, **181**, 309–316.

Marr, K.A., White, T.C., van Burik, J.A. and Bowden, R.A., (1997), Development of fluconazole resistance in *Candida albicans* causing disseminated infection in a patient undergoing marrow transplantation. *Clin Infect Dis*, **25**, 908–910.

Martin, E., Maier, F. and Bhakdi, S., (1994a), Antagonistic effects of fluconazole and 5-fluorocytosine on candidacidal action of amphotericin B in human serum. *Antimicrob Agents Chemother*, **38**, 1331–1338.

Martin, E., Stuben, A., Gorz, A., Weller, U. and Bhakdi, S., (1994b), Novel aspect of amphotericin B action: accumulation in human monocytes potentiates killing of phagocytosed *Candida albicans*. *Antimicrob Agents Chemother*, **38**, 13–22.

Marty, F.M., Cosimi, L.A. and Baden, L.R., (2004), Breakthrough zygomycosis after voriconazole treatment in recipients of hematopoietic stem-cell transplants. *N Engl J Med*, **350**, 950–952.

Mattner, F., Weissbrodt, H. and Strueber, M., (2004), Two case reports: fatal *Absidia corymbifera* pulmonary tract infection in the first postoperative phase of a lung transplant patient receiving voriconazole prophylaxis and transient bronchial *Absidia corymbifera* colonization in a lung transplant patient. *Scand J Infect Dis*, **36**, 312–314.

McCullough, M. and Hume, S., (1995), A longitudinal study of the change in resistance patterns and genetic relationship of oral *Candida albicans* from HIV-infected patients. *J Med Vet Mycol*, **33**, 33–37.

McGinnis, M.R., Pasarell, L., Sutton, D.A., Fothergill, A.W., Cooper, C.R., Jr. and Rinaldi, M.G., (1997), *In vitro* evaluation of voriconazole against some clinically important fungi. *Antimicrob Agents Chemother*, **41**, 1832–1834.

McNeil, M.M., Nash, S.L., Hajjeh, R.A., Phelan, M.A., Conn, L.A., Plikaytis, B.D. and Warnock, D.W., (2001), Trends in Mortality Due to Invasive Mycotic Diseases in the United States, 1980–1997. *Clin Infect Dis*, **33**, 641–647.

Meade, R.H., 3rd, (1979), Drug therapy reviews: clinical pharmacology and therapeutic use of antimycotic drugs. *Am J Hosp Pharm*, **36**, 1326–1334.

Medoff, G., Brajtburg, J., Kobayashi, G.S. and Bolard, J., (1983), Antifungal agents useful in therapy of systemic fungal infections. *Annu Rev Pharmacol Toxicol*, **23**, 303–330.

Mehta, R.T., Hopfer, R.L., Gunner, L.A., Juliano, R.L. and Lopez-Berestein, G., (1987), Formulation, toxicity, and antifungal activity *in vitro* of liposome-encapsulated nystatin as therapeutic agent for systemic candidiasis. *Antimicrob Agents Chemother*, **31**, 1897–1900.

Meletiadis, J., Mouton, J.W., Meis, J.F., Bouman, B.A. and Verweij, P.E., (2002), Comparison of the Etest and the sensititre colorimetric methods with the NCCLS proposed standard for antifungal susceptibility testing of *Aspergillus* species. *J Clin Microbiol*, **40**, 2876–2885.

Menichetti, F., (2004), How to improve the design of trials of antifungal prophylaxis among neutropenic adults with acute leukemia. *Clin Infect Dis*, **39** (Suppl 4), S181–S184.

Meunier, F., Aoun, M. and Bitar, N., (1992), Candidemia in immunocompromised patients. *Clin Infect Dis*, **14** (Suppl 1), S120–S125.

Michaelis, S. and Berkower, C., (1995), Sequence comparison of yeast ATP-binding cassette proteins. *Cold Spring Harb Symp Quant Biol*, **60**, 291–307.

Minari, A., Husni, R., Avery, R.K., Longworth, D.L., DeCamp, M., Bertin, M., Schilz, R., Smedira, N., Haug, M.T., Mehta, A. and Gordon, S.M., (2002), The incidence of invasive aspergillosis among solid organ transplant recipients and implications for prophylaxis in lung transplants. *Transpl Infect Dis*, **4**, 195–200.

Mio, T., Adachi-Shimizu, M., Tachibana, Y., Tabuchi, H., Inoue, S.B., Yabe, T., Yamada-Okabe, T., Arisawa, M., Watanabe, T. and Yamada-Okabe, H., (1997), Cloning of the *Candida albicans* homolog of *Saccharomyces cerevisiae GSC1/FKS1* and its involvement in beta-1,3-glucan synthesis. *J Bacteriol*, **179**, 4096–4105.

Moran, G.P., Sanglard, D., Donnelly, S.M., Shanley, D.B., Sullivan, D.J. and Coleman, D.C., (1998), Identification and expression of multidrug transporters responsible for fluconazole resistance in *Candida dubliniensis*. *Antimicrob Agents Chemother*, **42**, 1819–1830.

Mori, T., Matsumura, M., Kanamaru, Y., Miyano, S., Hishikawa, T., Irie, S., Oshimi, K., Saikawa, T. and Oguri, T., (1997), Myelofibrosis complicated by infection due to *Candida albicans*: emergence of resistance to antifungal agents during therapy. *Clin Infect Dis*, **25**, 1470–1471.

Moudgal, V., Little, T., Boikov, D. and Vazquez, J.A., (2005), Multiechinocandin- and multiazole-resistant *Candida parapsilosis* isolates serially obtained during therapy for prosthetic valve endocarditis. *Antimicrob Agents Chemother*, **49**, 767–769.

Mukherjee, P.K. and Chandra, J., (2004), Candida biofilm resistance. *Drug Resist Updat*, **7**, 301–309.

Mukherjee, P.K., Leidich, S.D., Isham, N., Leitner, I., Ryder, N.S. and Ghannoum, M.A., (2003), Clinical *Trichophyton rubrum* strain exhibiting primary resistance to terbinafine. *Antimicrob Agents Chemother*, **47**, 82–86.

Muller, F.M., Weig, M., Peter, J. and Walsh, T.J., (2000), Azole cross-resistance to ketoconazole, fluconazole, itraconazole and voriconazole in clinical *Candida albicans* isolates from HIV-infected children with oropharyngeal candidosis. *J Antimicrob Chemother*, **46**, 338–340.

Nascimento, A.M., Goldman, G.H., Park, S., Marras, S.A., Delmas, G., Oza, U., Lolans, K., Dudley, M.N., Mann, P.A. and Perlin, D.S., (2003), Multiple resistance mechanisms among *Aspergillus fumigatus* mutants with high-level resistance to itraconazole. *Antimicrob Agents Chemother*, **47**, 1719–1726.

National Committee for Clinical Laboratory Standards, (2002a), Reference Method for Broth Dilution Antifungal Susceptibility Testing of Filamentous Fungi. Approved standard M38-A. Wayne, PA, National Committee for Clinical Laboratory Standards.

National Committee for Clinical Laboratory Standards, (2002b), Reference Method for Broth Dilution Antifungal Susceptibility Testing of Yeast. Approved standard M27-A2. Wayne, PA, National Committee for Clinical Laboratory Standards.

National Committee for Clinical Laboratory Standards, (2004), Method for Antifungal Disk Diffusion Susceptibility Testing of Yeasts. Approved standard M44-A. Wayne, PA, National Committee for Clinical Laboratory Standards.

Ng, T.T. and Denning, D.W., (1993), Fluconazole resistance in *Candida* in patients with AIDS—a therapeutic approach. *J Infect*, **26**, 117–125.

Nolte, F.S., Parkinson, T., Falconer, D.J., Dix, S., Williams, J., Gilmore, C., Geller, R. and Wingard, J.R., (1997), Isolation and characterization of fluconazole- and amphotericin B-resistant *Candida albicans* from blood of two patients with leukemia. *Antimicrob Agents Chemother*, **41**, 196–199.

Oakley, K.L., Moore, C.B. and Denning, D.W., (1997), *In vitro* activity of SCH-56592 and comparison with activities of amphotericin B and itraconazole against *Aspergillus* spp. *Antimicrob Agents Chemother*, **41**, 1124–1126.

Odds, F.C., Brown, A.J. and Gow, N.A., (2003), Antifungal agents: mechanisms of action. *Trends Microbiol*, **11**, 272–279.

Odds, F.C., Van Gerven, F., Espinel-Ingroff, A., Bartlett, M.S., Ghannoum, M.A., Lancaster, M.V., Pfaller, M.A., Rex, J.H., Rinaldi, M.G. and Walsh, T.J., (1998), Evaluation of possible correlations between antifungal susceptibilities of filamentous fungi *in vitro* and antifungal treatment outcomes in animal infection models. *Antimicrob Agents Chemother*, **42**, 282–288.

Odds, F.C., Vranckx, L. and Woestenborghs, F., (1995), Antifungal susceptibility testing of yeasts: evaluation of technical variables for test automation. *Antimicrob Agents Chemother*, **39**, 2051–2060.

Oki, T., Hirano, M., Tomatsu, K., Numata, K. and Kamei, H., (1989), Cispentacin, a new antifungal antibiotic. II. *In vitro* and *in vivo* antifungal activities. *J Antibiot (Tokyo)*, **42**, 1756–1762.

Orozco, A.S., Higginbotham, L.M., Hitchcock, C.A., Parkinson, T., Falconer, D., Ibrahim, A.S., Ghannoum, M.A. and Filler, S.G., (1998), Mechanism of fluconazole resistance in *Candida krusei*. *Antimicrob Agents Chemother*, **42**, 2645–2649.

Osherov, N., May, G.S., Albert, N.D. and Kontoyiannis, D.P., (2002), Overexpression of Sbe2p, a Golgi protein, results in resistance to caspofungin in *Saccharomyces cerevisiae*. *Antimicrob Agents Chemother*, **46**, 2462–2469.

Ostrosky-Zeichner, L., (2004), Prophylaxis or preemptive therapy of invasive candidiasis in the intensive care unit? *Crit Care Med*, **32**, 2552–2553.

Ostrosky-Zeichner, L., Rex, J.H., Pappas, P.G., Hamill, R.J., Larsen, R.A., Horowitz, H.W., Powderly, W.G., Hyslop, N., Kauffman, C.A., Cleary, J., Mangino, J.E. and Lee, J., (2003), Antifungal susceptibility survey of 2,000 bloodstream *Candida* isolates in the United States. *Antimicrob Agents Chemother*, **47**, 3149–3154.

Paderu, P., Park, S. and Perlin, D.S., (2004), Caspofungin uptake is mediated by a high-affinity transporter in *Candida albicans*. *Antimicrob Agents Chemother*, **48**, 3845–3849.

Patel, R., (2000), Prophylactic fluconazole in liver transplant recipients: a randomized, double-blind, placebo-controlled trial. *Liver Transpl*, **6**, 376–379.

Perea, S., Lopez-Ribot, J.L., Kirkpatrick, W.R., McAtee, R.K., Santillan, R.A., Martinez, M., Calabrese, D., Sanglard, D. and Patterson, T.F., (2001), Prevalence of molecular mechanisms of resistance to azole antifungal agents in *Candida albicans* strains displaying high-level flu-conazole resistance isolated from human immunodeficiency virus-infected patients. *Antimicrob Agents Chemother*, **45**, 2676–2684.

Perea, S., Lopez-Ribot, J.L., Wickes, B.L., Kirkpatrick, W.R., Dib, O.P., Bachmann, S.P., Keller, S.M., Martinez, M. and Patterson, T.F., (2002), Molecular mechanisms of fluconazole resistance in *Candida dubliniensis* isolates from human immunodeficiency virus-infected patients with oropharyngeal candidiasis. *Antimicrob Agents Chemother*, **46**, 1695–1703.

Perea, S. and Patterson, T.F., (2002), Antifungal resistance in pathogenic fungi. *Clin Infect Dis*, **35**, 1073–1080.

Pereira, M., Felipe, M.S., Brigido, M.M., Soares, C.M. and Azevedo, M.O., (2000), Molecular cloning and characterization of a glucan synthase gene from the human pathogenic fungus *Paracoccidioides brasiliensis. Yeast*, **16**, 451–462.

Perfect, J.R., (2004a), Use of newer antifungal therapies in clinical practice: what do the data tell us? *Oncology (Williston Park)*, **18**, 15–23.

Perfect, J.R., (2004b), Management of invasive mycoses in hematology patients: current approaches. *Oncology (Huntingt)*, **18**, 5–14.

Perfect, J.R., Granger, D.L. and Durack, D.T., (1987), Effects of antifungal agents and gamma interferon on macrophage cytotoxicity for fungi and tumor cells. *J Infect Dis*, **156**, 316–323.

Petraitiene, R., Petraitis, V., Kelaher, A.M., Sarafandi, A.A., Mickiene, D., Groll, A.H., Sein, T., Bacher, J. and Walsh, T.J., (2005), Efficacy, plasma pharmacokinetics, and safety of icofungipen, an inhibitor of *Candida* isoleucyl-tRNA synthetase, in treatment of experimental disseminated candidiasis in persistently neutropenic rabbits. *Antimicrob Agents Chemother*, **49**, 2084–2092.

Petraitis, V., Petraitiene, R., Kelaher, A.M., Sarafandi, A.A., Sein, T., Mickiene, D., Bacher, J., Groll, A.H. and Walsh, T.J., (2004), Efficacy of PLD-118, a novel inhibitor of *Candida* isoleucyl-tRNA synthetase, against experimental oropharyngeal and esophageal candidiasis caused by fluconazole-resistant *C. albicans. Antimicrob Agents Chemother*, **48**, 3959–3967.

Peyron, F., Favel, A., Michel-Nguyen, A., Gilly, M., Regli, P. and Bolmstrom, A., (2001), Improved detection of amphotericin B-resistant isolates of *Candida lusitaniae* by Etest. *J Clin Microbiol*, **39**, 339–342.

Pfaller, M.A., Bale, M., Buschelman, B., Lancaster, M., Espinel-Ingroff, A., Rex, J.H., Rinaldi, M.G., Cooper, C.R. and McGinnis, M.R., (1995a), Quality control guidelines for National Committee for Clinical Laboratory Standards recommended broth macrodilution testing of amphotericin B, fluconazole, and flucytosine. *J Clin Microbiol*, **33**, 1104–1107.

Pfaller, M.A., Boyken, L., Hollis, R.J., Messer, S.A., Tendolkar, S. and Diekema, D.J., (2004), Clinical evaluation of a dried commercially prepared microdilution panel for antifungal susceptibility testing of five antifungal agents against *Candida* spp. and *Cryptococcus neoformans. Diagn Microbiol Infect Dis*, **50**, 113–117.

Pfaller, M.A. and Diekema, D.J., (2004), Twelve years of fluconazole in clinical practice: global trends in species distribution and fluconazole susceptibility of bloodstream isolates of *Candida. Clin Microbiol Infect*, **10** (Suppl 1), 11–23.

Pfaller, M.A., Diekema, D.J., Boyken, L., Messer, S.A., Tendolkar, S. and Hollis, R.J., (2003), Evaluation of the Etest and disk diffusion methods for determining susceptibilities of 235 bloodstream isolates of *Candida glabrata* to fluconazole and voriconazole. *J Clin Microbiol*, **41**, 1875–1880.

Pfaller, M.A., Diekema, D.J., Jones, R.N., Messer, S.A. and Hollis, R.J., (2002), Trends in antifungal susceptibility of *Candida* spp. isolated from pediatric and adult patients with bloodstream infections: SENTRY Antimicrobial Surveillance Program, 1997 to 2000. *J Clin Microbiol*, **40**, 852–856.

Pfaller, M.A., Diekema, D.J., Jones, R.N., Sader, H.S., Fluit, A.C., Hollis, R.J. and Messer, S.A., (2001), International surveillance of bloodstream infections due to *Candida* species: frequency of occurrence and in vitro susceptibilities to fluconazole, ravuconazole, and voriconazole of isolates collected from 1997 through 1999 in the SENTRY antimicrobial surveillance program. *J Clin Microbiol*, **39**, 3254–3259.

Pfaller, M.A., Jones, R.N., Doern, G.V., Fluit, A.C., Verhoef, J., Sader, H.S., Messer, S.A., Houston, A., Coffman, S. and Hollis, R.J., (1999), International surveillance of blood stream infections due to *Candida* species in the European SENTRY Program: species distribution and antifungal susceptibility including the investigational triazole and echinocandin agents. SENTRY Participant Group (Europe) *Diagn Microbiol Infect Dis*, **35**, 19–25.

Pfaller, M.A., Jones, R.N., Doern, G.V., Sader, H.S., Messer, S.A., Houston, A., Coffman, S. and Hollis, R.J., (2000), Bloodstream infections due to *Candida* species: SENTRY antimicrobial surveillance program in North America and Latin America, 1997–1998. *Antimicrob Agents Chemother*, **44**, 747–751.

Pfaller, M.A., Jones, R.N., Messer, S.A., Edmond, M.B. and Wenzel, R.P., (1998a), National surveillance of nosocomial blood stream infection due to *Candida albicans*: frequency of occurrence and antifungal susceptibility in the SCOPE Program. *Diagn Microbiol Infect Dis*, **31**, 327–332.

Pfaller, M.A., Marco, F., Messer, S.A. and Jones, R.N., (1998b), *In vitro* activity of two echinocandin derivatives, LY303366 and MK-0991 (L-743,792), against clinical isolates of *Aspergillus, Fusarium, Rhizopus*, and other filamentous fungi. *Diagn Microbiol Infect Dis*, **30**, 251–255.

Pfaller, M.A., Messer, S.A., Boyken, L., Rice, C., Tendolkar, S., Hollis, R.J., Doern, G.V. and Diekema, D.J., (2005), Global trends in the antifungal susceptibility of *Cryptococcus neoformans* (1990 to 2004) *J Clin Microbiol*, **43**, 2163–2167.

Pfaller, M.A., Messer, S.A., Karlsson, A. and Bolmstrom, A., (1998c), Evaluation of the Etest method for determining fluconazole susceptibilities of 402 clinical yeast isolates by using three different agar media. *J Clin Microbiol*, **36**, 2586–2589.

Pfaller, M.A., Rhine-Chalberg, J., Barry, A.L. and Rex, J.H., (1995b), Strain variation and antifungal susceptibility among bloodstream isolates of *Candida* species from 21 different medical institutions. *Clin Infect Dis*, **21**, 1507–1509.

Playford, E.G., Webster, A.C., Sorell, T.C. and Craig, J.C., (2004), Antifungal agents for preventing fungal infections in solid organ transplant recipients. *Cochrane Database Syst Rev*, CD004291.

Polak, A. and Scholer, H.J., (1975), Mode of action of 5-fluorocytosine and mechanisms of resistance. *Chemotherapy*, **21**, 113–130.

Polak, A., Scholer, H.J. and Wall, M., (1982), Combination therapy of experimental candidiasis, cryptococcosis and aspergillosis in mice. *Chemotherapy*, **28**, 461–479.

Posteraro, B., Sanguinetti, M., Sanglard, D., La Sorda, M., Boccia, S., Romano, L., Morace, G. and Fadda, G., (2003), Identification and characterization of a *Cryptococcus neoformans* ATP binding cassette (ABC) transporter-encoding gene, Cn*AFR1*, involved in the resistance to fluconazole. *Mol Microbiol*, **47**, 357–371.

Powderly, W.G., Kobayashi, G.S., Herzig, G.P. and Medoff, G., (1988), Amphotericin B-resistant yeast infection in severely immunocompromised patients. *Am J Med*, **84**, 826–832.

Prasad, R., De Wergifosse, P., Goffeau, A. and Balzi, E., (1995), Molecular cloning and characterization of a novel gene of *Candida albicans*, *CDR1*, conferring multiple resistance to drugs and antifungals. *Curr Genet*, **27**, 320–329.

Redding, S.W., Marr, K.A., Kirkpatrick, W.R., Coco, B.J. and Patterson, T.F., (2004), *Candida glabrata* sepsis secondary to oral colonization in bone marrow transplantation. *Med Mycol*, **42**, 479–481.

Redding, S.W., Pfaller, M.A., Messer, S.A., Smith, J.A., Prows, J., Bradley, L.L., Fothergill, A.W. and Rinaldi, M.G., (1997), Variations in fluconazole susceptibility and DNA subtyping of multiple *Candida albicans* colonies from patients with AIDS and oral candidiasis suffering one or more episodes of infection. *J Clin Microbiol*, **35**, 1761–1765.

Redding, S., Smith, J., Farinacci, G., Rinaldi, M., Fothergill, A., Rhine-Chalberg, J. and Pfaller, M., (1994), Resistance of *Candida albicans* to fluconazole during treatment of oropharyngeal candidiasis in a patient with AIDS: documentation by *in vitro* susceptibility testing and DNA subtype analysis. *Clin Infect Dis*, **18**, 240–242.

Reef, S.E. and Mayer, K.H., (1995), Opportunistic candidal infections in patients infected with human immunodeficiency virus: prevention issues and priorities. *Clin Infect Dis*, 21 (Suppl 1), S99–S102.

Revankar, S.G., Kirkpatrick, W.R., McAtee, R.K., Fothergill, A.W., Redding, S.W., Rinaldi, M.G. and Patterson, T.F., (1998), Interpretation of trailing endpoints in antifungal susceptibility testing by the National Committee for Clinical Laboratory Standards method. *J Clin Microbiol*, **36**, 153–156.

Rex, J.H., Cooper, C.R., Jr, Merz, W.G., Galgiani, J.N. and Anaissie, E.J., (1995a), Detection of amphotericin B-resistant *Candida* isolates in a broth-based system. *Antimicrob Agents Chemother*, **39**, 906–909.

Rex, J.H., Nelson, P.W., Paetznick, V.L., Lozano-Chiu, M., Espinel-Ingroff, A. and Anaissie, E.J., (1998), Optimizing the correlation between results of testing *in vitro* and therapeutic outcome *in vivo* for fluconazole by testing critical isolates in a murine model of invasive candidiasis. *Antimicrob Agents Chemother*, **42**, 129–134.

Rex, J.H., Pfaller, M.A., Barry, A.L., Nelson, P.W. and Webb, C.D., (1995b), Antifungal susceptibility testing of isolates from a randomized, multicenter trial of fluconazole versus amphotericin B as treatment of nonneutropenic patients with candidemia. NIAID Mycoses Study Group and the Candidemia Study Group. *Antimicrob Agents Chemother*, **39**, 40–44.

Rex, J.H., Pfaller, M.A., Galgiani, J.N., Bartlett, M.S., Espinel-Ingroff, A., Ghannoum, M.A., Lancaster, M., Odds, F.C., Rinaldi, M.G., Walsh, T.J. and Barry, A.L., (1997), Development of interpretive breakpoints for antifungal susceptibility testing: conceptual framework and analysis of *in vitro-in vivo* correlation data for fluconazole, itraconazole, and candida infections. Subcommittee on Antifungal Susceptibility Testing of the National Committee for Clinical Laboratory Standards. *Clin Infect Dis*, **24**, 235–247.

Rex, J.H., Pfaller, M.A., Rinaldi, M.G., Polak, A. and Galgiani, J.N., (1993), Antifungal susceptibility testing. *Clin Microbiol Rev*, 6, 367–381.

Rex, J.H., Pfaller, M.A., Walsh, T.J., Chaturvedi, V., Espinel-Ingroff, A., Ghannoum, M.A., Gosey, L.L., Odds, F.C., Rinaldi, M.G., Sheehan, D.J. and Warnock, D.W., (2001), Antifungal susceptibility testing: practical aspects and current challenges. *Clin Microbiol Rev*, **14**, 643–658, table of contents.

Rex, J.H., Rinaldi, M.G. and Pfaller, M.A., (1995c), Resistance of *Candida* species to fluconazole. *Antimicrob Agents Chemother*, **39**, 1–8.

Rex, J.H., Walsh, T.J., Sobel, J.D., Filler, S.G., Pappas, P.G., Dismukes, W.E. and Edwards, J.E., (2000), Practice guidelines for the treatment of candidiasis. Infectious Diseases Society of America. *Clin Infect Dis*, **30**, 662–678.

Richter, S.S., Galask, R.P., Messer, S.A., Hollis, R.J., Diekema, D.J. and Pfaller, M.A., (2005), Antifungal susceptibilities of *Candida* species causing vulvovaginitis and epidemiology of recurrent cases. *J Clin Microbiol*, **43**, 2155–2162.

Rodriguez-Tudela, J.L., Martin-Diez, F., Cuenca-Estrella, M., Rodero, L., Carpintero, Y. and Gorgojo, B., (2000), Influence of shaking on antifungal susceptibility testing of *Cryptococcus neoformans*: a comparison of the NCCLS standard M27A medium, buffered yeast nitrogen base, and RPMI-2% glucose. *Antimicrob Agents Chemother*, **44**, 400–404.

Rodriguez-Tudela, J.L. and Martinez-Suarez, J.V., (1995), Defining conditions for microbroth antifungal susceptibility tests: influence of RPMI and RPMI-2% glucose on the selection of endpoint criteria. *J Antimicrob Chemother*, **35**, 739–749.

Ryley, J.F., Wilson, R.G., Gravestock, M.B. and Poyser, J.P., (1981), Experimental approaches to antifungal chemotherapy. *Adv Pharmacol Chemother*, **18**, 49–176.

Saag, M.S., Graybill, R.J., Larsen, R.A., Pappas, P.G., Perfect, J.R., Powderly, W.G., Sobel, J.D. and Dismukes, W.E., (2000), Practice guidelines for the management of cryptococcal disease. Infectious Diseases Society of America. *Clin Infect Dis*, **30**, 710–718.

Safdar, A., van Rhee, F., Henslee-Downey, J.P., Singhal, S. and Mehta, J., (2001), *Candida glabrata* and *Candida krusei* fungemia after high-risk allogeneic marrow transplantation: no adverse effect of low-dose fluconazole prophylaxis on incidence and outcome. *Bone Marrow Transplant*, **28**, 873–878.

Safe, L.M., Safe, S.H., Subden, R.E. and Morris, D.C., (1977), Sterol content and polyene antibiotic resistance in isolates of *Candida krusei* , *Candida parakrusei*, and *Candida tropicalis*. *Can J Microbiol*, **23**, 398–401.

Sanglard, D., Ischer, F., Calabrese, D., Majcherczyk, P.A. and Bille, J., (1999), The ATP binding cassette transporter gene Cg*CDR1* from *Candida glabrata* is involved in the resistance of clinical isolates to azole antifungal agents. *Antimicrob Agents Chemother*, **43**, 2753–2765.

Sanglard, D., Ischer, F., Koymans, L. and Bille, J., (1998), Amino acid substitutions in the cytochrome P-450 lanosterol 14alpha-demethylase (*CYP51A1*) from azole-resistant *Candida albicans* clinical isolates contribute to resistance to azole antifungal agents. *Antimicrob Agents Chemother*, **42**, 241–253.

Sanglard, D., Ischer, F., Monod, M. and Bille, J., (1997), Cloning of *Candida albicans* genes conferring resistance to azole antifungal agents: characterization of *CDR2*, a new multidrug ABC transporter gene. *Microbiology*, **143** (Pt 2), 405–416.

Sanglard, D., Ischer, F., Parkinson, T., Falconer, D. and Bille, J., (2003), *Candida albicans* mutations in the ergosterol biosynthetic pathway and resistance to several antifungal agents. *Antimicrob Agents Chemother*, **47**, 2404–2412.

Sanglard, D., Kuchler, K., Ischer, F., Pagani, J.L., Monod, M. and Bille, J., (1995), Mechanisms of resistance to azole antifungal agents in *Candida albicans* isolates from AIDS patients involve specific multidrug transporters. *Antimicrob Agents Chemother*, **39**, 2378–2386.

Sanglard, D. and Odds, F.C., (2002), Resistance of *Candida* species to antifungal agents: molecular mechanisms and clinical consequences. *Lancet Infect Dis*, **2**, 73–85.

Santos, B. and Snyder, M., (2000), Sbe2p and sbe22p, two homologous Golgi proteins involved in yeast cell wall formation. *Mol Biol Cell*, **11**, 435–452.

Sar, B., Monchy, D., Vann, M., Keo, C., Sarthou, J.L. and Buisson, Y., (2004), Increasing *in vitro* resistance to fluconazole in *Cryptococcus neoformans* Cambodian isolates: April 2000 to March 2002. *J Antimicrob Chemother*, **54**, 563–565.

Saxen, H., Hoppu, K. and Pohjavuori, M., (1993), Pharmacokinetics of fluconazole in very low birth weight infants during the first two weeks of life. *Clin Pharmacol Ther*, **54**, 269–277.

Schaffner, A. and Bohler, A., (1993), Amphotericin B refractory aspergillosis after itraconazole: evidence for significant antagonism. *Mycoses*, **36**, 421–424.

Schaffner, A. and Frick, P.G., (1985), The effect of ketoconazole on amphotericin B in a model of disseminated aspergillosis. *J Infect Dis*, **151**, 902–910.

Schmitt, H.J., Bernard, E.M., Edwards, F.F. and Armstrong, D., (1991), Combination therapy in a model of pulmonary aspergillosis. *Mycoses*, **34**, 281–285.

Scholer, H.J., (1980), Flucytosine. In D.C.E. Speller, editor, *Antifungal Chemotherapy*. Chichester, UK, Wiley, pp. 35–106.

Schuetzer-Muehlbauer, M., Willinger, B., Krapf, G., Enzinger, S., Presterl, E. and Kuchler, K., (2003), The *Candida albicans* Cdr2p ATP-binding cassette (ABC) transporter confers resistance to caspofungin. *Mol Microbiol*, **48**, 225–235.

Seay, R.E., Larson, T.A., Toscano, J.P., Bostrom, B.C., O'Leary, M.C. and Uden, D.L., (1995), Pharmacokinetics of fluconazole in immune-compromised children with leukemia or other hematologic diseases. *Pharmacotherapy*, **15**, 52–58.

Sheehan, D.J., Hitchcock, C.A. and Sibley, C.M., (1999), Current and emerging azole antifungal agents. *Clin Microbiol Rev*, **12**, 40–79.

Singh, N., (2004), Antifungal prophylaxis in solid-organ transplant recipients: considerations for clinical trial design. *Clin Infect Dis*, **39** (Suppl 4), S200–S206.

Singh, N. and Yu, V.L., (2000), Prophylactic fluconazole in liver transplant recipients. *Ann Intern Med*, **132**, 843–844.

Siwek, G.T., Dodgson, K.J., de Magalhaes-Silverman, M., Bartelt, L.A., Kilborn, S.B., Hoth, P.L., Diekema, D.J. and Pfaller, M.A., (2004), Invasive zygomycosis in hematopoietic stem cell transplant recipients receiving voriconazole prophylaxis. *Clin Infect Dis*, **39**, 584–587.

Slavin, M.A., Osborne, B., Adams, R., Levenstein, M.J., Schoch, H.G., Feldman, A.R., Meyers, J.D. and Bowden, R.A., (1995), Efficacy and safety of fluconazole prophylaxis for fungal infections after marrow transplantation–a prospective, randomized, double-blind study. *J Infect Dis*, **171**, 1545–1552.

Smith, W.L. and Edlind, T.D., (2002), Histone deacetylase inhibitors enhance *Candida albicans* sensitivity to azoles and related antifungals: correlation with reduction in CDR and ERG upregulation. *Antimicrob Agents Chemother*, **46**, 3532–3539.

Sobel, J.D., (1985), Epidemiology and pathogenesis of recurrent vulvovaginal candidiasis. *Am J Obstet Gynecol*, **152**, 924–935.

Sobel, J.D., Wiesenfeld, H.C., Martens, M., Danna, P., Hooton, T.M., Rompalo, A., Sperling, M., Livengood, C., 3rd, Horowitz, B., Von Thron, J., Edwards, L., Panzer, H. and Chu, T.C., (2004), Maintenance fluconazole therapy for recurrent vulvovaginal candidiasis. *N Engl J Med*, **351**, 876–883.

St-Germain, G., (2001), Impact of endpoint definition on the outcome of antifungal susceptibility tests with *Candida* species: 24- versus 48-h incubation and 50 versus 80% reduction in growth. *Mycoses*, **44**, 37–45.

St-Germain, G., Laverdiere, M., Pelletier, R., Bourgault, A.M., Libman, M., Lemieux, C. and Noel, G., (2001), Prevalence and antifungal susceptibility of 442 *Candida* isolates from blood and other normally sterile sites: results of a 2-year (1996 to 1998) multicenter surveillance study in Quebec, Canada. *J Clin Microbiol*, **39**, 949–953.

Steinbach, W.J., Stevens, D.A. and Denning, D.W., (2003), Combination and sequential antifungal therapy for invasive aspergillosis: review of published *in vitro* and *in vivo* interactions and 6281 clinical cases from 1966 to 2001. *Clin Infect Dis*, **37** (Suppl 3), S188–S224.

Stone, E.A., Fung, H.B. and Kirschenbaum, H.L., (2002), Caspofungin: an echinocandin antifungal agent. *Clin Ther*, **24**, 351–377; discussion 329.

Sugar, A.M., Hitchcock, C.A., Troke, P.F. and Picard, M., (1995), Combination therapy of murine invasive candidiasis with fluconazole and amphotericin B. *Antimicrob Agents Chemother*, **39**, 598–601.

Swoboda, S.M., Merz, W.G. and Lipsetta, P.A., (2003), Candidemia: the impact of antifungal prophylaxis in a surgical intensive care unit. *Surg Infect (Larchmt)*, **4**, 345–354.

Takakura, S., Fujihara, N., Saito, T., Kudo, T., Iinuma, Y. and Ichiyama, S., (2004), National surveillance of species distribution in blood isolates of *Candida* species in Japan and their susceptibility to six antifungal agents including voriconazole and micafungin. *J Antimicrob Chemother*, **53**, 283–289.

Thompson, J.R., Douglas, C.M., Li, W., Jue, C.K., Pramanik, B., Yuan, X., Rude, T.H., Toffaletti, D.L., Perfect, J.R. and Kurtz, M., (1999), A glucan synthase *FKS1* homolog in *Cryptococcus neoformans* is single copy and encodes an essential function. *J Bacteriol*, **181**, 444–453.

Tritz, D.M. and Woods, G.L., (1993), Fatal disseminated infection with *Aspergillus terreus* in immunocompromised hosts. *Clin Infect Dis*, **16**, 118–122.

van Burik, J.A., Ratanatharathorn, V., Stepan, D.E., Miller, C.B., Lipton, J.H., Vesole, D.H., Bunin, N., Wall, D.A., Hiemenz, J.W., Satoi, Y., Lee, J.M. and Walsh, T.J., (2004), Micafungin versus fluconazole for prophylaxis against invasive fungal infections during neutropenia in patients undergoing hematopoietic stem cell transplantation. *Clin Infect Dis*, **39**, 1407–1416.

Vanden Bossche, H., Koymans, L. and Moereels, H., (1995), P450 inhibitors of use in medical treatment: focus on mechanisms of action. *Pharmacol Ther*, **67**, 79–100.

Vanden Bossche, H., Marichal, P. and Odds, F.C., (1994), Molecular mechanisms of drug resistance in fungi. *Trends Microbiol*, **2**, 393–400.

Vazquez, J.A., Lynch, M., Boikov, D. and Sobel, J.D., (1997), *In vitro* activity of a new pneumocandin antifungal, L-743,872, against azole-susceptible and -resistant Candida species. *Antimicrob Agents Chemother*, **41**, 1612–1614.

Venkateswarlu, K., Denning, D.W., Manning, N.J. and Kelly, S.L., (1996), Reduced accumulation of drug in *Candida krusei* accounts for itraconazole resistance. *Antimicrob Agents Chemother*, **40**, 2443–2446.

Venkateswarlu, K., Taylor, M., Manning, N.J., Rinaldi, M.G. and Kelly, S.L., (1997), Fluconazole tolerance in clinical isolates of *Cryptococcus neoformans*. *Antimicrob Agents Chemother*, **41**, 748–751.

Vigouroux, S., Morin, O., Moreau, P., Mechinaud, F., Morineau, N., Mahe, B., Chevallier, P., Guillaume, T., Dubruille, V., Harousseau, J.L. and Milpied, N., (2005), Zygomycosis after prolonged use of voriconazole in immunocompromised patients with hematologic disease: attention required. *Clin Infect Dis*, **40**, e35–e37.

Viudes, A., Peman, J., Canton, E., Salavert, M., Ubeda, P., Lopez-Ribot, J.L. and Gobernado, M., (2002), Two cases of fungemia due to *Candida lusitaniae* and a literature review. *Eur J Clin Microbiol Infect Dis*, **21**, 294–299.

Vonk, A.G., Netea, M.G., Denecker, N.E., Verschueren, I.C., van der Meer, J.W. and Kullberg, B.J., (1998), Modulation of the pro- and anti-inflammatory cytokine balance by amphotericin B. *J Antimicrob Chemother*, **42**, 469–474.

Walsh, T.J., Teppler, H., Donowitz, G.R., Maertens, J.A., Baden, L.R., Dmoszynska, A., Cornely, O.A., Bourque, M.R., Lupinacci, R.J., Sable, C.A. and dePauw, B.E., (2004), Caspofungin versus liposomal amphotericin B for empirical antifungal therapy in patients with persistent fever and neutropenia. *N Engl J Med*, **351**, 1391–1402.

Warnock, D.W., Burke, J., Cope, N.J., Johnson, E.M., von Fraunhofer, N.A. and Williams, E.W., (1988), Fluconazole resistance in *Candida glabrata*. *Lancet*, **2**, 1310.

Warnock, D.W., Johnson, E.M. and Rogers, T.R., (1998), Multi-centre evaluation of the Etest method for antifungal drug susceptibility testing of *Candida* spp. and *Cryptococcus neoformans*. BSAC Working Party on Antifungal Chemotherapy. *J Antimicrob Chemother*, **42**, 321–331.

Wheat, J., Marichal, P., Vanden Bossche, H., Le Monte, A. and Connolly, P., (1997), Hypothesis on the mechanism of resistance to fluconazole in *Histoplasma capsulatum* . *Antimicrob Agents Chemother*, **41**, 410–414.

Whelan, W.L., (1987), The genetic basis of resistance to 5-fluorocytosine in *Candida* species and *Cryptococcus neoformans*. *Crit Rev Microbiol*, **15**, 45–56.

Whelan, W.L. and Kerridge, D., (1984), Decreased activity of UMP pyrophosphorylase associated with resistance to 5-fluorocytosine in *Candida albicans*. *Antimicrob Agents Chemother*, **26**, 570–574.

White, M.H., (1998), Fluconazole and the changing epidemiology of candidemia. *Clin Infect Dis*, **27**, 233–234.

White, T.C., (1997), Increased mRNA levels of *ERG16*, *CDR*, and *MDR1* correlate with increases in azole resistance in *Candida albicans* isolates from a patient infected with human immunodeficiency virus. *Antimicrob Agents Chemother*, **41**, 1482–1487.

White, T.C., Holleman, S., Dy, F., Mirels, L.F. and Stevens, D.A., (2002), Resistance mechanisms in clinical isolates of *Candida albicans*. *Antimicrob Agents Chemother*, **46**, 1704–1713.

White, T.C., Marr, K.A. and Bowden, R.A., (1998), Clinical, cellular, and molecular factors that contribute to antifungal drug resistance. *Clin Microbiol Rev*, **11**, 382–402.

White, T.C., Pfaller, M.A., Rinaldi, M.G., Smith, J. and Redding, S.W., (1997), Stable azole drug resistance associated with a substrain of *Candida albicans* from an HIV-infected patient. *Oral Dis*, **3** (Suppl 1), S102–S109.

Wingard, J.R., (2004), Design issues in a prospective randomized double-blinded trial of prophylaxis with fluconazole versus voriconazole after allogeneic hematopoietic cell transplantation. *Clin Infect Dis*, **39** (Suppl 4), S176–S180.

Wingard, J.R., Merz, W.G., Rinaldi, M.G., Johnson, T.R., Karp, J.E. and Saral, R., (1991), Increase in *Candida krusei* infection among patients with bone marrow transplantation and neutropenia treated prophylactically with fluconazole. *N Engl J Med*, **325**, 1274–1277.

Wingard, J.R., Merz, W.G., Rinaldi, M.G., Miller, C.B., Karp, J.E. and Saral, R., (1993), Association of *Torulopsis glabrata* infections with fluconazole prophylaxis in neutropenic bone marrow transplant patients. *Antimicrob Agents Chemother*, **37**, 1847–1849.

Xiao, L., Madison, V., Chau, A.S., Loebenberg, D., Palermo, R.E. and McNicholas, P.M., (2004), Three-dimensional models of wild-type and mutated forms of cytochrome P450 14alpha-sterol demethylases from *Aspergillus fumigatus* and *Candida albicans* provide insights into posaconazole binding. *Antimicrob Agents Chemother*, **48**, 568–574.

Yoder, B.A., Sutton, D.A., Winter, V. and Coalson, J.J., (2004), Resistant *Candida parapsilosis* associated with long term fluconazole prophylaxis in an animal model. *Pediatr Infect Dis J*, **23**, 687–688.

Zaoutis, T.E., Foraker, E., McGowan, K.L., Mortensen, J., Campos, J., Walsh, T.J. and Klein, J.D., (2005), Antifungal susceptibility of *Candida* spp. isolated from pediatric patients: a survey of 4 children's hospitals. *Diagn Microbiol Infect Dis*, **52**, 295–298.

Chapter 11
An Anti-mutant Approach for Antimicrobial Use

Karl Drlica, J.-Y. Wang, Muhammad Malik, Tao Lu, Steven Park, Xinying Li, David S. Perlin, and Xilin Zhao

11.1 Introduction

Although a great deal is known about antimicrobial resistance mechanisms, the problem of resistance continues to grow. For example, methicillin-resistant *Staphylococcus aureus* is emerging as a community infection (Lieberman, 2003; Lowy, 2003), and global hot-spots of multidrug-resistant tuberculosis have appeared (Almeida et al., 2003; Coker, 2004; Espinal, 2003). In the case of fluoroquinolones, resistance is being observed even among highly susceptible organisms, such as *Escherichia coli* (Lautenbach et al., 2004), *Haemophilus influenzae* (Li et al., 2004a; Nazir et al., 2004), *Streptococcus agalactiae* (Kawamura et al., 2003; Wehbeh et al., 2005), and *Neisseria gonorrhoeae* (Dan, 2004). Whether new compounds can be identified and approved faster than susceptibility declines is uncertain (Projan and Shlaes, 2004).

Discussions of resistance generally fall into two categories. One deals with the acquisition of resistance by a given microbial population; the other addresses the dissemination of resistant pathogens from one patient to another. Although acquired and disseminated resistances are causally related, they differ fundamentally: the former is a dosing and environmental contamination problem, while the latter is largely an epidemiological issue. Since acquisition of resistance is an early step in a multistep process that can lead to dissemination, intervention at the acquisition stage is more likely to produce long-lasting solutions than efforts to halt late steps such as dissemination. Thus the present work focuses on acquisition of resistance.

Acquired resistance is not a new problem. In part of his early work on antimicrobial therapy, Paul Ehrlich noted resistant microbes. He also realized that resistant organisms could be controlled by raising antimicrobial doses (Ehrlich, 1913). The main cost of higher doses is increased side effects, which for many years were bypassed by the development of new compounds. As new compounds become more difficult to develop (Projan and Shlaes, 2004), we are likely to hear calls for use of higher drug concentrations (Croom and Goa, 2003; Feldman, 2004). But how high must antimicrobial concentrations be to significantly slow the acquisition of resistance?

One way to determine minimal anti-mutant concentrations derives from the mutant selection window hypothesis (Zhao and Drlica, 2001). This hypothesis

I. W. Fong and K. Drlica (eds.), *Antimicrobial Resistance and Implications for the Twenty-First Century.* © Springer 2008

postulates that subpopulations of single-step, resistant mutants are enriched and amplified when antimicrobial concentrations fall within a specific concentration range (mutant selection window). The upper boundary of the selection window is the concentration that blocks growth of all single-step mutants; the lower boundary is the concentration below which no mutant has a selective advantage. In a general sense, keeping antimicrobial concentrations above the selection window should restrict the expansion of resistant mutant subpopulations, just as keeping antimicrobial concentrations above MIC is expected to restrict the outgrowth of susceptible pathogens (Ambrose et al., 2002). The key difference between the selection window approach for antimicrobial chemotherapy and traditional approaches is that the window approach uses elevated drug concentrations that force the pathogen to acquire two concurrent resistance mutations for growth rather than one, thereby reducing the probability that mutant subpopulations will amplify.

We begin this final chapter by discussing general aspects of antimicrobial resistance to provide a context for the selection window hypothesis. Then we describe new tests of the hypothesis, including an extension to eukaryotic pathogens (yeasts). We expect that validation of the hypothesis through animal and clinical studies will eventually provide a way to determine anti-mutant dosing regimens by combining drug pharmacokinetics with in vitro susceptibility data, preferably from simple agar-plate measurements that can be made with many antimicrobial–pathogen combinations. To help guide these tests and refinements, we also outline some of the limitations of the window hypothesis. Phenotypic (non-mutational) resistance, although important clinically, is not discussed in detail because it is not directly subject to anti-mutant approaches.

11.2 Loss of Antimicrobial Susceptibility

Loss of antimicrobial susceptibility by genetic means can be divided into three stages. The first arises from entry of resistance genes into a bacterial population by horizontal transfer or by mutation. This stage is not easily manipulated with antimicrobials except by avoiding compounds that are inactivated by products of genes frequently carried on mobile genetic elements or by avoiding compounds that induce the SOS response (Cirz et al., 2005; Gudas and Pardee, 1976; Miller et al., 2004; Phillips et al., 1987), a process that increases error-prone DNA repair (Smith and Walker, 1998) and transfer of some types of integron (Beaber et al., 2004). The second stage, mutant enrichment (increase in the *proportion* of cells that are resistant) and mutant amplification (increase in the *number* of cells that are resistant), occurs during exposure of bacteria to antimicrobial agents. Acquisition of resistance can be manipulated by antimicrobial regimens, as discussed below. Mutant amplification during treatment is readily observed with animal models (Cirz et al., 2005; Jumbe et al., 2003; Wiuff et al., 2003) and is recognized as a problem for treatment of many bacterial diseases including tuberculosis (British Medical Research Council, 1950), pneumonia (Davidson et al., 2002; Fink et al., 1994; Fuller and Low, 2005) and bacteremia caused by *Enterobacter* species (Chow et al., 1991) and *S. aureus* (Sieradzki et al., 1999, 2003). However, with many other infections enrichment of mutant subpopulations may initially go undetected, since susceptibility surveillance is usually designed to examine only dominant members of infecting populations. Successive rounds of selective pressure eventually allow the mutants to dominate bacterial populations. At that point the effectiveness of the antimicrobial is diminished, and the bacterial isolate may be scored as resistant by clinical "breakpoint" criteria. Dissemination of such organisms from patient to patient constitutes the third stage of resistance. During disseminated resistance, also called primary resistance, the majority of pathogen cells recovered from a patient exhibits reduced susceptibility, making resistance readily detectable by clinical laboratory MIC measurement.

11.3 Overview of Strategies for Limiting Antimicrobial Resistance

The management of antimicrobial resistance can also be categorized. As pointed out above, resistant pathogens that are currently disseminating constitute one arm in the principle dichotomy of resistance (Figure 11.1a). Disseminating pathogens are most commonly controlled by improved hygiene (Anonymous, 2004; Pittet, 2003), patient and contact isolation (Jernigan et al., 1996; Kotilainen et al., 2003;

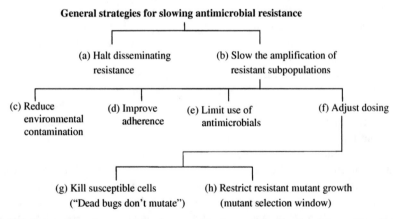

Fig. 11.1 Restricting resistance arising from genetic change. Situations involving resistant bacteria can be divided into two general classes. In one (**a**), the dominant members of the population are resistant. In such cases, the priority is to slow the dissemination of resistant mutants. In the other (**b**), only small subpopulations of resistant cells are present; application of antimicrobials exerts selective pressure that over time fosters selective amplification of new resistance. For the latter, general approaches are (**c**) reduction of environmental contamination, (**d**) improved adherence to approved antimicrobial regimens, (**e**) restricted clinical use of antimicrobials, such as limiting unnecessary prescriptions, and (**f**) adjustment of dosing regimens. Dosing strategies have evolved based on (**g**) removal of susceptible cells or (**h**) halting the growth of mutants, i.e., keeping antimicrobial concentrations above mutant susceptibility (mutant selection window).

Levy and Marshall, 2004), antimicrobial resistance surveillance (Felmingham, 2002; Sahm and Tenover, 1997), and shifting treatment to alternate compounds (Rice et al., 1996).

Restricting the acquisition of resistance by bacterial populations is the second arm in the management strategy (Figure 11.1b). Four approaches have evolved to control future resistance. One is reduction of environmental contamination by antimicrobials (Figure 11.1c) (Levy, 2002). This work has been extended to monitoring hospital wastewater, which can contain antimicrobials at concentrations high enough to enrich resistant pathogens (Kummerer and Henninger, 2003). Another strategy involves improving adherence to treatment regimens (Figure 11.1d) (Carey and Cryan, 2003; Pechere, 2001). Unfortunately, even good adherence may be inadequate when pathogen populations are large and mutation frequency is high, as has been observed in AIDS patients (Bangsberg et al., 2003, 2004). A third approach involves reducing the number of cases in which antimicrobials are prescribed (Figure 11.1e) (Levy, 2002; Levy and Marshall, 2004; Seppala et al., 1997). Widespread antimicrobial use is generally acknowledged as being responsible for the loss of antimicrobial susceptibility (Baquero et al., 2002; Cizman, 2003), and much of the use is thought to be unnecessary (Hecker et al., 2003). Efforts along these three lines have shown success (Aarestrup et al., 2001; Seppala et al., 1997), but many examples exist for which the prevalence of resistant isolates is still high (Drlica and Zhao, 2003) or growing (Bhavnani et al., 2005; Doern et al., 2005). Therefore, a fourth approach, adjustment of antimicrobial dosing (Figure 11.1f), has started to receive attention.

When antimicrobial regimens are optimized with respect to favorable patient outcome, effectiveness of a particular regimen is related to drug exposure experienced by the pathogen (Craig, 1998, 2001). That exposure is estimated through a relationship between drug pharmacokinetics and bacterial susceptibility. The overall goal has been to kill susceptible cells, since their death is expected to severely limit the generation of mutants (Figure 11.1g) (Blaser et al., 1987; Lipsitch and Levin, 1997; Zhanel et al., 2001). The essence of this strategy is expressed by the phrase "dead bugs don't mutate" (Dean, 2003; Stratton, 2003). Although this is an appealing phrase and treatments are often said to "eradicate" the infecting pathogen, antimicrobial action may not be sufficient for complete eradication. One reason is that susceptible bacterial populations frequently contain subpopulations that are phenotypically tolerant to the antimicrobial (Balaban et al., 2004) or enter a protected state upon antimicrobial treatment (Miller et al., 2004). Tolerant cells, which exhibit wild-type susceptibility upon re-growth, can amplify when drug concentrations drop below MIC. Another reason is that many infections, including abscesses, pneumonia, meningitis, tuberculosis, and urinary tract infections, generate such large bacterial burdens, above a billion cells (Bannon et al., 1998; Bingen et al., 1990; Fagon et al., 1990; Feldman, 1976; Jourdain et al., 1995; Mitchison, 1984; Wimberley et al., 1979), that resistant subpopulations are likely to exist before therapy begins. Even infections involving small bacterial burdens can contain resistant cells if mutators are present (Oliver et al., 2004) or if large numbers of patients are treated (when the probability of a resistant mutant occurring in a single patient, multiplied by the number of patients, equals or exceeds 1, a resistant mutant is likely to exist in one or more patients prior to treatment). Since drug concentrations sufficient to kill susceptible cells often fail to kill resistant mutants (Dong et al., 1998; Jumbe et al., 2003; Zhao et al., 1998), mutant subpopulations can expand even if drug concentrations do not drop below the MIC. Indeed, mathematical expressions have been derived that describe the rapid enrichment of mutants when resistant pathogens fail to be killed (Bonhoeffer and Nowak, 1997; Lipsitch and Levin, 1997; Muller and Bonhoeffer, 2003). Moreover, both in vitro and animal models establish experimentally that killing susceptible cells is not sufficient to block mutant enrichment (Gumbo et al., 2004; Jumbe et al., 2003; Tam et al., 2005). Thus traditional strategies aimed at killing susceptible bacteria are expected to gradually erode the utility of most antimicrobials by enriching mutants that can persist even when antimicrobial pressure is removed. High-level use of antimicrobials exacerbates this fundamental dosing problem.

A more stringent strategy, directed specifically against mutant subpopulations (Figure 11.1h), is required to block mutant amplification. A general approach is to use drug concentrations that allow growth of bacterial cells only if the cells acquire two or more concurrent resistance mutations, an event that is expected to occur rarely. This can be accomplished by combination therapy with two or more agents having different molecular targets (Iseman, 1994). The clearest example of successful combination therapy is seen with tuberculosis. Very large bacterial burdens and long periods of successive treatment with single agents cause resistance to develop often enough in individual tuberculosis patients that a favorable outcome is problematic. When multidrug therapy was instituted, the frequency at which resistance emerged dropped dramatically, from about 70% for streptomycin or isoniazid as monotherapy (British Medical Research Council, 1950; East African Hospitals and British Medical Research Council, 1960) to about 0.02% for multidrug combinations (Fox et al., 1999). In principle, monotherapy can also restrict mutant growth if antimicrobial concentrations are kept above the mutant selection window, which is described in the following section.

11.4 Mutant Selection Window

11.4.1 Definition of the Selection Window

In the 1990s Baquero suggested that dangerous concentration zones exist in which particular mutant types are selectively amplified (Baquero, 1990; Baquero and Negri, 1997a,b; Negri et al., 1994). The existence of selective concentration windows was demonstrated for low-level cefotaxime-resistant mutants, both in vitro and in animals, using susceptible bacterial populations to which a small number of resistant mutants had been added (Baquero et al., 1997; Negri and Baquero, 1998; Negri et al., 2000). Amplification of the mutants led to outgrowth of other mutants having even lower susceptibility. The latter were thought to have been present initially as a very small fraction of the bacterial population. These data led Baquero to postulate that the gradual loss of antimicrobial susceptibility is similar to hill climbing in which heights are achieved gradually: amplification of one mutant type facilitates the amplification of a second having slightly lower drug susceptibility (Baquero, 2001; Negri et al., 2000).

In the late 1990s, we discovered that fluoroquinolone concentration has a distinctive effect on the recovery of spontaneous mycobacterial mutants from drug-containing agar plates (Dong et al., 1999; Sindelar et al., 2000; Zhou et al., 2000). As drug concentration increases, the fraction of cells recovered as colonies drops, levels to a broad plateau, and then drops a second time (Figure 11.2). The initial drop arises from the inhibition of wild-type (susceptible) growth and occurs at concentrations approximated by MIC. The plateau is due to mutant subpopulations that are present at low frequencies (10^{-7}). Mutants isolated at low concentrations are predominantly non-target (non-gyrase) mutants (Figure 11.2, arrow A); at moderate quinolone concentrations several different gyrA (gyrase) mutants are recovered (Figure 11.2, arrow B). As the concentration increases, only the more protective gyrA mutations are observed (Figure 11.2, arrows C and D). When drug concentration is high enough to block the growth of the least susceptible, single-step mutant, colony recovery drops sharply a second time. The antimicrobial concentration at the second sharp drop, which is the MIC of the least susceptible, single-step mutant, is called the mutant prevention concentration (MPC). At concentrations above MPC, bacterial growth is expected to require two concurrent mutations, which are unlikely to occur. Although single mutants can still arise in bacterial populations, most, if not all, mutants fail to amplify when drug concentrations are kept above MPC; they are not recovered as colonies on agar plates.

It is important to emphasize that MPC, like MIC, is a bacteriostatic parameter because the antimicrobial agent is in the agar plates (lethal activity is measured by

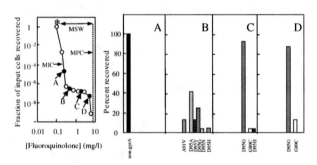

Fig. 11.2 Effect of fluoroquinolone concentration on recovery of mycobacterial colonies. *Mycobacterium smegmatis* was applied to agar plates containing the indicated concentrations of PD160788, an investigational fluoroquinolone, and after incubation colonies were recovered. Colonies were scored that grew when retested on the same drug concentration used to initially obtain the colony (*left panel*). DNA was isolated from colonies indicated by the labeled arrows (*solid circles*), and the nucleotide sequence for the quinolone-resistance-determining region (Friedman et al., 2001; Yoshida et al., 1990) was determined. Colonies were largely non-gyrase variants (*arrow A*), a variety of GyrA variants (*arrow B*), and almost all Asp-95 to Gly variants of GyrA (*arrows C and D*). Double-headed arrow indicates mutant selection window (MSW). Asterisk indicates bottom of the mutant selection window; MPC indicates the top of the selection window. Data taken from (Zhou et al., 2000) with permission of the publisher.

determining surviving cells by growth of colonies on drug-free agar). Thus cells applied to agar at MPC could still be alive even though they cannot form colonies. Indeed, with *E. coli*, viable cells do remain on agar at fluoroquinolone concentrations above MPC (Marcusson et al., 2005). These cells are persisting, wild-type cells rather than resistant mutants. Lethal activity with resistant mutants is considered below in Sections 5.1 and 6.

The response of mycobacterial growth to fluoroquinolone concentration extended Baquero's "dangerous zone" from a concentration range for a particular mutant to that of all single-step mutants. It also redefined the boundaries and led to the zones being collectively called the mutant selection window (Zhao and Drlica, 2001, 2002). The upper boundary of the window is MPC. The lower boundary, which is defined as the concentration below which selective pressure is absent, is seen experimentally as the beginning of the sharp drop in colony recovery that occurs at low drug concentration (asterisk in Figure 11.2). For convenience, we often approximate that value as the minimal concentration that blocks growth of 99% of the cells in the population ($MIC_{(99)}$). We emphasize that standard NCCLS values of MIC, which are frequently used to approximate the bottom of the selection window, overestimate that boundary because selective pressure exists below the standard MIC (Ciu et al., 2006).

11.4.2 Selection Window with Eukaryotic Pathogens

To extend the selection window idea to eukaryotic cells, we examined the effect of miconazole concentration on the recovery of spontaneous resistance mutants of *Candida albicans* and *Candida glabrata* from drug-containing agar. With a wild-type, laboratory strain of *C. albicans*, increasing the miconazole concentration caused the number of colonies recovered to drop sharply, level to a plateau, and drop sharply a second time (filled circles, Figure 11.3A). A single-step, miconazole-resistant mutant (strain KD2295), recovered from agar containing 5 mg/l miconazole, exhibited only a single, sharp drop in colony recovery when tested on miconazole-containing agar (open circles, Figure 11.3A). This single drop in colony recovery occurred at roughly the same miconazole concentration as the second drop observed with wild-type cells. These data are consistent with the second drop for wild-type cells occurring at the MPC (Dong et al., 1999).

A susceptible clinical isolate of *C. glabrata* behaved much like wild-type *C. albicans*: as miconazole concentration increased, colony recovery dropped sharply, leveled to a plateau, and dropped sharply a second time (filled circles, Figure 11.3B). With a resistant mutant of *C. glabrata* (strain CST56B), the drop in the mutant selection curve occurred at a high drug concentration (open circles, Figure 11.3B), much as seen with the laboratory mutant of *C. albicans* (Figure 11.3A).

For both yeast species examined, the effect of drug concentration on mutant recovery resembled that observed with mycobacteria when applied to fluoroquinolone-containing agar (Dong et al., 1999; Sindelar et al., 2000; Zhou et al., 2000). Thus, the concepts of mutant enrichment developed for bacterial

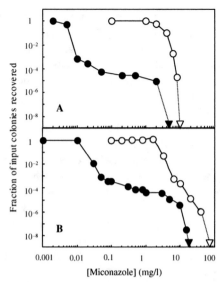

Fig. 11.3 Effect of miconazole concentration on the recovery of yeast colonies from agar plates. *Candida* isolates having an MIC < 1 mg/l for miconazole and fluconazole were grown to stationary phase in liquid medium, concentrated by centrifugation, and resuspended in water at about 10^9 cells/ml. Aliquots of dilutions containing up to 10^9 cells were then applied to agar plates containing various miconazole concentrations. After incubation at 30°C for 7 days (miconazole remained fully active in agar for at least 10 days), colonies emerged, were transferred to drug-free agar, and regrown. They were then retested on agar containing the miconazole concentration initially used for selection. The colonies were confirmed to be azole-resistant, indicating that the azole-resistance phenotype observed here is a stable, genetic phenomenon rather than an inducible or transient resistant phenotype that disappears upon subculture in the absence of drug (e.g., Marr et al., 2001). The fraction of input cells recovered as colonies was determined. *Panel A: Candida albicans* . Various numbers of cells of wild-type *C. albicans* (ATCC 90028, *solid circles*) or first-step mutant (KD2295, *open circles*) were applied to YPD agar containing the indicated concentrations of miconazole (the maximum cell density applied was 8×10^5 cells per cm^2 agar). Miconazole concentrations at which no colony was recovered when more than 10^9 cells were applied to agar plates are indicated by triangles (the symbol change emphasizes uncertainty involved in placing zero on a log plot). Replicate experiments showed similar results. *Panel B: Candida glabrata*. Various numbers of cells of wild-type *C. glabrata* (CST83B, solid circles) or a clinically resistant mutant (CST56B, open circles) were plated on YPD agar containing the indicated concentrations of miconazole. Triangles indicate miconazole concentrations at which no colony was recovered when more than 10^9 cells were applied to agar plates. With *C. glabrata* , small colonies were recovered after 11–16 days incubation; however, they did not regrow on drug-containing agar and were not scored as growth. Replicate experiments showed similar results.

systems (Zhao and Drlica, 2001) are likely to apply to both diploid (*C. albicans*) and haploid (*C. glabrata*) eukaryotic organisms. The levels of the plateaus shown in Figure 11.3 indicate that mutant subpopulations are present at a frequency of 10^{-4} to 10^{-5}. These data are consistent with previous work with *C. glabrata* (Cross et al., 2000). The high frequency of mutant recovery shown in Figure 11.3 suggests that resistance will readily develop during therapy unless infecting yeast populations are very small and/or azole concentrations at the site of infection are kept above MPC.

Topical miconazole concentrations are likely to exceed MPC during treatment (the concentration in ointments is typically 1% (w/v) or about 10 g/l); such high concentrations may be responsible for the low prevalence of resistance associated with the use of over-the-counter azole antifungals (Mathema et al., 2001). The methods described for miconazole can now be used to determine MPC for azoles used systemically to minimize the amplification of resistant mutants.

11.4.3 Complex Situations

The characteristic relationship between colony recovery and drug concentration seen for fluoroquinolones and mycobacteria (Figure 11.2) or for azoles and yeast (Figure 11.3) is also observed with other microbial species and other agents (e.g., erythromycin with *S. aureus* (Lu et al., 2003)). However, such is not always the case. Even for quinolones, examples have been found in which increasing drug concentration causes colony recovery to exhibit only a small inflection in the mutant recovery curve rather than a distinct plateau (Dong et al., 1999; Li et al., 2002). One reason for the absence of an obvious plateau is the presence of two drug targets (gyrase and topoisomerase IV) having similar drug sensitivity: drug concentrations that trap one enzyme on DNA (MIC) would be close to those that trap both (MPC). The mycobacteria tested exhibit a broad plateau (Dong et al., 1999; Sindelar et al., 2000; Zhou et al., 2000) because they have only gyrase as a fluoroquinolone target and because gyrase mutations are quite protective.

To test the explanation for steep mutant recovery curves, we measured the effect of a fluoroquinolone-resistance mutation on mutant recovery with *Streptococcus pneumoniae*. As shown in Figure 11.4, eliminating one quinolone target expands the inflection into a broad plateau. First-step mutations of *S. pneumoniae*, which may have little effect on MIC, can raise the MPC to where standard antimicrobial doses are unlikely to restrict the amplification of mutant subpopulations. The clinical implication is that a vast majority of isolates can appear to be susceptible by MIC-based break-point criteria, while many carry topoisomerase mutations (Doern et al., 2005) that are expected to increase the risk for treatment failure due to resistance (Fuller and Low, 2005; Gillespie et al., 2003; Li et al., 2002).

The presence of multiple drug targets with similar susceptibility explains why the selection curve, also called a population analysis profile (Sieradzki et al., 1999), drops sharply when *S. aureus* is applied to agar containing increasing concentrations of a β-lactam (penicillin) (Lu et al., 2003). That in turn explains why little difference is seen between the standard MIC and approximations of MPC for ceftriaxone, meropenem, imipenem, and ertapenem with another Gram-positive organism, *S. pneumoniae* (Hovde et al., 2003). A second reason for a sharp decline in colony recovery would be an exceedingly low mutation frequency. Examples of this situation have yet to be documented.

Situations have been found in which a broad plateau occurs without a second drop in colony recovery (Zhao and Drlica, 2002), i.e., the selection window is infinitely wide. Such cases, typified by rifampicin treatment of *E. coli*, are

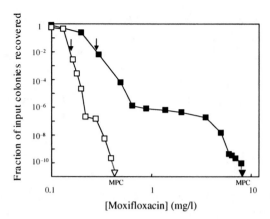

Fig. 11.4 Effect of a first-step mutation on the recovery of fluoroquinolone-resistant mutants. *S. pneumoniae* was applied to agar containing the indicated concentrations of moxifloxacin and then incubated to allow colony growth. The fraction of input colonies recovered is indicated for wild-type cells (*open squares*) and a *parC* resistance mutant (*filled squares*). MPC (zero colonies recovered when 10^{10} cfu were applied to plates) is indicated by *arrowheads*. Data taken from Li et al., (2002) with permission of the publisher.

interpreted as being due to our inability to achieve drug concentrations high enough to block mutant growth (single-step mutations are highly protective). Such drug–pathogen combinations are poorly suited to monotherapy (Binda et al., 1971; Dong et al., 2000).

Additional complexity occurs with β-lactams when resistance is due to phenotypic induction of protective enzymes (Minami et al., 1980) rather than to mutational derepression of those enzymes (Kuga et al., 2000). Such situations are not directly covered by the selection window hypothesis. If the induced resistance gene is borne on a plasmid that is initially present in a small fraction of the population, multidrug therapy is required to prevent plasmid-containing cells from selectively amplifying.

11.4.4 Estimating MPC

When a distinct plateau is absent, determining MPC from the shape of bacterial population analysis profiles is difficult. In these cases, the value of MPC has been approximated as the drug concentration at which no colony is recovered when more than 10^{10} cells are applied to agar plates (Dong et al., 1999; Li et al., 2002). Since most bacterial infections contain fewer than 10^{10} organisms, blocking the growth of 10^{10} cells should also block the growth of all spontaneous, single-step resistant subpopulations.

Although MPC has been estimated for a variety of compounds and several pathogens (Drlica, 2003; Hansen et al., 2006; Hermsen et al., 2005; Li et al., 2004b; Linde and Lehn, 2004; Marcusson et al., 2005; Metzler et al., 2004; Randall et al., 2004; Rodriguez et al., 2004a,b, 2005; Smith et al., 2004; Wetzstein, 2005), examining 10^{10} cells is not always straightforward. For example, with some situations, such as treatment of *S. aureus* with daptomycin or pradofloxacin (Silverman et al., 2001; Wetzstein, 2005), the size of the bacterial inoculum is likely to affect the measurement. With other species, such as *S. pneumoniae*, additional precautions are required to avoid autolysis. With this pathogen cells are harvested before autolysis occurs, and after concentration they are applied to many agar plates to achieve a total of 10^{10} cells examined for each drug concentration (Li et al., 2002).

Clinical isolates vary considerably in their history of antimicrobial exposure and thus in mutant subpopulations. Consequently, surveys with large numbers of microbial isolates are routinely used to characterize pathogen populations. To estimate MPC for large numbers of isolates, 10^9–10^{10} cells are applied to each agar plate of a series in which drug concentration differs by twofold increments (Blondeau et al., 2001). MPC is taken as the lowest concentration at which no growth is observed. With this survey method, the high cell density and high drug concentration increments preclude detection of individual colonies on sub-MPC plates (the population analysis method scores individual colonies by using large numbers of plates with smaller inocula). For quinolones, MPC determined by the survey method is close to MPC measured by population analysis with *Pseudomonas aeruginosa* (Hansen et al., 2006) and slightly higher with *S. pneumoniae* (Blondeau et al., 2001). The relative activity of different fluoroquinolones determined by the survey method is the same as observed by the population analysis method (Blondeau et al., 2004; Hansen et al., 2006).

11.4.5 Estimating MPC from MIC?

Early in the study of the selection window we noticed that MIC and MPC correlated very poorly ($r^2 = 0.39$) when linear regression was performed with closely related fluoroquinolones (Sindelar et al., 2000). Bacterial isolates were then found for which mutant MPC was disproportionately high or low relative to MIC (Linde and Lehn, 2004; Randall et al., 2004; Zhao and Drlica, 2002). The low correlation between MIC and MPC was recently emphasized by linear regression analysis with 20 clinical isolates of *E. coli* in which $r^2 = 0.58$ for ciprofloxacin (Marcusson et al., 2005); similar results were later obtained for a variety of quinolones with *P. aeruginosa*, *S. aureus*, and *S. pneumoniae* (Drlica et al., 2006). An example with *S. aureus* is shown in Figure 11.5. If poor correlation is a general phenomenon, MPC cannot be determined reliably from MIC for individual patients, nor can MIC-based criteria for anti-mutant dosing regimens be accurately applied to individual patients.

Low correlation between MIC and MPC can be attributed to a variety of factors. For example, some antimicrobials have more anti-mutant activity than others (Dong et al., 1998), which would make MPC disproportionately low. With respect to clinical isolates, some could have low-level resistance mutations that would affect MIC but not MPC. Others could have highly resistant subpopulations that would

Fig. 11.5 MPC as a function of MIC for ciprofloxacin with clinical isolates of *S. aureus*. MPC was determined under conditions in which isolated colonies could be detected when 10^{10} cells were applied to drug-containing agar at concentrations below MPC. r^2 was determined by linear regression analysis. Data are from Zhao et al., (2003).

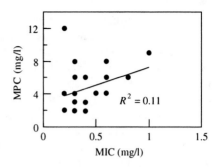

affect only MPC. Complete lack of correlation could even suggest that the antimicrobial has different molecular targets at the different concentrations used for MIC and MPC.

11.5 Experimental Support for the Selection Window Hypothesis

11.5.1 In Vitro Dynamic Models

Since the mutant selection window was defined using static drug concentrations with agar plates while antimicrobial concentration fluctuates during therapy, static and fluctuating concentrations needed to be related with respect to enrichment and amplification of resistant mutants. The Firsov–Zinner team set up an in vitro dynamic model using *S. aureus* such that fluoroquinolone concentrations were adjusted to be above MPC, below MIC, or between MIC and MPC. Then bacteria recovered from the model were examined for decreased susceptibility as an indicator of mutant amplification. Culture MIC increased when fluoroquinolone concentrations in the model were kept between MIC and MPC, but not when they were always above MPC or always below MIC (Firsov et al., 2003, 2004b) (Figure 11.6). A similar result was found with *S. pneumoniae* (Zinner et al., 2003), and the *S. aureus* work was confirmed by another laboratory (Campion et al., 2004, 2005a,b). The general principle has also been extended to vancomycin and daptomycin (Firsov et al., 2006). Thus static, agar plate values fit with results from in vitro dynamic models.

Longer exposure inside the selection window is expected to allow greater mutant amplification (Baquero and Negri, 1997b); however, adjustment is required for the position in the window (a much greater fraction of the population can amplify as resistant mutants when drug concentrations are in the lower portion of the window (Li et al., 2002; Zhou et al., 2000); also see Figure 11.2). Thus exposure for 20% of the time at the bottom of the window would have a different effect on mutant evolution than exposure for the same time near the top. This complexity is seen in dynamic, in vitro models in which time inside the window shows no simple relationship to outgrowth of mutants (Campion et al., 2004, 2005a,b). For example, with

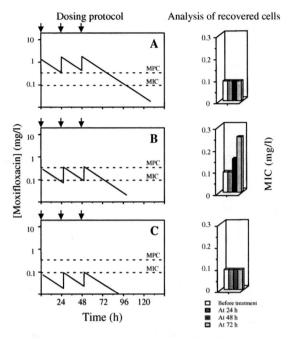

Fig. 11.6 Relationship between fluctuating fluoroquinolone concentration, the mutant selection window, and amplification of resistant mutants. An in vitro model was established with *S. aureus* in which moxifloxacin concentration was varied, as indicated in *Panels A, B, and C*, to be either above, within, or below the mutant selection window, which was determined separately with agar plates. Moxifloxacin was added at the times indicated by the *arrows*. Samples were removed at the times indicated under the *right panels*, and MIC was determined, as shown in histograms on the *right*. Data were taken from Firsov et al., (2003) with permission of the publisher.

S. aureus multiple rounds of fluctuating fluoroquinolone concentration can allow mutants exposed in the middle portion of the window to amplify enough to acquire a second mutation. The double mutant is under sufficient selective pressure to be enriched in a relatively short time. Since these double mutants have lower susceptibility than single mutants recovered in the upper portion of the window, the least susceptible mutant can appear to be selectively enriched in the middle rather than at the top of the window (Campion et al., 2004).

It is important to reiterate that MPC is a bacteriostatic parameter, which allows it to be applied to both bacteriostatic and bactericidal compounds. If a compound kills resistant mutants, MPC will overestimate the threshold needed to restrict mutant subpopulation outgrowth. Killing of mutants explains (1) why outgrowth of *S. aureus* mutants in an in vitro model is blocked even when levofloxacin is not above MPC at all times (Campion et al., 2005a), (2) why mutant growth is blocked with shorter times above MPC by the quinolone ABT492 than by levofloxacin (Firsov et al., 2004a), and (3) why enrichment of mutants requires ABT492 concentration to be inside the selection window twice as long as that of levofloxacin (fluoroquinolones with C-8–halogen or methoxy groups, such as the C-8–Cl compound ABT492, are disproportionately lethal for resistant mutants (Lu et al., 2001;

Zhao et al., 1998)). This overestimate due to killing of mutants is addressed in part by using PK/PD indices as a dosing guide, as discussed in Section 11.6.

11.5.2 Animal Models

The key question to be answered with animal systems is how well the boundaries of the selection window fit with agar plate determinations. We expect the fit to be good because MIC measured in vitro is close to that measured in vivo (Andes and Craig, 2002) and because MPC is simply MIC of the least susceptible, single-step mutant subpopulation. Since agar plate data are static while drug concentrations in animals fluctuate, comparisons require the pharmacokinetic considerations discussed above in Section 11.5.1. Three relevant studies address the general question of whether fluoroquinolone concentrations must remain above MPC throughout therapy to severely restrict amplification of resistant mutant subpopulations.

The first work involved infection of mouse thigh by *P. aeruginosa* treated with levofloxacin (Jumbe et al., 2003). Amplification of efflux mutants was restricted by doses of levofloxacin that produced an antimicrobial exposure of $AUC_{24}/MIC = 157$ (AUC_{24} is the area under the time–drug concentration curve in a 24 h period). To compare these data with in vitro studies, it is necessary to correct for host defense, which reduces the drug exposure necessary for a favorable outcome by 1.3–2 fold (Andes and Craig, 2002). That would make the comparable in vitro value of AUC_{24}/MIC range between 200 and 300, which is comparable to values observed by Firsov et al., (2003). Since MPC was not measured, estimates considering only efflux mutants are necessary. Mutant MIC, i.e., "efflux MPC," is about 8 mg/l (Jumbe et al., 2003). If we assume that the murine mutant-restricting AUC_{24}/MIC of 157 (Jumbe et al., 2003) also applies to humans, a 750 mg dose of levofloxacin, delivered once daily, would have an exposure of 157 for more than 60% of patients (Jumbe et al., 2003). However, that dose, which is expected to restrict mutant amplification, would keep drug concentration above efflux MPC for only 2 out of 24 h (http://www levaquin.com). According to this calculation, it is not necessary to keep fluoroquinolone concentrations above MPC throughout therapy to suppress mutant outgrowth. We note that MPC is about 9 mg/l when large bacterial populations and single-step *gyrA* mutants are considered (Hansen et al., 2006), which would cause the anti-mutant parameters in the mouse study (Jumbe et al., 2003) to be only slight underestimates.

A second animal report focused on moxifloxacin treatment of pneumococcal pneumonia in rabbits (Croisier et al., 2004; Etienne et al., 2004). When serum drug pharmacokinetics were controlled to simulate human conditions at various doses of moxifloxacin, wild-type cells were eradicated without detection of mutants. To facilitate mutant recovery, the selection window was expanded by infecting rabbits with a *parC* resistance mutant, which allowed *parC gyrA* double mutants to be recovered (Etienne et al., 2004) (Figure 11.4 shows how a *parC* resistance allele expands the selection window in vitro). When MPC of the *parC* strain was exceeded by free moxifloxacin concentration for more than half of the dosing period, no mutant was recovered. These data provide a second example in which it is not necessary to keep fluoroquinolone concentrations above MPC throughout therapy to block the amplification of resistant mutants, providing that drug concentrations are similar in serum and lung.

In the third model study, rabbits were infected with *S. aureus* placed inside a tissue cage (Cui et al., 2006). In this system a hollow plastic ball containing holes in its surface (whiffle ball) was implanted under the skin of a rabbit. After 4 weeks, connective tissue covered the surface of the ball, and the interior of the ball formed a compartment into which 10^{10} bacteria were injected. Levofloxacin was then administered orally, and both drug and bacterial concentrations inside the ball were monitored. MIC of the bacteria increased only when drug concentrations were inside the mutant selection window, which had been determined from agar plate data. The whiffle ball experiment establishes that the selection window has both upper and lower boundaries. Moreover, the amplification of resistant mutants was restricted when drug concentrations were above the MPC for only 20% of the dosing time.

The three animal studies indicate that it is unnecessary to keep fluoroquinolone concentration above MPC throughout therapy, probably because MPC is a bacteriostatic parameter while these compounds kill resistant mutants (Dong et al., 1998; Li et al., 2002). Corrections for this lethal activity are made empirically, as discussed below in Section 11.6.

11.5.3 Clinical Study

A central prediction of the selection window hypothesis is that drug concentrations inside the selection window allow the acquisition of resistance while eradicating susceptible cells. Concurrent amplification of resistant mutants and death of susceptible cells has been clearly demonstrated in vitro (Gumbo et al., 2004; Tam et al., 2005) and in an animal model (Jumbe et al., 2003). We examined the phenomenon clinically with a small number of tuberculosis patients ($n = 58$) shown to be colonized with rifampicin-susceptible *S. aureus* at the beginning of therapy (Liu et al., 2005). Four weeks of therapy that included rifampicin eradicated colonization in 53 (92%) of the patients. At the same time, rifampicin resistance was acquired in 5 (8%). Each resistant isolate had the same pulsed-field gel electrophoresis banding pattern and protein A polymorphism (*spa* type (Shopsin et al., 1999)) as its corresponding, initial, susceptible isolate. These genetic markers differed among the five resistant isolates. Thus resistance was acquired by selective amplification of mutant subpopulations rather than by dissemination. These data lead to the conclusion that treatments aimed only at eradicating susceptible cells (Figure 11.1g) are likely to be inadequate at blocking the acquisition of resistance. Although resistance may develop less frequently with other antimicrobial–pathogen combinations than with the *S. aureus*–rifampicin example, administration of millions of treatments will make even rare events significant, especially when resistant isolates subsequently disseminate.

11.6 Pharmacokinetic/Pharmacodynamic Indices and MPC

PK/PD indices (Mouton et al., 2005), which reflect the exposure of a pathogen to an antimicrobial, are commonly used to relate antimicrobial dosing regimens to efficacy (removal of susceptible bacterial populations). For example, empirical data indicate that the time at which antimicrobial concentration exceeds MIC ($t >$ MIC) is predictive of a favorable outcome for "time-dependent killers" and the area under the 24-h time–concentration curve divided by MIC (AUC_{24}/MIC) is predictive for "concentration-dependent killers" (Craig, 2002). Since MPC is the MIC of the least susceptible, single-step mutant, the empirical relationships are expected to apply to restricting resistant mutant outgrowth through replacement of MIC by MPC. In principle, anti-mutant dosing regimens for fluoroquinolones can be determined by measuring AUC_{24}/MPC. That value, determined in vitro, is about 22 for ciprofloxacin with *E. coli* (Olofsson et al., 2006). For animal studies we calculate AUC_{24}/MPC to be 20 (Jumbe et al., 2003) and 25 (Cui et al., 2006). Pharmacodynamic corrections to relate MIC, a bacteriostatic parameter, to bactericidal activity (Regoes et al., 2004) would apply to MPC to the extent that they apply to MIC.

Many early efforts to identify anti-mutant dosing regimens used the much more familiar AUC_{24}/MIC as an index. Mutant-restricting values of AUC_{24}/MIC for fluoroquinolones generally exceed 200 (Etienne et al., 2004; Firsov et al., 2003; Zinner et al., 2003). Since MIC correlates poorly with MPC (Figure 11.5 and (Drlica et al., 2006; Marcusson et al., 2005; Sindelar et al., 2000)), isolate-to-isolate variation in anti-mutant levels of AUC_{24}/MIC is likely to be high.

11.7 Stepwise Accumulation of Resistance Mutations

If multiple ways exist for a bacterium to reduce antimicrobial susceptibility and if resistance alleles have an additive effect, resistance can arise in a stepwise fashion (Eliopoulos et al., 1984; Shockley and Hotchkiss, 1970). To determine whether the mutant selection window rises with each successive mutation acquired by a bacterium, fluoroquinolone resistance of *H. influenzae* was examined (Li et al., 2004b). Wild-type cells were applied to ciprofloxacin-containing agar, and resistant cells containing mutations in *gyrA* were obtained. Application of one of these *gyrA* mutants to agar containing higher concentrations of ciprofloxacin allowed recovery of a second-step mutant with an additional mutation in *parC*. This mutant was applied to ciprofloxacin-containing agar, and a third-step mutant was obtained containing another mutation in *gyrA*. For each step in the selection process, the values of $MIC_{(99)}$ and MPC were determined, which defined the boundaries of the selection window (shaded regions, Figure 11.7).

With wild-type *H. influenzae*, the selection window is below the minimal serum concentration of ciprofloxacin measured with healthy volunteers (circles, Figure 11.7A); after the first-step *gyrA* mutation is acquired, the window rises to where it partially overlaps serum concentration (Figure 11.7B). Acquisition of the

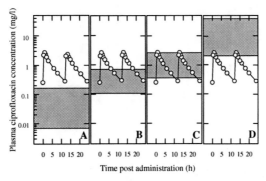

Fig. 11.7 Step-wise accumulation of fluoroquinolone resistance mutations and effect on the mutant selection window. Wild-type *H. influenzae* was applied to agar containing various concentrations of ciprofloxacin to determine MIC$_{(99)}$ and MPC, which were taken as the lower and upper boundaries of the mutant selection window, respectively (*shaded area of panel A*) (Li et al., 2004b). A first-step *gyrA* mutant was recovered and reapplied to agar containing various concentrations of ciprofloxacin to determine MIC$_{(99)}$ and MPC for the first-step mutant (*shaded area in panel B*). A second-step *gyrA parC* double mutant was recovered, and its selection window (*shaded area in panel C*) was determined as above. A third-step *gyrA parC gyrA* triple mutant was recovered and used to determine its selection window (*shaded area, panel D*). To relate the selection window changes to drug pharmacokinetics, ciprofloxacin concentration in human plasma (*circles*) is shown (samples were taken from healthy volunteers who had been administered 500 mg oral doses twice daily for 7 days; data were taken over the course of 24 h on day 7 of treatment, as reported in Gonzalez et al., (1984)).

second-step *parC* mutation causes the window to fully overlap serum concentration (Figure 11.7C). With the third-step *gyrA parC gyrA* mutant, the window exceeds serum concentration (Figure 11.7D). These changes in window position relative to therapeutic drug concentration are expected to cause each successive mutation to be more readily fixed in the population due to the diminished time that mutant growth is restricted and to the diminished killing of mutant cells. Thus it is not surprising that isolates from patients can contain many resistance mutations (Li et al., 2004a).

A variety of ways have also been found in which low-level, non-target resistance mutations can affect subsequent enrichment of mutant subpopulations. One is represented by mutator mutants in which mutation frequencies can be elevated up to a thousand fold (Miller, 1996; Oliver et al., 2004). These mutations are expected to raise the plateau of the population analysis profile. Another is seen in overexpression of efflux genes (Markham and Neyfakh, 1996). Efflux overexpression mutations are expected to shift the population analysis profiles to higher drug concentrations, thereby making it more likely that high drug concentrations will fall inside the selection window rather than above it. A third type of non-target mutation has been found in *S. pneumoniae*. These mutations, which are still genetically undefined, confer low-level protection from fluoroquinolones without affecting efflux and without altering the amino acid sequence of the GyrA, GyrB, ParC, or ParE proteins (X. Li and K. Drlica, unpublished observations). One of these mutants, recovered at a frequency of 10^{-5} (arrow in Figure 11.8), produced a population

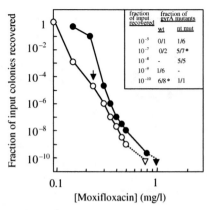

Fig. 11.8 Effect of fluoroquinolone concentration on the recovery of single-step, resistant mutants of *S. pneumoniae*. A wild-type population of *S. pneumoniae*, strain ATCC49619 (*open circles*) was grown to late log phase and then distributed at various dilutions to agar plates containing the indicated concentrations of moxifloxacin as described in Li et al., (2002). After incubation, colonies were regrown on drug-free agar and retested for growth on the selected concentration of moxifloxacin. The fraction of colonies recovered relative to the number of input colony-forming units is shown as a function of moxifloxacin concentration. The triangle indicates no colony recovered at that drug concentration and bacterial load. A non-target mutant, KD2113, was recovered under the conditions shown by the unlabeled *arrow*. A liquid culture was grown, applied to moxifloxacin-containing agar, and colony recovery (*filled circles*) was determined as with wild-type cells. *Inset* compares the recovery of *gyrA* mutants from wild-type (wt) and non-target mutants (nt mut) at various points on the population analysis profile expressed as a fraction of input cells recovered. Asterisks indicate the highest fraction of mutants recovered in which *gyrA* mutants represent the majority.

analysis profile showing increased MIC but little difference from wild-type cells at high moxifloxacin concentration (compare circles, Figure 11.8). When second-step mutant subpopulations were recovered from the non-target mutant at various moxifloxacin concentrations (solid circles), *gyrA* mutants were obtained at a frequency 1000 times higher than observed with wild-type cells (inset, Figure 11.8). Thus a small change in MIC can correlate with a large effect on the type of mutant selected.

11.8 Clinical Implications of the Mutant Selection Window Hypothesis

The most important concept emerging from the selection window hypothesis is that traditional dosing regimens are flawed with respect to resistance. They tend to seek the lowest drug dose required for clinical efficacy, thereby facilitating the acquisition of resistance by placing drug concentrations inside the selection window. A corollary of the hypothesis is that keeping drug concentrations above the window will severely restrict the amplification of mutant subpopulations. This requirement can be relaxed for compounds that kill resistant mutants. For such agents, anti-mutant doses are determined empirically as AUC_{24}/MPC. Conversely, if dose is fixed, compounds can be ranked for anti-mutant activity by AUC_{24}/MPC.

Another general statement emerges from the stepwise acquisition of resistance seen with ciprofloxacin and *H. influenzae*. The data in Figure 11.7 indicate that acquiring the first mutation is rate-limiting because the selection window is far below the serum drug concentration, while that is not the case for subsequent mutations. The same principle is likely to apply with other bacteria, especially when the first-step mutation causes a large increase in the upper boundary of the selection window, as seen with a fluoroquinolone-resistance mutation of *S. pneumoniae* (Figure 11.4) and with a multi-antibiotic resistance (MAR) mutation of *Salmonella enterica* (Randall et al., 2004). Thus blocking the initial amplification of first-step mutants is important to anti-mutant dosing strategies regardless of whether the mutation confers clinical resistance. Treatment of *S. pneumoniae* with levofloxacin may eventually serve as an example, since an increase in topoisomerase mutants, which was predicted from selection window considerations (Tillotson et al., 2001), has been attributed to the use of this agent (Bhavnani et al., 2005; Doern et al., 2005). Increasing the levofloxacin dose from 500 to 750 mg may slow the acquisition of resistance.

The relationship between MPC and pharmacokinetics can be used to determine whether monotherapy is likely to restrict amplification of mutant subpopulations or whether combination therapy is required. Fluoroquinolone treatment of *P. aeruginosa* infections serves as an example. For ciprofloxacin and levofloxacin, MPC is 3 and 9 mg/l, respectively (Hansen et al., 2006). Approved treatments of 750 mg twice daily (ciprofloxacin) or once daily (levofloxacin) generate an AUC_{24}/MPC of 10.5 and 10, respectively, using published values of AUC (http://www.ciprousa.com and http://www.levaquin.com). These numbers are below the 22 value estimated for restricting outgrowth with *E. coli* (Olofsson et al., 2006), another Gram-negative pathogen. Dropping the mutant-restrictive AUC_{24}/MPC by half due to immunocompetence (Andes and Craig, 2002) would still make monotherapy marginal for these compounds. Other situations are also expected to require combination therapy. For example, mutations that confer very high levels of protection, such as rifampicin resistance in *S. aureus* (Zhao and Drlica, 2002) and most forms of resistance in *Mycobacterium tuberculosis* (Dong et al., 2000), extend the selection window to such high drug concentrations that achievable concentrations tend to fall inside the window.

Combination therapy is also required when resistance genes enter a bacterial population at high frequency, such as from horizontal transfer of integrons and plasmids (Hall, 1997; Michael et al., 2004) or from elevated mutation frequency (Chopra et al., 2003; Oliver et al., 2004). In these cases, bacterial populations may contain double or triple mutant subpopulations prior to drug treatment, thereby requiring higher monotherapy drug concentrations than are readily achieved.

MPC can also be used to evaluate indications for new compounds. For example, the quinolone garenoxacin exhibits potent activity against *S. aureus,* and some investigators suggested that it might even be useful with ciprofloxacin-resistant *S. aureus* (Discotto et al., 2001; Fung-Tomc et al., 2001; Jones et al., 2001; Weller et al., 2002). However, determination of MPC with ciprofloxacin-resistant isolates suggested that garenoxacin resistance would be acquired as quickly as ciprofloxacin resistance with fully susceptible isolates (Acar and Goldstein, 1997; Blumberg, 1991), because in both cases serum drug concentrations fall inside the mutant selection window for much of the dosing period (Zhao et al., 2003). These same in vitro measurements of selection window dimensions indicate that garenoxacin treatment of ciprofloxacin-susceptible isolates might allow little mutant amplification if drug concentrations at the site of mutant amplification are at least as high as serum concentrations (published serum drug concentrations are above MPC for the entire dosing period) (Gajjar et al., 2003; Zhao et al., 2003). These predictions have yet to be tested in vivo.

11.9 Assumptions and Limitations of the Selection Window

Since the upper boundary of the selection window, the MPC, marks the concentration at which a cell must acquire two concurrent mutations for growth, a central assumption is that such a double mutant will be present at a much lower frequency than a single mutant. One type of support for this assumption is the dramatic reduction in the frequency of mutant recovery observed when S. aureus is treated with two compounds having different targets (Zhao and Drlica, 2002), even when both compounds are in the same class (Strahilevitz and Hooper, 2005). Single compounds that have a similar activity for two targets also greatly lower the recovery of resistant mutants (Ince et al., 2002). The converse is also seen. When a single compound has two targets, removal of one target through the introduction of a resistance mutation facilitates the recovery of resistant mutants, even when the introduced mutation has little effect on MIC (Li et al., 2002; Zhao et al., 1997). Thus the assumption that multiple concurrent mutations arise in the same cell less often than single mutations is well founded.

Application of the selection window hypothesis is based on the assumption that the drug susceptibility of mutant subpopulations is similar in vivo and in vitro and that mutation frequencies measured in vitro reflect those occurring in vivo. If susceptibility were lower or if mutation frequency were higher in vivo, drug concentration thresholds for restricting mutant growth would need to be higher than currently estimated. Since fluoroquinolone MPC measured in vitro can be used as a mutant-restriction threshold in three animal models, initial tests have been passed. However, additional animal studies with many pathogen–antimicrobial combinations are required to fully justify these assumptions.

Another assumption in applying data collected in vitro is that host defenses are ignored. A study with mice suggests that immunocompetence could reduce mutant-restrictive thresholds by roughly 50% (Andes and Craig, 2002). Thus using in vitro data without correction for host defense leads to a conservative estimate of the thresholds with most patients.

An important limitation of the hypothesis is that the threshold required to restrict amplification of resistant mutants depends on the size of the treated pathogen population. If that value is calculated by considering multiple patients, which can number in the millions with some bacterial infections, keeping drug concentration above MPC is expected to slow, but never prevent, the selective amplification of mutants—the occurrence of two concurrent resistance mutations in the same cell is statistically likely (double mutants would be enriched despite drug concentrations being kept above MPC or AUC_{24}/MPC being kept high). This numerical consideration emphasizes that application of the selection window hypothesis is *not* expected to replace efforts designed to limit antimicrobial consumption (Figure 11.1c, e). Likewise, MPC cannot be considered formally as a *resistance* prevention concentration.

Another limitation concerns determination of drug pharmacokinetics, since keeping drug concentrations above MPC or having a high value of AUC_{24}/MPC are central tasks of the approach. Serum pharmacokinetics with healthy volunteers are generally available; however, obtaining patient data may not be straightforward, and pharmacokinetics at the site of infection are often difficult to determine. Moreover, some of the bacterial population may be sequestered in sites that are protected from drug action. Such pharmacodynamic heterogeneity helps account for the evolution of resistance with multidrug therapy of tuberculosis (Lipsitch and Levin, 1998). These issues, plus person-to-person pharmacokinetic variation, emphasize the importance of developing new compounds with very narrow selection windows ($MIC_{(99)} \approx MPC$) (Zhao and Drlica, 2002), since such compounds should have reduced ability to allow mutant amplification regardless of pharmacokinetic properties.

Finally, it is important to reiterate that the mutant selection window hypothesis applies only to genetically based resistance and is most useful for resistance arising step-wise from spontaneous chromosomal mutations. The hypothesis does not apply to phenotypically "induced" resistance or to the protection of an entire bacterial population through the secretion of degradative enzymes by a few resistant cells. Resistance that is gained horizontally, such as through plasmid vectors, requires special consideration because the frequency of transmission may be high (10^{-5} to 10^{-1} (Alonso et al., 2005; Luo et al., 2005; Oppegaard et al., 2001)). That would cause the resistant subpopulation to be a large fraction of an infecting population, perhaps large enough to make it likely that a plasmid-containing cell carries a second-step resistance allele. If so, drug concentrations would need to be high enough to force the cell to acquire three or more mutations for growth. That is unlikely to be achieved by monotherapy. If the plasmid-borne resistance allele confers full protection from the agent, plasmid-bearing cells will amplify at all concentrations of the antimicrobial, which also precludes monotherapy. Thus horizontally transmitted resistance requires multidrug therapy even when the bulk of the bacterial population is deemed susceptible by break-point criteria.

11.10 Concluding Remarks

A current lament concerning resistance is that antimicrobial use guidelines are too vague, consisting of little more than a plea for using the agents wisely (Burke, 2003; Peterson and Dalhoff, 2004). For situations in which the amplification of resistant mutants can be described as hill climbing, the mutant selection window hypothesis leads to quantitative strategies for making dosing decisions: by requiring bacteria to obtain two or more mutations for growth, the organisms are forced to climb a steep cliff. Traditional efforts that focus on killing susceptible cells require microbes to acquire only one mutation for growth and rely on host defenses to halt mutant amplification. Those strategies allow the acquisition of resistance even while eradicating the susceptible population (Liu et al., 2005).

So far, the selection window hypothesis has had little effect on clinical practice. One reason is that MPC has been measured for only a few antimicrobial–pathogen combinations, and clinical laboratories do not routinely measure the parameter. Standardization is still required, as are less labor-intensive assays. Unfortunately, MPC is not reliably estimated from determinations of MIC (e.g., Figure 11.5). Another, and perhaps more important reason for limited use is that anti-mutant strategies involve using doses that are higher than that generally needed to cure. Consequently, anti-mutant strategies are expected to cause more side-effects for patients, who at the individual level may see little benefit from the higher doses. Thus implementing anti-mutant strategies has costs. The size of these costs relative to the benefits of longer antimicrobial life-spans is unknown.

Acknowledgments We thank Caroline Logan for technical assistance and Joyce de Azavedo, Alexander Firsov, Marila Gennaro, and Bruce Levin for many critical comments, most of which led to changes in the manuscript. The work was supported by NIH grants AI35257 and AI063431.

References

Aarestrup, F., Seyfarth, A., Emborg, H., Pedersen, K., Hendriksen, R. and Bager, F., (2001), Effect of abolishment of the use of antimicrobial agents for growth promotion on occurrence of antimicrobial resistance in fecal enterococci from food animals in Denmark. *Antimicrob Agents Chemother*, **45**, 2054–2059.

Acar, J. and Goldstein, F., (1997), Trends in bacterial resistance to fluoroquinolones. *Clin Infect Dis*, **24**, S67-S73.

Almeida, D., Rodrigues, C., Udwadia, Z., Lalvani, A., Gothi, G., Mehta, P. and Mehta, A., (2003), Incidence of multidrug-resistant tuberculosis in urban and rural India and implications for prevention. *Clin Infect Dis*, **36**, e152–e154.

Alonso, G., Baptista, K., Ngo, T. and Taylor, D., (2005), Transcriptional organization of the temperature-sensitive transfer system from the IncHI1 plasmid R27. *Microbiology*, **35**, 3563–3573.

Ambrose, P., Zoe-Powers, A., Russo, R., Jones, D. and Owens, R., (2002), Utilizing pharmacodynamics and pharmacoeconomics in clinical and formulary decision making. In C. Nightingale, T. Murakawa, and P. Ambrose, editors, *Antimicrobial Pharmacodynamics in Theory and Clinical Practice*. New York, Marcel Dekker, pp. 385–409.

Andes, D. and Craig, W., (2002), Pharmacodynamics of the new fluoroquinolone gatifloxacin in murine thigh and lung infection models. *Antimicrob Agents Chemother*, **46**, 1665–1670.

Anonymous (2004) Hospital hygiene sweeps resistance under the carpet. *The Lancet Infect Dis*, **4**, 713.

Balaban, N., Merrin, J., Chait, R., Kowalik, L. and Leibler, S., (2004), Bacterial persistence as a phenotypic switch. *Science*, **305**, 1622–1625.

Bangsberg, D., Charlebois, E., Grant, R., Holodniy, M., Deeks, S., Perry, S., Conroy, K., Clark, R., Guzman, D., Zolopa, A. and Moss, A., (2003), High levels of adherence do not prevent accumulation of HIV drug resistance mutations. *AIDS*, **17**, 1925–1932.

Bangsberg, D.R., Moss, A.R. and Deeks, S.G., (2004), Paradoxes of adherence and drug resistance to HIV antiretroviral therapy. *J Antimicrob Chemother*, **53**, 696–699.

Bannon, J., Hatem, M. and Noone, M., (1998), Anaerobic infections of the urinary tract: are they being missed? *J Clin Pathol*, **51**, 709–710.

Baquero, F., (1990), Resistance to quinolones in Gram-negative microorganisms: mechanisms and prevention. *Eur Urol*, 17 (Suppl 1), 3–12.

Baquero, F., (2001), Low-level antibacterial resistance: a gateway to clinical resistance. *Drug Resist Updat*, **4**, 93–105.

Baquero, F., Baquero-Artigao, G., Canton, R. and Garcia-Rey, C., (2002), Antibiotic consumption and resistance selection in *Streptococcus pneumoniae*. *J Antimicrob Chemother*, 50 (Suppl S2), 27–37.

Baquero, F. and Negri, M., (1997a), Selective compartments for resistant microorganisms in antibiotic gradients. *Bioessays*, **19**, 731–736.

Baquero, F. and Negri, M., (1997b), Strategies to minimize the development of antibiotic resistance. *J Chemother*, 9 (Suppl), 29–37.

Baquero, F., Negri, M., Morosini, M. and Blazquez, J., (1997), The antibiotic selective process: concentration-specific amplification of low-level resistant populations. *Ciba Found Symp*, **207**, 93–105.

Beaber, J.W., Hochhut, B. and Waldor, M.K., (2004), SOS response promotes horizontal dissemination of antibiotic resistance genes. *Nature*, **427**, 72–74.

Bhavnani, S.M., Hammel, J., Jones, R. and Ambrose, P., (2005), Relationship between increased levofloxacin use and decreased susceptiblity of *Streptococcus pneumoniae* in the United States. *Diagn Microbiol Infect Dis*, **51**, 31–37.

Binda, G., Domenichini, A., Gottardi, A., Orlandi, B., Ortelli, E., Pacini, B. and Fowst, G., (1971), Rifampicin, a general review. *Arzneim Forsch*, **21**, 1908–1977.

Bingen, E., Lambert-Zechovsky, N., Mariani-Kurkdjian, P., Doit, C., Aujard, Y., Fournerie, F. and Mathieu, H., (1990), Bacterial counts in cerebrospinal fluid of children with meningitis. *Eur J Clin Microbiol Infect Dis*, **9**, 278–281.

Blaser, J., Stone, B., Groner, M. and Zinner, S., (1987), Comparative study with enoxacin and netilmicin in a pharmacodynamic model to determine importance of ratio of antibiotic peak concentration to MIC for bactericidal activity and emergence of resistance. *Antimicrob Agents Chemother*, **31**, 1054–1060.

Blondeau, J., Hansen, G., Metzler, K. and Hedlin, P., (2004), The role of PK/PD parameters to avoid selection and increase of resistance: mutant prevention concentration. *J Chemother*, Suppl 3, 1–19.

Blondeau, J., Zhao, X., Hansen, G. and Drlica, K., (2001), Mutant prevention concentrations (MPC) or fluoroquinolones with clinical isolates of *Streptococcus pneumoniae*. *Antimicrob Agents Chemother*, **45**, 433–438.

Blumberg, (1991), Rapid development of ciprofloxacin resistance in methicillin-susceptible and-resistant *Staphylococcus aureus*. *J Infect Dis*, **63**, 1279–1285.

Bonhoeffer, S. and Nowak, M., (1997), Pre-existence and emergence of drug resistance in HIV-1 infection. *Proc R Soc Lond B Biol Sci*, **264**, 631–637.

British Medical Research Council, (1950), Treatment of pulmonary tuberculosis with streptomycin and para-amino-salicylic acid. *Br Med J*, **2**, 1073–1085.

Burke, J.P., (2003), Infection control—a problem for patient safety. *N Engl J Med*, **348**, 651–656.

Campion, J., Chung, P., McNamara, P., Titlow, W. and Evans, M., (2005a), Pharmacodynamic modeling of the evolution of levofloxacin resistance in *Staphylococcus aureus*. *Antimicrob Agents Chemother*, **49**, 2189–2199.

Campion, J., McNamara, P. and Evans, M., (2005b), Pharmacodynamic modeling of ciprofloxacin resistance in *Staphylococcus aureus*. *Antimicrob Agents Chemother*, **49**, 209–219.

Campion, J., McNamara, P. and Evans, M.E., (2004), Evolution of ciprofloxacin-resistant *Staphylococcus aureus* in *in vitro* pharmacokinetic environments. *Antimicrob Agents Chemother*, **48**, 4733–4744.

Carey, B. and Cryan, B., (2003), Antibiotic misuse in the community—a contributor to resistance? *Irish Med J*, **96**, 43–45.

Chopra, I., O'Neill, A. and Miller, K., (2003), The role of mutators in the emergence of antibiotic-resistant bacteria. *Drug Resist Updat*, **6**, 137–145.

Chow, J., Fine, M., Shlaes, D., Quinn, J., Hooper, D., Johnson, M., Ramphal, R., Wagener, M., Miyashiro, D. and Yu, V., (1991), *Enterobacter* bacteremia: clinical features and emergence of antibiotic resistance during therapy. *Ann Intern Med*, **115**, 585–590.

Cirz, R., Chin, J., Andes, D., Crecy-Lagard, V., Craig, W. and Romesberg, F.E., (2005), Inhibition of mutation and combating the evolution of antibiotic resistance. *Plos Biol*, **3**, 1024–1033.

Cizman, M., (2003), The use and resistance to antibiotics in the community. *Int J Antimicrob Agents*, **21**, 297–307.

Coker, R.J., (2004), Multidrug-resistant tuberculosis: public health challenges. *Trop Med Int Health*, **9**, 25–40.

Craig, W., (1998), Pharmacokinetic/pharmacodynamic parameters: rationale for antibacterial dosing of mice and men. *Clin Infect Dis*, **26**, 1–12.

Craig, W., (2001), Does the dose matter? *Clin Infect Dis*, **33**, S233–S237.

Craig, W.A., (2002), Pharmacodynamics of antimicrobials: general concepts and applications. In C. Nightingale, T. Murakawa and P. Ambrose, editors, *Antimicrobial Pharmacodynamics in Theory and Clinical Practice*. New York, Marcel Dekker, pp. 1–22.

Croisier, D., Etienne, M., Piroth, L., Bergoin, E., Lequeu, C., Portier, H. and Chavanet, P., (2004), In vivo pharmacodynamic efficacy of gatifloxacin against *Streptococcus pneumoniae* in an experimental model of pneumonia: impact of the low levels of quinolone resistance on the enrichment of resistant mutants. *J Antimicrob Chemother*, **54**, 640–647.

Croom, K.F. and Goa, K.L., (2003), Levofloxacin: a review of its use in the treatment of bacterial infections in the United States. *Drugs*, **63**, 2769–2802.

Cross, E., Park, S. and Perlin, D.S., (2000), Cross-resistance of clinical isolates of *Candida albicans* and *Candida glabrata* to over-the-counter azoles used in the treatment of vaginitis. *Microb Drug Res*, **6**, 155–161.

Cui, J., Liu, Y., Wang, R., Tong, W., Drlica, K. and Zhao, X., (2006), The mutant selection window demonstrated in rabbits infected with *Staphylococcus aureus*. *J* Infect Dis, **194**, 1601–1608.

Dan, M., (2004), The use of fluoroquinolones in gonorrhoea: the increasing problem of resistance. *Expert Opin Pharmacother*, **5**, 829–854.

Davidson, R., Cavalcanti, R., Brunton, J., Bast, D.J., deAzavedo, J.C., Kibsey, P., Fleming, C. and Low, D.E., (2002), Resistance to levofloxacin and failure of treatment of pneumococcal pneumonia. *N Engl J Med*, **346**, 747–750.

Dean, N., (2003), Encouraging news from the antibiotic resistance front. *Chest*, **124**, 423–424.

Discotto, L.F., Lawrence, L., Denbleyker, K. and Barrett, J.F., (2001), *Staphylococcus aureus* mutants selected by BMS-284756. *Antimicrob Agents Chemother*, **45**, 3273–3275.

Doern, G.V., Richter, S., Miller, A., Miller, N., Rice, C., Heilmann, K. and Beekmann, S., (2005), Antimicrobial resistance among *Streptococcus pneumoniae* in the United States: have we begun to turn the corner on resistance to certain antimicrobial classes? *Clin Infect Dis*, **41**, 139–148.

Dong, Y., Xu, C., Zhao, X., Domagala, J. and Drlica, K., (1998), Fluoroquinolone action against mycobacteria: effects of C8 substituents on bacterial growth, survival, and resistance. *Antimicrob Agents Chemother*, **42**, 2978–2984.

Dong, Y., Zhao, X., Domagala, J. and Drlica, K., (1999), Effect of fluoroquinolone concentration on selection of resistant mutants of *Mycobacterium bovis* BCG and *Staphylococcus aureus*. *Antimicrob Agents Chemother*, **43**, 1756–1758.

Dong, Y., Zhao, X., Kreiswirth, B. and Drlica, K., (2000), Mutant prevention concentration as a measure of antibiotic potency: studies with clinical isolates of *Mycobacterium tuberculosis*. *Antimicrob Agents Chemother*, **44**, 2581–2584.

Drlica, K., (2003), The mutant selection window and antimicrobial resistance. *J Antimicrob Chemother*, **52**, 11–17.

Drlica, K. and Zhao, X., (2003), Controlling antibiotic resistance: strategies based on the mutant selection window. In I.W. Fong and K. Drlica, editors, *Re-emergence of Established Microbial Pathogens in the 21st Century*. New York, Kluwer Academic/Plenum Publishers, pp. 295–331.

Drlica, K., Zhao, X., Blondeau, J. and Hesje, C., (2006), Low correlation between minimal inhibitory concentration (MIC) and mutant prevention concentration (MPC). *Antimicrob Agents Chemother*, **50**, 403–404.

East African Hospitals and British Medical Research Council, (1960), Comparative trial of isoniazid alone in low and high dosage and isoniazid plus PAS in the treatment of acute pulmonary tuberculosis in East Africa. *Tubercle*, **40**, 83–102.

Ehrlich, P., (1913), Chemotherapeutics: scientific principles, methods and results. *Lancet*, **ii**, 445–451.

Eliopoulos, G., Gardella, A. and R. Moellering, J., (1984), In vitro activity of ciprofloxacin, a new carboxyquinoline antimicrobial agent. *Antimicrob Agents Chemother*, **25**, 331–335.

Espinal, M.A., (2003), The global situation of MDR-TB. *Tuberculosis*, **83**, 44–51.

Etienne, M., Croisier, D., Charles, P.-E., Lequeu, C., Piroth, L., Portier, H., Drlica, K. and Chavanet, P., (2004), Effect of low-level resistance on subsequent enrichment of fluoroquinolone-resistant *Streptococcus pneumoniae* in rabbits. *J Infect Dis*, **190**, 1472–1475.

Fagon, J., Chastre, J., Trouillet, J., Domart, Y., Dombret, M., Bornet, M. and Gibert, C., (1990), Characterization of distal bronchial microflora during acute exacerbation of chronic bronchitis. *Am Rev Respir Dis*, **142**, 1004–1008.

Feldman, C., (2004), Clinical relevance of antimicrobial resistance in the management of pneumococcal community-acquired pneumonia. *J Lab Clin Med*, **143**, 269–283.

Feldman, W., (1976), Concentrations of bacteria in cerebrospinal fluid of patients with bacterial meningitis. *J Pediatr*, **88**, 549–552.

Felmingham, D., (2002), The need for antimicrobial resistance surveillance. *J Antimicrob Chemother*, **50** (Suppl S1), 1–7.

Fink, M., Snydman, D., Niederman, M., Leeper, K., Johnson, R., Heard, S., Wunderink, R., Caldwell, J., Schentag, J., Siami, G., Zameck, R., Haverstock, D., Reinhart, H., Echols, R. and Group, S.P.S., (1994), Treatment of severe pneumonia in hospitalized patients: results of a multicenter, randomized, double-blind trial comparing intravenous ciprofloxacin with imipenem–cilastatin. *Antimicrob Agents Chemother*, **38**, 547–557.

Firsov, A., Smirnova, M., Lubenko, I., Vostrov, S., Portnoy, Y., and Zinner, S., (2006), Testing the mutant selection window hypothesis with *Staphylococcus aureus* exposed to daptomycin and vancomycin in an *in vitro* dynamic model. *J Antimicrob Chemother*, **58**, 1185–1192

Firsov, A., Vostrov, S., Lubenko, I., Arzamastsev, A., Portnoy, Y. and Zinner, S., (2004a), ABT492 and levofloxacin: comparison of their pharmacodynamics and their abilities to prevent the selection of resistant *Staphylococcus aureus* in an *in vitro* dynamic model. *J Antimicrob Chemother*, **54**, 178–186.

Firsov, A., Vostrov, S., Lubenko, I., Drlica, K., Portnoy, Y. and Zinner, S., (2003), In vitro pharmacodynamic evaluation of the mutant selection window hypothesis: four fluoroquinolones against *Staphylococcus aureus*. *Antimicrob Agents Chemother*, **47**, 1604–1613.

Firsov, A., Vostrov, S., Lubenko, I., Zinner, S. and Portnoy, Y., (2004b), Concentration-dependent changes in the susceptibility and killing of *Staphylococcus aureus* in an in vitro dynamic model that simulates normal and impaired gatifloxacin elimination. *Int J Antimicrob Agents*, **23**, 60–66.

Fox, W., Elklard, G. and Mitchison, D., (1999), Studies on the treatment of tuberculosis undertaken by the British Medical Research Council Tuberculosis Units, 1946–1986, with relevant subsequent publications. *Int J Tuberc Lung Dis*, **3**, S231–S279.

Friedman, S.M., Lu, T. and Drlica, K., (2001), A mutation in the DNA gyrase A gene of *Escherichia coli* that expands the quinolone-resistance-determining region. *Antimicrob Agents Chemother*, **45**, 2378–2380.

Fuller, J.D. and Low, D.E., (2005), A review of *Streptococcus pneumoniae* infection treatment failures associated with fluoroquinolone resistance. *Clin Infect Dis*, **41**, 118–121.

Fung-Tomc, J., Valera, L., Minassian, B., Bonner, D. and Gradelski, E., (2001), Activity of the novel des-fluoro(6) quinolone BMS-284756 against methicillin-susceptible and -resistant staphylococci. *J Antimicrob Chemother*, **48**, 735–748.

Gajjar, D., Bello, A., Ge, Z., Christopher, L. and Grasela, D., (2003), Multiple-dose safety and pharmacokinetics of oral garenoxacin in healthy subjects. *Antimicrob Agents Chemother*, **47**, 2256–2263.

Gillespie, S.H., Voelker, L.L., Ambler, J., Traini, C. and Dickens, A., (2003), Fluoroquinolone resistance in *Streptococcus pneumoniae*: evidence that *gyrA* mutations arise at a lower rate and that mutation in *gyrA* or *parC* predisposes to further mutation. *Microb Drug Res*, **9**, 17–24.

Gonzalez, M., Uribe, F., Moisen, S., Fuster, A., Selen, A., Welling, P. and Painter, B., (1984), Multiple-dose pharmacokinetics and safety of ciprofloxacin in normal volunteers. *Antimicrob Agents Chemother*, **26**, 741–744.

Gudas, L.J. and Pardee, A.B., (1976), DNA synthesis inhibition and the induction of protein X in *Eschericia coli*. *J Mol Biol*, **101**, 459–477.

Gumbo, T., Louie, A., Deziel, M., Parsons, L., Salfinger, M. and Drusano, G., (2004), Selection of a moxifloxacin dose that suppresses drug resistance in *Mycobacterium tuberculosis* by use of an in vitro pharmacodynamic infection model and mathematical modeling. *J Infect Dis*, **190**, 1642–1651.

Hall, R.M., (1997), Mobile gene cassettes and integrons: moving antibiotic resistance genes in gram-negative bacteria. *Ciba Found Symp*, **207**, 192–202.

Hansen, G., Zhao, X., Drlica, K. and Blondeau, J., (2006), Mutant prevention concentration for ciprofloxacin and levofloxacin with *Pseudomonas aeruginosa*. *Int J Antimicrob Agents*, **27**, 120–124.

Hecker, M., Aron, D., Patel, N., Lehmann, M. and Donskey, C., (2003), Unnecessary use of antimicrobials in hospitalized patients. *Arch Intern Med*, **163**, 972–978.

Hermsen, E., Hovde, L., Konstantinides, G. and Rotschafer, J., (2005), Mutant prevention concentration of ABT-492, levofloxacin, moxifloxacin, and gatifloxacin against three common respiratory pathogens. *Antimicrob Agents Chemother*, **49**, 1633–1635.

Hovde, L., Rotschafer, S., Ibrahim, K., Gunderson, B., Hermsen, E. and Rotschafer, J., (2003), Mutation prevention concentration of ceftriaxone, meropenem, imipenem, and ertapenem against three strains of *Streptococcus pneumoniae*. *Diagn Microbiol Infect Dis*, **45**, 265–267.

Ince, D., Zhang, X., Silver, C., and Hooper, D.C., (2002), Dual targeting of DNA gyrase and topoisomerase IV: target interactions of garenoxacin (BMS-284756, T-3811ME), a new desfluoroquinolone. *Antimicrob Agents Chemother*, **46**, 3370–3380.

Iseman, M., (1994), Evolution of drug-resistant tuberculosis: a tale of two species. *Proc Natl Acad Sci USA*, **91**, 2428–2429.

Jernigan, J., Titus, M., Groschel, D., Getchell-White, S. and Farr, B., (1996), Effectiveness of contact isolation during a hospital outbreak of methicillin-resistant *Staphylococcus aureus*. *Am J Epidemiol*, **143**, 496–504.

Jones, R., Pfaller, M., Stilwell, M. and SENTRY-Antimicrobial-Surveillance-Program-Participants-Group, (2001), Activity and spectrum of BMS 284756, a new des-F (6) quinolone, tested against strains of ciprofloxacin-resistant Gram-positive cocci. *Diagn Microbiol Inf Dis*, **39**, 133–135.

Jourdain, B., Novara, A., Joly-Guillou, M., Dombret, M., Calvat, S., Trouillet, J., Gilbert, C. and Chastre, J., (1995), Role of quantitative cultures of endotracheal aspirates in the diagnosis of nosocomial pneumonia. *Am J Respir Crit Care Med*, **152**, 241–246.

Jumbe, N., Louie, A., Leary, R., Liu, W., Deziel, M., Tam, V., Bachhawat, R., Freeman, C., Kahn, J., Bush, K., Dudley, M., Miller, M. and Drusano, G., (2003), Application of a mathematical model to prevent in vivo amplification of antibiotic-resistant bacterial populations during therapy. *J Clin Invest*, **112**, 275–285.

Kawamura, Y., Fujiwara, H., Mishima, N., Tanaka, Y., Tanimoto, A., Ikawa, S., Itoh, Y. and Ezaki, T., (2003), First *Streptococcus agalactiae* isolates highly resistant to quinolones with point mutations in *gyrA* and *parC*. *Antimicrob Agents Chemother*, **47**, 3605–3609.

Kotilainen, P., Routamaa, M., Peltonen, R., Oksi, J., Rintala, E., Meurman, O., Lehtonen, O., Eerola, E., Salmenlinna, S., Vuopio-Varkila, J. and Rossi, T., (2003), Elimination of epidemic methicillin-resistant *Staphylococcus aureus* from a university hospital and district institutions, Finland. *Emerg Infect Dis*, **9**, 169–175.

Kuga, A., Okamoto, R. and Inoue, M., (2000), *ampR* gene mutations that greatly increase class C β-lactamase activity in *Enterobacter cloacae*. *Antimicrob Agents Chemother*, **44**, 561–567.

Kummerer, K. and Henninger, A., (2003), Promoting resistance by the emission of antibiotics from hospitals and households into effluent. *Clin Microbiol Infect*, **9**, 1203–1214.

Lautenbach, E., Strom, B., Nachamkin, I., Bilker, W., Marr, A., Larosa, L. and Fishman, N., (2004), Longitudinal trends in fluoroquinolone resistance among Enterobacteriaceae isolates from inpatients and outpatients, 1989–2000; differences in the emergence and epidemiology of resistance across organisms. *Clin Infect Dis*, **38**, 655–662.

Levy, S., (2002), The 2000 Garrod lecture: factors impacting on the problem of antibiotic resistance. *J Antimicrob Chemother*, **49**, 25–30.

Levy, S.B. and Marshall, B., (2004), Antibacterial resistance worldwide: causes, challenges, and responses. *Nat Med*, **10**, S122-S129.

Li, X., Mariano, N., Rahal, J.J., Urban, C.M. and Drlica, K., (2004a), Quinolone-resistant *Haemophilus influenzae* in a long-term care facility: nucleotide sequence characterization of alterations in the genes encoding DNA gyrase and DNA topoisomerase IV. *Antimicrob Agents Chemother*, **48**, 3570–3572.

Li, X., Mariano, N., Rahal, J.J., Urban, C.M. and Drlica, K., (2004b), Quinolone-resistant *Haemophilus influenzae*: determination of mutant selection window for ciprofloxacin, garenoxacin, levofloxacin, and moxifloxacin. *Antimicrob Agents Chemother*, **48**, 4460–4462.

Li, X., Zhao, X. and Drlica, K., (2002), Selection of *Streptococcus pneumoniae* mutants having reduced susceptibility to levofloxacin and moxifloxacin. *Antimicrob Agents Chemother*, **46**, 522–524.

Lieberman, J., (2003), Appropriate antibiotic use and why it is important: the challenges of bacterial resistance. *Pediatr Infect Dis J*, **22**, 1143–1151.

Linde, H.-J. and Lehn, N., (2004), Mutant prevention concentration of nalidixic acid, ciprofloxacin, clinafloxacin, levofloxacin, norfloxacin, ofloxacin, sparfloxacin or trovafloxacin for *Escherichia coli* under different growth conditions. *J Antimicrob Chemother*, **53**, 252–257.

Lipsitch, M. and Levin, B., (1997), The population dynamics of antimicrobial chemotherapy. *Antimicrob Agents Chemother*, **41**, 363–373.

Lipsitch, M. and Levin, B., (1998), Population dynamics of tuberculosis treatment: mathematical models of the roles of non-compliance and bacterial heterogeneity in the evolution of drug resistance. *Int. J. Tuberc Lung Dis*, **2**, 187–199.

Liu, Y., Cui, J., Wang, R., Wang, X., Drlica, K. and Zhao, X., (2005), Selection of rifampicin-resistant *Staphylococcus aureus* during tuberculosis therapy: concurrent bacterial eradication and acquisition of resistance. *J Antimicrob Chemother*, **56**, 1172–1175.

Lowy, F.D., (2003), Antimicrobial resistance: the example of *Staphylococcus aureus*. *J Clin Invest*, **111**, 1265–1273.

Lu, T., Zhao, X., Li, X., Drlica-Wagner, A., Wang, J.-Y., Domagala, J. and Drlica, K., (2001), Enhancement of fluoroquinolone activity by C-8 halogen and methoxy moieties: action against a gyrase resistance mutant of *Mycobacterium smegmatis* and a gyrase-topoisomerase IV double mutant of *Staphylococcus aureus*. *Antimicrob Agents Chemother*, **45**, 2703–2709.

Lu, T., Zhao, X., Li, X., Hansen, G., Blondeau, J. and Drlica, K., (2003), Effect of chloramphenicol, erythromycin, moxifloxacin, penicillin, and tetracycline concentration on the recovery of resistant mutants of *Mycobacterium smegmatis* and *Staphylococcus aureus*. *J Antimicrob Chemother*, **52**, 61–64.

Luo, H., Wan, K. and Wang, H., (2005), High-frequency conjugation system facilitates biofilm formation and pAMβ1 transmission by *Lactococcus lactis*. *Appl Environ Microbiol*, **71**, 2970–2978.

Marcusson, L., Olofsson, S., Lindgren, P., Cars, O. and Hughes, D., (2005), Mutant prevention concentration of ciprofloxacin for urinary tract infection isolates of *Escherichia coli*. *J Antimicrob Chemother*, **55**, 938–943.

Markham, P. and Neyfakh, A., (1996), Inhibition of the multidrug transporter NorA prevents emergence of norfloxacin resistance in *Staphylococcus aureus*. *Antimicrob Agents Chemother*, **40**, 2673–2675.

Marr, K., Lyons, C., Ha, K., Rustad, T. and White, T., (2001), Inducible azole resistance associated with a heterogeneous phenotype in *Candida albicans*. *Antimicrob Agents Chemother*, **45**, 52–59.

Mathema, B., Cross, E., Dun, E., Park, S., Dedell, J., Slade, B., Williams, M., Riley, L., Chaturvedi, V. and Perliln, D.S., (2001), Prevalence of vaginal colonization by drug-resistant *Candida* species in college-age women with previous exposure to over-the-counter azole antifungals. *Clin Infect Dis*, **33**, e23-e27.

Metzler, K., Hansen, G., Hedlin, P., Harding, E., Drlica, K. and Blondeau, J., (2004), Comparison of minimal inhibitory and mutant prevention concentrations of 4 fluoroquinolones: methicillin-susceptible and -resistant *Staphylococcus aureus. Int J Antimicrob Agents*, **24**, 161–167.

Michael, C.A., Gillings, M., Holmes, A., Hughes, L., Andrew, N., Holley, M. and Stokes, H., (2004), Mobile gene cassettes: a fundamental resource for bacterial evolution. *Am Nat*, **164**, 1–12.

Miller, C., Thomsen, L., Gaggero, C., Mosseri, R., Ingmer, H. and Cohen, S., (2004), SOS response induction by β-lactams and bacterial defense against antibiotic lethality. *Science*, **305**, 1629–1631.

Miller, J., (1996), Spontaneous mutators in bacteria: insights into pathways of mutagenesis and repair. *Annu. Rev. Microbiol.* **50**, 625–643.

Minami, S., Yotsuji, A., Inoue, M. and Mitsuhashi, S., (1980), Induction of β -lactamase by various β-lactam antibiotics in *Enterobacter cloacae. Antimicrob Agents Chemother*, **18**, 382–385.

Mitchison, D.A., (1984), Drug resistance in mycobacteria. *Br Med Bull*, **40**, 84–90.

Mouton, J., Dudley, M., Cars, O., Derendorf, H. and Drusano, G., (2005), Standardization of pharmacokinetic/pharmacodynamic (PK/PD) terminology for anti-infective drugs: an update. *J Antimicrob Chemother*, **55**, 601–607.

Muller, V. and Bonhoeffer, S., (2003), Mathematical approaches in the study of viral kinetics and drug resistance in HIV-1 infection. *Curr Drug Targets Infect Disord*, **3**, 329–344.

Nazir, J., Urban, C., Mariano, N., Burns, J., Tommasulo, B., Rosenberg, C., Segal-Maurer, S. and Rahal, J., (2004), Quinolone-resistant *Haemophilus influenzae* in a long-term care facility: clinical and molecular epidemiology. *Clin Infect Dis*, **38**, 1564–1569.

Negri, M. and Baquero, F., (1998), In vitro selective concentrations of cefapime and ceftazidime AmpC beta-lactamase hyperproduce *Enterobacter cloacae* variants. *Clin Microbiol Infect*, **4**, 585–588.

Negri, M., Morosini, M., Loza, E. and Baquero, F., (1994), In vitro selective antibiotic concentrations of β-lactams for penicillin-resistant *Streptococcus pneumoniae* populations. *Antimicrob Agents Chemother*, **38**, 122–125.

Negri, M.-C., Lipsitch, M., Blazquez, J., Levin, B.R. and Baquero, F., (2000), Concentration-dependent selection of small phenotypic differences in TEM β-lactamase-mediated antibiotic resistance. *Antimicrob Agents Chemother*, **44**, 2485–2491.

Oliver, A., Levin, B., Juan, C., Baquero, F. and Blazquez, J., (2004), Hypermutation and the preexistence of antibiotic-resistant *Pseudomonas aeruginosa* mutants: implications for susceptibility testing and treatment of chronic infections. *Antimicrob Agents Chemother*, **48**, 4226–4233.

Olofsson, S., Marcusson, L., Komp-Lindgren, P., Hughes, D. and Cars, O., (2006), Selection of ciprofloxacin resistance in *Escherichia coli* in an in vitro kinetic model: relation between drug exposure and mutant prevention concentration. *J Antimicrob Chemother*, **57**, 1116–1121.

Oppegaard, H., Steinum, T. and Wasteson, Y., (2001), Horizontal transfer of a multi-drug resistance plasmid between coliform bacteria of human and bovine origin in a farm environment. *Appl Environ Microbiol*, **67**, 3732–3734.

Pechere, J.-C., (2001), Patients' interviews and misuse of antibiotics. *Clin Infect Dis*, **33** (Suppl 3), S170–S173.

Peterson, L.R. and Dalhoff, A., (2004), Toward target prescribing: will the cure for antimicrobial resistance be specific, directed therapy through improved diagnostic testing? *J Antimicrob Chemother*, **53**, 902–905.

Phillips, I., Culebras, E., Moreno, F. and Baquero, F., (1987), Induction of the SOS response by new 4-quinolones. *J Antimicrob Chemother*, **20**, 631–638.

Pittet, D., (2003), Hand hygiene: improved standards and practice for hospital care. *Curr Opin Infect Dis*, **16**, 327–335.

Projan, S. and Shlaes, D., (2004), Antibacterial drug discovery: is it all downhill from here? *Clin Microbiol Infect*, **10** (Suppl 4), 18–22.

Randall, L., Cooles, S., Piddock, L. and Woodward, M., (2004), Mutant prevention concentrations of ciprofloxacin and enrofloxacin for *Salmonella enterica*. *J Antimicrob Chemother*, **54**, 688–691.

Regoes, R., Wiuff, C., Zappala, R., Garner, K., Baquero, F. and Levin, B., (2004), Pharmacodynamic functions: a multiparameter approach to the design of antibiotic treatment regimens. *Antimicrob Agents Chemother*, **48**, 3670–3676.

Rice, L., Eckstein, E., DeVente, J. and Shlaes, D., (1996), Ceftazidime-resistant *Klebsiella pneumoniae* isolates recovered at the Cleveland Department of Veterans Affairs Medical Center. *Clin Infect Dis*, **23**, 118–124.

Rodriguez, J., Cebrian, L., Lopez, M., Ruiz, M., Jimenez, I. and Royo, G., (2004a), Mutant prevention concentration: comparison of fluoroquinolones and linezolid with *Mycobacterium tuberculosis*. *J Antimicrob Chemother*, **53**, 441–444.

Rodriguez, J., Cebrian, L., Lopez, M., Ruiz, M. and Royo, G., (2004b), Mutant prevention concentration: a new tool for choosing treatment in nontuberculous mycobacterial infections. *Int J Antimicrob Agents*, **24**, 352–356.

Rodriguez, J., Cebrian, L., Lopez, M., Ruiz, M. and Royo, G., (2005), Usefulness of various antibiotics against *Mycobacterium avium-intracellulare*, measured by their mutant prevention concentration. *Int J Antimicrob Agents* **25**, 221–225.

Sahm, D. and Tenover, F., (1997), Surveillance for the emergence and dissemination of antimicrobial resistance in bacteria. *Infect Dis Clin North Am*, **11**, 767–783.

Seppala, H., Klaukka, T., Vuopio-Varkila, J., Maotiala, A., Helenius, H. and Lager, K., (1997), The effect of changes in the consumption of macrolide antibiotics on erythromycin resistance in group A streptococci in Finland. *N Engl J Med*, **337**, 441–446.

Shockley, T. and Hotchkiss, R., (1970), Stepwise introduction of transformable penicillin resistance in pneumococcus. *Genetics* **64**, 397–408.

Shopsin, B., Gomez, M., Montgomery, S., Smith, D.H., Waddington, M., Dodge, D.E., Bost, D., Riehman, M., naidich, S. and Kreiswirth, B.N., (1999), Evaluation of protein A gene polymorphic region DNA sequencing for typing of *Staphylococcus aureus* strains. *J Clin Microbiol*, **37**, 3556–3563.

Sieradzki, K., Leski, T., Dick, J., Borio, L., and Tomasz, A., (2003), Evolution of a vancomycin-intermediate *Staphylococcus aureus* strain in vivo: multiple changes in the antibiotic resistance phenotypes of a single lineage of methicillin-resistant *S. aureus* under the impact of antibiotics administered for chemotherapy. *J Clin Microbiol*, **41**, 1687–1693.

Sieradzki, K., Roberts, R., Haber, S. and Tomasz, A., (1999), The development of vancomycin resistance in a patient with methicillin-resistant *Staphylococcus aureus* infection. *N Engl J Med*, **340**, 517–523.

Silverman, J., Oliver, N., Andrew, T. and Li, T., (2001), Resistance studies with daptomycin. *Antimicrob Agents Chemother*, **45**, 1799–1802.

Sindelar, G., Zhao, X., Liew, A., Dong, Y., Zhou, J., Domagala, J. and Drlica, K., (2000), Mutant prevention concentration as a measure of fluoroquinolone potency against mycobacteria. *Antimicrob Agents Chemother*, **44**, 3337–3343.

Smith, B. and Walker, G., (1998), Mutagenesis and more: *umuDC* and the *Escherichia coli* SOS response. *Genetics*, **148**, 1599–1610.

Smith, H., Walters, M., Hisanaga, T., Zhanel, G. and Hoban, D., (2004), Mutant prevention concentrations for single-step fluoroquinolone-resistant mtuants of wild-type, efflux-positive, or *parC* or *gyrA* mutation-containing *Streptococcus pneumoniae* isolates. *Antimicrob Agents Chemother*, **48**, 3954–3958.

Strahilevitz, J. and Hooper, D.C., (2005), Dual targeting of topoisomerase IV and gyrase to reduce mutant selection: direct testing of the paradigm by using WCK-1734, a new fluoroquinolone, and ciprofloxacin. *Antimicrob Agents Chemother*, **49**, 1949–1956.

Stratton, C., (2003), Dead bugs don't mutate: susceptibility issues in the emergence of bacterial resistance. *Emerg Infect Dis*, **9**, 10–16.

Tam, V., Louie, A., Deziel, M., Liu, W., Leary, R. and Drusano, G., (2005), Bacterial-population responses to drug-selective pressure: examination of garenoxacin's effect on *Pseudomonas aeruginosa*. *J Infect Dis*, **192**, 420–428.

Tillotson, G., Zhao, X. and Drlica, K., (2001), Fluoroquinolones as pneumococcal therapy: closing the barn door before the horse escapes. *The Lancet Infect Dis*, **1**, 145–146.

Wehbeh, W., Rojas-Diaz, R., Li, X., Mariano, N., Grenner, L., Segal-Maurer, S., Tommasulo, B., Drlica, K., Urban, C. and Rahal, J., (2005), Fluoroquinolone-resistant *Streptococcus agalactiae*: epidemiology and mechanism of resistance. *Antimicrob Agents Chemother*, **49**, 2495–2497.

Weller, T., Andrews, J., Jevons, G. and Wise, R., (2002), The in vitro activity of BMS-2847676, a new des-fluorinated quinolone. *J Antimicrob Chemother*, **49**, 177–184.

Wetzstein, H.-G., (2005), Comparative mutant prevention concentrations of pradofloxacin and other veterinary fluoroquinolones indicate differing potentials in preventing selection of resistance. *Antimicrob Agents Chemother*, **49**, 4166–4173.

Wimberley, N., Faling, L. and Bartlett, J., (1979), A fiberoptic bronchoscopy technique to obtain uncontaminated lower airway secretions for bacterial culture. *Am Rev Respir Dis*, **119**, 337–343.

Wiuff, C., Lykkesfeldt, J., Svendsen, O. and Aarestrup, F., (2003), The effects of oral and intramuscular administration and dose escalation of enrofloxacin on the selection of quinolone resistance among *Salmonella* and coliforms in pigs. *Res Vet Sci*, **75**, 185–193.

Yoshida, H., Bogaki, M., Nakamura, M. and Nakamura, S., (1990), Quinolone resistance-determining region in the DNA gyrase *gyrA* gene of *Escherichia coli*. *Antimicrob Agents Chemother*, **34**, 1271–1272.

Zhanel, G.G., Walters, M., Laing, N. and Hoban, D.J., (2001), In vitro pharmacodynamic modelling simulating free serum concentrations of fluoroquinolones against multidrug-resistant *Streptococcus pneumoniae*. *J Antimicrob Chemother*, **47**, 435–440.

Zhao, X. and Drlica, K., (2001), Restricting the selection of antibiotic-resistant mutants: a general strategy derived from fluoroquinolone studies. *Clin Infect Dis*, **33** (Suppl 3), S147–S156.

Zhao, X. and Drlica, K., (2002), Restricting the selection of antibiotic-resistant mutants: measurement and potential uses of the mutant selection window. *J Infect Dis*, **185**, 561–565.

Zhao, X., Eisner, W., Perl-Rosenthal, N., Kreiswirth, B. and Drlica, K., (2003), Mutant prevention concentration for garenoxacin (BMS-284756) with ciprofloxacin-susceptible and ciprofloxacin-resistant *Staphylococcus aureus*. *Antimicrob Agents Chemother*, **47**, 1023–1027.

Zhao, X., Wang, J.-Y., Xu, C., Dong, Y., Zhou, J., Domagala, J. and Drlica, K., (1998), Killing of *Staphylococcus aureus* by C-8-methoxy fluoroquinolones. *Antimicrob Agents Chemother*, **42**, 956–958.

Zhao, X., Xu, C., Domagala, J. and Drlica, K., (1997), DNA topoisomerase targets of the fluoroquinolones: a strategy for avoiding bacterial resistance. *Proc Natl Acad Sci USA*, **94**, 13991–13996.

Zhou, J.-F., Dong, Y., Zhao, X., Lee, S., Amin, A., Ramaswamy, S., Domagala, J., Musser, J.M. and Drlica, K., (2000), Selection of antibiotic resistance: allelic diversity among fluoroquinolone-resistant mutations. *J Infect Dis*, **182**, 517–525.

Zinner, S., Lubenko, I., Gilbert, D., Simmons, K., Zhao, X., Drlica, K. and Firsov, A., (2003), Emergence of resistant *Streptococcus pneumoniae* in an in vitro dynamic model that simulates moxifloxacin concentration in and out of the mutant selection window: related changes in susceptibility, resistance frequency, and bacterial killing. *J Antimicrob Chemother*, **52**, 616–622.

Index

Printed in the United States
94955LV00001B/169/A

9 780387 724171